The University of Chicago School Mathematics Project

Authors

John W. McConnell
Susan Brown
Susan Eddins
Margaret Hackworth
Leroy Sachs
Ernest Woodward
James Flanders
Daniel Hirschhorn
Cathy Hynes
Lydia Polonsky
Zalman Usiskin

About the Cover
Graphed here is the path of a baseball that is lobbed from
a height of 4 feet and attains a height of 12 feet after traveling a horizontal distance
of 10 feet. The path is part of a parabola.

Scott, Foresman and Company
Editorial Offices: Glenview, Illinois Regional Offices: Sunnyvale, California •
Atlanta, Georgia • Glenview, Illinois • Oakland, New Jersey • Dallas, Texas

Acknowledgments

Authors

John W. McConnell
Instructional Supervisor of Mathematics, Glenbrook South H.S., Glenview, IL

Susan Brown
Mathematics Teacher, York H.S., Elmhurst, IL

Susan Eddins
Mathematics Teacher, Illinois Mathematics and Science Academy, Aurora, IL

Margaret Hackworth
Mathematics Supervisor, Pinellas County Schools, Largo, FL

Leroy Sachs
Mathematics Teacher (retired), Clayton H.S., Clayton, MO

Ernest Woodward
Professor of Mathematics, Austin Peay State University, Clarksville, TN

James Flanders
UCSMP

Daniel Hirschhorn
UCSMP

Cathy Hynes
Mathematics Teacher, The University of Chicago Laboratory Schools

Lydia Polonsky
UCSMP

Zalman Usiskin
Professor of Education, The University of Chicago

Editorial Development and Design

Scott, Foresman staff, Publishers Services, Incorporated, Kristin Nelson Design, Jill Ruter Design

We wish to acknowledge the generous support of the **Amoco Foundation** and the **Carnegie Corporation of New York** in helping to make it possible for these materials to be developed and tested.

UCSMP Production and Evaluation

Series Editors: Zalman Usiskin, Sharon Senk (Michigan State University)
Managing Editor: Natalie Jakucyn
Technical Coordinator: Susan Chang
Director of Evaluations: Sandra Mathison (State University of New York, Albany)
Assistants to the Director: Penelope Flores, David Matheson, Catherine Sarther

A list of schools that participated in the research and development of this text may be found on page *iii*.

ISBN: 0-673-45263-8

We cannot thank everyone who helped us on this book by name. We wish particularly to acknowledge Carol Siegel, who coordinated the use of these materials in schools, and Peter Bryant, Dan Caplinger, Janine Crawley, Kurt Hackemer, Michael Herzog, Maryann Kannappan, Mary Lappan, Teresa Manst, and Victoria Ritter of our technical staff.

The following teachers taught preliminary versions of this text, participated in the pilot and formative research, and contributed many ideas to help improve the text:

Ed Brennan
George Washington High School
Chicago Public Schools

Patricia Doliboa
Clearwater High School
Clearwater, Florida

Monica Hatfield
Morton East High School
Cicero, Illinois

Marie Hill
Glenbrook South High School
Glenview, Illinois

Martha Huberty
Aptakisic Junior High School
Prairie View, Illinois

Jacquie Jensen
O'Neill Junior High School
Downers Grove, Illinois

Rob Johan
O'Neill Junior High School
Downers Grove, Illinois

Joe Lee
Parkway West Junior High School
Chesterfield, Missouri

Edythe Olshan
Von Steuben Mathematics
and Science Academy
Chicago Public Schools

Janet Ramser
Northeast High School
Clarksville, Tennessee

Paula Rossino
Disney Magnet School
Chicago Public Schools

Candace Schultz
Wheaton-Warrenville Middle School
Wheaton, Illinois

Chris Senorski
Austin Academy
Chicago Public Schools

Barbara Simak
McClure Junior High School
Western Springs, Illinois

Mary Szczypta
Elk Grove High School
Elk Grove, Illinois

George Zerfass
Glenbrook South High School
Glenview, Illinois

The following schools used an earlier version of UCSMP *Algebra* in a nationwide study. Their comments, suggestions, and performance guided the changes made for this version.

Rancho San Joaquin High School
Lakeside Middle School
Irvine High School
Irvine, California

Mendocino High School
Mendocino, California

Lincoln Junior High School
Lesher Junior High School
Blevins Junior High School
Fort Collins, Colorado

Bacon Academy
Colchester, Connecticut

Rogers Park Junior High School
Danbury, Connecticut

Hyde Park Career Academy
Bogan High School
Chicago, Illiinois

Morton East High School
Cicero, Illinois

Springman Junior High School
Glenview, Illinois

Carl Sandburg Junior High School
Winston Park Junior High School
Palatine, Illinois

Fruitport High School
Fruitport, Michigan

Taylor Middle School
Roosevelt Middle School
Van Buren Middle School
Albuquerque, New Mexico

Crest Hills Middle School
Shroder Paideia
Walnut Hills High School
Cincinnati, Ohio

Easley Junior High School
Easley, South Carolina

R.C. Edwards Junior High School
Central, South Carolina

Liberty Middle School
Liberty High School
Liberty, South Carolina

Glen Hills Middle School
Glendale, Wisconsin

Robinson Middle School
Mapledale Middle School
Milwaukee, Wisconsin

We wish to acknowledge and thank the many other schools and students who have used earlier versions of these materials. We wish also to acknowledge the contribution of the text *Algebra Through Applications with Probability and Statistics,* by Zalman Usiskin (NCTM, 1979), developed with funds from the National Science Foundation, to some of the conceptualizations and problems used in this book.

UCSMP Algebra

The University of Chicago School Mathematics Project (UCSMP) is a long-term project designed to improve school mathematics in grades K-12. UCSMP began in 1983 with a 6-year grant from the Amoco Foundation. Additional funding has come from the Ford Motor Company, the Carnegie Corporation of New York, the National Science Foundation, the General Electric Foundation, GTE, and Citicorp.

The project is centered in the Departments of Education and Mathematics of the University of Chicago, and has the following components and directors:

Resources	Izaak Wirszup, Professor Emeritus of Mathematics
Primary Materials	Max Bell, Professor of Education
Elementary Teacher Development	Sheila Sconiers, Research Associate in Education
Secondary	Sharon L. Senk, Assistant Professor of Mathematics and Education, Syracuse University (on leave)
	Zalman Usiskin, Professor of Education
Evaluation	Larry Hedges, Professor of Education
	Susan Stodolsky, Professor of Education

From 1983-1987, the director of UCSMP was Paul Sally, Professor of Mathematics. Since 1987, the director has been Zalman Usiskin.

The text *Algebra* was developed by the Secondary Component (grades 7-12) of the project, and constitutes the second year in a six-year mathematics curriculum devised by that component. As texts in this curriculum complete their multi-stage testing cycle, they are being published by Scott, Foresman and Company. The schedule for first publication of the texts follows. Titles for the last two books are tentative.

Transition Mathematics	spring, 1989
Algebra	spring, 1989
Geometry	spring, 1990
Advanced Algebra	spring, 1989
Functions, Statistics, and Trigonometry, with Computers	spring, 1991
Precalculus and Discrete Mathematics	spring, 1991

A first draft of *Algebra* was written and piloted during the 1985-86 school year. After a major revision, a field trial edition was used in about ten schools in 1986-87. A second revision was given a comprehensive nationwide test during 1987-88. Results are available by writing UCSMP. The Scott, Foresman and Company edition is based on improvements suggested by the authors, editors, and some of the many teacher and student users of earlier editions.

Comments about these materials are welcomed. Address queries to Mathematics Product Manager, Scott, Foresman and Company, 1900 East Lake Avenue, Glenview, Illinois 60025, or to UCSMP, The University of Chicago, 5835 S. Kimbark, Chicago, IL 60637.

UCSMP *Algebra* is designed for a first-year course in algebra. It differs from other books for this course in six major ways. First, it has **wider scope** in content. It integrates geometry, statistics, and probability into the algebra. These topics are not isolated as separate units of study or enrichment. They are employed to motivate, justify, extend, and otherwise enhance important concepts of algebra.

Second, **reading and problem solving** are emphasized throughout. Students can and should be expected to read this book. The explanations were written for students and tested with them. The first set of questions in each lesson is called "Covering the Reading." The exercises guide students through the reading and check their coverage of critical words, rules, explanations, and examples. The second set of questions is called "Applying the Mathematics." These questions extend student understanding of the principles and applications of the lesson. To further widen student horizons, "Exploration" questions are provided in every lesson.

Third, there is a **reality orientation** towards both the selection of content and the approaches allowed the student in working out problems. Algebra is rich in applications and problem solving. Being able to do algebra is of little ultimate use to individuals unless they can apply that content. Real-life situations motivate algebraic ideas and provide the settings for practice of algebra skills. The variety of content of this book permits lessons on problem-solving strategies to be embedded in application settings.

Fourth, fitting the reality orientation, students are expected to use current **technology**. Calculators are assumed throughout this book because virtually all individuals who use mathematics today find it helpful to have them. Scientific calculators are recommended because they use an order of operations closer to that found in algebra and have numerous keys that are helpful in understanding concepts at this level. Computer exercises present important representations of the language and algorithms of algebra. To help the student develop a sense of when technology is appropriate, many lessons contain questions requiring mental computation.

Fifth, **four dimensions of understanding** are emphasized: skill in carrying out various algorithms; developing and using mathematical properties and relationships; applying mathematics in realistic situations; and representing or picturing mathematical concepts. We call this the SPUR approach: **S**kills, **P**roperties, **U**ses, **R**epresentations.

Sixth, the **instructional format** is designed to maximize the acquisition of both skills and concepts. The book is organized around lessons meant to take one day to cover. Ideas introduced in a lesson are reinforced through "Review" questions in the immediately succeeding lessons. This daily review feature allows students several nights to learn and practice important concepts and skills. The lessons themselves are sequenced into carefully constructed chapters. At the end of each chapter, a carefully focused Progress Self-Test and Chapter Review, keyed to objectives in all the dimensions of understanding, are then used to solidify performance of skills and concepts from the chapter so that they may be applied later with confidence. Finally, to increase retention, important ideas are again reviewed in "Review" questions of later chapters.

CONTENTS

With algebra, you can describe patterns of all kinds, work with formulas, discuss unknowns in problems, quickly graph ideas, and write computer programs. Algebra can be considered to be the language of mathematics. The first goal of UCSMP *Algebra* is to introduce you to this wonderful and rich language.

You will find algebra different from arithmetic. Sometimes algebra is harder, but often it's easier because there is less computation. Also, you will learn that graphs can describe many of the ideas in algebra, and these will help you.

For this book, you will need a ruler (to draw and measure along lines), with both centimeter and inch markings, and graph paper. It is best if the ruler is made of transparent plastic.

You will also need a scientific calculator in many places in this book, beginning in Chapter 1. Scientific calculators differ widely in the range of keys they have. If you are going to buy or borrow a calculator, it should have the following keys: $\boxed{x^y}$ or $\boxed{y^x}$ (powering), $\boxed{\sqrt{x}}$ (for square root), $\boxed{x!}$ (factorial), $\boxed{\pm}$ or $\boxed{+/-}$ (for negative numbers), $\boxed{\pi}$ (pi), and $\boxed{1/x}$ (reciprocals), and it should write very large or very small numbers in scientific notation. We recommend a *solar-powered* calculator so that you do not have to worry about batteries, though some calculators have batteries which can last for many years and work in dim light. A good calculator can last for many years.

There is another important goal of this book: to assist you to become able to learn mathematics on your own, so that you will be able to deal with the mathematics you see in newspapers, magazines, on television, on any job, and in school. The authors, who are all experienced teachers, offer the following advice.

1. You cannot learn much mathematics just by watching other people do it. You must participate. Some teachers have a slogan:

    Mathematics is not a spectator sport.

2. You are expected to read each lesson. Read slowly, and keep a pencil with you as you check the mathematics that is done in the book. Use the Glossary or a dictionary to find the meaning of a word you do not understand.

3. You are expected to do homework every day while studying from this book, so put aside time for it. Do not wait until the day before a test if you do not understand something. Try to resolve the difficulty right away and ask questions of your classmates or teacher. You are expected to learn many things by reading, but school is designed so that you do not have to learn everything by yourself.

4. If you cannot answer a question immediately, don't give up! Read the lesson again; read the question again. Look for examples. If you can, go away from the problem and come back to it a little later.

We hope you join the many thousands of students who have enjoyed this book. We wish you much success.

Basic Concepts

About 3500 years ago, and Egyptian wrote the hieroglyphics shown below. They tell how to find the area of a rectangle with length 10 units and width 2 units.

Today's description in English can be shorter.

The area of a rectangle equals its length times its width.

We can shorten the English statement by using symbols for equals and times.

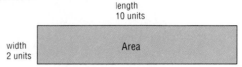

length
10 units

width
2 units

Area

$$\text{Area of a rectangle} = \text{length} \times \text{width}$$
$$= 10 \text{ units} \times 2 \text{ units}$$
$$= 20 \text{ square units}$$

This statement can be abbreviated still more by using variables.

$$A = \ell w$$

By using variables, the area relationship is described in a very concise way. The formula $A = \ell w$ is much shorter than the hieroglyphics or the description in English. For many people, formulas are clearer and easier to use than words.

Variables are basic to the language of algebra. Formulas are one way in which algebra makes things easier to understand and use. In this chapter, you will study ideas fundamental to variables, to algebra, and to this book.

Numbers in Algebra

There is an old *rule of thumb* for estimating your distance from a flash of lightning. First, determine the number of seconds between the flash and the sound of thunder. Divide this number by 5. The result is the approximate distance in miles. For example, if you count 10 seconds between the flash and the sound of thunder, you are $\frac{10}{5}$, or 2, miles from the lightning. In the language of algebra, if s is the number of seconds, then $\frac{s}{5}$ is your distance (in miles) from the lightning. The letter s is a variable. A **variable** is a letter or other symbol that can be replaced by any number (or other object) in some set.

In this situation the letter s was chosen because s is the first letter of "seconds." The full formula is:

$$n = \frac{s}{5}, \text{ where } n = \text{distance in miles}; s = \text{time in seconds.}$$

The sentence $n = \frac{s}{5}$ is called an **equation** because it uses the mathematical verb "=." Other verbs are shown below.

$\neq$	is not equal to		$\approx$	is approximately equal to
$<$	is less than		$>$	is greater than
$\leq$	is less than or equal to		$\geq$	is greater than or equal to

A sentence with one of the other verbs is called an **inequality**. For instance, the sentence $\frac{1}{2} < \frac{5}{8}$ is an inequality. A sentence with a variable is called an **open sentence**. $n = \frac{s}{5}$ is an open sentence. Open sentences are powerful mathematical statements because they allow you to symbolize many patterns and relationships in an efficient, organized way. A **solution** to a sentence is a replacement for the variable that makes the statement true.

Example 1 Which of the numbers 7, 8, or 9 is a solution to the open sentence

$$3 \cdot x + 15 = 4 \cdot x + 6?$$

Solution Try 7. Does $3 \cdot 7 + 15 = 4 \cdot 7 + 6$?
No, $36 \neq 34$.
Try 8. Does $3 \cdot 8 + 15 = 4 \cdot 8 + 6$?
No, $39 \neq 38$
Try 9. Does $3 \cdot 9 + 15 = 4 \cdot 9 + 6$?
Yes, $42 = 42$.

So 9 is a solution. The numbers 7 and 8 are not solutions.

In later chapters you will learn methods for finding solutions to equations. For now, you may have to solve equations by using trial and error.

When an open sentence has more than one solution, it may be easiest to describe the solutions with a graph or an inequality.

Example 2 Water will remain ice for all temperatures less than 0° Celsius.
a. Write an open sentence describing this situation.
b. Graph all the points which make the sentence true.

Solution
a. Let T represent the temperature of the water. Then water is ice if $T < 0°$. ($0° > T$ would also be a correct statement.)
b. Graph $T < 0°$. First mark $T = 0°$ with an open circle to indicate that 0° does not make $T < 0°$ true. Then draw a heavy arrow to the left to represent all points less than 0°.

Example 3 Carl's car gets at least 19.3 miles per gallon.
a. Write an algebraic sentence which describes this situation.
b. Graph the solutions to the sentence on a number line.

Solution
a. If m is the number of miles per gallon, m can be 19.3 or larger.

$$m \geq 19.3$$

b. Mark a dot on 19.3 and draw a heavy arrow to the right. (To find 19.3, you can separate the interval between 19 and 20 into ten equal parts.) The arrow indicates that numbers like 19.4, 19.35, 20, 50.2 and $70\frac{1}{2}$ all make the sentence $m \geq 19.3$ true. The closed circle means that 19.3 itself makes the sentence true.

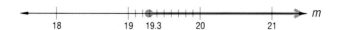

An inequality compares numbers. Often it is easier to compare numbers in decimal form than fraction form.

■ ■ ■ ■ ■ ■ ■ ■

Example 4   Write an inequality that compares $\frac{1}{3}$ and $\frac{3}{8}$.

Solution   A fraction indicates a division. Carry out that division to convert the fraction to a decimal.

$$\frac{1}{3} = 1 \div 3 \qquad\qquad \frac{3}{8} = 3 \div 8$$
$$= .\overline{3} \approx .333 \qquad\qquad = .375$$

In decimal form, $.\overline{3}$ is seen to be smaller than .375.

$$\text{So, } \frac{1}{3} < \frac{3}{8}.$$

A correct answer using the "is greater than" sign is $\frac{3}{8} > \frac{1}{3}$.

Questions

Covering the Reading

These questions check your understanding of the reading. If you cannot answer a question you should go back to the lesson to help you find the answer.

1. You clock the time between lightning and thunder as 8 seconds. How far away was the lightning?

2. What is a variable?

In 3–5, tell whether what is written is a sentence.
3. $5 \cdot x + 3 < 2$ 4. $8 - 2 \cdot y$ 5. $-5 \leq r$

6. What is the symbol for "is approximately equal to"?

7. Which of the numbers 5, 9, 11 are solutions to $x \leq 9$?

8. Which of the numbers 5, 6, or 7 is a solution of $2 \cdot y + 3 = 4 \cdot y - 9$?

In 9 and 10, **a.** write an open sentence to describe the situation; **b.** graph all the points that make the sentence true.

9. Dan ran more than 3 miles.

10. The temperature was below 10°F all day.

11. *Multiple choice*   In Example 3, the inequality shown is $m \geq 19.3$. Which of the following means the same thing?
(a) $m \approx 19.3$ (b) $19.3 \leq m$ (c) $19.3 < m$

In 12 and 13, write an inequality to compare the two numbers.

12. $\frac{5}{8}, \frac{4}{7}$ **13.** $\frac{5}{6}, \frac{17}{20}$

Applying the Mathematics

These questions extend your understanding of the content of the lesson. You should study the examples and explanations if you cannot get an answer. Check your answers with the Selected Answer section in the back of the book.

14. Before 1985, the average household in the United States contained more than 1.9 persons.
a. Write an inequality about the date using the variable d.
b. Write an inequality about the average number of persons per household using the variable p.

In 15 and 16, write an inequality to describe each graph. Use the variable that is next to the graph.

15.

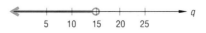

16.

17. *Multiple choice* $z \geq 100$ is the same as
  (a) $z \leq 100$.   (b) $100 \geq z$.   (c) $100 \leq z$.

18. Which of the values 4, 8, or 16 solves $5 \cdot x < 40$?

In 19–21, use this information.
The mixed number $2\frac{1}{4} = 2 + \frac{1}{4}$.

  Since $\frac{1}{4} = .25$ you can write

  $$2\frac{1}{4} = 2 + .25$$
  $$= 2.25$$

Write the mixed number as a decimal.

19. $3\frac{3}{8}$ **20.** $1\frac{1}{9}$ **21.** $112\frac{14}{15}$

22. Order from smallest to largest: $1\frac{1}{2}$, 1.52, $\frac{2.9}{2}$.

23. Let $d = 3$. Find a value of n so that:
  a. $\frac{n}{d} > 1$ **b.** $\frac{n}{d} < 1$ **c.** $\frac{n}{d} = 1$

24. Copy and complete so the statement describes the graph.

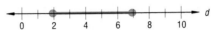

  $\underline{\quad ? \quad} \leq d \leq \underline{\quad ? \quad}$

25. Which of the numbers ⁻3, ⁻2, 7, and ⁻7 are included in the graph below?

26. Graph $y > \frac{2}{3}$ on a number line.

27. Write in order from smallest to largest: $\frac{7}{10}$, $\frac{2}{3}$, $\frac{3}{4}$.

Often it is quicker and more convenient to do problems in your head. Punching calculator keys for simple problems is time-consuming and may lead to careless mistakes. Do not use your calculator or work with paper and pencil on these problems. Just write an answer.

28. Compute in your head.
  a. $10 \cdot 3.7$ **b.** $1\frac{1}{2} \cdot 2$ **c.** $4 \cdot \$2.25$

These questions ask you to explore mathematics topics related to the chapter. Often, these questions require that you use dictionaries and other books. Sometimes they will ask you to perform an experiment. Many exploration questions have more than one correct answer.

29. Find a fraction between $\frac{11}{20}$ and $\frac{14}{25}$.

In Lesson 1-1, number lines were used to graph solutions to algebraic sentences. In this lesson, number lines are used to organize information and make sense out of data.

To illustrate a way to organize data on a number line, Mr. Tobias asked the students in his algebra class how many letters were in each of their last names. Here are the students' responses in the order in which they were given.

6, 6, 8, 4, 5, 6, 5, 4, 4, 5, 9, 7, 11, 6, 6, 5, 5, 9, 4, 8, 5

To picture this information, Mr. Tobias made a **dot frequency diagram**. First he drew a horizontal number line. For each student's response, he placed a dot above that number on the line.

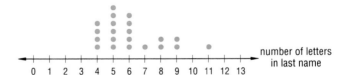

The number of dots above each number records how often that response occurred. From the drawing you can answer several questions about Mr. Tobias's class.

■ ■ ■ ■ ■ ■ ■ ■ ■

Example 1   Consider the dot frequency diagram above.

   a. How many students' names have 6 letters?
   b. How many students' names have more than 8 letters?
   c. Would a response of 0 have made sense?
   d. What is the mean (or average) number of letters per last name?

Solution

a. There are five dots above the number 6 on the dot frequency diagram, so five students have 6-letter last names.

b. There are three dots to the right of 8 on the number line. Three students have last names of more than 8 letters.

c. No. Every last name must have at least one letter.

d. Recall from your previous work that the mean is calculated as follows.

$$\text{mean} = \frac{\text{sum of the numbers}}{\text{number of numbers}} =$$

$$\frac{6 + 6 + 8 + 4 + 5 + 6 + 5 + 4 + 4 + 5 + 9 + 7 + 11 + 6 + 6 + 5 + 5 + 9 + 4 + 8 + 5}{21}$$

$$= \frac{128}{21}$$

$$\approx 6.1$$

The **mean** is a single number that represents a set of data. Numbers that represent sets of data are called **statistics**. Two other statistics used to describe data are the **median** and the **mode**. These statistics describe the middles or centers of the data. Statisticians call them measures of **central tendency.**

Statistics of Central Tendency:

The mean is the average of a set of numbers.

The median is the middle value of a set of numbers when the numbers are ranked in order.

The mode is the most frequently occurring number or category in a set of numbers or categories.

Example 2   Refer again to the responses in Mr. Tobias's class.

a. What is the mode for the set of data?

b. What is the median for the set of data?

Solution

a. The mode is the most frequently occurring value. The dot frequency diagram shows the most frequent response in Mr. Tobias's class was 5. The mode is therefore 5.

b. The median is the middle value of a set of numbers when the numbers are ranked in order. To calculate the median, first list the 21 values in order. You can use the dot frequency diagram to do this. Then cut the list into halves.

4, 4, 4, 4, 5, 5, 5, 5, 5, 5, $\boxed{6}$, 6, 6, 6, 6, 7, 8, 8, 9, 9, 11

There are 21 numbers, so the 11th number is the one in the middle. It is the number 6, so 6 is the median.

To find the median in Example 2, we counted until we found the middle value. If there are an even number of values, there is no single middle value. You must then find the *two* values in the middle. The mean of these is the median of the whole set of values.

Example 3   Sharma has test scores of 50, 60, 100, 70, 100, 60, 60, 100, 40, and 100.

   a. If s represents a test score, for how many of Sharma's tests was $s \leq 60$?

   b. What is her median score?

   c. What is her mean score?

   d. What is the mode?

   e. Which statistic gives her the highest score?

   Solution   A dot frequency diagram helps to organize the data.

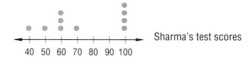

Sharma's test scores

   a. $s \leq 60$ means a score of 60 or lower. Sharma had one score of 40, one of 50, and three of 60. So Sharma had five tests for which $s \leq 60$.

   b. List the scores in order from the diagram. There are an even number of scores, so locate the middle two.

$$40, 50, 60, 60, \boxed{60, 70,} \ 100, 100, 100, 100$$

   Find the mean of the middle scores. $\dfrac{60 + 70}{2} = 65$
   Her median score is 65.

   c. To find the mean, first add the scores. The sum is 740. There are 10 scores, so the mean is $\frac{740}{10} = 74$.

   d. The mode is 100, the most common value.

   e. The mode gives her the highest score. (The median gives her the lowest score.)

It is possible for a set of data to have more than one mode. For example, there are two modes for the set of scores 50, 70, 20, 70, 60, 40, and 20. Both 70 and 20 occur the same number of times. On the other hand, if no one piece of data occurs most often, the set of data has no mode. The set of scores 65, 90, 80, 60, 75, 77, and 56 has no mode.

1. The dot frequency diagram below shows the number of students who received certain scores on an algebra test.

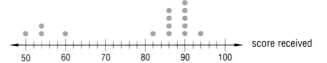

   a. How many students received a score of 50?
   b. How many students received a score of more than 80?
   c. Find the mean or average score for the class.

2. The __?__ is the most frequently occurring number or category in a set of numbers or categories.

3. The middle value of a set of scores is 65. This is the __?__ for the set of scores.

4. Seven members of a boxing team compete as flyweights. Their weights in pounds are: 112, 110, 111, 108, 107, 98, and 98. Find the median for these weights.

5. For one quarter Sheila has scores of 68, 65, 80, 75, 40, 60, 62, and 80 in social studies. What is the **a.** mean; **b.** median; and **c.** mode of these scores? **d.** Which statistic gives her the highest measure of her scores?

6. Give a set of data that has no mode.

7. Shown below are the daily high temperature readings in degrees Fahrenheit for the first 27 days of January, 1979 in Sterling, Alaska.

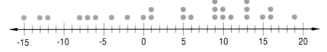

   a. If the daily high temperature is T degrees, for how many days was $T < 0$?
   b. For how many days was $T > $ -10?
   c. Give the mode or modes for this data.
   d. Find the median.

8. In Question 4, does the median or mode give a better idea of the weights of the team members?

9. Consider the mean, median and mode for the scores in Question 5. In your opinion, which statistic (the mean, median, or mode) is the *worst* description of how Sheila did that quarter?

10. The weekly salaries for workers in a small business are $160, $140, $140, $200, $150, $300, $150, $140 and $500.
    a. Make a dot frequency diagram for this data.
    b. Find the mean, median and mode.

c. The company president wants to impress a prospective employee. Would the president be more likely to use the mean, median or mode to describe the workers' salaries?

d. The employees' union is negotiating for a raise. Would it be more likely to use the mean, median, or mode to describe the workers' salaries?

Review

Every lesson from here on contains review questions which give practice on ideas presented in earlier lessons. Lessons in parentheses after the questions indicate where the idea was first presented. If you cannot do a review question, look back at the indicated lesson. Some skills provide practice on ideas of previous courses.

11. The temperatures at which water is steam are all those greater than 100 °Celsius.
   a. Write an inequality describing this situation.
   b. Graph the inequality. *(Lesson 1-1)*

12. Write an inequality that has the same meaning as -15 < y. *(Lesson 1-1)*

13. Put these numbers in order from least to greatest. *(Previous course)*

$$-10, 6.8, 5, -4, -15, 0, -2$$

14. The Jets lost 11 yards on their first play and gained 7 on their second play. What was their net gain or loss? *(Previous course)*

In 15 and 16, add. *(Previous course)*

15. -11 + 7

16. -7 + -5

17. For the first five days it was open, Harold's Electronics gave away pens to customers. The cost was $70.00 per day. What was the cost for all five days? *(Previous course)*

In 18 and 19, multiply. (Recall that a positive times a positive is positive; a negative times a negative is positive; and a positive times a negative is negative.) *(Previous course)*

18. 5 · -70

19. -10 · -4

20. Remember that x^2 means $x \cdot x$. Find **a.** 7^2 **b.** 10^2 **c.** 2^3
   (Previous course)

In 21–23, multiply. *(Previous course)*

21. $\frac{2}{7} \cdot \frac{3}{5}$

22. $\frac{4}{9} \cdot \frac{3}{10}$

23. $\frac{2}{5} \cdot 10$

24. Compute in your head. *(Previous course)*
   a. $\frac{1}{3}$ of 9   **b.** 25% of 20

Exploration

25. Mathematical terms are often borrowed from ordinary vocabulary. Explain the relationship between the following phrases and the statistical meaning of the underlined word.
   a. <u>median</u> strip of a highway; **b.** pie a la <u>mode</u>

Intervals and Estimates

An **interval** is the set of numbers between two numbers a and b, possibly including a and b. The numbers a and b are called the **endpoints** of the interval. Intervals occur often in real situations and are described in a variety of ways.

One way to describe an interval is with its middle point (midpoint) and a distance from that point. For instance, a pollster surveyed voters and reported: "Accurate to within 3%, I found that 58% of voters plan to vote for candidate A."

The graph below shows the percent of voters who could be expected to vote for candidate A.

The percentage of voters who plan to vote for A can be 3% higher or lower than 58%. It can range from 55% to 61%. This interval can be described as

$$58\% \pm 3\%.$$

The symbol $\pm$ is read "plus or minus." Use the + sign to get the upper endpoint.

$$58\% + 3\% = 61\%$$

Use the − sign to get the lower endpoint.

$$58\% - 3\% = 55\%$$

Notice that the midpoint of the interval is the average of the endpoints. $58 = \dfrac{55 + 61}{2}$

A second way to describe an interval is by giving its endpoints as in the sentence: "You must be between 18 and 35 to enter the military." But the English language presents problems. The word *between* doesn't tell you whether 18 and 35 are included. An inequality is more precise. One way to interpret "between 18 and 35" is with the compound sentence $18 \leq a \leq 35$, where a represents a person's age. This interval is called a **closed interval** because it includes its endpoints.

Another interpretation of "between 18 and 35" does not include the endpoints. The interval $18 < a < 35$ is an **open interval.**

If an interval is described by words rather than symbols, you may have to decide from the situation if it is closed or open. In some situations, you may have a choice.

An interval can include one endpoint and not the other. Here is a graph of the positive numbers less than or equal to 10. $0 < x \leq 10$

The endpoints are the numbers 0 and 10. 10 is included in the graph, but 0 is not.

The interval from the minimum to the maximum value of a set of data is useful for summarizing data on dot frequency diagrams.

Example 1   Shown below are the ages of U.S. presidents when they were first inaugurated. Describe the interval from the minimum to the maximum age using: **a.** an inequality; **b.** a graph; **c.** words.

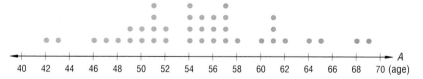

Solution

a. The maximum age is 69, the minimum is 42. The interval from minimum to maximum is $42 \le A \le 69$.

b.

42   69

(Technically, we might graph with dots from 42 to 69, but it is easier to draw a bar.)

c. Presidents' ages at inauguration range from 42 to 69.

The word *range* in solution c above indicates the interval. The **length of the interval** is also called its **range** and is found by subtracting the minimum value from the maximum value.

range = maximum value − minimum value

The range is a statistic which gives you an idea of the *spread* of data. In Example 1, the range is 69 years − 42 years = 27 years.

If an interval is open, use its endpoints as the maximum and minimum values.

Death Valley,
California

Example 2   Find the range of the interval $3.5 < w < 18.07$.

Solution   range = $18.07 - 3.5 = 14.57$

Notice the two uses of range in the next example. When you are asked for the range in this book, you should give a number as an answer.

Example 3   Elevations in the state of California range from 86 meters below sea level (in Death Valley) to 4418 meters above sea level (on top of Mt. Whitney).

a. Graph the elevations in California on a vertical number line.
b. Calculate the range of the interval.

Solution

elevation E
(meters)

4418   5000
         4000
         3000
         2000
         1000
-86   0 (sea level)
         -1000

a. The question translates to graphing the inequality

$$-86 \leq E \leq 4418,$$

where E is an elevation in California. E can be any real number. The inequality is graphed at the left.

b. The maximum value is 4418 m. The minimum value is -86 m.

range = maximum − minimum
       = 4418 − (-86)
       = 4418 + 86
       = 4504

The range of elevation in California is 4504 meters.

Questions

Covering the Reading

1. What is an interval?

2. Graph the closed interval 29 ± 4.

3. A poll accurate within 5% showed that 47% of the voters plan to vote for Mamie.
   a. Describe this interval using the "$\pm$" symbol.
   b. What are the endpoints of this interval?

4. **a.** Find the midpoint of the open interval from 100 to 180. **b.** Graph the interval.

5. Use the following information. "Children ages five through twelve pay half price."
   a. Rewrite this using an inequality.
   b. Is this a closed or open interval?
   c. What is the range of the interval?

6. *Multiple choice*   The graph   [graph: -2 -1 0 1 2 3] x represents:

   (a) $-2 \leq x \leq 2$ (b) $-2 < x < 2$ (c) $-2 \leq x < 2$ (d) $-2 < x \leq 2$

7. Suppose 15 students in a class were asked how much television they watched last week. The dot frequency diagram below shows their responses. Describe the interval from minimum to maximum time using **a.** an inequality, **b.** a graph, and **c.** words.

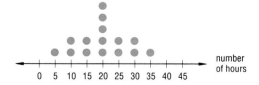

number of hours

0   5   10   15   20   25   30   35   40   45

8.

m is the coordinate of the midpoint of the interval graphed above.
a. What is the value of m?
b. How much larger is 14 than m?
c. How much smaller is 8 than m?
d. Describe the interval using the "±" symbol.

9. Consider the interval graphed here.

a. Describe the interval using an inequality.
b. What is the midpoint of this interval?

In 10 and 11, graph the interval.
10. The best golfers had scores in the 70s.

11. Children under the age of 5 ride free.

In 12 and 13, consider the closed interval $\frac{1}{2} \pm \frac{1}{8}$.
12. What are the endpoints of the interval?

13. Describe this interval using an inequality.

14. A machinist is making metal rods for lamps. The metal rods are to be 1.25" in diameter with an allowable error of at most .005." (The number .005" is called the *tolerance*.) Give an interval for the possible widths w of the rods.

15. A person makes more than $1800 a month. Let d be the amount made.
a. What inequality describes this situation? *(Lesson 1-1)*
b. Graph the inequality. *(Lesson 1-1)*

In 16 and 17, a survey was taken of the number of times the students in Ms. Lawson's class went swimming last month. Here are the data. Each number represents the number of swimming trips for a single student.

3	1	0	2	0
1	4	6	2	0
0	1	0	3	5

16. Which is the most appropriate number line for the data? *(Lesson 1-2)*

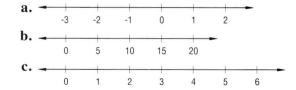

17. **a.** Construct a dot frequency diagram of the data using the appropriate number line from Question 16.
    b. What is the mode of the data?
    c. What is the median of the data? *(Lesson 1-2)*

18. Calculate $328 - x$ when $x = 2.56$. *(Previous course)*

19. In 1983, the typical American teenager watched TV 15% of the time. *(Previous course)*
    a. How many hours would this represent out of a 24-hour day?
    b. How many hours would this represent out of the hours in one week?
    c. How many days does this equal in a year?

20. A band performed at a party and earned $100, which the seven members want to share. How much will each person get, rounded to the nearest cent? *(Previous course)*

Skill sequences are questions intended to maintain and extend skills which you already have and will need in later chapters.

21. *Skill sequence*   Find the sum. *(Previous course)*
    a. $\frac{4}{5} + \frac{9}{5}$ **b.** $\frac{4}{5} + \frac{9}{10}$ **c.** $\frac{4}{5} + \frac{9}{11}$

22. Compute in your head. *(Previous course)*
    a. $5 + \text{-}9$ **b.** $\text{-}5 + \text{-}9$ **c.** $\text{-}5 + 9$

23. Compute in your head. *(Previous course)*
    a. $7 \cdot \text{-}8$ **b.** $\text{-}7 \cdot 8$ **c.** $\text{-}7 \cdot \text{-}8$

Exploration

24. Each of the words below stands for an interval of time. How long is the interval?
    a. decade   **b.** century   **c.** millenium
    d. fortnight   **e.** sennight   **f.** lustrum

25. There is a third meaning of *range* suggested by Example 3. What is that meaning?

26. Why did the nomad put a tent on his stove?

27. Which presidents are represented by the endpoints of the interval in Example 1?

1-4

Order of Operations

Finding the value of an expression is called **evaluating** the expression. **Numerical expressions** like $6 + 3^2 - 1$ combine numbers. If an expression includes one or more variables, like $4 + 3x$, it is called an **algebraic expression.** No matter which type of expression you are evaluating, if it involves several operations you must be careful to do them in the correct order.

Rules for Order of Operations

1. First do operations within parentheses or other grouping symbols.
2. Within grouping symbols or if there are no grouping symbols:
   a. Do all powers from left to right.
   b. Do all multiplications and divisions from left to right.
   c. Do all additions and subtractions from left to right.

To evaluate an algebraic expression you must have values to substitute for the variables. It is important to understand that $3x$ means $3 \cdot x$ and $5(A + B)$ means $5 \cdot (A + B)$.

Example 1 Evaluate $4 + 3x$ **a.** when $x = 9$ and **b.** when $x = -1$.

Solution

a. Let $x = 9$.   Then $4 + 3x$ $= 4 + 3 \cdot 9$ Substitute 9 for x.

   $= 4 + 27$ Multiply first.

   $= 31$ Add.

b. Let $x = -1$.   Then $4 + 3x$ $= 4 + 3 \cdot -1$ Substitute -1 for x.

   $= 4 + -3$

   $= 1$

The value of $4 + 3x$ depends upon what value is used for x.

Example 2 Evaluate $7n^3$ when $n = 2$.

Solution Substitute 2 for n. Do the power *before* the multiplication.

$$7 \cdot 2^3 = 7 \cdot 8$$
$$= 56$$

Suppose you did not follow the correct order and multiplied 7 and 2 first, then raised to the third power. Your answer would be 2744 instead of 56 and would be incorrect.

Most scientific calculators use the same order of operations as algebra. Though we do Examples 1 and 2 without calculators, because the numbers are easy, the following sequences will check them.

Example 1: 4 $\boxed{+}$ 3 $\boxed{\times}$ 9 $\boxed{=}$ $\boxed{ 31}$

Example 2: 7 $\boxed{\times}$ 2 $\boxed{x^y}$ 3 $\boxed{=}$ $\boxed{ 56}$

If you have never used a scientific calculator, see Appendix A.

■ ■ ■ ■ ■ ■ ■ ■ ■

Example 3   Evaluate $\dfrac{5(A + B)}{2}$ when $A = 3.4$ and $B = 7.2$.

Solution   Substitute 3.4 for A and 7.2 for B.

Then $\dfrac{5(A + B)}{2} = \dfrac{5(3.4 + 7.2)}{2}$

$\phantom{Then \dfrac{5(A + B)}{2}} = \dfrac{5(10.6)}{2}$ Work inside the parentheses.

$\phantom{Then \dfrac{5(A + B)}{2}} = \dfrac{53}{2}$ Remove parentheses. 5(10.6) means 5 · 10.6.

$\phantom{Then \dfrac{5(A + B)}{2}} = 26.5$

Most computer languages follow the rules for order of operations. In **BASIC** (Beginner's All-Purpose Symbolic Instruction Code), the arithmetic symbols are + for addition, − for subtraction, * for multiplication, / for division, ^ for powering and ( ) for grouping. In computer symbols $5(A + B)$ is written $5 * (A + B)$. 2^3 is written $2 ^ 3$.

■ ■ ■ ■ ■ ■ ■ ■ ■

Example 4   Use a computer to evaluate $3(4x - 5) + y$ when $x = 3.1$ and $y = 0.6$.

Solution   Here is a computer program written in BASIC to evaluate the expression $3(4x - 5) + y$.

```
10 PRINT "ANSWER TO EXAMPLE 4"
20 LET X = 3.1
30 LET Y = 0.6
40 PRINT 3 * (4 * X − 5) + Y
50 END
```

Run the program. The computer should print

ANSWER TO EXAMPLE 4
22.8

A shortcut to writing the program in Example 4 is to do the substitutions in your head and type PRINT 3 ∗ (4 ∗ 3.1 − 5) + 0.6

When you press the return key, you should see 22.8

On a computer, when you evaluate expressions which are fractions, you must watch grouping carefully. You see that when you evaluate $\frac{10 + 4}{2 + 5}$ by hand, you get $\frac{14}{7} = 2$. If you type PRINT 10+4/2+5 you see 17 as the answer. The slash (/) is *not* a grouping symbol. The computer interprets $10 + 4 / 2 + 5$ as $10 + (4/2) + 5$. To get the correct interpretation, you must show your intended grouping with parentheses: $(10 + 4)/(2 + 5)$.

■ ■ ■ ■ ■ ■ ■

Example 5 Use a computer to evaluate $\left(\dfrac{x + 10.7}{y + 4}\right)^7$ when $x = 3.1$ and $y = 0.6$.

Solution 1 To a computer the expression must be
((x + 10.7) / (y + 4))^7.
Substitute in your head and type PRINT ((3.1 + 10.7) / (0.6 + 4))^7

Press the return key and see **2187**

Solution 2 Run the same program that is in Example 4 but change lines 10 and 40.

```
10 PRINT "ANSWER TO EXAMPLE 5"
20 LET X = 3.1
30 LET Y = 0.6
40 PRINT ((X + 10.7) / (Y + 4))^7
50 END
```

Questions

In 1–3, identify each expression as numerical or algebraic.

1. $\dfrac{8(7) + 2}{12}$

2. $\dfrac{2x}{4 - 8}$

3. $a^2 + b^2$

In 4–8, evaluate.

4. $10^2 \cdot 7^2$

5. $12 - 2 \cdot 4$

6. $3(10 - 6)^3 + 15$

7. $\dfrac{a + 2b}{5}$ when $a = 11.6$ and $b = 9.2$

8. $.5 * H * (A + H)$ when $H = 32$ and $A = .7$

9. If you type PRINT $8+6/3-2$ what will a computer print?

10. What will the computer print on the screen when this program is run?
```
10 PRINT "ANSWER TO QUESTION 10"
20 LET A = 4.5
30 LET B = 2.6
40 PRINT 2 * (4 * A - 3) + B
50 END
```

11. Using the programs in the lesson as patterns, write a computer program or statement to evaluate $\left(\dfrac{m + 12.5}{n + 3}\right)^{10}$ when $m = 2.4$ and $n = 0.2$.

Applying the Mathematics

In 12–14, translate into BASIC.

12. $6xy$

13. $5x + 3y^{10}$

14. $\dfrac{7x + y}{x + 7y}$

15. The perimeter of the hexagon at the right is $4a + 2b$. Find the perimeter when $a = 10$ mm and $b = 35$ mm.

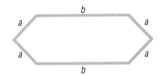

16. Begin with the innermost parentheses and work toward the outer parentheses. $6 - (5 - (4 - (3 - 2)))$.

17. $\dfrac{1.4x - 2.3y}{4xy} = \underline{\ ?\ } \div \underline{\ ?\ }$.

18. Let $d =$ number of days, $t =$ number of trips, and $m =$ number of meals.
   a. Write an expression representing the cost of a vacation at Hotel St. Jacques using d, t, and m.
   b. Find the cost when $d = 3$, $t = 2$, and $m = 5$.

HOTEL
ST. JACQUES

$49 per day
$15 per sightseeing trip
$12 per meal

19. Graph the interval $-4 \leq y \leq 6$. *(Lesson 1-3)*

20. Graph the closed interval 25 ± 4. *(Lesson 1-3)*

21. *P* is the sale price of a record whose list price is $7.98. *(Lesson 1-3)*
  a. Describe the possible values of *P* with an inequality.
  b. Graph.

22. Which of the numbers -4, 2, 6 make the sentence $15 + n + {}^-11 > 0$ true? *(Lesson 1-1)*

23. Mrs. Marlowe gave her students a quiz. Their scores are shown below.

```
30  80  70  60   70
40  80  80  90  100
80  90  60  80   70
```

  a. Construct a dot frequency diagram using the scores.
  b. Find the mode of the scores.
  c. Find the median of the scores.
  d. Find the mean of the scores.
  e. Find the range of scores.
  f. On the basis of these statistics, are Mrs. Marlowe's students doing well? *(Lessons 1-2, 1-3)*

24. *Skill sequence*   Compute. *(Previous course)*
  a. $\frac{2}{3} + \frac{4}{3}$ **b.** $\frac{2}{3} + \frac{4}{9}$ **c.** $\frac{2}{3} + \frac{4}{13}$

25. *Skill sequence*   If $x \cdot y = 85.7$, give
  a. $^-x \cdot y$ **b.** $x \cdot {}^-y$ **c.** xy *(Previous course)*

26. Compute in your head. *(Previous course)*
  a. $.8 \cdot 5$ **b.** $.8 \cdot .5$ **c.** $.8 \cdot 5.5$

27. *Multiple choice*   Which of the following are sentences? *(Lesson 1-1)*
  (a) Ten is more than the number.
  (b) Ten more than the number.
  (c) Ten is less than the number.
  (d) Ten less than the number.

28. Copy and complete with decimal answers. *(Previous course)*
  a. 5.8 hundred = $5.8 \cdot 100 = \underline{\ ?\ }$
  b. 5.8 thousand = $5.8 \cdot \underline{\ ?\ } = \underline{\ ?\ }$
  c. 5.8 million = $\underline{\ ?\ } \cdot \underline{\ ?\ } = \underline{\ ?\ }$

29. Molly has one and three-eighths of a yard of ribbon and uses half of it to make a bow. Multiply $\frac{1}{2} \cdot 1\frac{3}{8}$ to find half of $1\frac{3}{8}$. *(Previous course)*

30. A computer can change the value of a variable. Find out what a computer will print when given the following instructions.
```
10 LET X = 52
20 LET X = 4 + 2 * X
30 PRINT X
40 END
```

Almost everyone, at some time or another, has been in a lot of traffic. Perhaps you have been in a traffic jam on your way to an amusement park, or to a sports event, or to a concert. In big cities, adults may be in traffic jams every day on their way to work. Have you ever wondered how many cars are on the road?

An estimate to the answer can be calculated using the sentence below. Here N is the estimated number of cars on the road if they are traveling at safe distances from one another.

$$N = \frac{20Ld}{600 + s^2}$$

N = estimated number of cars
L = number of lanes of road
d = length of road (in feet)
s = average speed of the cars (in miles per hour)

This sentence is a *formula* for the number of cars *in terms of* the number of lanes of the road, the length of the road, and the average speed of the cars. A **formula** is a sentence in which one variable is given in terms of other variables and numbers. In this formula, N is given in terms of L, d, and s.

Example 1 How many cars are there on a 1-mile stretch of a 2-lane highway if the cars are going at an average speed of about 30 miles per hour?

Solution First determine the value of each variable. Here the only difficulty is that the length of the road is given in miles but is needed in feet. Recall that there are 5280 feet in a mile.

$$L = 2 \text{ lanes}$$
$$d = 5280 \text{ feet}$$
$$s = 30 \text{ miles per hour}$$

Now replace the variables by their values.

$$N = \frac{20Ld}{600 + s^2}$$
$$N = \frac{20 \cdot 2 \cdot 5280}{600 + 30^2}$$

Evaluate using the order of operations. A calculator helps here.

$$N = \frac{211200}{1500}$$

$$N \approx 141$$

There are about 141 cars on the road.

The formula shows that the number of cars on a road depends quite a bit on the average speed of the cars. When cars go faster there should be a greater distance between them for safety.

Formulas do not necessarily work for all possible values of the variables. In the formula above, for example, it is meaningless to consider negative numbers for any of the variables and L should be a small whole number. All the values which *may* be meaningfully substituted for a variable make up the **domain** or **replacement set** of the variable. In Example 1, the domains for s and d are the positive real numbers, and the domain for L is the set of whole numbers excluding zero.

Example 2 In some bowling leagues, bowlers who average under 200 can get handicaps added to their score. The handicap H of a bowler, with average A, is often found by using the formula $H = .8(200 - A)$.

  a. What is the handicap of a person whose average score is 145?
  b. What would a bowler's final score be for a game in which his or her actual score was 145?
  c. What happens when an average score greater than 200 is substituted into the formula?
  d. What is the domain of A?

Solution
  a. Substitute 145 in the formula for A and follow the order of operations.
  $$H = .8(200 - 145)$$
  $$= .8(55)$$
  $$= 44$$
  b. Add the handicap to the actual score. $145 + 44 = 189$
  c. If $A > 200$, then $200 - A$ is negative. (This gives a negative handicap, which is ignored.)
  d. $0 \le A \le 200$. Since the average must be less than or equal to 200 in order for this formula to work, the domain of A is the set of numbers satisfying $0 \le A \le 200$.

A computer program can be used to evaluate formulas. In the next example, the programming statement INPUT is used. An **INPUT** statement makes the computer pause and wait for you to type in a value. The computer signals you by flashing its cursor (the symbol that shows you where you are on the screen). On some computers you may see a "?" printed on the screen as well.

Example 3 The circumference C of a circle can be estimated using the formula

$$C = \pi d$$

where d is the diameter of the circle and $\pi \approx 3.14159$.

a. Write a computer program to find C, given d.

b. Estimate the circumference of a circle with diameter 15.5 feet.

Solution

a.

```
10 PRINT "EVALUATE CIRCUMFERENCE FORMULA"
20 PRINT "GIVE DIAMETER"
30 INPUT DIAM
40 LET CIRC = 3.14159 * DIAM
50 PRINT "DIAMETER", "CIRCUMFERENCE"
60 PRINT DIAM, CIRC
70 END
```

Note: Using a comma in the PRINT statements makes the computer print in columns.

b. When the program is run, the lines EVALUATE CIRCUMFERENCE FORMULA and GIVE DIAMETER will automatically appear. Now type 15.5. When you press the RETURN or ENTER key you will see:

```
DIAMETER CIRCUMFERENCE
15.5              48.694645
```

Questions

Covering the Reading

In 1–4, use the formula for the number of cars on a road.

1. In this formula, N is given in terms of __?__.

2. About how many cars are on a 2-mile stretch of a 3-lane highway if the average speed of the cars is 50 mph?

3. About how many cars are on a 1.5 mile part of a 4-lane highway if the average speed of the cars is 20 mph?

4. In this formula, give the domain for: **a.** L, **b.** d, and **c.** s.

5. What is the formula for the handicap H of a bowler whose average score is A?

6. What is the handicap of a bowler whose average score is 120?

7. *Multiple choice*   A bowler with which average is not entitled to a handicap?
(a) 95   (b) 145   (c) 195   (d) 205

8. Substitute 200 for A in the handicap formula. Explain what you get.

9. What does the INPUT statement make a computer do?

10. Using a __?__ in a PRINT statement makes a computer print in columns.

11. What is the circumference of a circle with diameter 7.3 in.?

Applying the Mathematics

12. a. Complete this program which computes area and perimeter of rectangles. Use these formulas: Area $= LW$, Perimeter $= 2L + 2W$.

```
10 PRINT "AREA AND PERIMETER OF A RECTANGLE"
20 INPUT "LENGTH";L
30 INPUT "WIDTH";W
40 LET A = _?_
50 LET P = _?_
60 PRINT "AREA","PERIMETER"
70 PRINT A, P
80 END
```

b. What will the computer find for the area and perimeter when $L = 52.5$ and $W = 38$?

13. The force you feel when riding an elevator is caused by the acceleration of the elevator acting against the pull of gravity.

$$f = \frac{wa}{g}, \text{ where}$$

$f =$ force in pounds
$w =$ your weight in pounds
$a =$ acceleration of elevator in feet per second per second
$g =$ pull of gravity, usually taken to be 32 ft per second per second (32 ft/sec/sec)

a. What is the force when you weigh 120 lbs and the acceleration of the elevator is 5 ft/sec/sec? Use $g = 32$.
b. What happens to the force f as the acceleration of the elevator is increased?
c. On whom does the elevator exert a greater force, a heavy person or a light person?

In 14 and 15, an adult's normal weight w (in pounds) can be estimated by the formula $w = \frac{11}{2}h - 220$ when his or her height h (in inches) is known.

14. Estimate the normal weight of a person who is 6 feet tall.

15. a. According to this formula, what is the normal weight of a person who is 40 inches tall?
b. Explain your answer to part a by relating it to the domain of the variable h.

In 16 and 17, Fahrenheit temperature T can be approximated by counting the chirps C a cricket makes in one minute, and applying the formula

$$T = \tfrac{1}{4}C + 37.$$

16. A cricket chirps 200 times per minute. Estimate the temperature.

17. Estimate the temperature when a cricket chirps 180 times per minute.

Review

In 18–20, evaluate. *(Lesson 1-4)*

18. $(11 - 7)^3$ **19.** $5^2 \cdot 5^3$ **20.** $57 - 3 \cdot 11$

21. What will the computer print on its screen when this program is run? *(Lesson 1-4)*

```
10 PRINT "ANSWER TO QUESTION 21"
20 LET M = 4.56
30 LET N = 3
40 PRINT 4 * M − 9/N
50 END
```

22. Evaluate $28 - 3(x - 4)$ when $x = 9$. *(Lesson 1-4)*

23. Write a computer program or statement that will print the value of $x + y + 3z$ when $x = 5.7$, $y = 2.006$, and $z = 51.46$. *(Lesson 1-4)*

24. Multiply 1234.5678 by 100,000 in your head. *(Previous course)*

25. *Skill sequence*   Find the sums. Write your result as a fraction or whole number. *(Previous course)*
a. $\frac{3}{4} + \frac{1}{4}$ **b.** $\frac{3}{8} + \frac{1}{4}$ **c.** $\frac{3}{10} + \frac{1}{4}$

26. *Skill sequence*
a. What is 16% of 24?
b. What is 16% of 2400?
c. Suppose 16% of the 24 million people aged 12–18 in the U.S. like jazz. How many people is this? *(Previous course)*

27. The real number π has been calculated to more than 29.36 million decimal places. Write 29.36 million as a single number without a decimal point. *(Previous course)*

Exploration

28. One of the world's most famous formulas was discovered by Albert Einstein in 1905. It is $E = mc^2$.
a. What do E, m, and c stand for?
b. What physical phenomenon does the formula describe?

1-6

Sets

A **set** is a collection of objects called **elements.** Usually the elements are put together for a purpose. Sets are found outside of mathematics, but with different names.

When a set is called	An element is often called
herd of dairy cattle	cow
team	player
committee	member
the U.S. Senate	senator
class	student
place setting	plate

A set often has different properties than its elements. For example, a team in baseball can win the World Series, but a player cannot. A cow moos, but a herd cannot. The Senate can pass legislation, but a senator cannot. A plate might have a radius, a place setting does not.

The standard symbols used for a set are braces {...}, and commas go between the elements. When Lulu, Mike, Nell, Oscar, Paula, and Quincy are the six members of a committee C, you could write
$$C = \{\text{Lulu, Mike, Nell, Oscar, Paula, Quincy}\}.$$

The order of naming elements makes no difference. {Oscar, Mike, Lulu, Nell, Paula, Quincy} is the same committee C. Two sets are **equal** if they have the same elements.

Some frequently used sets in mathematics are the set of whole numbers indicated by the letter W; the set of integers, indicated by I; and the set of real numbers, indicated by R.

W = The set of **whole numbers:** 0, 1, 2, 3, ... Other examples of whole numbers are 100; 1990; five; $\frac{16}{2}$; and 7,000,000.

I = The set of **integers** includes the whole numbers and their opposites. A list could begin 0, 1, -1, 2, -2, 3, -3, ... Other examples of integers are -17.00; $\frac{8}{2}$; 2001; and negative one thousand.

R = The set of **real numbers** consists of all numbers that can be represented as finite or infinite decimals. These include positive and negative numbers, whole numbers, zero, fractions, and decimals themselves. Examples of real numbers are 5; 100,000; -0.0042; $-3\frac{1}{3}$; 0; π; $.\overline{13}$; and $\sqrt{2}$.

Notice that all whole numbers are also integers, and all integers are also real numbers.

The set of real numbers possesses a property not held by the sets of whole numbers or integers. For instance, between 12.6 and 12.7 is the number 12.68. Even if you do not know the numbers, you can still find a number between them. One of the numbers between x and y is their mean, $\frac{x + y}{2}$. Because of the **Density Property**, the set of real numbers is an appropriate domain to use when a variable stands for a measure.

Density Property of Real Numbers:

Between any two real numbers are many other real numbers.

The three sets mentioned on pages 30–31 are often used as domains for variables. The domain of a variable in an open sentence is the set of numbers that might make sense as solutions. The **solution set** is the set of numbers from the domain that actually are solutions.

■ ■ ■ ■ ■ ■ ■

Example 1 Let x = number of people at a meeting.
  a. Give a reasonable domain for x.
  b. If there were more than 20 people, write this as an algebraic sentence using x.
  c. Graph the solution set to part b.

  Solution
  a. Since x is a count of the number of people at the meeting, it does not make sense to have x be a fraction or negative. The domain for x is the set of whole numbers.
  b. $x > 20$
  c. You must show all the whole numbers greater than 20. The three dots next to the graph mean "and so on."

Example 2

Let w = weight of a roast.

a. Give a reasonable domain for w.
b. The roast weighs less than six pounds. Write this as an inequality using w.
c. Graph the solution set to part b.

Solution

a. The roast can weigh a fraction of a pound, but not a negative number. The domain could be the set of positive real numbers.
b. $w < 6$, or $0 < w < 6$ (read "w is greater than 0 and less than 6.") Either is acceptable, since we know w is positive.
c.

Example 3

Let t = temperature outside.

a. Give a reasonable domain for t.
b. It is colder than 6° outside. Write this as an inequality using t.
c. Graph the solution set to part b.

Solution

a. A temperature can be a fraction and it can be positive or negative. The domain is the set of real numbers.
b. $t < 6$
c.

The situation determines whether you should use dots on its graph or whether you should indicate the numbers with a bar. Generally when a situation involves counting, the situation is called **discrete** and is graphed as dots. When a situation requires a measurement, like weight or time, it is called **continuous** and is graphed with a bar.

Sometimes, however, you may use a bar to describe discrete data. "From 34,000,000 to 40,000,000 people are expected to watch the next World Series game on TV" is a situation involving a discrete number of people. To graph the 6,000,001 dots (including endpoints) representing discrete people is time-consuming and the graph looks continuous anyway. Therefore the graph to describe this situation is a bar.

When a set S is discrete, you may want to know the number of elements in it. The symbol **N(S)** stands for the number of elements in S. N({3, 6, 12, 24}) = 4 because the set has 4 elements.

There is a set which has no elements in it. It is called the **empty set** or **null set.** The symbol { } can be used to refer to this set. The Danish letter ø is also used. The set ø might refer to

> the set of points of intersection of two parallel lines,
> the set of U.S. Presidents under 35 years
>   of age,
> the set of all the living dinosaurs,
> the solution set of $x + 2 = x + 4$,

or many other things. N({ }) = N(ø) = 0

Questions

Covering the Reading

1. The objects in a set are called __?__.

In 2–5, a set and an element are given. **a.** State whether the set is discrete or continuous. **b.** Name another element in the set.

2. set: baseball team   element: pitcher

3. set: family   element: mother

4. set: {2, 11, -6}   element: -6

5. set of real numbers   element: 2

6. Which of the following sets are equal?
$A = \{0, 2, -5\}$ $B = \{2, -5\}$ $C = \{-5, 0, 2\}$

In 7–9, tell whether the number is an element of W, I, or R. (It is possible that a number belongs to more than one of these sets.)

7. -10   **8.** $\frac{6}{2}$ **9.** 0.5

In 10–12, let S be the solution set for the sentence. **a.** Give the solution set. **b.** Find N(S).

10. $-5m = 25$ where the domain is R.

11. $-3 < y < 3$ where the domain is W.

12. $-3 < y < 3$ where the domain is I.

13. State the Density Property of Real Numbers.

In 14–16, write a real number that is between the two given numbers.

14. 3.4 and 3.5   **15.** *a*, *b*   **16.** -45 and -45.1

17. Let *E* = elevation of land in Alaska.
  a. Give a reasonable domain for *E*.
  b. Elevations in Alaska range from sea level to 6194 meters (Mt. McKinley). Write this as an algebraic sentence involving *E*.
  c. Write and graph the solution set to part b.

18. Let *h* = height of a dinosaur.
  a. Give a reasonable domain for *h*.
  b. Heights of adult *Tyrannosaurus rex* skeletons range from about 12 m to 15 m. Write this as an algebraic sentence involving *h*.
  c. Write and graph the solution set to part b.

19. a. What does ø represent?
  b. Give a situation where ø might be used.

Applying the Mathematics

In 20–22, tell whether the set is discrete or continuous.

20. The police department will hire only people who are more than 63 inches tall. *T* is the set of acceptable heights.

21. The Hubbard family has more than 5 children. *C* is the set of possible numbers of children.

22. *P* is the set of prime numbers.

In 23 and 24, use the following two sets.

$$C = \{-4, -2, 0, 2, 4\} \qquad D = \{2, 4, 6, 8\}$$

23. Which elements are in both sets *C* and *D*?

24. Which elements appear in set *C* or set *D* or both?

In 25–27, graph the elements of the set on a number line.

25. solution set to $-42 < x \le -11$ where *x* is a real number

26. {0, 1, 2, 3, 4, 5, ...}

27. the set of even whole numbers not greater than 10

28. What is N(R) if $R = \{7, 8, 9, \ldots, 18\}$?

29. Using $\{2, 5, 7, 9\}$ as a domain, find the solution set of $4a + 1 > 20$.

30. Give two real numbers that are between $\frac{3}{10}$ and $\frac{1}{3}$.

Review

31. Shoe size s and foot length F (in inches) for men are related by the formula $s = 3F - 24$. What is the shoe size of someone whose foot is $10\frac{1}{2}$ inches long? *(Lesson 1-5)*

32. *Multiple choice*　　Which expression summarizes the following directions? Begin with a number n. Find the product of n and 12.4. Subtract 11. Multiply the result by 6. *(Lesson 1-4)*
(a) $n \cdot 12.4 - 11 \cdot 6$　　　(b) $n \cdot (12.4 - 11) \cdot 6$
(c) $(n \cdot 12.4) - (11 \cdot 6)$　　(d) $(n \cdot 12.4 - 11) \cdot 6$

33. Insert parentheses around one of the subtractions so that the sentence is true. *(Lesson 1-4)*
$35 - 20 - 7 = 22$

34. Use this graph. *(Lesson 1-3)*

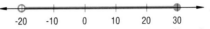

a. Copy and complete to describe the interval: ___?___ $< x \leq$ ___?___.
b. What is the range of the interval?

35. Evaluate in your head when $m = 3$. *(Lesson 1-4)*
a. $-2m$　　　　　**b.** m^2
c. $2 + 2m$　　　　**d.** $m + -12$

Exploration

36. Collections of animals frequently are given special names. Match the group name with the correct animal as in "school of fish."

Group name	Animal
cloud	ants
colony	bees
exaltation	crows
gaggle	fish
hive	foxes
leap	geese
mob	gnats
murder	kangaroos
pride	larks
nest	leopards
school	lions
skulk	nightingales
watch	oxen
yoke	vipers

DENNIS THE MENACE

"THIS WEEK WE'RE STUDYING REAL DEEP STUFF, LIKE: A BUNCH OF SHEEP IS A FLOCK ... AND A FLOCK OF FLOWERS IS A BUNCH."

37. Name three sets outside of mathematics that are not mentioned in this lesson. For each, indicate what an element is usually called.

LESSON

1-7

Probability

Probability measures how likely it is that something could happen. Suppose you toss an ordinary die once. There are six different faces which could show on the top of the die.

The result of tossing a die is called an **outcome.** There are six different outcomes: 1, 2, 3, 4, 5, 6. If each outcome is assumed to occur as often as any other outcome, the outcomes are called **equally likely** or **random.** Thus, the probability of tossing 4 is $\frac{1}{6}$, since 4 is one out of the six outcomes.

■ ■ ■ ■ ■ ■ ■ ■

Example 1 Hurricanes happen often enough in Florida that public officials must have evacuation plans ready. These plans must take into account whether people are likely to be at home, work, or school. If a hurricane hits, what is the probability that it occurs on Sunday when there are the fewest people in schools and offices?

Solution Sunday is one out of the seven days of the week. If a hurricane occurs, the probability that it hits on Sunday is $\frac{1}{7}$.

Finding probabilities is often a matter of counting. Suppose you draw one card from an ordinary deck of 52 playing cards. What is the probability that it is a 5? In the situation "selecting a card from a deck" there are 52 equally-likely outcomes. Of these, "picking a 5" is not a single outcome, but a set of 4 different outcomes:

5♣ (5 of clubs)
5♦ (5 of diamonds)
5♥ (5 of hearts)
5♠ (5 of spades)

Since there are four ways to draw a 5 out of a possible 52 ways to choose a card from the deck, the probability of "picking a 5 from a deck of cards" is $\frac{4}{52}$. Here "choosing a 5" is called a *success*. "Not choosing a 5" is called a *failure*. To compute the **probability of a success,** you divide the number of outcomes that are successes by the total number of outcomes.

36

Definition:

Suppose a situation has T possible equally-likely outcomes of which S are successes. Then the probability of a success is $\frac{S}{T}$.

An **event** is a set of outcomes. Tossing a coin once has two possible outcomes, heads or tails. Many probabilities can be calculated simply by counting both the number of outcomes in the event and the number of all possible outcomes.

Example 2   A number is selected randomly from {1, 2, 3, 4, 5, 6, 7, 8, 9}. What is the probability that the number is prime?

Solution   Remember that a prime number is a whole number greater than 1 that is divisible only by 1 and itself.

$$1 \ \textcircled{2}\textcircled{3} \ 4 \ \textcircled{5} \ 6 \ \textcircled{7} \ 8 \ 9$$

The four ways to select a prime number are circled. The total number of outcomes is 9. Since the number is selected randomly, the probability that it is a prime is $\frac{4}{9}$.

When all outcomes of a situation like "tossing a die" have the same probability, the die is said to be *fair* or *unbiased*. In this lesson, dice, coins, and cards are assumed to be fair.

Example 3   When you roll a die, what is the probability that you will get
  a. an even number?
  b. a 7?
  c. a number less than 7?

Solution   The list of possible, equally-likely outcomes is 1, 2, 3, 4, 5, 6.
  a. Three of the numbers, 2, 4, and 6 are even.
     The probability of tossing an even number is $\frac{3}{6}$, or $\frac{1}{2}$.
  b. 7 is not a possible outcome. The number of successes is 0.
     The probability of tossing a 7 is $\frac{0}{6}$, or 0.
  c. There are six successes: 1, 2, 3, 4, 5, 6. So the probability is $\frac{6}{6}$, or 1.

Notice in Example 3 that the event in part b is impossible and its probability is zero. This is the smallest number that a probability can be. The largest probability possible is 1. This happens in cases like part c that always happen. If P is a probability, then $0 \leq P \leq 1$.

Outcomes are not always equally likely. Consider the probability that you will find $1,000,000 on your way home today. There are two outcomes: finding the money and not finding it. This suggests that the probability of each is $\frac{1}{2}$. But "finding $1,000,000 today" and "not finding $1,000,000 today" are not equally-likely outcomes. The probability of finding the money is near zero; the probability of not finding it is near 1. You cannot use the probability formula $P = \frac{S}{T}$ in this situation.

Sometimes whether or not events are equally likely depends on how you look at them. For example, in many games you add numbers that you roll on two dice. There are eleven possible sums: 2, 3, 4, 5, 6, 7, 8, 9, 10, 11, and 12. However, these are not equally likely. A sum of 12 occurs less frequently than a sum of 7 or 8. However, you can overcome this problem by considering all 36 possible pairs of numbers rather than the 11 possible sums. The possible pairs are equally likely.

Example 4   Suppose two fair dice are thrown once. What is the probability that the sum of the dots is 7?

Solution   Each die has 6 sides. The 36 possible outcomes are shown in the array below.
The 36 possible outcomes are equally likely, so the formula $P = \frac{S}{T}$ can be used. Here $T = 36$. The outcomes that give a sum of 7 are circled. There are six successes, so $S = 6$.

$$P = \frac{S}{T} = \frac{6}{36}$$
$$= \frac{1}{6}$$

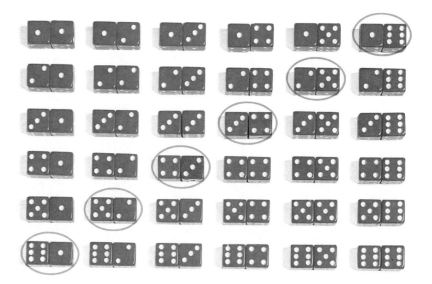

Questions

Covering the Reading

1. For a situation with 400 outcomes occurring randomly, what is the probability of a single outcome?

2. If a hurricane strikes, what is the probability that it hits on a week-day?

3. If T outcomes are equally likely, and S are successes, what is the probability of a success?

In 4 and 5, a single die is tossed once. Find each of the following:

4. the probability of tossing a 5

5. the probability that the die shows a number less than 3

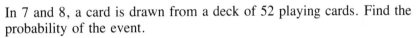

6. Carlotta mails an entry to a magazine sweepstakes. She says: "The probability of winning is $\frac{1}{2}$, since either I will win or I won't." What is wrong with this argument?

In 7 and 8, a card is drawn from a deck of 52 playing cards. Find the probability of the event.

7. selecting the 7 of clubs

8. selecting a queen

9. A number is drawn randomly from the set {1, 2, 3, 4, 5}. What is probability that the number is
   a. an even number?
   b. not an even number?
   c. a number less than 10?

10. A pair of dice is tossed once.
   a. Show all the successful outcomes of getting a sum of 9.
   b. Find the probability of getting a 9.

Applying the Mathematics

In 11 and 12, a pair of dice is tossed once. For each event, **a.** list the set of successful outcomes; **b.** find the probability.

11. a sum greater than 9

12. a sum which is not greater than 3

13. You are holding a raffle. You want the chances of winning to be 1 in 10. If 80 people enter the raffle, how many prizes do you need?

14. If a coin is flipped once, what is the probability that it lands on heads?

15. A dime and a nickel are tossed once. The four possible outcomes are in the array below. The outcome HT means that the nickel came up heads and the dime came up tails. If the coins are fair, find:
  a. the probability that both coins show heads.
  b. the probability that one coin shows a head and one shows a tail.

```
              dime
            H      T
       ┌─────────────────
     H │   HH     HT
nickel │
     T │   TH     TT
```

16. A dime, a nickel, and a penny are tossed once.
  a. Write the eight possible outcomes.
  b. If the coins are fair, find the probability that all three coins are heads.
  c. If the coins are fair, find the probability that only one coin shows heads.

17. Tell why each of the following cannot represent a probability.
  a. $\frac{50}{40}$ **b.** $\frac{7}{-14}$ **c.** 2

18. Bo T. Fell, one of the world's greatest algebra teachers, teaches only 3 of the 12 algebra classes at his school. If you were at his school and students were assigned randomly to classes, what is the probability you would not get Mr. Fell?

19. If a letter is picked randomly from the alphabet, what is the probability it is between I and U (not including them)?

Review

20. a. Write an inequality to describe this graph. *(Lesson 1-6)*

  b. What is the domain?

21. The domain for g is $\{-3, 0, 5\}$. For each element of the domain, evaluate $10g^2$. *(Lesson 1-4, 1-6)*

22. a. Using I as a domain, graph $y > -1$ on one number line, and $y < 7$ on another.
  b. What values appear on both graphs? *(Lessons 1-1,1-6)*

23. Give a real number that is between x and y. *(Lesson 1-6)*

24. Often 15% of a restaurant bill is left for a tip. If a bill was $24.89, then what would be the tip? *(Previous course)*

25. *Multiple choice*   A theater has the following prices:

Infants (under 3)   -   free
Children (under 12)   - $2.50
Adults   - $5.00

Which graph shows the ages of people who will pay $2.50? *(Lesson 1-3)*

(a)

(b)

(c)

26. A certain pitcher has gotten a hit 20% of her times at bat. If the pitcher has batted 80 times, how many hits does she have? *(Previous course)*

27. Compute in your head, given that $34 \cdot 651 = 22{,}134$. *(Previous course)*
   a. $.01 \cdot 34 \cdot 651$ **b.** $.001 \cdot 34 \cdot 651$ **c.** $.0001 \cdot 34 \cdot 651$

28. Round 14.57052 to: **a.** the nearest tenth; **b.** the nearest hundredth; **c.** the nearest thousandth. *(Previous course)*

29. At Lowe High School, 240 students are in a musical group. If there are 1600 students in the school, what percent of the students are in a musical group? *(Previous course)*

30. When outcomes are not equally likely, the situation is called *biased*. Biased coins are sometimes called ''weighted'' or ''two-headed.'' What are some names for **a.** a biased pair of dice; **b.** a biased deck of cards?

1-8

Relative Frequency and the Equal Fractions Property

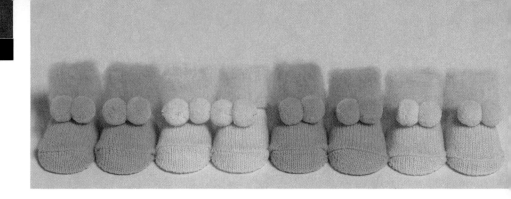

A **relative frequency** can be used to estimate how likely it is that a certain thing will happen. Consider the following question. How likely is it that a newborn baby will be a boy? One way to approach this problem is to use available data. In the U.S. in 1980, of about 3,612,258 children born, 1,852,616 were boys. The relative frequency of boys to births is found by dividing the number of boys by the total number of births.

$$\text{relative frequency} = \frac{\text{number of boys born}}{\text{total number of births}}$$

$$= \frac{1,852,616}{3,612,258} \approx 0.513 \approx 51.3\%$$

The definition of relative frequency is similar to the definition of probability.

Relative frequency of an event:

$$\frac{\text{number of times the event occurred}}{\text{total number of possible occurrences}}$$

Relative frequencies are sometimes given as decimals. A batter who has made 24 hits out of 90 times at bat is reported to have a "batting average" of $\frac{24}{90} \approx .267$. Survey results are typically reported in percent form. If a survey of 2000 adults shows that 1142 own a car, then the relative frequency of car owners is $\frac{1142}{2000} = .571 \approx 57\%$.

Relative frequencies and probabilities are not the same. A relative frequency always results from an actual experiment or data. A probability is always calculated from assumptions or is just assigned. In the example at the beginning of this lesson, the assumption might be that $\frac{1}{2}$, or 50%, of newborns are boys, but a more accurate probability would be 51.3%. However, we often use actual experiments to estimate the probability of an event.

Example 1 Suppose that 1025 people responded to the question "In which month do you feel the happiest?" with the following results:

January	19	June	138	November	24
February	12	July	153	December	182
March	33	August	47	No Choice	136
April	71	September	61		
May	96	October	53		

Estimate the probability that a randomly selected person would choose December as his or her happiest month.

Solution Calculate the relative frequency of those who said that December was their favorite month.

$$\text{relative frequency} = \frac{182}{1025} \approx 0.178$$

So, an estimate for the probability that a randomly selected person would select December is 0.178, or about 18%.

Suppose you asked eight friends the question, "Which do you prefer, chicken or hamburgers?" If two of your friends said "chicken," then the relative frequency of "chicken" answers is $\frac{2}{8}$ = .25 = 25%. Now suppose a marketing company asked the same question of 1200 people. If 300 of these people said "chicken," the relative frequency of "chicken" answers would be $\frac{300}{1200}$. Their relative frequency is the same as yours, since $\frac{2}{8} = \frac{300}{1200}$.

But people would certainly have more faith in 300 out of 1200 than in 2 out of 8. We might assign a probability of 0.25 to liking chicken more than hamburgers.

You know $\frac{2}{8}$ and $\frac{300}{1200}$ are equal, because they equal the same decimal, 0.25. But also, one fraction can be converted to the other.

$$\frac{2}{8} = \frac{2 \cdot 150}{8 \cdot 150} = \frac{300}{1200}$$

You have often rewritten fractions in this way. This is an instance of a general pattern we call the **Equal Fractions Property.**

Equal Fractions Property:

If $k \neq 0$ and $b \neq 0$, $\frac{a}{b} = \frac{ak}{bk}$.

The Equal Fractions Property can be used to simplify fractions. Very often, it helps to think of it "backwards," as $\frac{ak}{bk} = \frac{a}{b}$.

■ ■ ■ ■ ■ ■ ■ ■

Example 2   Simplify $\dfrac{34}{51}$.

Solution   Look for common factors. $\dfrac{34}{51} = \dfrac{2 \cdot 17}{3 \cdot 17}$

Apply the Equal Fractions Property. $= \dfrac{2}{3}$

The Equal Fractions Property can also be used to rewrite fractions involving variables. For instance, $\dfrac{2mx}{3my} = \dfrac{2x}{3y}$ and $\dfrac{5x}{6w} = \dfrac{5xy}{6wy}$.

Note: Here and elsewhere, when a fraction is written, we assume its denominator is not zero.

■ ■ ■ ■ ■ ■ ■ ■

Example 3   Simplify $\dfrac{2rt}{12rs}$.

Solution   Look for common factors. $\dfrac{2rt}{12rs} = \dfrac{2 \cdot r \cdot t}{6 \cdot 2 \cdot r \cdot s}$

$= \dfrac{t}{6s}$

The Equal Fractions Property can be used to help interpret information you encounter in the media.

■ ■ ■ ■ ■ ■ ■ ■

Example 4   A newspaper reports that its telephone survey of households showed 40% favored building a sports stadium. Give some possibilities of the number of households surveyed and the number favoring a sports stadium.

Solution   You must find pairs of reasonable numbers. Each pair must form a ratio that is equal to 40%. $40\% = .40 = \frac{40}{100}$. So one possibility is that 40 favored the stadium out of 100 households surveyed.

The equal fractions property says that $\dfrac{40}{100} = \dfrac{40k}{100k}$ where $k \neq 0$.
One possibility is that $k = 6$.

Then $\dfrac{40}{100} = \dfrac{40 \cdot 6}{100 \cdot 6} = \dfrac{240}{600}$.

So another possibility is that 600 households were surveyed and 240 favored the sports stadium.

It is also possible that fewer than 100 households were surveyed.

$$\dfrac{40}{100} = \dfrac{2 \cdot 20}{5 \cdot 20} = \dfrac{2}{5}$$

Perhaps only 5 were surveyed, and 2 favored the stadium.

The table below summarizes some important similarities and differences between relative frequency and probability.

Relative Frequency	Probability
1. Between 0 and 1, inclusive	Between 0 and 1, inclusive.
2. A relative frequency of 0 means that an event did not occur.	A probability of 0 indicates that an event is impossible.
3. A relative frequency of 1 means that an event occurred every time.	A probability of 1 indicates that an event must happen.
4. The more often an event occurred relative to the possible number of occurrences, the closer its relative frequency is to 1.	The more likely an event is, the closer its probability is to 1.

Questions

Covering the Reading

1. In 1970, of 3,791,258 children born in the U.S., 1,915,378 were boys.
   a. What was the relative frequency of boys to births in 1970?
   b. Using this information, estimate the probability of a male birth in 1970. (Give your answer as a percent.)
   c. How does this estimate compare to the estimate calculated with the data from 1980?
   d. Estimate the probability of a female birth in 1970.

2. Relative frequencies are often used to __?__ probabilities that certain events will happen.

In 3 and 4, use the data of Example 1. Give the relative frequency for a person responding:

3. October is his or her happiest month. (Write as a decimal and as a percent.)

4. Her or his happiest month begins in winter (January, February, or March).

5. State the Equal Fractions Property.

6. In the fraction $\frac{x}{y}$, what value can y not have?

7. Simplify:
   a. $\frac{25}{85}$
   b. $\frac{42x}{420x}$
   c. $\frac{36}{9y}$

8. a. Simplify $\frac{2xy}{3ay}$. b. Rewrite $\frac{2x}{3a}$ with a denominator of $3ya$.

9. In a survey of a neighborhood, 60% of those surveyed expressed a need for greater police protection.
   a. If 100 were surveyed, how many said they needed more police protection?
   b. If 500 people were surveyed, how many said they needed more police protection?
   c. If 5 people were surveyed, how many said they needed more police protection?

10. If the probability of an event is 1, what do you know about that event?

11. If the relative frequency of an event is 0, is the event impossible?

Multiple choice   In 12 and 13, find the relative frequency not equal to the other three.

12. (a) $\frac{9}{11}$ (b) $\frac{99}{121}$ (c) $\frac{90}{100}$ (d) $\frac{450}{550}$

13. (a) $\frac{100}{260}$ (b) $\frac{35}{91}$ (c) $\frac{38}{100}$ (d) $\frac{500}{1300}$

14. According to a survey in 1981, one out of four teenagers who worked earned more than $200 a month. Use the Equal Fractions Property to write three possibilities for the total number of teenagers surveyed and the number who earned more than $200.

15. Event X has a probability of $\frac{3}{4}$. Event Y has a probability of $\frac{1}{2}$. Event Z has a probability of $\frac{2}{3}$.
    a. If you could find actual relative frequencies of these events, which event would you expect to have the largest relative frequency?
    b. Event Y has the smallest probability. Must it have the smallest relative frequency?

In 16 and 17, consider the following. A tire company tests 50 tires to see how long they last under typical road conditions. The results they obtained are shown at the right.

Mileage until worn out	Number of tires
10,000–14,999	1
15,000–19,999	3
20,000–24,999	6
25,000–29,999	15
30,000–34,999	14
35,000–39,999	7
40,000 or more	4

16. What is the relative frequency of a tire lasting less than 25,000 miles? (Write as a fraction.)

17. What is the relative frequency of a tire lasting at least 10,000 miles?

18. Since about 51% of people born in the U.S. are male, why aren't 51% of all living people in the U.S. male?

19. *Multiple choice*   An event occurred c times out of t possibilities. The relative frequency of the event was 30%. Which is true?
    (a) $\frac{c}{t} = 0.3$ (b) $\frac{t}{c} = 0.3$
    (c) $ct = 0.3$ (d) $t - c = 0.3$

20. *Multiple choice* Of the people surveyed, $\frac{4}{9}$ thought the American League team would win the World Series. If 36n people were surveyed, how many thought the American League team would win?
(a) $\frac{4}{9}$n (b) 4n (c) 9n (d) 16n

21. What is y if $\frac{6y}{27} = \frac{2}{3}$? (Hint: $\frac{2}{3}$ is how many twenty-sevenths?)

22. *True or false* When $x = 1$, then $\frac{4 + x}{2 + x} = \frac{2}{1}$.

23. In a certain school, out of 50 girls surveyed,
    5 girls played both basketball and softball;
    12 girls played basketball but not softball;
    15 girls played softball but not basketball.
What percentage of girls surveyed played softball?

Basketball   Softball

Review

24. Imagine tossing two fair dice once. Refer to Example 4 of Lesson 1-7 for a list of all possible outcomes. Find each of the following:
a. P(sum of 8), i.e., the probability of getting a sum of 8
b. P(sum not equal to 8)
c. P(sum less than 13) *(Lesson 1-7)*

25. A card is drawn randomly from a deck of 52 playing cards. Find each of the following:
a. P(7 of hearts)   **b.** P(a heart)   **c.** P(a king) *(Lesson 1-7)*

26. One fifth of the students will not go to the dance. Of these, nine tenths have to work. What is the fraction of all the students who will not go to the dance because they have to work? *(Previous course)*

27. *Skill sequence* What does each equal? *(Previous course)*
a. 3.95 from 12.8
b. 1.28 minus .395
c. $39.95 less $1.52

28. Order from largest to smallest. *(Previous course, Appendix B)*
$6.5 \cdot 10^{14}$ $7.2 \cdot 10^{15}$ $9.4 \cdot 10^{13}$

29. What is N(ø) + N({6, 7, 8,}) − N({2, 4})? *(Lesson 1-6)*

In 30–32, determine in your head which is larger. *(Lesson 1-1)*
30. $2\frac{1}{2}$ or 2.45   **31.** $\frac{7}{2}$ or $\frac{7}{3}$ **32.** $\frac{3}{4}$ or $\frac{4}{5}$

Exploration

33. In Example 4 of Lesson 1-7, two fair dice were thrown once. The probability of getting a sum of 7 was calculated to be $\frac{1}{6}$.
a. Try the corresponding experiment. Find a pair of dice and actually toss them 72 times. Record the results.
b. What relative frequency of getting a 7 do you get?
c. For any sums from 2 through 12, does any relative frequency you get match the corresponding probability?

Summary

Algebra is a language of numbers, operations, and variables. In this chapter, three important uses of variables are discussed. Variables describe number relationships, as in $k > 19.5$ or $-3 < g < 0$. Variables are a shorthand used in formulas like $A = \ell w$. And variables may represent unknowns, as in the open sentence $x + 5 = 8$.

Graphs help in understanding variables and numbers. Data can be pictured on a graph with a dot frequency diagram. Solutions to inequalities and intervals can be graphed on a number line.

Like any other language, algebra has its rules. Among these are the rules for order of operations: work in parentheses, then powers, then multiplications or divisions, then additions or subtractions. Scientific calculators and computers usually follow the same rules.

The language of sets helps with many ideas in this chapter. The domain of a variable is the set of values it may take. The solution set to a sentence is the set of values making that sentence true. An event is a set of outcomes.

Both relative frequencies and probabilities indicate the likelihood of an event. Relative frequencies are calculated from actual data. The relative frequency of an event E is the fraction of times E actually occurs out of the total number of possible occurrences. Probabilities are either taken from relative frequencies or calculated by assuming things about outcomes. If outcomes are assumed to occur randomly, the probability of an event E is the number of ways E can occur divided by the number of all possible outcomes.

Vocabulary

Below are the most important terms and phrases for this chapter. You should be able to give a general description and a specific example of each.

Lesson 1-1
variable
open sentence, solution
equation, inequality
$=, \neq, <, \leq, \approx, >, \geq$

Lesson 1-2
dot frequency diagram
statistics
mean, median, mode, central tendency

Lesson 1-3
interval, endpoints
open interval, closed interval
$\pm$, range, length of interval

Lesson 1-4
evaluating the expression
order of operations
numerical expression, algebraic expression
BASIC, PRINT, LET

Lesson 1-5
formula
domain, replacement set
INPUT

Lesson 1-6
set, element, {...}, equal sets
whole numbers, integers, real numbers
Density Property of Real Numbers
solution set, discrete, continuous
N(S)
empty set, null set, { }, ø

Lesson 1-7
outcome, equally likely or random
probability of a success
event

Lesson 1-8
relative frequency
Equal Fractions Property

Progress Self-Test

Take this test as you would take a test in class. Use a ruler and calculator. Then check your work with the solutions in the Selected Answers section in the back of the book.

In 1–3, evaluate each expression.

1. $2(a + 3b)$ when $a = 3$ and $b = 5$.

2. $5 \cdot 6^n$ when $n = 4$.

3. $\dfrac{p + t^2}{p + t}$ when $p = 5$ and $t = 2$.

4. Which is largest, $\dfrac{7}{12}$, $\dfrac{13}{22}$, or $\dfrac{3}{5}$?

5. Simplify: $\dfrac{63abc}{28b}$.

6. The last four times Kevin ran the 100-meter dash his times were 13.1, 12.6, 12.7, and 12.3 seconds. Find the mean of his times.

7. Dennis earned $6, $4.50, $10.00, $5.00, $5.00, and $6 for six nights of babysitting. Find the median for these amounts.

8. Here is a program to be run by a computer. What will be printed?

```
10 PRINT "ANSWER TO QUES. 8"
20 LET M = 6.3
30 LET N = 1.4
40 PRINT M * N
50 END
```

9. The Humbert family car goes from 240 to 336 miles on a tank of gas. Graph this interval on a horizontal number line.

10. A road has a 25 mph speed limit. A person is driving at S mph and is speeding. Express the possible values of S with an inequality.

In 11 and 12, W = the set of whole numbers, I = the set of integers, R = the set of real numbers, and P = the set of positive real numbers.

11. Which one of these sets is a reasonable domain for L, the length of an animal?

12. -5 is an element of which two of these sets?

13. A poll accurate to within 3% showed 67% of the registered voters planned to vote. What are the maximum and minimum percentages of voters planning to vote?

14. Which of the numbers 2, 5, and 8 makes the open sentence $4y + 7 = 2y + 23$ true?

15. Give three values for x that make $3x < 24$ true.

16. The formula $c = 20(n - 1) + 25$ gives the cost of first-class postage in 1988. In the formula, c is the cost in cents and n is the weight of the mail rounded up to the nearest ounce. What does it cost to mail a 3.2 ounce letter?

17. The area of a circle is given by the formula $A = \pi r^2$. To the nearest square meter, what is the area of a circle with radius 3 meters? (Use $\pi \approx 3.14159$.)

In 18 and 19, a survey was done to determine the cost of granola bars. The prices (in cents per ounce) of 21 brands are shown on the dot frequency diagram below.

cents per ounce

18. a. How many brands cost 12¢ per ounce?

b. How many brands cost 15¢ per ounce or more?

19. Describe the range of costs using an inequality.

20. In 1981, a survey of 100,000 U.S. citizens showed 2284 born before 1901. What was the relative frequency of a 1981 U.S. citizen being born before 1901?

21. There are 300 tickets in a raffle drawing for a 10-speed bike. If you buy 8 tickets, what is the probability you will win the bike?

22. A letter from the alphabet is chosen randomly. What is the probability it is a letter in the word CATS?

23. Alain is in the finals of a bowling tournament. He will face the winner of the match between Donna and Sarah. Alain has beaten Donna 6 games out of 9, while he has beaten Sarah 3 games out of 5. Which opponent is Alain more likely to beat?

24. Let S be the solution set for $x < 8$ where the domain is W. Graph S on a number line.

25. If $G = \{-5, -4, -3, \ldots, 5\}$, what is $N(G)$?

Chapter Review

Questions on **SPUR** Objectives

SPUR stands for **S**kills, **P**roperties, **U**ses, and **R**epresentations.
The Chapter Review questions are grouped according to the
SPUR Objectives for this chapter.

SKILLS deal with following a procedure to get an answer.

■ **Objective A.** *Operate with, and compare fractions. (Lessons 1-1, 1-8)*

1. Write an inequality to compare $\frac{1}{6}$ and $\frac{3}{20}$.

2. Which is larger, $2\frac{3}{8}$ or $\frac{7}{3}$?

3. Which is larger, $\frac{3}{5}$ or $\frac{3}{4}$?

4. Which is smaller, 1.7 or $1\frac{2}{3}$?

In 5–6, use the Equal Fractions Property to simplify fractions.

5. $\dfrac{11pq}{12mp}$ 6. $\dfrac{49ab}{14b}$

In 7 and 8, write as a single fraction.

7. $\frac{1}{5} + \frac{2}{3} - \frac{3}{4}$ 8. $2\frac{1}{6} + 1\frac{1}{3}$

■ **Objective B.** *Find solutions to open sentences using trial and error. (Lesson 1-1)*

9. Which of the numbers 3, 4, or 5 is a solution to $2x + 13 = 3x + 9$?

10. Using $\{1, 4, 7\}$ as a domain, give the solution set of $7x - 13 < 2x$.

11. Give three values for y that make $2y \geq 150$ true.

12. Solve for z: $2 + z = 3$.

■ **Objective C.** *Evaluate numerical and algebraic expressions. (Lesson 1-4)*

13. Evaluate $-2p$ when $p = 3.5$.

14. Evaluate $4x^2$ when $x = 1$.

15. Evaluate $35 + 5 \cdot 2$.

16. *True or false* $2 - 3/5 + 6 = (2 - 3)/(5 + 6)$.

17. Find the value of $4(p - q)$ when $p = 13.8$ and $q = 5.4$.

18. Evaluate $5(M - N)$ when $M = \frac{2}{3}$ and $N = \frac{1}{5}$.

19. Evaluate $2b^3$ when $b = 5$.

20. Evaluate $\left(\dfrac{n}{4}\right)^2$ when $n = 36$.

■ **Objective D.** *Give the output for or write short computer programs and statements that evaluate expressions. (Lessons 1-4, 1-5)*

In 21 and 22, what will the computer print if the statement is entered?

21. PRINT 8 * 3 − 28/2

22. PRINT -1 * (1.5 − 2.3)

23. If the program below is run, what will be printed after 2 is entered as input?

```
10 PRINT "ANSWER TO QUESTION 23"
20 PRINT "ENTER A NUMBER"
30 INPUT M
40 PRINT 3 + M/2 + M
50 END
```

24. The area of a circle can be found by using the formula $A = \pi r^2$ where r is the radius of the circle and $\pi \approx 3.14159$.

   a. Write a computer program to find A given $r = 12.5$. (Hint: πr^2 is the same as $\pi \cdot r \cdot r$.)

   b. What is the area of a circle with radius 12.5 cm?

PROPERTIES deal with the principles behind the mathematics.

■ **Objective E.** *Read and interpret set language and notation. (Lesson 1-6)*

25. Let B = the solution set to $-2 < n < 4$, where n is an integer.
    a. List the elements of B.
    b. What is N(B)?

26. Let P = {10, 20, 30, 40, 50} and Q = the solution set to $x < 34$. Which elements are in both sets P and Q?

27. *True or false* $\frac{2}{1}$ is an element of the set of whole numbers.

28. a. What is ø called?   b. What is N(ø)?

29. If S = the set of states in the USA, what is N(S)?

In 30 and 31, use these sets.
A = {-3, 0, 3, 6} B = {6, 3, 0, -3}
C = {-3, 3, 6} D = {-4, 0, 4, 8}

30. Which of the sets are equal?

31. Find N(D) − N(C) + N(B).

■ **Objective F.** *Find and interpret the probability of an event when outcomes are assumed to occur randomly. (Lesson 1-7)*

32. A number is selected randomly from {2, 4, 6, 8, 10, 12}. What is the probability that the number is divisible by 5?

33. A card is picked randomly from a standard deck of 52 playing cards. What is the probability that the card is a seven?

34. Two fair dice are thrown once. What is the probability of a sum of 6?

35. Event A has a probability of 0.3. Event B has a probability of $\frac{4}{9}$. Event C has a probability of 33%. Which event is
    a. most likely to happen?
    b. least likely to happen?

USES deal with applications of mathematics in real situations.

■ **Objective G.** *In real situations, choose a reasonable domain for a variable. (Lesson 1-6)*

In 36–39, choose a domain for the variable from these sets:

    real numbers   whole numbers
    positive real numbers   integers

Tell whether the domain is discrete or continuous.

36. d, the distance that a jogger runs

37. roast beef is being ordered for n people

38. P is the altitude of a point on the surface of the earth.

39. the weight w of a molecule

■ **Objective H.** *Calculate the mean, median, and mode for a set of data. (Lesson 1-2)*

In 40 and 41, use the following information. Nine kids were asked how much they paid for their jeans. They paid:
    $22, $10.50, $14, $12, $12, $17, $38, $25 and $33.

40. Find the mean, median, and mode for these prices.

41. *True or false*   The mode does not give an appropriate description of the amount the majority of kids paid.

In 42 and 43, the ages of participants in an exercise class are 26, 49, 23, 52, 49, 23, 33, and 39.

42. Find the mean age of the participants.

43. *True or false*   The median age is an integer.

■ **Objective I.** *Evaluate formulas in real situations. (Lesson 1-5)*

44. The percent discount p on an item is given by the formula
$$p = 100\left(1 - \frac{n}{g}\right)$$
where g is the original price and n is the new price. Find the percent discount on a pair of jeans whose price is reduced from $20 to $15.

45. If a can of orange juice is h centimeters high and has a bottom radius of r centimeters, then the volume equals $\pi r^2 h$ cubic centimeters. What is the volume, to the nearest cubic centimeter, of a can 12 centimeters high with a radius of 6 centimeters?

In 46 and 47, use this information. The cost c of carpeting a room is given by $c = p\left(\dfrac{lw}{9}\right)$, where

p is the price of the carpeting per square yard and

l and w are the length and width of the room in feet.

46. Find the cost of carpeting a 12' by 15' room with carpeting that sells for $19.95 per square yard.

47. At $5.99 per square yard, what is the cost of carpeting an 8' by 30' deck?

■ **Objective J.** *Use relative frequency to determine information about surveys.* *(Lesson 1-8)*

48. One in four people refused to answer a market survey. Give three possibilities for the number of people surveyed and the number of people who refused to answer.

49. A survey found three fifths of amateur violinists to be women. If 400 violinists were surveyed, how many were women?

50. At a shopping mall m people were asked about a certain product. w people liked the product. What is the relative frequency of people who liked the product?

51. In 1980, France had a reported population of 54,652,000. There were 9,619,000 people over the age of 60. What is the relative frequency of people 60 or under? (Give your answer as a percent.)

52. One in about 86 births in the U.S. results in twins.
   a. What is the relative frequency of having twins?
   b. What is an estimate of the probability of having twins?

53. *Multiple choice* An event occurred t times out of p possibilities. The relative frequency of the event was $\frac{1}{8}$. Which is true?
   (a) $\dfrac{t}{p} = \dfrac{1}{8}$ (b) $\dfrac{p}{t} = \dfrac{1}{8}$
   (c) $\dfrac{t}{p + t} = \dfrac{1}{8}$ (d) $\dfrac{p}{p + t} = \dfrac{1}{8}$

REPRESENTATIONS deal with pictures, graphs, or objects that illustrate concepts.

■ **Objective K.** *Use graphs or symbols to describe intervals.* *(Lessons 1-1, 1-3)*

54. Kim bought stamps for less than $25. On a number line, graph what she might have spent.

55. Consider the sign pictured.
   a. Express the interval of legal speeds as an inequality using s to represent speed.
   b. Graph all legal speeds.

SPEED LIMIT **65** / **45** MINIMUM

56. Graph the closed interval 35 ± 1.5.

57. Graph the solution set to $y \geq 19$,
   a. if y is a real number;
   b. if y is an integer.

In 58 and 59, choose from the three graphs below.

58. Which of these could be a graph of the solutions to $57 < n \leq 62$?

59. Which graph represents the statement "There are from 57 to 62 students with green eyes in the school"?

(a)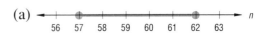

(b)

(c)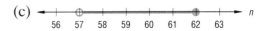

■ **Objective L.** *Interpret dot frequency graphs.* *(Lesson 1-2)*

60. A class of 16 students was asked how many hours of exercise they get a week. The dot frequency diagram below shows their responses.

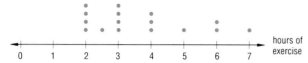

a. Describe the range of the hours with an inequality.
b. What is the median number of hours exercised?

c. What is the mean number of hours exercised?
d. Give the modes of the responses.

61. Below are the record high temperatures *T* in degrees Fahrenheit in the 50 states and the District of Columbia through 1986.

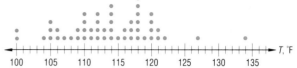

a. How many states had a record high of 105°?
b. In how many states was $T \le 110°$?
c. Give the mode for this data.

REFRESHER

Chapter 2, which discusses addition in algebra, assumes that you have mastered certain objectives in your previous mathematics courses. Use these questions to check your mastery.

A. Add any positive numbers or quantities.

1. $3.5 + 4.3 =$

2. $122.4 + 11 + .16 =$

3. $3.024 + 7.9999 =$

4. $1\frac{1}{2} + 2\frac{1}{4} =$

5. $\frac{2}{3} + 8\frac{1}{3} =$

6. $\frac{2}{5} + \frac{1}{6} + \frac{3}{7} =$

7. $6\% + 12\% =$

8. $20\% + 11.2\% =$

9. $11 \text{ cm} + .03 \text{ cm} =$

10. $.4 \text{ km} + 1.9 \text{ km} =$

11. $2' \ 3'' + 9'' =$

12. $6' + 11'' + 4'' =$

13. $30 \text{ oz} + 8 \text{ lb} =$

14. $4 \text{ lb } 13 \text{ oz} + 2 \text{ lb } 12 \text{ oz} =$

B. Add positive and negative integers.

15. $30 + \text{-}6$

16. $\text{-}11 + \text{-}4$

17. $\text{-}1 + \text{-}1 + 3$

18. $\text{-}99 + 112$

19. $\text{-}2 + 4 + \text{-}6$

20. $8 + \text{-}8 + \text{-}8 + \text{-}8$

C. Graph ordered pairs on the coordinate plane.

21. $(4, 3)$

22. $(5, \text{-}2)$

23. $(\text{-}2, 4)$

24. $(\text{-}3, \text{-}1)$

25. $(0, 4)$

26. $(0, \text{-}2)$

27. $(\text{-}3, 0)$

28. $(1, 0)$

29. $(0, 0)$

D. Solve equations of the form $x + a = b$, where *a* and *b* are positive integers.

30. $x + 3 = 11$

31. $9 + z = 40$

32. $665 + w = 1072$

33. $7 = m + 2$

34. $2000 = n + 1461$

35. $472 = 173 + s$

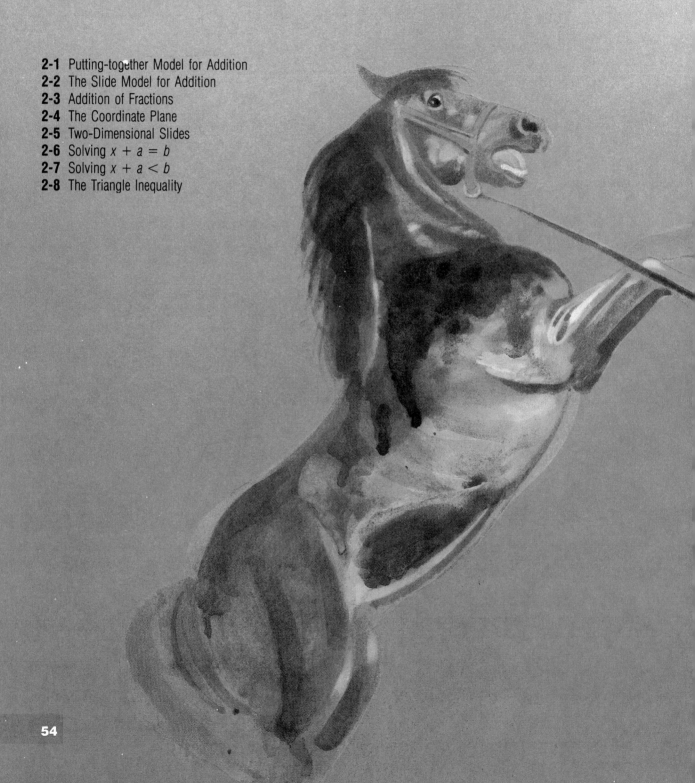

CHAPTER 2

Addition in Algebra

You often hear about automobile accidents. But life before automobile travel was not perfectly safe either. One hundred years ago, horses were a major means of transportation. However, travel on horseback could also be dangerous. Powerful kicks by the horse caused major injuries, even death. This was so serious a problem that the Prussian government kept records of how many men were killed by horse-kick in the Prussian army. A corps (pronounced core) was a major subdivision of the army that had about 10,000 men. Below are the data on eleven corps for 10 years.

Addition is used to total the numbers in each *row* across to find how many deaths there were each year. Addition is used to total the numbers in each *column* to find how many deaths there were in each corps for the 10-year period. The total of the columns is 79; this is also the total of the rows. In the decade 1880–1889, 79 men in the eleven listed corps were killed by horse-kick.

Number of Deaths by Horse-kick in the Prussian Army from 1880-1889 for 11 Corps

Year	I	II	III	IV	V	VI	VII	VIII	IX	X	XI	Total
1880	3	2	1	1	1				2	1	4	15
1881			2	1			1		1			5
1882	2					1		1	1	2	1	8
1883		1	2		1	2	1		1		3	11
1884		1					1			2		4
1885						1			2		1	4
1886	1			1	1	1			1		1	6
1887	1	2	1			3	2	1	1		1	12
1888	1	1			1	1					1	5
1889		1	1		1	1			1	2	2	9
Total	8	8	7	3	5	10	5	2	10	7	14	79

LESSON

2-1

Putting-together Model for Addition

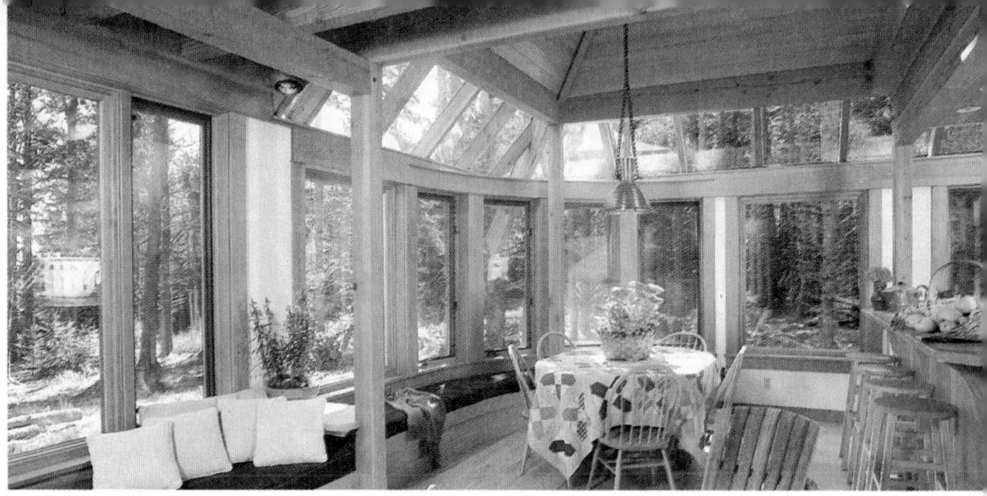

Morgan Windows, Inc., manufactures custom-designed windows for houses. They have two plants, one in New York, and one in New Orleans. The company can keep track of its progress through charts and graphs. A **bar graph** with separate bars for the two plants presents a way to compare the two facilities. This is the graph shown at the left. You can see that the profits for the New York plant went down over the five-year period. The profits for the New Orleans plant increased until 1987 and then decreased in 1988.

Morgan Windows, Inc. 1984-1988 Profits

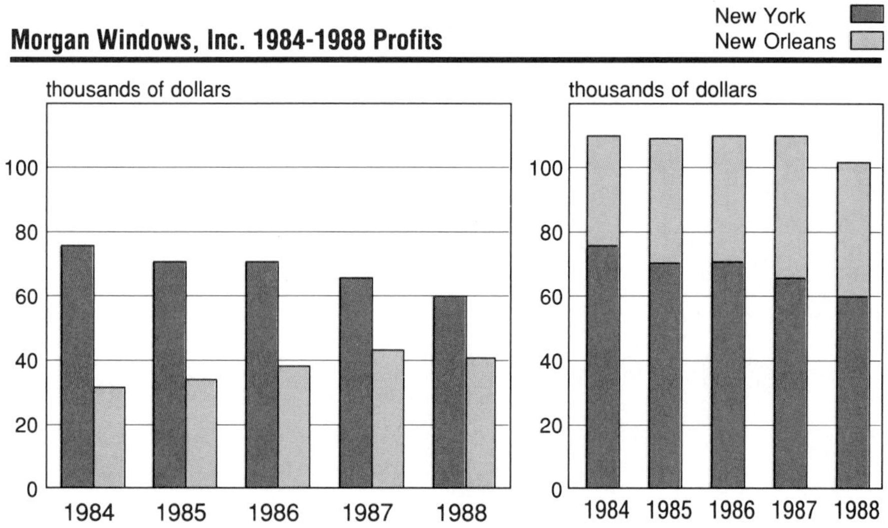

To determine whether overall profits were increasing or decreasing, the bar graph was rearranged into the **stacked bar graph** on the right. Each bar for New Orleans was placed on top of the bar representing New York for the same year. You can see that profits for the company were nearly constant until 1988.

To make this stacked bar graph, the two profits for each year shown in the left graph are added. For instance, in 1988, New York had a profit of $60,000 and New Orleans had a profit of $41,000. In total, they had a profit of $101,000. So the bar for 1988 is drawn to the 101 mark.

A **model** for an operation is a general pattern that categorizes many of the uses of the operation. The stacked bar graph illustrates an important model for addition.

Putting-together Model for Addition:

A quantity x is put together with a quantity y *with the same units*. If there is no overlap, then the result is the quantity

$$x + y.$$

Example 1 Ms. Argonaut is given a daily allowance of $115.00 by her company for her travel expenses on a business trip. She spends $45.95 for her hotel room. She estimates that her automobile costs will be about $4.00 for gas and $7.50 for parking. Let E be the amount still available for other expenses. Write an equation to show how E is related to the other quantities.

Solution The different expenses do not overlap, so the putting-together model applies. The total allowance is $115.00. So,

$$49.95 + 4.00 + 7.50 + E = 115.00.$$

The next example involves sets and the putting-together model. Notice that the sets contain no common elements. If there were overlap, the putting-together model would have to be modified.

Example 2 Consider $A = \{1, 2, 3, ..., 10\}$ and $B = \{100, 101, 102, ..., 199\}$. If C is a set including all the elements from either A or B, find $N(C)$, the number of elements in set C.

Solution Since there is no overlap in A and B, $N(C) = N(A) + N(B)$. $N(A) = 10$ and $N(B) = 100$, so $N(C) = 10 + 100 = 110$.

When two or more quantities are put together, it does not make any difference which comes first. If the stacked bar graph on the previous page had the New York bars on top of the New Orleans bars, the total heights would be the same. For 1988,

$$\$41{,}000 + \$60{,}000 = \$60{,}000 + \$41{,}000.$$

In Example 2, $N(A) + N(B) = N(B) + N(A).$

Example 2 gives instances of a general pattern named by Francois Servois in 1814. He used the French word "commutatif," which means "switchable." The English name is **commutative property.**

Commutative Property of Addition:

For any real numbers a and b,

$$a + b = b + a.$$

Sometimes you must change the order of numbers being added before you can carry out an addition.

Example 3 In making a square base for a coffee machine, there is to be $2\frac{1}{2}$ cm on each side of the circular warming tray. If the coffee pot is d cm in diameter, write a formula for the side of the base, b, in terms of d.

Solution First draw a picture.

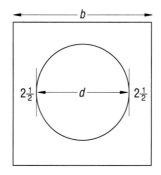

Putting together the distances gives

$$b = 2\frac{1}{2} + d + 2\frac{1}{2}.$$

The expression $2\frac{1}{2} + d + 2\frac{1}{2}$ can be simplified. There are three numbers to add. The order of operations says to do additions from left to right. So the above equation means

$$b = (2\frac{1}{2} + d) + 2\frac{1}{2}.$$

The commutative property allows you to switch the first two terms.

$$= (d + 2\frac{1}{2}) + 2\frac{1}{2}$$

But we would like to associate the middle term $2\frac{1}{2}$ with the other $2\frac{1}{2}$, not with d. The **associative property of addition** enables you to do this.

$$= d + (2\frac{1}{2} + 2\frac{1}{2})$$
$$= d + 5$$

Notice that $d + 5$ is simpler than $(2\frac{1}{2} + d) + 2\frac{1}{2}$. This is one reason for knowing the properties. The associative property was given its name by the Irish mathematician Sir William Rowan Hamilton.

Associative Property of Addition:

For any real numbers a, b, and c,

$$(a + b) + c = a + (b + c).$$

Both the commutative property and the associative property have to do with changing order. The commutative property says you can change the order of the *numbers* being added. The associative property says you can change the order of *additions* by *regrouping* the numbers. Together these properties enable you to add as many numbers as you please, in any order. This is why the totals of rows and columns in the Prussian horse-kick table on page 55 are the same.

Questions

Covering the Reading

1. What is a model for an operation?

2. State the putting-together model for addition.

In 3 and 4, refer to the graphs of the Morgan Windows, Inc., profits.

3. In which year did both the New York and New Orleans plants of Morgan Windows, Inc., show a decrease in profits?

4. Approximately what were the combined profits of both plants in 1984?

5. Refer to Example 1. On another trip, Ms. Argonaut spent $52.50 for a hotel room, $23.75 for food, and $20 for a bus ticket. Write an equation to show how she accounts for these amounts and the amount still available for other expenses from her $115 daily allowance.

6. Let $E = \{3, 4, 5, 6, 7, 8\}$, $F = \{10, 11, 12, \ldots, 20\}$, and $G = $ the set of numbers in either set E or set F. Find N(E), N(F), and N(G).

7. Refer to Example 3. What size base would you need if the diameter of the pot was **a.** 12 cm? **b.** 10 cm?

8. The commutative property for addition gets its name from a French word meaning __?__.

9. Using the associative property involves __?__ the numbers.

10. Simplify $28 + k + 30$.

Questions 11–14 are *multiple choice*. Which addition property is illustrated?
(a) commutative only (b) both commutative and associative
(c) associative only (d) neither commutative nor associative

11. $2L + 2W = 2W + 2L$

12. $(2x + 3) + 4 = 2x + (3 + 4)$

13. The total for rows equals the total for columns in the Prussian horse-kick example.

14. $[N(X) + N(Y)] + N(Z) = N(X) + [N(Z) + N(Y)]$

15. Rita got on a scale and weighed 50 kg. She took her cat in her arms and the scale went up x kg. It then read 54 kg. Write an addition equation relating x, 50, and 54.

16. Write an inequality relating the three numbers mentioned in the following situation. The bill for lunch came to $10.50. Manuel had M dollars. Nancy had $4.25. Together they did not have enough to pay the bill.

17. Let $E = \{0, 2, 4, 6, 8\}$, and $F = \{0, 5, 10\}$. Suppose G is a set including all the elements in either E or F. Tell why $N(G) \neq N(E) + N(F)$.

18. Joe and his friend Dave together earned $25 on a bike-a-thon. Joe and his sister Sarah together earned $35 on the bike-a-thon. How much did the three of them earn altogether?

19. Write an expression for the perimeter of this triangle.

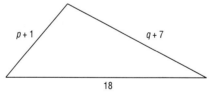

20. a. Write an equation to express the area of the largest rectangle in terms of the areas of the smaller rectangles.

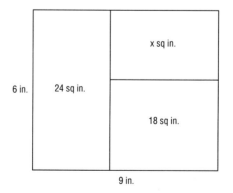

b. Find x. Use trial and error if necessary.

21. Use the associative and commutative properties to add mentally:

$$49.95 + 59.28 + .05 + .72$$

22. Find values of a, b, and c that show that $(a - b) - c = a - (b - c)$ is *not* always true. This indicates that, in general, subtraction is not associative.

23. In each situation, tell whether "followed by" is a commutative operation.
a. Putting on your socks, followed by putting on your shoes.
b. Putting cream in your coffee, followed by putting sugar in your coffee.
c. Writing on the blackboard, followed by erasing the blackboard.
d. Make up an example of your own. Tell whether it is commutative.

Review

24. There are 12 National League (NL) baseball fields and 14 American League (AL) fields. Six NL fields are natural grass and eleven AL fields are artificial grass. What is the relative frequency of a major league ball park having natural grass? *(Lesson 1-8)*

25. The Carter family has 35 classical CDs, 8 CDs of musicals, 10 folk CDs, and 15 rock-and-roll CDs. If Ben Carter picks a CD randomly, what is the probability the disk is classical? *(Lesson 1-7)*

26. Does $x + 14.7 = 20.3$ when $x = 5.6$? *(Lesson 1-4)*

27. *Skill sequence*   Simplify. *(Lesson 1-8)*
a. $\dfrac{14}{21}$ **b.** $\dfrac{14xy}{21xz}$ **c.** $\dfrac{30ab}{45bn}$

28. *Multiple choice*   During the championship basketball game, the Mustangs led by as many as 5 points and were down by as many as 3 points. If p is the point spread of the two teams, which graph best describes the game situation? *(Lesson 1-3)*

(a)    (b)

(c) 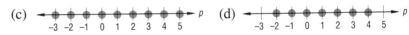   (d)

29. Rewrite as a simple fraction *(Previous course)*
a. $8\frac{2}{3}$ **b.** $-8\frac{2}{3}$ **c.** $5\frac{3}{10}$

Exploration

30. When $\frac{1}{3}$ cup of sugar is added to $\frac{2}{3}$ cup of coffee, the result does not fill a container that holds one cup. Why not?

31. The discussion about horse-kicks in the Prussian Army may be hard to believe, but it is true and uses real data.
a. In what continent was Prussia?
b. What country or countries today cover the land which was once Prussia?

The Slide Model for Addition

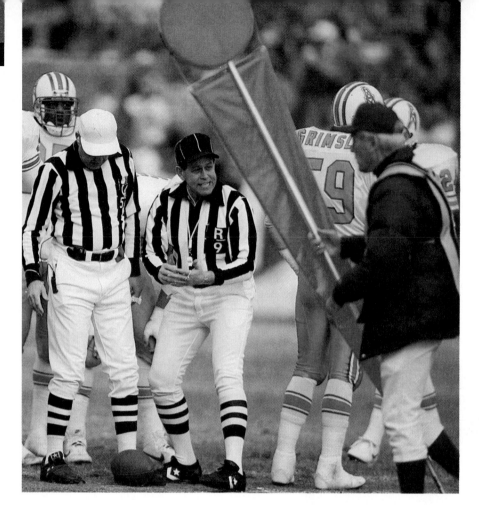

Many situations use both positive and negative numbers. You may add gains and losses in a football game, deposits and withdrawals in savings accounts, profits and losses in business endeavors, and so on. You can consider positive and negative numbers as being in opposite directions. Here is a situation that leads to the addition problem -4 + 6.

Example 1 A football team lost 4 yards on the first play and gained 6 yards on the second play. What is the net gain of these two plays?

Solution Think of -4 as a slide to the left. Think of 6 as a slide to the right. The net result is a slide to the right of 2 units. This means a net gain of 2 yards.

$$-4 + 6 = 2$$

Check The two plays can be represented on a number line where 0 is the "line of scrimmage."

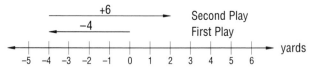

The lower arrow, from 0 to -4, indicates a loss of 4 yards. The upper arrow, from -4 to 2, represents a gain of 6 yards.

In the **slide model for addition,** positive numbers are shifts or slides in one direction. Negative numbers are slides in the opposite direction. The $+$ sign means "followed by." The sum indicates the net result of the two slides.

Slide Model for Addition:

If a slide x is followed by a slide y, the result is the slide $x + y$.

The slide model gives a method for visualizing some properties of real numbers. Suppose you deposit $480 in a bank. If you make no withdrawal or deposit, the amount you will have is $480 + 0$ or $480. You can think of this as a slide of 480 followed by a slide of 0. A slide of zero does not change a position at all. Adding 0 to a number keeps the *identity* of that number. So 0 is called the **additive identity.**

Additive Identity Property:

For any real number a,

$$a + 0 = 0 + a = a.$$

What happens when you have one slide followed by another which is the same length, but in the opposite direction? The result is always 0.

$$7 + \text{-}7 = 0 \qquad\qquad 5 \cdot 10^{32} + \text{-}5 \cdot 10^{32} = 0$$
$$\text{-}11.2 + 11.2 = 0 \qquad\qquad \text{-}\tfrac{3}{5} + \tfrac{3}{5} = 0$$

Below is the graph of $7 + \text{-}7$, which illustrates this result.

The numbers 7 and -7 are called **opposites** or **additive inverses** of each other. The opposite of any real number x is written **-x.**

Property of Opposites:

For any real number a,
$$a + -a = 0.$$

If $a = -60$, then $-a$ is the opposite of -60, or $-a = 60$. So $-a$ is not always a negative number. *When a is negative, -a is positive.*

The numbers a and $-a$ are additive inverses, as are $-a$ and $-(-a)$. A number has only one additive inverse, so $-(-a) = a$. We call this the opposite-of-an-opposite property, or for short, the **Op-op Property.**

Op-op Property:

For any real number a,
$$-(-a) = a.$$

These properties apply to all numbers: positive, negative, or zero.

Example 2 Simplify $(-8 + y) + -4$.

Solution First change the order.

$(-8 + y) + -4 = (y + -8) + -4$ Commutative Property of Addition
$\qquad\qquad\quad = y + (-8 + -4)$ Associative Property of Addition

Slides verify that $-8 + -4 = -12$.

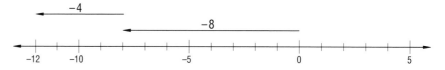

So $(-8 + y) + -4 = y + -12$.

Check For any particular value of y, $(-8 + y) + -4$ must give the same value as $y + -12$. We try $y = 7$ and follow order of operations.

Then $(-8 + y) + -4 = (-8 + 7) + -4 = -1 + -4 = -5$
$\qquad\qquad y + -12 = 7 + -12 = -5.$

Since both expressions have a value of -5, they are equal.

You should find that with practice you can do some simplifying steps in your head. Often we write more steps than are needed so that you can follow closely. Ask your teacher how many steps you should write.

Number lines can show units. In Example 1, the units on the number line represent yards. The example shows that -4 yards + 6 yards = 2 yards. Sometimes the number line is in multiples of a number. The Morgan Windows, Inc., bar graphs in the last lesson were labeled in thousands. If the units or multiples are of the same number or quantity, their sum can be calculated.

Example 3 Simplify $7s + -2s$.

> **Solution** Mark off a number line into segments of length s and then do the slide.

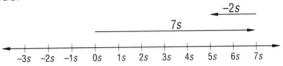

> The slide ends at $5s$. So $7s + -2s = 5s$.

The expressions $7s$ and $-2s$ are called **terms.** A term is either a single number or a product of numbers and variables. When the variables in the terms are the same, they are called **like terms** and the addition can be performed. This simplification is called **adding like terms**. Exponents of variables must match for terms to be like. The terms $5x^2$ and $9x^2$ are like terms. So, $5x^2 + 9x^2 = 14x^2$, just as 5 square inches + 9 square inches = 14 square inches. However, the expression $13t + -4a$ cannot be simplified because its terms are not like terms. Neither can $3 + 8x$ or $2x + x^5$ be simplified. Nor can 6 meters + 2 centimeters be simplified, unless you convert one unit to the other.

Adding Like Terms Property:

> For any real numbers a, b, and x,
> $$ax + bx = (a + b)x$$

Adding like terms is a special use of a more general property known as the Distributive Property. Your teacher may want you to call it by that name.

1. On the first play, a football team gained 7 yards. On the second play, the team lost 3 yards.
  a. Represent this situation on a number line.
  b. What was the net yardage for the two plays?

2. State the slide model for addition.

3. Add: **a.** -10 + -13 **b.** 9x + -11x **c.** -172 + 29

In 4 and 5, simplify.
  4. 3 + -4 + x
  5. -2 + y + -9

6. a. 0 + -10 = __?__.
  b. Why is zero called the additive identity?

7. Another name for an additive inverse is __?__.

8. What is the additive inverse of -x?

9. Give a value of n for which -n is positive.

10. Simplify -(-7).

In 11–13, an instance of what property is given?
  11. x + -x = 0
  12. -y = 0 + -y
  13. 15x + 14x = 29x

14. Check the answer in Example 3 by substituting 8 for s in 7s + -2s = 5s.

15. Check that 3 + 8x is not the same as 11x by substituting 4 for x.

16. Describe a real situation that leads to -9 + 9 = 0.

In 17 and 18, write the addition expression suggested by each situation.
17. The temperature goes up $x°$, then falls 3°, then rises 5°.

18. A person withdraws d dollars, deposits c dollars, and deposits b dollars.

19. Mary collects and trades old records. She has 40 Bing Crosby records. If she sells n of them and buys twice as many as she sells, how many will she have in all?

20. Display the number 3.14 on your calculator. Press the $\boxed{+/-}$ twice.
  a. What number appears in the display?
  b. What property has been checked?

21. If p = -8 and q = -10, then -(p + q) = __?__.

22. If Andy's age is A, write an expression for his age:
  a. 3 years from now
  b. 4 years ago

In 23–27, simplify.

23. $x + b + 0 + \text{-}x + \text{-}b + \text{-}c$

24. $\text{-}(\text{-}(\text{-}4))$ **25.** $(17 + \text{-}35) + (36 + \text{-}18)$

26. $\text{-}7t^2 + 22t^2$ **27.** $11a + 3a - 4$

28. a. Simplify $a + 2b + 3b + 4$.
    b. Check your answer by substituting 6 for a and 3 for b.

Review

29. Write a simplified expression for the area of the figure shown. *(Lesson 2-1)*

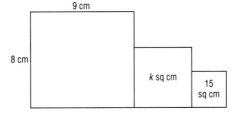

30. Fran had m dollars in savings. She received n dollars for her allowance. She then had more than enough money to buy a basketball for $48. Express this using an inequality. *(Lessons 1-3, 2-1)*

31. In chapter 1, you saw the formula for the force f that you feel when riding in an elevator:
$$f = \frac{wa}{g}$$
where w = your weight, a = acceleration of the elevator, and $g \approx 32$ feet per second per second (gravity).
    a. In the language BASIC, F = _?_. *(Lesson 1-4)*
    b. What force does a 90-pound person feel riding in an elevator accelerating at 4.5 feet per second per second? Round your answer to the nearest tenth of a pound. *(Lesson 1-5)*

32. Freezing temperatures are those at or below 32 degrees Fahrenheit.
    a. Graph all possibilities for freezing temperatures on a vertical number line. *(Lesson 1-1)*
    b. Write an inequality representing all freezing temperatures in the Fahrenheit scale using x as the variable. *(Lesson 1-6)*

33. If T is the set of planets in the solar system, then N(T) = _?_. *(Lesson 1-6)*

34. The radius of the earth is about 6.37 million meters. Write this quantity in the following ways:
    a. in decimal form without the word million *(Previous course)*
    b. in scientific notation *(Appendix B)*

Exploration

35. Negative numbers appear on TV in many situations. What real situation might each number represent?
    a. -5.32 in stock market averages
    b. -9 in rocket launches
    c. -3 in golf

LESSON

2-3

Addition of Fractions

The slide model for addition can help you picture addition of fractions. For instance, the slide model confirms that $\frac{1}{4} + \frac{3}{4} = 1$.

What does this tell you about adding fractions? It shows you that $\frac{1}{4} + \frac{3}{4} \neq \frac{1+3}{4+4} = \frac{4}{8}$. You do *not* add numerators and denominators. In fact, you add the numerators and keep the same denominator, called the **common denominator.**

Adding Fractions Property:

For all real numbers a, b, and c, with $c \neq 0$,

$$\frac{a}{c} + \frac{b}{c} = \frac{a+b}{c}.$$

Sometimes expressions involving fractions give a result which can be simplified.

Example 1 Simplify $\dfrac{-9 + 3b}{b} + \dfrac{9}{b}$.

Solution Since the denominators are the same, the Adding Fractions Property can be applied.

$$\frac{-9 + 3b}{b} + \frac{9}{b} = \frac{-9 + 3b + 9}{b}$$

$$= \frac{3b}{b} \qquad \text{Adding Like Terms Property,}$$
$$\qquad\qquad\qquad \text{Additive Identity Property}$$

$$= 3 \qquad \text{Equal Fractions Property}$$

The expression equals 3, regardless of the value of b.

Check We substitute 2 for b in the original expression and follow order of operations

$$\frac{-9 + 3 \cdot 2}{2} + \frac{9}{2} = \frac{-9 + 6}{2} + \frac{9}{2} = \frac{-3}{2} + \frac{9}{2} = \frac{6}{2} = 3. \text{ It checks.}$$

The Adding Fractions Property is quite similar to the Adding Like Terms Property. Instead of like terms, the denominators are the same. Both are special uses of a more general Distributive Property, which you will study in Chapter 6.

The rule for adding fractions is more complicated when the denominators are not the same. To picture adding $\frac{1}{2}$ and $\frac{1}{3}$, think of finding the coordinate of P.

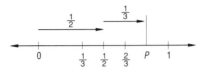

The difficulty lies in the fact that P does not lie on one of the tick marks. The interval between 0 and 1 must be divided into intervals which can measure both $\frac{1}{2}$ and $\frac{1}{3}$. Dividing it into sixths will work, since $\frac{1}{2}$ and $\frac{1}{3}$ can both be expressed as sixths. By the Equal Fractions Property, $\frac{1}{2} = \frac{3}{6}$ and $\frac{1}{3} = \frac{2}{6}$.

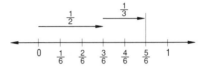

So $\frac{1}{2} + \frac{1}{3} = \frac{3}{6} + \frac{2}{6} = \frac{5}{6}$. Changing the interval on the number line to sixths illustrates the process of finding a common denominator.

The process of adding fractions with different denominators involves three steps. First, find a common denominator. Then use the Equal Fractions Property to rewrite the fractions with the common denominator. Last, add the numerators. Write this sum over the denominator. The next example involves mixed numbers, which have both whole number and fraction parts.

Example 2   Find the perimeter of the triangle below.

$1\frac{3}{4}''$ $1\frac{1}{5}''$ $2\frac{1}{6}''$

Solution 1   Change each mixed number to a fraction, then add.
$$1\frac{3}{4} + 2\frac{1}{6} + 1\frac{1}{5} = \frac{7}{4} + \frac{13}{6} + \frac{6}{5}$$

A common denominator for $\frac{7}{4}$, $\frac{13}{6}$, and $\frac{6}{5}$ is 60.

$$= \frac{105}{60} + \frac{130}{60} + \frac{72}{60}$$

$$= \frac{307}{60}$$

$$= 5\frac{7}{60} \text{ inches}$$

Solution 2   The perimeter is $1\frac{3}{4} + 2\frac{1}{6} + 1\frac{1}{5}$.

Think of each mixed number as a sum. For example, $1\frac{3}{4} = 1 + \frac{3}{4}$.

$1\frac{3}{4} + 2\frac{1}{6} + 1\frac{1}{5} = 1 + \frac{3}{4} + 2 + \frac{1}{6} + 1 + \frac{1}{5}$

Group the whole numbers together and the fractions together.

$$= 1 + 2 + 1 + \frac{3}{4} + \frac{1}{6} + \frac{1}{5}$$

$$= 4 + \frac{45}{60} + \frac{10}{60} + \frac{12}{60}$$

A common denominator for $\frac{3}{4}$, $\frac{1}{6}$, and $\frac{1}{5}$ is 60.

$$= 4 + \frac{67}{60}$$

$$= 4 + 1\frac{7}{60}$$

$$= 5\frac{7}{60} \text{ inches}$$

Check   Change the fractions to decimals. $1\frac{3}{4} = 1.75$, $2\frac{1}{6} \approx 2.17$, $1\frac{1}{5} = 1.2$, $5\frac{7}{60} \approx 5.12$. Is $1.75 + 2.17 + 1.2 \approx 5.12$? Yes.

The next example uses negative numbers in fraction form. Notice that $-\frac{9}{8}$ and $\frac{-9}{8}$ represent the same number.

Example 3   On Monday, McDonald's stock rose $\frac{3}{4}$ point; on Tuesday it fell $1\frac{1}{8}$. What was the net change?

Solution 1   A loss of $1\frac{1}{8}$ is a change of $-1\frac{1}{8}$. The answer can be found by computing $\frac{3}{4} + -1\frac{1}{8}$. You can use a common denominator of 8.

$$\frac{3}{4} + -1\frac{1}{8} = \frac{6}{8} + -1\frac{1}{8} = \frac{6}{8} + \frac{-9}{8} = \frac{6 + -9}{8} = -\frac{3}{8}.$$

The stock went down $\frac{3}{8}$ of a point.

Instead of rewriting the fractions with a common denominator, there is a formula that can be used when adding two fractions. You may find the formula helpful. It is developed by letting $\frac{a}{b}$ and $\frac{c}{d}$ stand for any two fractions. Then $b \neq 0$ and $d \neq 0$. A common denominator is bd. By the equal fractions property $\frac{a}{b} = \frac{ad}{bd}$ and $\frac{c}{d} = \frac{bc}{bd}$.

$$\frac{a}{b} + \frac{c}{d} = \frac{ad}{bd} + \frac{bc}{bd}$$

$$= \frac{ad + bc}{bd}$$

This formula is used in the first solution of Example 2.

Example 4   Write $\frac{1}{4} + \frac{x}{3}$ as a single fraction.

Solution 1   Use the formula. $\frac{1}{4} + \frac{x}{3} = \frac{1 \cdot 3 + 4 \cdot x}{12}$

$$= \frac{3 + 4x}{12}$$

Solution 2   Use the common denominator 12.

$$\frac{1}{4} + \frac{x}{3} = \frac{1 \cdot 3}{4 \cdot 3} + \frac{4 \cdot x}{4 \cdot 3}$$

$$= \frac{3 + 4x}{12}$$

Check   Substitute a number for x. We use $x = 2$. Then does

$$\frac{1}{4} + \frac{x}{3} = \frac{3 + 4x}{12}?$$

Does $\frac{1}{4} + \frac{2}{3} = \frac{3 + 4 \cdot 2}{12}?$ Yes, both sides equal $\frac{11}{12}$.

Questions

Covering the Reading

1. To add two fractions by adding the numerators, what must be true of the denominators?

2. What common denominator can you use to help in adding $\frac{1}{3}$ and $\frac{7}{4}$?

3. Use the slide model of addition to show $\frac{1}{4} + \frac{3}{8} = \frac{5}{8}$.

4. In Example 4, the final answer was $\frac{3 + 4x}{12}$. Show by substituting $x = 2$ that $\frac{7x}{12}$ is *not* the same as $\frac{3 + 4x}{12}$.

5. Tell what your calculator gives for the following computations of $-\frac{5}{8}$ and $\frac{-5}{8}$: $5 \div 8 = \pm$ and $5 \pm \div 8 =$.

In 6–9, simplify.

6. $\frac{-3}{5} + \frac{1}{5}$

7. $1\frac{1}{2} + 2\frac{2}{3} + \frac{1}{3}$

8. $\frac{3}{4} + \frac{-2}{x}$

9. $\frac{6y + 11}{2y} + \frac{-11}{2y}$

Applying the Mathematics

10. *True or false* $\frac{a}{b} + \frac{c}{d} = \frac{a+c}{b+d}$.

11. Write as a single fraction.

a. $\frac{x}{5} + \frac{x}{3}$

b. $\frac{5}{x} + \frac{3}{x}$

c. $\frac{f}{g} + \frac{3}{5}$

d. $\frac{b}{c} + \frac{3}{x}$

12. a. Calculate $1\frac{3}{4} + 2\frac{1}{2} + {}^-5\frac{7}{8}$.

  b. Check part a by changing the fractions to decimals and adding on a calculator.

13. Add $\dfrac{k}{2} + \dfrac{2k}{3} + \dfrac{k}{4}$.

14. In a tug-of-war, team A pulled the rope $3\frac{1}{2}$ ft toward its side, then team B pulled it $1\frac{2}{3}$ ft toward its side. After another minute, team B pulled the rope $4\frac{5}{6}$ ft more in its direction. What is the status of the middle of the rope at this point?

15. Simplify $(\frac{1}{3} + {}^-\frac{1}{3}) + (f + \frac{1}{3})$.

Review

16. a. What is the mean of $\dfrac{x}{2}$ and $\dfrac{x}{3}$?

  b. Give a real number that is between $\dfrac{x}{2}$ and $\dfrac{x}{3}$. *(Lessons 1-2, 1-6)*

17. *Multiple choice* Which of the following *must* be a negative number? *(Lesson 2-2)*

  (a) $-x$ (b) $-(-3)$ (c) $-(-(-\frac{1}{3}))$ (d) $-(-a)$

18. Write an expression to describe each situation. *(Lesson 2-2)*

  a. Jamie's temperature went up $2°$, then down $y°$, then down $2.4°$.
  b. Marie climbed u meters up the hill then d meters back down.
  c. Stanley earned $7e$ dollars, then paid back $5e$ dollars to one friend while collecting c dollars from another.

19. Complete the following: *(Previous course)*

Percent	Decimal	Fraction
40%	= .40	= $\frac{2}{5}$
60%	= _?_	= $\frac{3}{5}$
5%	= .05	= _?_
_?_	= _?_	= $\frac{1}{10}$

20. An acute angle has a measure between $0°$ and $90°$. Graph the interval on a number line. *(Lesson 1-3)*

21. Give an example of a situation that could be represented by the graph at the left. *(Lesson 1-3)*

22. There are two children in the Windsor family. What is the probability that one is a boy and one is a girl? *(Lesson 1-7)*

23. Evaluate $4x^2 - 5y$, when $x = 3$ and $y = 5$. *(Lesson 1-4)*

24. Match each phrase with an equivalent expression. *(Previous course)*

  a. three times a number   **i.** $n + 3$
  b. three more than a number   **ii.** $3n$
  c. the sum of three and a number **iii.** n^3
  d. the quotient of a number and 3
  e. a number to the third power   **iv.** $\dfrac{n}{3}$

25. In the eleven-year period from 1977 through 1987, Edwin Moses won 122 consecutive victories, including an Olympic gold medal, in the 400-meter hurdles. What was his mean number of victories per year? *(Lesson 1-2)*

26. *Multiple choice*   Which is *not* equal to the others? *(Lesson 1-2)*
(a) $1\frac{2}{5}$ (b) 1.4   (c) $\frac{14}{10}$ (d) 1.\overline{4}

27. The pairs of lines in these drawings are called __?__. *(Previous course)*

28. Give the coordinates of each ordered pair on this graph. *(Previous course)*

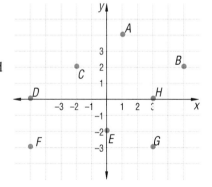

Exploration

29. a. Use your calculator to compute each sum.
   i. $\frac{1}{2} + \frac{1}{4} + \frac{1}{8}$ **ii.** $\frac{1}{2} + \frac{1}{4} + \frac{1}{8} + \frac{1}{16}$

   iii. $\frac{1}{2} + \frac{1}{4} + \frac{1}{8} + \frac{1}{16} + \frac{1}{32}$

b. If you could do an infinite addition problem, what do you think you would get as an answer to
$\frac{1}{2} + \frac{1}{4} + \frac{1}{8} + \frac{1}{16} + \frac{1}{32} + \frac{1}{64} + \ldots$?

The Coordinate Plane

Coordinate graphs of the plane make it possible to picture relationships between pairs of numbers. They can display a lot of information and show trends in a small space. Here is an example of what you can do with such a graph.

Mrs. Hernandez, a teacher, read the headline in the newspaper at the left. To test whether the headline is true, Mrs. Hernandez asked her class to keep track of the time they spent at home studying and the time they spent watching TV. The next day, she had her students record their times on the board. The table below shows what her students reported.

Time Spent on TV and Homework (minutes)

Student	TV	Homework	Student	TV	Homework
Alex	60	30	Jim	120	75
Beth	0	60	Kerry	30	45
Carol	120	30	Lawanda	120	45
David	90	90	Meg	150	60
Evan	210	0	Nancy	180	15
Frank	150	30	Ophelia	60	120
Gary	0	90	Paula	90	75
Harper	90	60	Quincy	60	45
Irene	120	0			

To see if there is a connection between the time spent watching television and the time spent studying, Mrs. Hernandez graphed these data on a two-dimensional **coordinate graph.** She drew perpendicular number lines called **axes.** The point with coordinate 0 was the same on the axes. This point is called the **origin.**

Mrs. Hernandez then located a point for each student by starting at the origin and moving right the number of minutes spent watching TV. Then she moved up the number of minutes spent on homework. She drew a point on the graph and coded it with the student's initial.

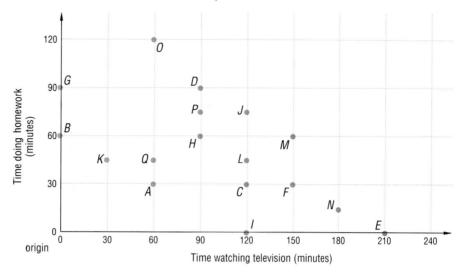

Paula's dot is at 90 right, 75 up since she spent 90 minutes watching TV and 75 minutes on homework. This point can be expressed as the ordered pair (90, 75) and is labeled as point *P* on the graph.

A two-dimensional coordinate graph is needed for these data since each response involves two numbers. This kind of graph is called a **scattergram.**

Example 1 Refer to the table and the scattergram on pages 74–75.
   a. Responses of how many students are shown on the scattergram?
   b. How many students reported doing homework for exactly 90 minutes?
   c. How many students watched at least 120 minutes of television?
   d. Who is represented by the ordered pair (60, 120)?

Solution
   a. There are 17 student responses because there are 17 data points on the scattergram.
   b. Two students (Gary, David) did exactly 90 minutes of homework because there are two data points on the horizontal line for 90.
   c. The line for 120 minutes of TV is in the middle of the graph. Look for points on this line or to the right of it. There are 8 such points, so 8 students watched at least 120 minutes of TV.
   d. The point at O, 60 right, 120 up, represents Ophelia.

From the pattern of dots, Mrs. Hernandez's class decided that the headline was generally true. As students watched more TV, they did less homework.

A scattergram can show trends that are not obvious from a table.

Example 2

United States "exports" are goods made in the U.S. and sold in another country. United States "imports" are goods made in another country and sold in the U.S. "Net exports" are calculated by subtracting imports from exports. They tell whether money is flowing in or out of a country.

a. Graph the information in this table.

b. In which year was the decline from the previous year greatest?

c. Is there an obvious trend from which to predict the future?

United States Net Exports of Goods and Services (billions of dollars)

Year	Amount
1975	23
1976	10
1977	-10
1978	-10
1979	-5
1980	10
1981	14
1982	0
1983	-37
1984	-99
1985	-103

Solution

a. The variable time is usually graphed on the horizontal axis. On the vertical axis, use a scale that allows numbers from -103 to 23. The table can be viewed as points named (1975, 23), (1976, 10), and so on.

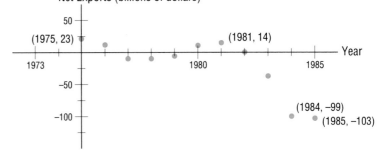

b. Read from left to right, when the dots go down, there is a decline. The biggest decline occurred in 1984.

c. From 1981 to 1985 there was a decline. That is one trend. But the years 1975 to 1981 show that trends can change from declines to increases. There is no obvious trend from which to predict the future.

Examples 1 and 2 use graphs to describe discrete situations. Example 3 displays a continuous situation with a graph.

76

Example 3   Here is a graph indicating the speed Harold Hooper traveled as he drove from home to work. Invent a story that explains the graph.

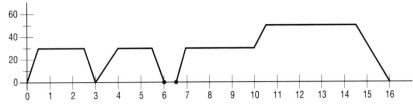

Time measured from Harold's house (minutes)

Solution   Some key times are 3 minutes after he had left home, between 6 and $6\frac{1}{2}$ minutes after he left, and at 10 minutes after he left. He was going 0 miles per hour at the three-minute mark. He may have stopped for a stop sign. He was stopped for about $\frac{1}{2}$ minute beginning at the six-minute mark. This could have resulted from Harold's stopping at a stoplight. The change of speed at 10 minutes could be explained by a change of speed limit from 30 mph to 55 mph. (Of course, other stories are possible.)

Questions

Covering the Reading

1. The number lines used in a coordinate graph are called __?__ and share a point called the __?__.

In 2–4, refer to Example 1.

2. Which ordered pair describes Alex's responses?

3. **a.** According to the graph, how many students did not watch any television?
   b. How much time did each of these students spend on homework?

4. What is the trend in this graph?

In 5–7, refer to Example 2.

5. Why are the numbers on the vertical axis given in billions of dollars?

6. When were the net exports the greatest?

7. How much less were the net exports in 1985 than in 1975?

Applying the Mathematics

8. Refer to the table and scattergram on pages 74 and 75.
   a. What is the median TV time?
   b. What is the median homework time?
   c. Which student's times are closest to the ordered pair of medians found in parts a and b?

9. Refer to Example 3. What speed was the mode for Harold's trip to work?

Questions 10 and 11 are *multiple choice*. Which of the following situations is represented by the graph?
(a) The distance traveled in h hours at 50 mph
(b) The distance traveled in h hours at 75 mph
(c) The distance you are from home if you started 150 miles from home and traveled at 50 mph

10.
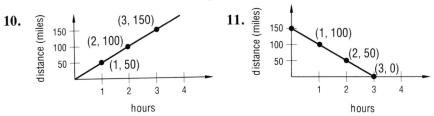

11.

12. *Multiple choice* Which situation is represented by the graph at left?
(a) The cost d in dollars of h hamburgers at a cost of $1 each.
(b) The cost d in dollars of h hamburgers at a cost of $2 each.
(c) The change d that you could get back from a $10 dollar bill, if you purchased h hamburgers at a cost of $2 each.

In 13–15, use the graph below.

13. On the average, how many minutes did a worker have to work in 1940 to buy one dozen eggs?

14. In general, would you say it took a greater or lower percentage of a person's time to buy the four items in 1970 than in 1930?

15. *Multiple choice* In dollars, the cost of the items in 1977 as compared with 1930 is
(a) more   (b) less   (c) can't tell

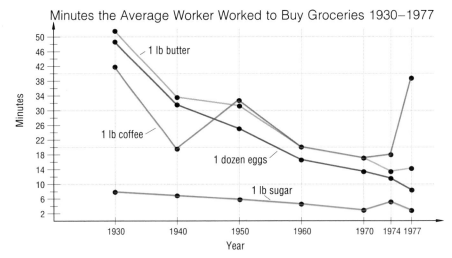

Minutes the Average Worker Worked to Buy Groceries 1930–1977

16. a. Make a coordinate graph showing the following set of points:
{(2, 3), (4, 5), (6, 7), (8, 9), (10, 11)}
b. If the pattern continues, what is the missing coordinate in (100, __?__)?
c. If the pattern continues, what is the missing coordinate in (m, __?__)?

17. a. Add. $-7 + 2 + -5 + 9 + 8 + -3$
b. Add. $-7m + 2m + -5m + 9m + 8m + -3m$ *(Lesson 2-2)*

18. The value of IBM stock went down $2\frac{3}{8}$ points on July 30. On July 31 and August 1 it went up $\frac{3}{8}$ of a point each day. Find the net change in IBM stock over this three-day period. *(Lesson 2-3)*

19. Three distinct coins are tossed once. What is the probability that all three will land heads up? *(Lesson 1-7)*

In 20–22, write as a single fraction. *(Lesson 2-3)*

20. $\frac{x}{3} + -\frac{2}{5}$ **21.** $\frac{2}{a} + \frac{7}{5a}$ **22.** $\frac{3}{p} + -\frac{2}{q}$

23. Suppose $h_1 + h_2 = 0$ and $h_1 = 17.3$.
a. What is the value of h_2?
b. What property does this illustrate? *(Lesson 2-2)*

24. Lucy lost two pounds, gained p pounds, then lost q pounds. Write an addition expression representing the change in Lucy's weight. *(Lesson 2-1)*

25. m∠PQR is 140°. Write an equation that relates 140, $x + 7$, and y. *(Lesson 2-1, Previous course)*

26. Write a set S with prime numbers as elements where N(S) = 5. *(Lesson 1-6)*

27. To estimate the number of bricks N needed in a wall, some bricklayers use the formula $N = 7LH$, where L and H are the length and height of the wall in feet. About how many bricks would a bricklayer need for a wall 8.5 feet high and 24.5 feet long? *(Lesson 1-5)*

28. To compute Jane's salary, use the formula $s = 4.50h + 6.75t$, where h is regular hours worked and t represents overtime. How much did she earn this week with 20 regular hours and 5 hours overtime? *(Lesson 1-5)*

29. Evaluate $-3x^3$ when $x = -1.4$. *(Lesson 1-4)*

30. Evaluate in your head when $t = 4$. *(Lesson 1-4)*
a. $t + -9$ **b.** $-3t$ **c.** $-2t^2$

31. a. Do a survey of at least 10 of your friends. Ask them how much time they spent watching TV and how much they spent doing homework yesterday.
b. Plot your results on a scattergram.
c. Do your data agree with the newspaper headline in this lesson?

2-5

Two-Dimensional Slides

You probably have played video games in which a character moves across the video screen. Programmers of games move the characters by first imagining them on a coordinate plane. In diagram 2 below, the *center* of the character's face is at the point (4,3).

1. how you see the screen

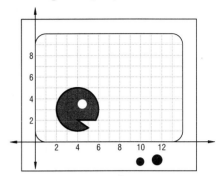

2. how the programmer sees the screen

One basic movement of a character is a **two-dimensional slide.** The movement from the original position or **preimage** to the final position or **image** is shown in diagram 3. The arrow shows the path the center takes. The programmer models this movement as a horizontal slide followed by a vertical slide, as shown in diagram 4.

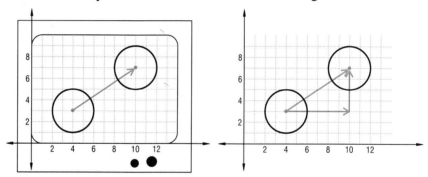

3. a slide of the figure

4. how the programmer sees the slide

In diagram 4, the slide is 6 units to the right and 4 units up. To slide a point 6 to the right, you must add 6 to the first coordinate of the point. To slide 4 units up, add 4 to the second coordinate. The center of the face was originally (4, 3). The new center is (4 + 6, 3 + 4), which is (10, 7).

To slide the entire preimage, a general pattern for sliding *any* point on the figure is needed. It is customary to call the general point (x, y). The first coordinate is called the **x-coordinate.** The horizontal axis is labeled x and called the **x-axis.** The second coordinate is called the **y-coordinate.** The vertical axis is labeled y and called the **y-axis.** If a preimage point is (x, y), then the image point after a slide 6 units to the right and 4 units up is (x + 6, y + 4).

Using negative numbers, you can indicate a slide left or down.

Example 1   Find the image of the point $A = (12, 5)$ after a slide 10 units to the left and 8 units down.

Solution   Add -10 to the first coordinate and -8 to the second coordinate. The image is $(12 + -10, 5 + -8)$, or $(2, -3)$. On the graph, the point A' (read "A prime") is the image of the point A.

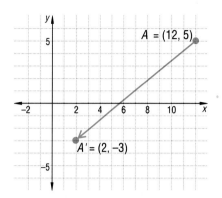

Check   To check, graph the preimage $(12, 5)$. Slide it 10 units to the left and 8 units down. It ends at the point $(2, -3)$.

The more points the figure has, the easier it may be to see the slide.

Example 2   In $\triangle ABC$, let $A = (-2, 4)$, $B = (-1, 7)$, and $C = (2, 3)$. Slide the figure 1 unit to the right and 6 units down.

Solution   Add 1 to each x-coordinate and -6 to each y-coordinate. Graph the image $\triangle A'B'C'$.

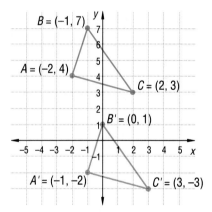

Check $\triangle ABC$ looks like it has been slid 1 unit to the right and 6 units down to get $\triangle A'B'C'$. (That is why this use of addition is called a two-dimensional slide.)

Example 3 A preimage point (x, y) is moved 7 units up. What is its image?

Solution (x, y) may be any point on the plane. But no matter what the values of x and y are, to move 7 units up, add 7 to the second coordinate. The image is $(x, y + 7)$. (There is no move left or right, so nothing is added to the first coordinate.)

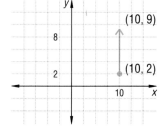

Check Substitute values for x and y and graph. Suppose $x = 10$ and $y = 2$. The preimage is then $(10, 2)$. The image is $(10, 2 + 7)$ or $(10, 9)$. Is $(10, 9)$ seven units above $(10, 2)$? The graph shows this, so it checks.

The x-axis and the y-axis divide the coordinate plane into four **quadrants.** The four quadrants are named I, II, III, and IV as shown below.

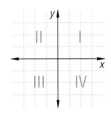

Questions

Covering the Reading

1. The first coordinate of a point is also called its __?__-coordinate. The second coordinate is also called its __?__-coordinate.

2. When you slide a figure, the original figure is called the __?__ and the resulting figure is called its __?__.

3. Write the image of $(-2, -1)$ after a slide 1.5 units to the left and 6 units up.

4. Find the image of $(0, 0)$ after a slide 45 units up.

5. Draw a coordinate plane and graph the point $P = (3, 5)$ and its image after a slide of 2 units to the right and 2 units down.

6. A preimage is $(-3, 1.5)$. Graph the preimage and its image after a slide 0.5 unit to the left and 4 units up.

7. Let point $P = (x, y)$.
   a. Write the image of P after a slide 3 units to the right and 7 units down.
   b. Check your answer to part a by picking values for x and y and graphing.

82

8. a. Copy the triangle *PQR* with
   P = (-5, -2), *Q* = (-7, -5),
   and *R* = (-3, -8).
   b. On the same
   axes, graph the
   image of this
   figure by finding
   the image of
   each vertex after
   a slide of 3 units
   to the right and
   4 units up.

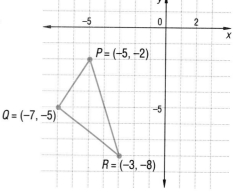

9. After a slide 3 units right and 9 units up, an image is (7, -1). What are the coordinates of its preimage? (Hint: You must work backwards.)

10. A point is (7, 2) and its image is (15, -4). Describe the slide: __?__ units to the (left or right) and __?__ units (up or down).

11. One route Tony can take to get
to school is by going 2 blocks
east, 4 blocks north, and
another 2 blocks east. Name
three other routes Tony can
take to get from his house to
school.

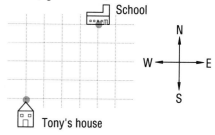

12. Examine the two-dimensional slide below.
   a. Under this slide, the image of any point is __?__ units right and __?__ units above the preimage.
   b. Under this slide, the image of (x, y) is (x + __?__, y + __?__).

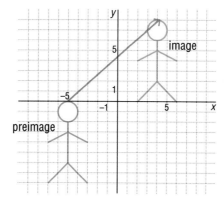

13. A can of orange juice sells for $.59 at the grocery store. On a coordinate graph, make *cost* the unit on one axis and *number of cans* the unit on the other axis. Plot a graph showing the cost of 1, 2, 3, 4, 5, and 6 cans. *(Lesson 2-4)*

In 14 and 15, simplify. *(Lesson 2-3)*

14. $\frac{3}{5} + (-2\frac{3}{10}) + 1\frac{1}{2}$

15. $\frac{2}{7} + \frac{x}{3}$

16. John wants $\frac{1}{2}$ of a pizza. Sue wants $\frac{1}{3}$ of the same pizza, and Anne wants $\frac{1}{6}$ of the same pizza. Will one pizza be large enough for the three of them? *(Lesson 2-3)*

17. The Ceramco Co. showed a loss of $5 million in 1983 and a profit of $3.2 million in 1984. What was the net profit or loss over the two years? *(Lesson 2-2)*

In 18–21, simplify. *(Lessons 2-2, 2-3)*

18. $-x + 14 + x$

19. $\frac{1}{2} + \frac{1}{3} + \frac{1}{2}$

20. $-2.5y + 3.5y$

21. 40% of 100

22. *Skill sequence* If $2y + 1 = 8$, evaluate. *(Lesson 1-4)*
   a. $3(2y + 1)$ **b.** $(2y + 1) + 10$ **c.** $2y + 26$

23. a. Which property tells you segments $\overline{AB}$ and $\overline{CD}$ below have the same length?
   b. Express that length in terms of x and y. *(Lesson 2-1)*

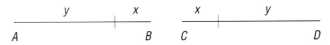

24. An air conditioning unit with a high energy efficient ratio (EER) gives more cooling with less electricity. To find the EER of a unit, divide the BTU (British Thermal Unit) number by the number of watts. The higher the EER, the more efficient the air conditioner.

$$EER = \frac{BTU}{watts}$$

   a. Find the EER to the nearest tenth for an air conditioner having BTU = 12,600 and watts = 1315.
   b. Find the EER to the nearest tenth for an air conditioner having BTU = 5000 and watts = 850.
   c. Which air conditioner, in part a or part b, is more efficient? *(Lesson 1-5)*

25. $P = \{$all integers less than $0\}$. Graph P. *(Lesson 1-6)*

26. Refer to the information about the Prussian Army on page 55 of this chapter. Here are the data for a twelfth corps.

1880	'81	'82	'83	'84	'85	'86	'87	'88	'89	Total
d	0	4	0	1	0	d	2	1	0	14

What is the value of d? *(Lessons 2-1, 2-2)*

Exploration

27. In 1980, the U.S. center of population was 1/4 mile west of De Soto, Missouri. According to American Demographics, the center of population moves 58 feet west and 29 feet south each day.
   a. At this rate, how far will it have moved by the year 2020?
   b. Where will it be? (You will need to consult a map.)

Solving
$x + a = b$

You have been solving some simple equations by trial and error. For more complex equations, guessing at solutions is not easy and is too slow. It is important, therefore, to have a more systematic way of solving equations. Here is the idea behind a property that is particularly helpful in solving equations.

Suppose Tina's age is T years and Robert's age is R years. If Tina and Robert are the same age, then

$$T = R.$$

Eight years from now, Tina's age will be $T + 8$ years and Robert's age will be $R + 8$ years. They will still be the same age, so

$$T + 8 = R + 8.$$

Similarly, Tina's age three years ago was $T + {}^-3$ and Robert's age was $R + {}^-3$. Thus,

$$T + {}^-3 = R + {}^-3$$

since they would have been the same age then as well.

The general property used is the **Addition Property of Equality.**

Addition Property of Equality:

> For all real numbers a, b, and c:
> if $a = b$,
> then $a + c = b + c$.

In solving an equation, this property indicates that you can add any number c to both sides of the equation without changing its solutions. In the equation $a = b$, a and b may represent algebraic expressions.

∎ ∎ ∎ ∎ ∎ ∎ ∎∎

Example 1   Solve: $x + {}^-126 = 283$.

Solution 1   Beginners put in all the steps.

$(x + {}^-126) + 126 = 283 + 126$	Addition Property of Equality (126 is added to both sides.)
$x + ({}^-126 + 126) = 409$	Associative Property of Addition (The additions can be done in any order.)
$x + 0 = 409$	Property of Opposites (${}^-126$ and 126 are opposites.)
$x = 409$	Additive Identity Property (Replace $x + 0$ with x.)

Check Substitute 409 for x in the original equation. Does $409 + -126 = 283$? Yes. So 409 is the solution.

Solution 2 Experts do some work mentally and may write fewer steps.

$$x + -126 = 283$$
$$x = 283 + 126 \qquad \text{Addition Property of Equality}$$
$$x = 409$$

Check Experts still check. This is done like the check in Solution 1.

All the steps were shown in Solution 1 to Example 1 to illustrate the properties that justify this process. Like the expert, you do not always need to include all steps in solving an equation. Directions in the problem and your teacher's instructions will guide you in choosing what steps to include.

The key to solving equations is knowing what should be done to both sides. In Example 1, 126 is added to both sides because it is the opposite of -126. So the left side simplifies to just x. For this type of equation there is only one step to remember.

To solve an equation of the form

$$x + a = b$$

add $-a$ to both sides and simplify.

In an equation, the variable can be on either side. And, since addition is commutative, the variable may be first or second on that side. In each of the four equations below, -47 can be added to both sides to find the same solution, 51.

$$x + 47 = 98$$
$$47 + x = 98$$
$$98 = x + 47$$
$$98 = 47 + x$$

To solve more complicated addition equations, you can use the commutative and associative properties to simplify one side to the form $x + a$.

Example 2   Solve $-10 = -3.2 + (x + 8.2)$

Solution 1   For beginners

$-10 = (-3.2 + x) + 8.2$	Associative Property of Addition
$-10 = (x + -3.2) + 8.2$	Commutative Property of Addition
$-10 = x + (-3.2 + 8.2)$	Associative Property of Addition
$-10 = x + 5$	
$-10 + -5 = (x + 5) + -5$	Addition Property of Equality
$-10 + -5 = x + (5 + -5)$	Associative Property of Addition
$-15 = x + 0$	Property of Opposites
$-15 = x$	Additive Identity Property

Check   Substitute -15 for x on the right side of the original equation.

$$-10 = -3.2 + (-15 + 8.2)$$
$$-10 = -3.2 + -6.8$$
$$-10 = -10$$

Since -15 makes the equation true, -15 is the solution.

Solution 2   For experts

$-10 = -3.2 + (x + 8.2)$	Commutative and Associative
$-10 = x + 5$	Properties of Addition
$-10 + -5 = x$	Addition Property of Equality
$-15 = x$	

Check   The check is identical to that in Solution 1.

Many equations with addition can be solved mentally. In working with complicated numbers or in rearranging formulas, however, the Addition Property of Equality can be particularly useful.

Example 3   The perimeter p of a triangle with sides of lengths a, b, and c is given by the formula $p = a + b + c$. Solve this equation for c.

Solution   To solve for c means to rewrite the equation so that c is alone on one side.

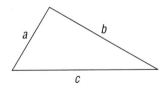

$$p = a + b + c$$

Add $-a$ to each side.
$$-a + p = -a + a + b + c$$
$$-a + p = 0 + b + c$$
$$-a + p = b + c$$

Add $-b$ to each side.
$$-b + -a + p = -b + b + c$$

Simplify again.
$$-b + -a + p = 0 + c$$
$$-b + -a + p = c$$

Most people prefer to put the variable for which they are solving on the left side of the equation. Thus, $c = -b + -a + p$

Check When $a = 10$, $b = 15$, and $c = 20$, then $p = 45$. Substitute these values in the answer. Does $20 = -15 + -10 + 45$? Yes.

■ ■ ■ ■ ■ ■ ■

Example 4 Solve for y. $3x + y = 11x$

Solution Add $-3x$ to both sides. $-3x + 3x + y = -3x + 11x$
Combine terms. $y = 8x$

Check Substitute $8x$ for y. Does $3x + 8x = 11x$? Yes.

When you solve equations, we *strongly recommend* that you arrange your work so that the equal signs of each line are directly below each other (as the examples show). This helps to avoid confusion.

Questions

Covering the Reading

In 1–4, suppose your age is A and a friend's age is B. What does each sentence mean?

1. $A = B$

2. $A + 4 = B + 4$

3. $A + -6 = B + -6$

4. If $A = B$, then $A + C = B + C$.

5. State the Addition Property of Equality.

In 6 and 7, consider these steps in solving the equation $-173 + a = 209$.
    a. $-173 + a = 209$
    b. $173 + -173 + a = 173 + 209$
    c. $0 + a = 382$
    d. $a = 382$

6. What property was used to get from step a to step b?

7. What property was used to get from step c to step d?

In 8 and 9, **a.** What number should be added to each side when solving the equation? **b.** Find the solution. **c.** Check your answer.

8. $m + 42 = 87$ 9. $-12 + y = -241$

10. Solve: $8 = -5 + (x + 2)$

11. Let p be the perimeter of a triangle. Let a, b, and c represent the lengths of the sides of the triangle.
    a. What is a formula for p in terms of a, b, and c?
    b. Solve the formula for b.
    c. Find b if $a = 21$, $c = 47$, and $p = 101$.

12. Consider the equation $g + 9h = 85h$.
  a. Solve for g.
  b. Check your work by letting $h = 2$ and solving the resulting equation.

Applying the Mathematics

13. *Multiple choice*   Which equation does not have the same solution as the others?
  (a) $15 + x = \text{-}3$ (b) $\text{-}3 + x = 15$
  (c) $x + 15 = \text{-}3$ (d) $\text{-}3 = x + 15$

14. Hank has $5.27. A solar calculator costs $14. Let n be how much more he needs.
  a. What addition equation could be solved to find out how much more money he needs?
  b. Solve this equation.

In 15 and 16, **a.** simplify the left side. **b.** Solve and check.

15. $C + 48 + \text{-}5 = 120$

16. $x + 3 + 5 + \text{-}x + 8 + x = 29$

In 17 and 18, **a.** find the solution. **b.** Check your answer.

17. $3\frac{1}{4} + x = 10\frac{1}{2}$ **18.** $15.2 = f + 2.15$

19. Solve for d: $a + d = c$.

20.a.

20.b.

20. Pictured at left top is a balance scale. On the left side of the top scale is 1 box and 2 one-kilogram weights. They balance with 5 kilograms on the right side. If b is the weight of the box, the situation is described by the equation $b + 2 = 5$.
  a. How much does the box weigh?
  b. Describe the situation for the bottom scale using an equation.

In 21 and 22, write an equation, solve and check.

21. The temperature was 25°F yesterday and now it is -12°F. Let c be the change in temperature. By how much has the temperature changed?

22. On June 1, Carlos's savings account showed a balance of $4347.59. During the next month he deposited a total of $752.85 and withdrew $550.00. On July 1, he asked the bank to tell him how much was in the account. The teller said $4574.14 including interest. How much interest had been earned during June?

Review

23. Triangle $D'E'F'$ is a slide image of triangle DEF. $D = (0, 0)$, $E = (1, 4)$, $F = (3, 6)$ and $F' = (5, 2)$.
  a. Describe the slide choosing the appropriate directions: __?__ units (left or right) and __?__ units (up or down).
  b. What are the coordinates of D' and E'? *(Lesson 2-5)*

Cost and Ratings for Popular Strawberry Ice Creams		
Brand	**Cost per Serving**	**Rating**
Berry N'ice	57	3
Perfect Parfait	55	5
Delicious	52	5
Merry Berry	49	-1
Sundae Special	46	3
Gourmet	43	2
Fabulous Flavors	26	1
Bon Appetit	22	2
Betty's Best	19	3
Select	18	-1
Mmm Good	18	1
Creamy Creations	17	-2
Mix-In Magic	17	-2
I. Scream	16	3
Ambrosia	16	-1
Tasty Treat	12	-2
Nuts and Berries	12	-2
Sweet Swirl	12	-3

In 24–29, a consumer research organization evaluated the flavor and texture of popular strawberry ice creams. The rating scale had a maximum possible score of 6 (flavor and texture very good) and a minimum possible of -6 (very bad). See the chart at the left for data. A scattergram is shown below. *(Lesson 2-4)*

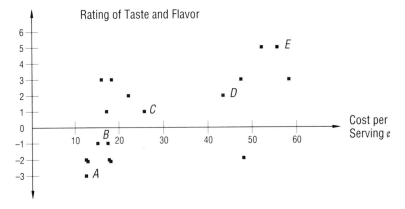

24. Which point (*A*, *B*, *C*, *D*, or *E*) represents Gourmet?

25. Which point represents Select?

26. What is the range for cost ?

27. Which ice cream seems to be a poor value; that is, it costs a lot but has a low rating?

28. Is there a tendency for ice cream which costs more to taste better?

29. Find the median cost for the 18 brands.

30. Estimate the answers in your head. *(Previous course)*
    a. $496 - 308$ **b.** $983 \div 102$ **c.** 29^2

Exploration

31. Here is a computer program.

```
10 PRINT "SOLVE AN EQUATION X + A = B"
20 PRINT "GIVE A AND B"
30 INPUT A, B
```

The semicolons in lines 40 and 50 below instruct the computer to print the following character on the same line as the previous one with one space between them.

```
40 PRINT "THE SOLUTION TO X + ";A;" = ";B
50 PRINT "IS X = ";B+(-A)
60 END
```

    a. What will the computer print after this program is run with the input 3, 2?
    b. Change this computer program so that it will solve $x + 5 + 7 = 2$. Use the command INPUT *A,B,C*.
    c. Run this program.

LESSON

2-7

Solving
$x + a < b$

Suppose your age is x and an *older* friend's age is y. Then

$$x < y.$$

Five years from now you will still be younger.

$$x + 5 < y + 5$$

In general, j years from now you will be younger than your friend.

$$x + j < y + j$$

In the same way, k years ago you were younger.

$$x + -k < y + -k$$

These examples illustrate the **Addition Property of Inequality.**

Addition Property of Inequality:

For all real numbers a, b, and c,
if $\quad\quad a < b$,
then $a + c < b + c$.

The Addition Property of Inequality allows you to add the same number to both sides of an inequality. So you can solve an inequality in the same way that you solved an equation.

To solve an inequality of the form

$$x + a < b,$$

add $-a$ to both sides and simplify.

Example 1   Solve: $x + 31 < 42$

Solution   Add -31 to each side. $x + 31 + \text{-}31 < 42 + \text{-}31$
Now simplify. $x + 0 < 11$
                                                                      $x < 11$

Note that it's just like solving an equation.

Check   The answer $x < 11$ means that any number less than 11 will work in the original sentence $x + 31 < 42$. Since this inequality has infinitely many solutions, you cannot check the answer by substituting a single number. You must do two things.

Step 1: Check the number 11 by substituting it in the original inequality. It should make both sides equal. Does $11 + 31 = 42$? Yes.

Step 2: Pick some number that works in the answer $x < 11$. This number should also work in the original inequality. We pick the number 7.
Is $7 + 31 < 42$? Yes, $38 < 42$.

Since both steps worked, $x < 11$ is the solution to $x + 31 < 42$.

Inequalities often have many solutions. Because all the solutions in Example 1 cannot be listed, we either write them in the form of a simpler inequality, $x < 11$, or we graph them.

Here is a graph of $x < 11$.

Inequalities with the $>$ sign are solved just like those with the $<$ sign. Here is why. Suppose your age is a, and a friend's age is b. If you are older than your friend now, you will be older c years from now. So, if $a > b$, then $a + c > b + c$. The same idea works for $\leq$ and $\geq$.

Thus, sentences with $=$, $<$, $>$, $\leq$, or $\geq$ that involve addition can all be solved in the same way.

You can add the same number to both sides of an equation or inequality without affecting the solutions.

Example 2 You can vote in government elections in the U.S. if you are 18 years old or over. Joan will be able to vote in an election three years from now. How old is she now?

Solution Let J be Joan's age. Three years from now her age is $J + 3$. Since she will be able to vote, she will be at least 18.

$$J + 3 \geq 18$$

Add -3 to each side. $$J \quad\;\; \geq 15$$

Joan is at least 15 years old.

Check
Step 1. Try 15. If Joan is now 15. Will she be able to vote in 3 years? Yes.
Step 2. Try a number that works in $J \geq 15$. We use 28. If Joan is 28, will she be able to vote in 3 years? Yes.

Notice in Example 2 that $J = 15$ would not be the correct answer because it does not also tell you that she may be older than 15. $J \geq 15$ is the best way to describe her possible age. $J \geq 15$ is also read ''Joan is *at least* 15.'' In the next example, some steps are omitted to simplify the work.

Example 3 Graph all solutions to $-87 \geq x + 6$.

Solution Use the Addition Property of Inequality. Add -6 to both sides.

$$-87 + \text{-}6 \geq x + 6 + \text{-}6$$
$$-93 \geq x$$

Check
Step 1: Does $-87 = -93 + 6$? Yes.
Step 2: Try a number that is a solution to $-93 \geq x$, such as -100. Does -100 work in the original sentence? Yes, because $-87 \geq -100 + 6$.

The expression $-93 \geq x$ can be read several ways. One way is $x \leq -93$. Another way is ''x is *at most* -93.''

Questions

Covering the Reading

In 1–3, suppose your age is x and an older friend's age is y.

1. What inequality relates x and y?

2. What inequality relates the ages fifty years from now?

3. What inequality relates the ages six years ago?

In 4 and 5, solve.

4. $-7 + x \leq 2$

5. $y + 4.6 \geq 4.79$

6. For the inequality $12 < x + 8$, Miko got the answer $x < 4$.
 a. Substitute 4 in the original inequality. Does it make the two sides equal?
 b. Choose some number that works in $x < 4$. Does it make the original inequality true?
 c. Is Miko's answer correct?

7. *Multiple choice*   Which inequality represents the situation, Frank is at least 50 years old?
 (a) $E < 50$ (b) $E > 50$ (c) $E \leq 50$ (d) $E \geq 50$

8. *Multiple choice*   Which inequality represents the situation, Rosa is at most 29 years old?
 (a) $R < 29$ (b) $R > 29$ (c) $R \leq 29$ (d) $R \geq 29$

9. Liz will be able to vote in an election 2 years from now.
 a. Write an inequality whose solutions tell Liz's possible ages.
 b. Solve.

10. Given are steps in the solution of the inequality $t + 18 < -3$. Name the property that supports each step.
 Step 1: $t + 18 + -18 < -3 + -18$
 Step 2: $t + 0 \quad < -21$
 Step 3: $t \quad\quad < -21$

Applying the Mathematics

11. *True or false*   The graph of $x + 3 < 17$ is the same as the graph of $14 > x$.

12. Solving $46 + x < -39$ you get $x < -85$. Use this information to solve $46 + x > -39$ in your head.

In 13 and 14, **a.** answer the question. **b.** Graph the solutions.

13. Your class is planning a picnic for the first graders. There is $162 to spend for refreshments and prizes. The food costs $129. You do not have to spend all the money. How much can be spent for prizes?

14. The temperature is now 31 °C. The high temperature record for this date is 37 °C. How much does the temperature have to change to break the record?

15. Solve and check: $\frac{1}{2} + (x + \frac{3}{4}) \geq 3$.

16. Solve for n: $18m + n \geq 19m$.

Review

In 17–19, solve in your head. *(Lesson 2-6)*
17. $3 + x = 3$ **18.** $10 = y + 4$ **19.** $-3 = z + 2\frac{1}{2}$

In 20 and 21, solve using the Addition Property of Equality. *(Lesson 2-6)*
20. $a + -11.2 = 24$ **21.** $9 + B = -11.6$

22. Solve $-11 = (2 + y) + -7$. *(Lesson 2-6)*

In 23 and 24, use the rectangle at the right.
$x + y = 90°$, and $a + b = 90°$.

23. If $x = 30°$, find y. *(Lesson 2-6)*

24. Solve the equation
$a + b = x + y$ for b. *(Lesson 2-6)*

25. Find the image of the point $P = (3, 6)$ after a slide of 4 units to the right and 6 units up. *(Lesson 2-5)*

26. Here is a list of the average monthly temperatures (in °F) in Los Angeles for one year.
     57   58   59   62   65   68
     73   74   73   68   63   58
a. Find the mean of the monthly temperatures.
b. Find the median. *(Lesson 1-2)*

27. Copy and complete the ordered pairs below.
     $(0, \underline{\ ?\ })$, $(1, \underline{\ ?\ })$, $(2, \underline{\ ?\ })$, $(3, \underline{\ ?\ })$, $(4, \underline{\ ?\ })$, $(5, \underline{\ ?\ })$
a. Complete so that the second coordinate of each ordered pair is 2 less than the first coordinate.
b. Graph these points on a coordinate plane. *(Lesson 2-5)*

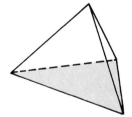

28. Simplify $\frac{-3}{4} + \frac{x^2}{5}$. *(Lesson 2-3)*

29. Darrell has a pair of 4-sided dice like the one at left. The sides are numbered 1 to 4 and all outcomes are equally likely. If the two dice are rolled once, what is the probability of rolling a sum of 6 on the two dice? (The face that is counted is the one that is down.) *(Lesson 1-7)*

Exploration

30. Give five inequalities whose solutions can be described by $x < 24$.

The Triangle Inequality

In geometry, capital letters usually name points. The distance from point A to point B is written AB. This is not multiplication because points cannot be multiplied. The line segment connecting A and B is written $\overline{AB}$.

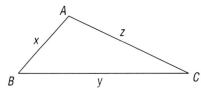

Suppose A, B, and C are 3 points. If A, B, and C lie on the same straight line, and B is between A and C, then $AC = AB + BC$. That is, at right, $AC = x + y$. You put together the smaller lengths to get the larger length.

However, A, B, and C might be the 3 vertices of a triangle.

To determine the possible values for AC, you can apply an important relationship called the **triangle inequality.**

Triangle Inequality: The sum of the lengths of two sides of any triangle is greater than the length of the third side.

In $\triangle ABC$ drawn above, this relationship means three inequalities are true.

$$x + y > z \qquad x + z > y \qquad y + z > x$$

■ ■ ■ ■ ■ ■ ■

Example 1 Suppose two sides of a triangle have lengths 15 and 22. What are the possible lengths of the third side?

Solution 1 Let $x = 15$ and $y = 22$. Substitute into the Triangle Inequality to find the possible values of z.

$$15 + 22 > z \text{ and } 15 + z > 22 \text{ and } 22 + z > 15$$

Solve each inequality.

$$37 > z \text{ and } \qquad z > 7 \quad \text{and} \qquad z > \text{-}7$$

The first two inequalities show that when two sides of a triangle are 15 and 22, the third side must be shorter than 37 but longer than 7. This can be written as $7 < z < 37$. This expression can be read several ways. One way is "x is greater than 7 *and* x is less than 37." Another is "x is *between* 7 and 37." (The third inequality shows that it must also be longer than -7. But that is obvious since length is always positive.)

Solution 2 Draw pictures. Let $AB = 15$ and $BC = 22$. Think of a hinge connected at point B. $\overline{AB}$ is fixed. $\overline{BC}$ moves.

Hinge wide open.
AC is almost 37.

Hinge partially open.
AC can be any number between 7 and 37.

Hinge almost closed.
AC is near 7.

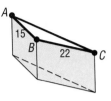

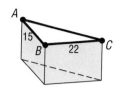

The hinge can open so that A, B, and C are on the same line and are no longer vertices of a triangle. Then $AC = 37$.

When the hinge is fully closed, A, B, and C are on the same line and are not vertices of a triangle. Then $AC = 7$.

So AC must have a value less than when the hinge is completely open (37) and greater than when the hinge is closed (7). Hence $7 < AC < 37$.

If you know nothing about A, B, and C, then either $AB + BC = AC$ (they are on the same line) or $AB + BC > AC$ (they are vertices of a triangle). So $AB + BC \geq AC$ is always true.

Example 2 Amy lives one mile from school and 0.4 mile from Larry. How far does Larry live from school?

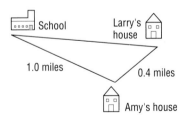

Solution Let d be the distance from school to Larry's house. Then, by the Triangle Inequality,

$$1.0 + 0.4 \geq d \text{ and } 0.4 + d \geq 1.0 \text{ and } 1.0 + d \geq 0.4.$$

There are equal signs in the inequalities since the houses and the school could be on the same road. Solve each inequality.

$$1.4 \geq d \quad \text{and} \quad d \geq 0.6 \quad \text{and} \quad d \geq \text{-}0.6$$

$1.4 \geq d$ means Larry can live no more than 1.4 miles from school, $d \geq 0.6$ means the distance Larry lives from school must be at least 0.6 mile. Since distance is always positive, $d \geq \text{-}0.6$ yields no new information. So Larry lives from 0.6 to 1.4 miles from school. The solution is given by the interval

$$0.6 \leq d \leq 1.4.$$

Questions

Covering the Reading

1. In the drawing below, *A* is on $\overline{BC}$. How long is *BC*?

2. If *M* is between *P* and *Q* on a line, *PM* = 5, and *PQ* = 7, what is *MQ*?

3. State the Triangle Inequality.

4. Refer to the triangle at the right. Copy and complete.
   a. $k + n > $?
   b. $n + m > $?
   c. ? $ + $? $ > n$

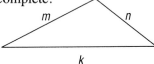

5. Refer to the triangle at the right.
   a. Name three inequalities that involve *x*.
   b. Solve each inequality for *x*.
   c. *x* must be less than __?__.
   d. *x* must be greater than __?__.
   e. __?__ $< x <$ __?__.

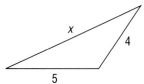

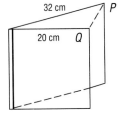

6. Two metal plates are joined by a hinge as in the drawing at the left.
   a. Name the three inequalities *PQ* must satisfy.
   b. *PQ* can be no shorter than __?__ cm.
   c. *PQ* can be no longer than __?__ cm.

7. Suppose Larry lives 1.3 km from school and 0.8 km from Amy. The distance Amy lives from school must be greater than or equal to __?__ but less than or equal to __?__.

Applying the Mathematics

8. a. If two sides of a triangle have lengths $100x^2$ and $75x^2$, the third side can have any length between __?__ and __?__.
   b. If two sides of a triangle have lengths *a* and *b*, with $a > b$, then the third side *c* can have any length between __?__ and __?__.

9. Why is there no triangle with sides of lengths 1 cm, 2 cm, and 4 cm?

10. By air, it is 1061 miles from Miami to St. Louis and 1724 miles from St. Louis to Seattle. Use only this information to answer the questions.
   a. What is the shortest possible air distance from Miami to Seattle?
   b. What is the longest possible air distance?
   c. Graph the possible distances on a number line.

11. Sirius, the brightest star in the nighttime sky, is 8.7 light-years from Earth. Procyon, a bright star near Sirius, is 11.3 light-years from Earth. Let *m* be the distance between Sirius and Procyon. What values can *m* have?

12. Betty can walk to school in 25 minutes. She can walk to her boyfriend's house in 10 minutes. Let *t* be the length of time for her boyfriend to walk to school. If they walk at the same rate, what are the possible values of *t* ?

In 13–15, solve. *(Lesson 2-6)*

13. $\frac{4}{3} = x + \frac{1}{6}$ **14.** $-423 = -234 + y$ **15.** $z + (2 + -z) + 6 = z$

16. Ben had S dollars in his savings account. He made deposits of $28.75 and $36.57. His balance was then $137.48.
   a. Relate these four quantities in an addition equation.
   b. Solve for S. *(Lesson 2-6)*

In 17 and 18, simplify each expression. *(Lesson 2-3)*

17. $-(-p^3 + p^3)$ **18.** $b + c + d + (-c)$

19. *Skill sequence* If $4x - 7y = 3$, evaluate. *(Lesson 1-4)*
   a. $5(4x - 7y)$
   b. $(4x - 7y)^2$
   c. $6(4x - 7y) - (4x - 7y)$

20. An estate was divided among four heirs in the following manner:
   Von   45%
   Winnie   30%
   Xenia   15%
   Yuri   10%
   If the estate was worth $85,000 how much did each heir receive?
   (Previous course)

21. Which is larger, 45% of 85,000 or 85% of 45,000? *(Previous course)*

22. What is $\frac{1}{2}$% of 400? *(Previous course)*

23. A squirrel slid down 50 cm and then climbed up an unknown amount c, then slid down 20 cm, winding up 210 cm above its starting position. Find c. *(Lesson 2-6)*

24. The perimeter of a triangle is 15 cm. The sides all have different integer lengths. How many different combinations of sides are possible? (Hint: The answer may be fewer than you think. Use trial and error and the Triangle Inequality.)

Summary

In algebra as in arithmetic, addition is a basic operation. This chapter discusses sets, properties, uses, equations, inequalities, and graphs related to addition.

The most frequent applications of addition occur in situations which are represented by a putting-together or a slide. Putting-together occurs when quantities that do not overlap are combined. A slide occurs when you start with a quantity and go higher or lower by a given amount. Slides help picture addition of integers, addition of fractions, and combining like terms.

The properties of addition can be verified through their uses. For example, putting-together quantities in a different order yields the same sum,

so addition is commutative. Other properties are mentioned below, in the Vocabulary.

Graphing provides a picture that can help clarify solutions to a problem or trends in data. If the relationship between two quantities is being considered, a coordinate graph in a plane is needed to show both values. A two-dimensional slide can be represented as a combination of a horizontal and vertical slide on a coordinate graph.

The simplest sentences to solve are of the form $x + a = b$ or $x + a < b$. The first step in solving each is to add $-a$ to both sides. Then simplify. One simple application of inequalities is in finding possible lengths of third sides in a triangle.

Vocabulary

Below are the most important terms and phrases for this chapter. You should be able to give a general description and specific examples of each.

Lesson 2-1
bar graph, stacked bar graph
Putting-together Model for Addition
Commutative Property of Addition
Associative Property of Addition

Lesson 2-2
Slide Model for Addition
additive identity
Additive Identity Property
opposite, additive inverse
Property of Opposites
Op-op Property
terms, like terms
Adding Like Terms Property

Lesson 2-3
common denominator
Adding Fractions Property

Lesson 2-4
coordinate graph
axes
origin
scattergram

Lesson 2-5
two-dimensional slide
preimage
image
x-coordinate, y-coordinate
x-axis, y-axis
quadrant

Lesson 2-6
Addition Property of Equality

Lesson 2-7
Addition Property of Inequality

Lesson 2-8
Triangle Inequality

Progress Self-Test

Take this test as your would take a test in class. You will need graph paper. Then check your work with the solutions in the Selected Answers section in the back of the book.

1. What property of addition is illustrated by $(x + y) + {}^-4 = x + (y + {}^-4)$?

2. Write an equation to show how x and y are related to the given numbers in the rectangle below.

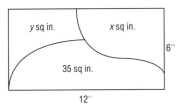

3. The temperature falls $10°$ then rises $d°$. Let $t°$ be the change in temperature. Write an equation relating these numbers.

In 4–10, simplify.

4. $^-11 + 85 + {}^-47$

5. $x + 5 + x + {}^-8 + {}^-x$

6. $\dfrac{2}{n} + \dfrac{m}{n}$

7. $\dfrac{3}{4} + \dfrac{3}{5} + \left(\dfrac{-3}{10}\right)$

8. $^-8\frac{1}{2} + \left(^-3\frac{1}{3}\right)$

9. $^-(^-(^-p))$

10. $3x + 4y + {}^-5x$

11. *Multiple choice*   Which of the following equals $\dfrac{a}{b} + \dfrac{c}{b}$?

(a) $\dfrac{a + c}{b + b}$ (b) $\dfrac{ab + bc}{b}$ (c) $\dfrac{a + c}{2b}$ (d) $\dfrac{a + c}{b}$

12. Marcia went on a diet. Below are the changes in her weight.

First week	Second week	Third week
lost 4 lb	gained $1\frac{1}{2}$ lb	lost 3 lb

a. Write an equation to find the change she must have the fourth week for a 10-lb overall loss.

b. Solve this equation.

13. What property is being illustrated? If $y + 11 = 3$, then $y + 15 = 7$.

In 14–17, solve.

14. $x + {}^-4 = 12$

15. $^-3.5 + a > 10.2$

16. $\frac{5}{2} + y = \frac{17}{4}$

17. $4 > 3 + (z + {}^-10)$

18. Solve for b: $b + a = 100$.

19. Is $^-100$ an element of the solution set of $15 \le x + 87$?

20. Graph the solutions of $^-4 \ge 9 + x$.

21. Write, but do not solve, an inequality for this situation. Vern needs at least $150 to buy a puppy. He has saved $47.50. How much more money m does he need?

22. The perimeter of the triangle is 43. Find b when $a = 9$ and $c = 21$.

23. Carla's age is C. What was her age 7 years ago?

24. Find the image of $(5, {}^-2)$ after a slide of 4 units to the left and 5 units up.

25. At the right, $\triangle A'B'C'$ is the image of $\triangle ABC$ under a slide. Under this slide, find B', the image of B.

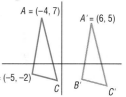

In 26 and 27, Wynken lives 30 km from Blynken. Blynken lives 12 km from Nod.

26. If x is the distance from Wynken to Nod, what inequalities must x satisfy?

27. What is the smallest possible value of x?

In 28–30, boys and girls fill a bus which seats 45 students. Let b be the number of boys on the bus and g be the number of girls.

28. Write an equation using b and g.

29. Find g in each ordered pair showing a possible combination (b, g) of boys and girls: $(10, \underline{\ ?\ })$, $(20, \underline{\ ?\ })$, $(23, \underline{\ ?\ })$, $(44, \underline{\ ?\ })$.

30. On a coordinate graph, plot points with b on the horizontal axis and g on the vertical axis.

Chapter Review

Questions on **SPUR** Objectives

SPUR stands for **S**kills, **P**roperties, **U**ses, and **R**epresentations.
The Chapter Review questions are grouped according to the
SPUR Objectives for this chapter.

SKILLS deal with the procedures used to get answers.

■ **Objective A.** *Add fractions. (Lesson 2-3)*

In 1–8, write as a single fraction.

1. $\frac{1}{2} + \frac{1}{3} + \frac{1}{4}$

2. Add. $\frac{1}{2} + \frac{2}{3} + \frac{1}{3} + \frac{-3}{4}$

3. $\frac{-7}{10} + -.35$

4. $-1\frac{3}{8} + (-4\frac{7}{10})$

5. $\frac{x}{3} + \frac{y}{3}$

6. $\frac{30}{a} + \frac{10}{a}$

7. $\frac{m}{n} + \frac{p}{q}$

8. $\frac{x}{5} + -\frac{3}{2}$

■ **Objective B.** *Combine like terms. (Lesson 2-2)*

In 9–16, simplify.

9. $8 + 4 + -2 + -6 + 1$

10. $8x + -3x + 10x$

11. $4a + 2b + -5 + 8a$

12. $2x + 7x + 8 + 5 + 3x$

13. $t + 1 + 2t + 2$

14. $-3m + 4m + -m$

15. $5s + 6b + -5s$

16. $13x^2 + -7.5x + 8x^2$

 $21x^2$

■ **Objective C.** *Solve and check equations of the form $a + x = b$. (Lesson 2-6)*

In 17–26, solve and check.

17. $2 + m = 12$

18. $x + -11 = 12$

19. $2.5 = t + 3.1$

20. $\frac{1}{3} + a + \frac{3}{4} = 5$

21. $21{,}625 + m = 29{,}112$

22. $(3 + n) + -11 = -5 + 4$

23. $-2 + y = -3$

24. Solve for p: $a + p = c$.

25. Solve for r: $15p + r = 20p$

26. Solve for t: $4s + 3s = t + 2s$

■ **Objective D.** *Solve inequalities of the form $a + x < b$. (Lesson 2-7)*

27. $x + 3 > 8$

28. $-28 > y + 22$

29. $-2 + (5 + x) > 4$

30. Solve for r: $d \le r + 45$

31. For the inequality $15 + x < 20$, Brian got the answer $x < 35$. Check whether Brian is correct.

PROPERTIES deal with the principles behind the mathematics.

■ **Objective E.** *Identify properties of addition. (Lessons 2-1, 2-2, 2-3)*

In 32–37, an instance of what property of addition is given?

32. $2(L + W) = 2(W + L)$

33. $x + 0 + 3 = x + 3$

34. $4 + (28 + -16) + -23 = 4 + 28 + (-16 + -23)$

35. $-(-31) = 31$

36. $\frac{2}{3} + \frac{5}{3} = \frac{7}{3}$

37. $8x + -13x = -5x$

Objective F. *Identify and apply properties used in solving equations and inequalities. (Lessons 2-6, 2-7)*

In 38 and 39, an instance of what property is given?

38. If $t + 18 < \text{-}3$, then $t + 18 + \text{-}18 < \text{-}3 + \text{-}18$.

39. You can add 64 to both sides of an equation without affecting the solutions.

40. Hillary adds -14 to both sides of $x + \text{-}7 = 14$. What sentence results?

41. *Multiple choice*   Which inequality has the same solution set as $3 < x \le 15$?
(a) $15 < x \le 3$ (b) $15 > x \ge 3$
(c) $15 \ge x > 3$ (d) $3 \le x < 15$

Objective G. *Use the Triangle Inequality to determine possible lengths of sides of triangles. (Lesson 2-8)*

In 42 and 43, use the Triangle Inequality to write the three inequalities which must be satisfied by lengths of sides in the triangle.

42.

43.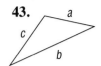

In 44 and 45, find the possible values for y.

44.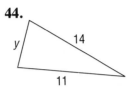

45.

USES deal with applications of mathematics in real situations.

Objective H. *Use the models of addition to form and solve sentences involving addition. (Lessons 2-1, 2-2, 2-6, 2-7)*

46. Two children wish to buy a $50 present for their parents. If one child has saved $5, how much does the other child need to have saved?
  a. Do this problem in your head.
  b. Imagine that you couldn't do the problem mentally. Write an equation that might help answer the question.

47. If the temperature is -11°C, by how much must it increase to become 13°C?

48. The two largest milk-producing states are Wisconsin and California. One year, Wisconsin produced 2.246 billion pounds of milk. Together these states produced 3.652 billion pounds of milk. How much milk was produced in California that year?

49. Mark has $5.40 and would like to buy a pair of jeans for $26. He earns d dollars

babysitting and $7.50 for mowing the lawn, but still does not have enough money. What sentence relates $5.40, $26, $7.50, and d?

50. Eli needs $5 more for a concert ticket. How much must he earn to go to the concert and have at least $4 to spend on bus fare and food?

51. The temperature was T_1 degrees. It changed by C degrees. Now it is more than T_2 degrees. Give a sentence relating T_1, C, and T_2.

52. Represent the perimeter of the quadrilateral below as a simplified algebraic expression.

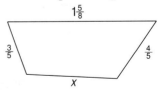

53. Bessie's stock rose $1\frac{1}{4}$ points Monday, rose $\frac{1}{8}$ point Tuesday, and fell $\frac{3}{4}$ point Wednesday. What was the overall change in her stock for these three days?

Objective I. *Apply the Triangle Inequality relationship in real situations. (Lesson 2-8)*

54. It is 346 miles from El Paso to Phoenix and 887 miles from Dallas to Phoenix. From only this information, what can you say about the distance from Dallas to El Paso?

55. Malinda lives 20 minutes by train from Roger and 30 minutes by train from Charles. By train, how long would it take to get from Roger's place to Charles's place? (Assume all trains go at the same rate.)

REPRESENTATIONS deal with pictures, graphs, or objects that illustrate concepts.

Objective J. *Graph solutions to inequalities on a number line. (Lesson 2-7)*

In 56–59, graph all solutions to the inequality.

56. $12 + y \leq 48$

57. $-66 > z + 20$

58. $1.5 + C - 6.2 < 12.1$

59. $\frac{1}{5} + \frac{2}{10} \geq \frac{3}{5} + p$

Objective K. *Plot points and interpret information on a coordinate graph. (Lesson 2-4)*

60. Draw a graph to illustrate these data (year, the U.S. population per square mile). Label the axes. (1800, 6.1), (1850, 7.9), (1900, 25.6), (1950, 50.7), (1980, 62.6)

In 61 and 62, the graph below shows the height of a boy's head from the ground as he rides in a ferris wheel. *(Lesson 2-4)*

61. Where is the boy (top, bottom, halfway up) after 40 seconds on the ride?

62. After everyone is on, how many times does the ferris wheel go around before it begins to let people off?

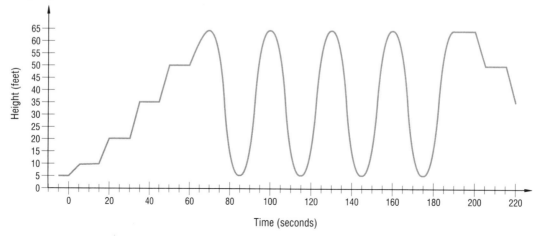

In 63–65, use the graph below.

63. Which state had more people in 1930, Texas or Ohio?

64. In which decade was there a time when the populations in Ohio and Texas were the same?

65. In which ten-year interval did Ohio show the greatest increase in population?

■ **Objective L.** *Interpret two-dimensional slides on a coordinate graph.* *(Lesson 2-5)*

66. Find the image of (2, -4) after a slide of 40 units to the left and 60 units up.

67. Find the image of (x, y) after a slide of 4 units to the right and 10 units down.

68. Find a and b when $(4 + a, 9 + b) = (3, 17)$.

69. After a slide, the image of C is $C' = (6,4)$. Graph the image of $\triangle ABC$ below by finding the image of each vertex.

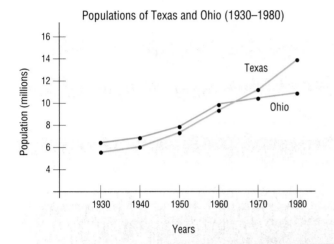

Populations of Texas and Ohio (1930–1980)

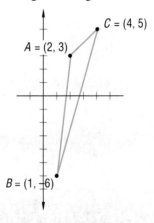

REFRESHER

Chapter 3, which discusses subtraction in algebra, assumes that you have mastered certain objectives in your previous mathematics studies. Use these optional questions to check your mastery.

A. Subtract positive and negative integers.

1. $40 - 200$
2. $76 - 79$
3. $-2 - 6$
4. $-12 - -11$
5. $111 - -88$
6. $-2 - -3$

B. Measure angles.

7. *Multiple choice*
The measure of angle V is
(a) between 0° and 45°
(b) between 45° and 90°
(c) greater than 90°

8. *Multiple choice*
The measure of angle W is
a. between 0° and 45°
b. between 45° and 90°
c. greater than 90°

9. Measure $\angle V$ to the nearest degree.

10. Measure $\angle W$ to the nearest degree.

11. Draw an angle whose measure is 110°.

C. Solve simple equations of the form $x - a = b$ when a and b are positive integers.

12. $x - 40 = 11$

13. $878 = y - 31$

14. $w - 64 = 49$

15. $100 = z - 402$

Subtraction in Algebra

Here is a puzzle originally done with matchsticks. Given are six incorrect equations involving adding or subtracting whole numbers written as Roman numerals. Can you correct each by repositioning just one match?

Subtraction has uses more important than puzzles. Any quantities related by addition (3, 4, and 7, for example) are also related by subtraction. In this chapter you will review what you know about subtraction and apply that knowledge to solving equations and inequalities, to counting problems, to graphing, and to geometry.

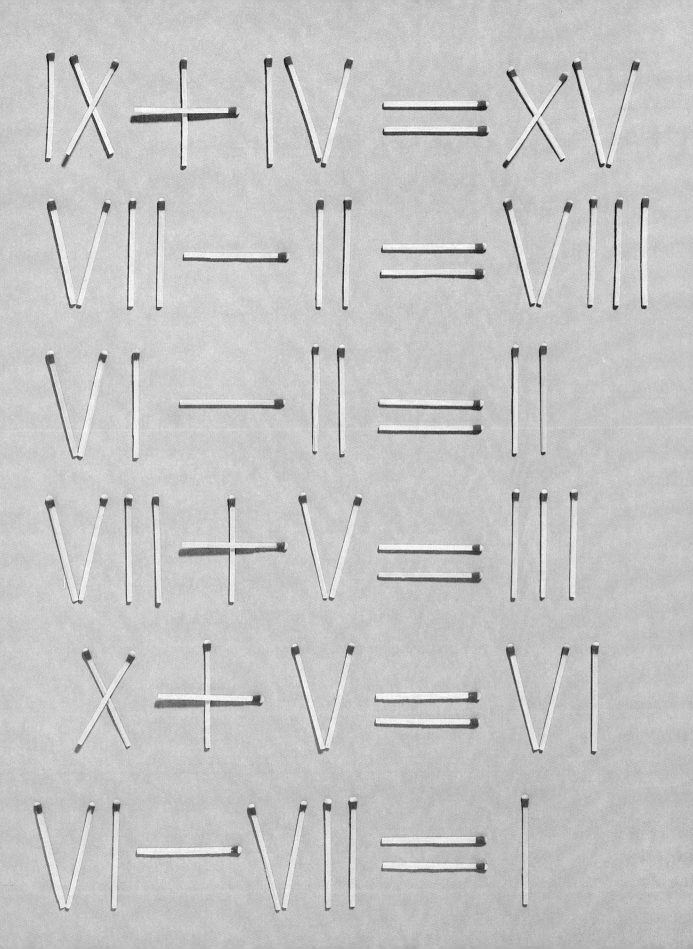

3-1

Subtraction of Real Numbers

You learned how to subtract with questions like this one.

Question: If you have 7 pennies and take 2 away, how many pennies are left?

Answer: $7 - 2 = 5$

The same question can be answered by adding: $7 + -2 = 5$. So $7 - 2 = 7 + -2$. Here is another question that can be answered either by adding or by subtracting.

■ ■ ■ ■ ■ ■ ■

Example 1   A temperature is -5°C. It falls 3°. What is the new temperature?

> **Solution 1**   Begin with -5. Subtract 3.
> $-5 - 3 = -8$ The new temperature is -8 °C.
>
> **Solution 2**   Begin with -5. Add -3.
> $-5 + -3 = -8$ The new temperature is -8 °C.

These instances show that subtracting a number is the same as adding its opposite. The general pattern is the **algebraic definition of subtraction.**

Algebraic Definition of Subtraction:

> For all real numbers a and b,
> $$a - b = a + -b.$$

Changing a subtraction problem into an addition problem is especially helpful when the numbers are not integers.

Example 2   Evaluate $x - y$ when $x = 7.31$ and $y = -5.62$.

Solution   Substitution gives $x - y = 7.31 - {-5.62}$.
The opposite of -5.62 is 5.62, so by the algebraic definition of subtraction

$$= 7.31 + 5.62$$
$$= 12.93.$$

Check   Use a calculator.

7.31 $\boxed{-}$ 5.62 $\boxed{+/-}$ $\boxed{=}$
You should see 12.93 displayed.

Caution: Subtraction is *not* associative. A computation like $3 - 9 - 1$ will not give the same answer if you do $3 - 9$ first as when you do $9 - 1$ first. You must follow the order of operations and work from left to right. (Do $3 - 9$ first.) However, you can gain flexibility by changing the subtractions to adding the opposites. $3 + {-9} + {-1}$. Now either addition can be done first, and the answer is -7.

Example 3   During the week of June 10, 1985, the price of IBM stock changed as follows:

Monday	Tuesday	Wednesday	Thursday	Friday
up $\frac{1}{4}$	down $1\frac{5}{8}$	down $5\frac{1}{4}$	down $2\frac{1}{8}$	up $2\frac{1}{2}$

Find the net change in the price of IBM stock for the week.

Solution   The net change is figured by combining all of the daily changes for the week. Up means add. Down means subtract.

$$\text{net change} = \tfrac{1}{4} - 1\tfrac{5}{8} - 5\tfrac{1}{4} - 2\tfrac{1}{8} + 2\tfrac{1}{2}$$

By changing each subtraction to adding the opposite, the numbers may be added in any order.

$$= \tfrac{1}{4} + {-1\tfrac{5}{8}} + {-5\tfrac{1}{4}} + {-2\tfrac{1}{8}} + 2\tfrac{1}{2}$$

Now use the commutative and associative properties of addition. Group the positives and the negatives.

$$= (\tfrac{1}{4} + 2\tfrac{1}{2}) \;+\; ({-1\tfrac{5}{8}} + {-5\tfrac{1}{4}} + {-2\tfrac{1}{8}})$$
$$= \quad 2\tfrac{3}{4} \quad + \quad {-9}$$
$$= {-6\tfrac{1}{4}}$$

The net change is down $6\frac{1}{4}$.

Check   Convert to decimals and use a calculator.
Net change $= 0.25 - 1.625 - 5.25 - 2.125 + 2.5 = -6.25$

You could keep the whole number parts separate from the fraction parts in Example 3. But be careful. Whereas $1\frac{5}{8}$ means $1 + \frac{5}{8}$, so $-1\frac{5}{8}$ means $-1 - \frac{5}{8}$ or $-1 + -\frac{5}{8}$.

Example 4 Simplify $-10 - (-y)$.

Solution By the algebraic definition of subtraction,

$$-10 - (-y) = -10 + y$$
$$= y + -10$$
$$= y - 10$$

Check Let $y = 25$. Then $-10 - (-y) = -10 - (-25) = -10 + 25 = 15$. Also, $y - 10 = 25 - 10 = 15$.

Example 5 Simplify $5x + 3y - 2 - x$.

Solution Think of x as $1x$. Use the algebraic definition of subtraction.
$$5x + 3y - 2 - 1x = 5x + 3y + -2 + -1x$$
Combine like terms. $\qquad\qquad\qquad = 4x + 3y + -2$
Apply the algebraic definition of subtraction. $\quad = 4x + 3y - 2$

Questions

Covering the Reading

1. The temperature is 12 °F. It falls 15°.
   a. Write an addition expression to find the new temperature.
   b. Write a subtraction expression to find the new temperature.

2. State the algebraic definition of subtraction.

In 3–5, apply the algebraic definition of subtraction to rewrite the subtraction as an addition.

3. $-2 - 7$ 4. $28 - -63$ 5. $x - (-d)$

6. During one week, the price of World Wide Widget stock changed as follows. Find the net change.

Monday	Tuesday	Wednesday	Thursday	Friday
down $2\frac{1}{4}$	up $\frac{1}{2}$	down $\frac{1}{8}$	down $\frac{1}{2}$	up 2

In 7 and 8, **a.** rewrite each subtraction as an addition; **b.** evaluate the expression.

7. $\frac{3}{5} - -\frac{7}{10}$ 8. $-3 - 4 - -7 - -11$

9. a. *True or false* $(3 - 9) - 1 = 3 - (9 - 1)$
b. What property is or is not verified in part a?

In 10 and 11, calculate.
10. $20 - 4 - 3$ **11.** $-7 - 30 - 20$

12. Write the key sequence to do $-73 - -91$ on your calculator:
a. using the $\boxed{-}$ key; **b.** using the $\boxed{+}$ key.

In 13 and 14, simplify.
13. $10p - 2q + 4 + 8q$ **14.** $-2a - 3a + 4b - b$

Applying the Mathematics

15. Evaluate $3 - x^2$ when $x = 5$.

16. Evaluate $-x - y$ when $x = -12$ and $y = 2$.

17. Evaluate $a - y - b$ when $a = -1$, $b = 2$, and $y = -3$.

18. What problem does this key sequence answer? $8.37 \boxed{+/-} \boxed{-} 7.01 \boxed{=}$

19. Mr. Whittaker's doctor advised him to lose weight. The changes in Mr. Whittaker's weight were:

First Week	Second Week	Third Week	Fourth Week
lost 4 lb	lost 3 lb	lost 3 lb	gained 5 lb

a. Write an expression for the net change using addition.
b. Write an expression for the net change using subtraction.
c. What was the net change for the four weeks ?

20. Let t be Toni's weight. Let f be Fred's weight. Suppose $t - f = 35$.
a. Find a value for t and a value for f such that $t - f = 35$.
b. Who is heavier, Toni or Fred?
c. Use values for t and f from part a. Find the value of $f - t$.

21. Find all possible values for x using the clues.

Choices for x: -7, -4, -3, -1, 1, 3, 4, 7
Clue 1: $x > 0 - 2$
Clue 2: $x < 4 - -1$
Clue 3: $x \neq 2 - -1$
Clue 4: $-x \neq -4$

22. Evaluate $a - b$ and $b - a$ when
a. $a = 5$ and $b = -1$
b. $a = 1$ and $b = 3$
c. $a = -2$ and $b = 0$
d. $a = -3$ and $b = -6$
e. From part a to part d, does subtraction seem to be commutative?

23. a. Simplify $-x - 2x - 3x - 4x - 5x$.
b. Simplify $y - y - 2y - 3y - 4y - 5y$.
c. Simplify $6x - (5x - (4x - (3x - 32)))$. Begin with the innermost parentheses and work toward the outer parentheses.
d. Simplify $x - (3x - 6x)$.

24. Find and graph the solution set: $3 \leq x + -4$. *(Lesson 2-7)*

25. *Skill sequence* Solve. *(Lesson 2-6)*
 a. $-3 = x + (-7)$
 b. $2.7 + y = 3.4$
 c. $z + (-1\frac{1}{4}) = -\frac{3}{4}$

26. *Multiple choice* Tell which graph best fits each description. *(Lesson 2-4)*
 a. The number of people in a restaurant from 6 A.M. to 6 P.M.
 b. The number of people in a school from 6 A.M. to 6 P.M.
 c. The number of people in a hospital from 6 A.M.to 6 P.M.

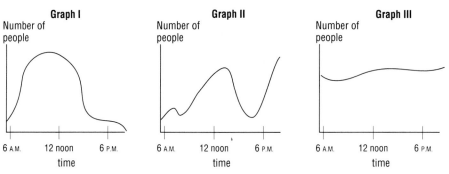

27. Evaluate in your head when $b = 20$. *(Lesson 1-4)*
 a. $-3 + b$ **b.** $\frac{1}{2}b$ **c.** $2b^2$

28. Event A has a probability of 66%, event B has a probability of $\frac{3}{5}$, and event C has a probability of 0.7. Which event is most likely to happen? *(Lesson 1-7)*

29. Solve for y: $2x + y + 5x = 28x + 3$. *(Lesson 2-6)*

30. The puzzles in the chapter opener are written with Roman numerals. Match each Roman numeral at left with its decimal form at right. *(Previous course)*
 a. L   1
 b. M   5
 c. C   10
 d. D   50
 e. V   100
 f. X   500
 g. I   1000

31. Roman numerals were the most common way of writing numbers in Western Europe from about 100 B.C. to A.D. 1700. Even in this century, they have often been used to name years and on timepieces. Match each year with its corresponding Roman numeral.
 a. the year Christopher Columbus first sailed for the New World   **i.** MDCCCLXXXV
 b. the year the character "Mickey Mouse" first appeared in a cartoon   **ii.** MCDXXXI
 c. the year the first gasoline-powered automobile appeared   **iii.** MCDXCII
 d. the year Joan of Arc was killed   **iv.** MCMXXVIII

Models for Subtraction

Pine Island had an area of 27.8 square miles. During a hurricane, 1.6 square miles of beach were washed away. The area of the island left was 27.8 − 1.6, or 26.2 square miles. This illustrates an important model for subtraction.

Take-Away Model for Subtraction:

If a quantity y is taken away from an original quantity x, the quantity left is $x - y$.

The take-away model leads to algebraic expressions involving subtraction.

Example 1 In baseball, the batter hits into a playing field. The foul lines form a 90° angle. Suppose on a hard-hit ground ball each of the four infielders can cover an angle of about 13°. The pitcher can cover about 6°. How much of the infield is left for the hitter to hit through?

shortstop

2nd baseman

1st baseman

3rd baseman

pitcher
6°

13°

13°

13°

13°

foul

foul

batter

Solution The infielders and the pitcher can cover 13° + 13° + 13° + 13° + 6° = 58° of the hitting region. Subtract 58° from the 90° region: 90° − 58° = 32°. A hit can occur in about 32° of the field.

If each infielder could cover 15°, with the pitcher covering 6°, then by the same procedure, there would be 24° of the infield for a ball to get through. In general, if all infielders can cover d degrees and the pitcher can cover P degrees, then there are $90 - 4d - P$ degrees remaining for a hard-hit ground ball to get through.

A second model for subtraction is used when two quantities are compared.

Comparison Model for Subtraction:

> The quantity $x - y$ tells how much quantity x differs from the quantity y.

You saw one application of the comparison model for subtraction in Lesson 1-3 when you computed the range of a set of numbers. Recall that the interval from the minimum to the maximum is called the range. It is found by subtracting the minimum value from the maximum value. The range tells you how much the maximum differs from the minimum.

Example 2 The greatest recorded difference in temperature during a single day was in Browning, Montana, in January, 1916. During one 24-hour period, the low temperature was -56°F and the high temperature was 44°F. What was the range of the temperature?

Solution Compare the numbers by subtracting.

$$\text{range} = \text{maximum} - \text{minimum}$$
$$= 44 - (-56)$$
$$= 44 + 56$$
$$= 100$$

The range of the temperatures was 100°.

Examples 3 and 4 use the comparison model with variables to describe relationships.

Example 3 Andrea is A years old. Her younger brother, Tim, is 12 years old. How much older is Andrea than Tim?

Solution To compare the ages, use the comparison model. Subtract the numbers. Therefore Andrea is $A - 12$ years older than her brother.

Check If Andrea is 15, she is 3 years older than her brother. If $A = 15$, then $A - 12 = 3$, which checks.

Example 4

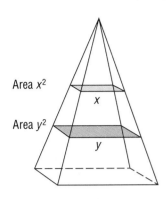

Area x^2

Area y^2

The Transamerica Building in San Francisco has the shape of a pyramid. Each floor is approximately square. Let x be the side of one floor and let y be the side of a lower floor. Write an expression to compare the areas of the two floors.

Solution The two floors have areas x^2 and y^2. Since floor y has the larger area, the difference in area is $y^2 - x^2$.

In Example 4, be careful to use order of operations. Take powers before subtracting. If $y = 50$ meters and $x = 30$ meters, the difference in areas is $50^2 - 30^2$ square meters, which is $2500 - 900$, or 1600 square meters.

Questions

Covering the Reading

1. State the take-away model for subtraction.

2. In the baseball example, if each infielder could cover $14°$ and the pitcher $7°$, how much of the infield is there for ground balls to get through?

3. State the comparison model for subtraction.

4. A most unusual recorded temperature change occurred in Spearfish, South Dakota on January 22, 1943. Over a period of two minutes the temperature rose from $-4°$ F to $45°$ F. What was the range of temperature?

In 5–6, use Example 3.

5. Andrea's older cousin Sam is 18. How much older is Sam than Andrea?

6. Suppose Andrea is 17 years old. How much older is she than her brother Tim?

7. Suppose one side of a floor in the Transamerica Building has length 40 meters. Let b meters be the length of a side of a higher floor. Write an expression for the difference in the areas of the two floors.

Applying the Mathematics

8. The Arnold Company buys a piece of property which has an area of 14,000 square feet. On it the company builds a store with area S square feet and a parking lot with area 2580 square feet. Write an expression for the area left for the lawn.

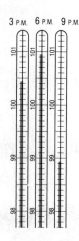

3 P.M. 6 P.M. 9 P.M.

In 9 and 10, Bernie's age is B, John's age is $B - 3$, and Robin's age is $B - 7$.

9. **a.** Who is older, Robin or Bernie? **b.** How much older?

10. **a.** Who is older, John or Robin? **b.** How much older?

11. These thermometers at the left show a hospital patient's temperature at three times.
    a. What was the change from 3 P.M. to 6 P.M.?
    b. What was the change from 3 P.M. to 9 P.M.?

12. Some students studied a chapter for which they took a pretest and a posttest.
    a. Complete the table below. The change may be negative.

Student	Pretest	Posttest	Change
Chui, L.	57	65	8
Fields, S.	43	41	?
Ivan, J.	63	?	5
Washington, C.	?	51	-3

   b. What is the median change?
   c. What is the mean of the changes?

13. In New York City, the sun's rays make a $72\frac{1}{2}°$ angle with the ground at noon on the first day of summer (about June 21). On the first day of winter (about December 21), the angle is $25\frac{1}{2}°$. By how much do the angles differ?

14. Neanderthal man lived in Europe, North Africa, and Central Asia from about 200,000 B.C. to about 30,000 B.C. How long was this?

15. A carpenter has a board that is x feet long. He cuts a 3-foot piece from it. Write an expression for the length of the remaining piece.

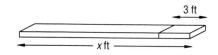

3 ft

x ft

16. At a carnival, John estimates that there are 325 marbles in a barrel. Maria's estimate is 500. The actual count is 422. Who has a closer estimate?

17. What is the difference in area between the two rectangles at the right?

3 x

x

7

Review

In 18-22, simplify. *(Lesson 3-1)*

18. $42{,}531 - 36{,}195 - (-14{,}259)$

19. $-8.7 - 16.03$

20. $-p - (-q)$

21. $-7ab + 2a - 5b - 6ab - 4a + b$

22. $y + (6 - (2.7y + 0.4y))$

23. E is 4 units to the left of A. F is n units to the right of A.
  a. What is the coordinate of E?
  b. Write an expression for the coordinate of F. *(Lesson 2-2)*

24. Four instances of a pattern are given below. Describe the general pattern using two variables. *(Previous course)*

$$9^2 - 4^2 = (9 + 4)(9 - 4)$$
$$31^2 - 29^2 = (31 + 29)(31 - 29)$$
$$7^2 - 8^2 = (7 + 8)(7 - 8)$$
$$3.5^2 - 2.5^2 = (3.5 + 2.5)(3.5 - 2.5)$$

25. A formula in a computer program is $J = 3 * I + 1$. Find the value of J when $I = 12$. *(Lesson 1-4)*

26. In Tinytown, the only gas station is 1 mile from the elementary school and 0.4 mile from the high school. Explain why it is impossible for the two schools to be 2 miles apart. *(Lesson 2-8)*

27. The image of (x, y) is $(7, -1)$. It results from a slide four units to the right and two units down. Find x and y. *(Lesson 2-5)*

In 28 and 29, use this information: At Central High, all graduates must take either earth science or biology. There were G graduates last year. E of them took only earth science. B of them took only biology. A of them took both. Suppose a science reporter picks at random a graduate to interview. *(Lesson 1-7)*

28. What is the probability that the student took both biology and earth science?

29. What is the probability the student took earth science but not biology?

30. *Skill sequence* Solve. *(Lesson 2-7)*
  a. $x - 8 > 5$
  b. $16 + x > 1 - 0$
  c. $-28.3 > x + 17.5$

31. The area of a right triangle is $\frac{1}{2}$ the product of its legs. Find the areas in your head.
(Lesson 1-4)

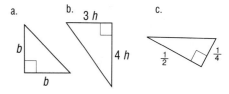

Exploration

32. The number of years between historical events can be calculated by subtracting the years in which the events occurred. You will be within a year of being correct. (December 31, 1985 and January 1, 1986 are one day apart, not one year apart.)
  a. How many years apart are the Declaration of Independence and the end of the Civil War?
  b. How many years are between the death of Archimedes in 222 B.C. and the death of Julius Caesar in A.D. 43. (Be careful! There was no year 0.)
  c. Nicholas and Nicole have birth years one year apart. What is the smallest and largest possible difference in their birth days?

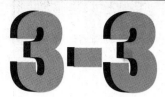

3-3

Solving $x - a = b$ and $x - a < b$

The range of temperatures recorded on the planets of our solar system is 682°C. The minimum temperature of -220°C was recorded on Pluto. The maximum was recorded on Venus. To find the maximum recorded temperature you can use a subtraction equation.

Let $T =$ the maximum temperature recorded. Substitute into the formula for range.

$$\text{range} = \text{maximum} - \text{minimum}$$
$$682 = T - (-220)$$

Notice that you can change the subtraction to addition.

$$682 = T + 220$$

Solve as you would an addition equation. Add -220 to each side.

$$682 + {-220} = T + 220 + {-220}$$
$$462 = T$$

The maximum recorded temperature is 462°C.

The subtraction equation above was changed to one involving addition. You can do this with *any* sentence involving subtraction. Then you can use all you know about solving addition sentences.

■ ■ ■ ■ ■ ■ ■ ■

Example 1 Hometown Bank and Trust requires a minimum balance of $1500 for free checking. If Mr. Archer can withdraw $3276 and still have free checking, how much is in his account?

Solution Let M be the amount of money in Mr. Archer's account. Then the amount left after the withdrawal is $M - 3276$. This quantity must be greater than or equal to $1500.

$$M - 3276 \geq 1500$$

Change subtraction to addition. $M + {-3276} \geq 1500$

Add 3276 to both sides of the inequality. $M + {-3276} + 3276 \geq 1500 + 3276$
$$M \geq 4776$$

The answer, $M \geq 4776$, means that Mr. Archer has to have at least $4776 in his account in order to keep free checking after a withdrawal of $3276.

Check As with any inequality, check the answer with two steps.

Step 1: Check 4776 by substituting in the original inequality.
Does $4776 - 3276 = 1500$? Yes.

Step 2: Check some number that works in $M \geq 4776$. We choose 5000.
Is $5000 - 3276 \geq 1500$?
Yes, since $1724 \geq 1500$.

In Example 2, each line of the solution includes the reason for that step.

Example 2 Solve $-1 > x - 4.2$ and graph its solutions.

Solution

$-1 > x + -4.2$	algebraic definition of subtraction
$-1 + 4.2 > x + -4.2 + 4.2$	Addition Property of Inequality
$-1 + 4.2 > x + 0$	Property of Opposites
$3.2 > x$	Additive Identity Property

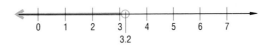

Check Step 1: Check 3.2. This should make the sides equal.
Does $-1 = 3.2 - 4.2$? Yes.

Step 2: Check a value smaller than 3.2, like 0.
Is $-1 > 0 - 4.2$?
Is $-1 > -4.2$? Yes, it checks.

To summarize: in order to solve sentences of the form
$$x - a = b \quad \text{or} \quad x - a < b,$$
rewrite as $x + -a = b$ or $x + -a < b$,
and solve as you would any addition sentences.

Questions

Covering the Reading

1. To solve $x - a = b$, first rewrite the sentence as $x +$ __?__ $= b$, and then solve like any addition sentence.

2. The range of temperatures recorded on Mercury one day was 797°F. The minimum temperature recorded that day was -23°F. Using the definition of range,
   a. write an equation to find Mercury's maximum recorded temperature that day;
   b. solve to find the maximum.

In 3–5, **a.** rewrite as an addition sentence and **b.** solve.

3. $s - 1240 = 20{,}300$

4. $5.3 = w - {-4.1}$

5. $x - {-60} < 140$

6. Refer to Example 1. How much would be in Mr. Archer's account if he can withdraw $4582 and still have free checking?

7. The steps in the solution of the inequality $K - 25 < 755$ are given. State the reason for each step.
   Step 1: $K + {-25} < 755$ **a.** __?__
   Step 2: $K + {-25} + 25 < 755 + 25$ **b.** __?__
   Step 3: $K + 0 < 755 + 25$ **c.** __?__
   Step 4: $K < 780$ **d.** __?__

8. Graph all solutions to $-6 > x - 4$.

Applying the Mathematics

9. Write, but do not solve, an inequality for the following problem: Houston Investment provides free travelers' checks to customers having at least $1000 in a savings account. Marilyn Pulowski wants to withdraw $2768.00 for a trip. How much should be in her account so that she can make the withdrawal and still qualify for free travelers' checks?

10. The range of temperatures that have been used in laboratory experiments is about $8.2 \cdot 10^6$ kelvins. The minimum recorded lab temperature is $5 \cdot 10^{-8}$ kelvins. Write a subtraction equation which you could use to solve for the maximum recorded temperature M. You do not have to solve this sentence.

In 11 and 12, solve for x.

11. $x - 35 \leq y$ 12. $8x - 3r - 7x = r$

13. The formula $p = s - c$ relates profit, selling price, and cost. Here the formula is solved for s. Give a reason for each step.

Step 1: $p = s + -c$ **a.** ___?___
Step 2: $p + c = (s + -c) + c$ **b.** ___?___
Step 3: $p + c = s + (-c + c)$ **c.** ___?___
Step 4: $p + c = s + 0$ **d.** ___?___
Step 5: $p + c = s$ **e.** ___?___

Review

In 14–15, simplify. *(Lessons 1-4, 3-1)*

14. $5 - (4 - (3 - (2 - 1)))$

15. $1 - (2 - (3 - (4 - 5)))$

16. $m - (2m - 4m)$

17. Subtract and check with a calculator. *(Lesson 3-1)*

$120\frac{2}{5} - 19\frac{9}{10}$

In 18 and 19, solve for q and graph the solution set. Do *not* use a calculator.

18. $q + 21 \leq 11$ *(Lesson 2-7)*

19. $3 + q - 12 > 4 - -18$ *(Lessons 2-7, 3-3)*

20. Solve for a: $a + -2b = 10$. *(Lesson 2-6)*

21. Four instances of a pattern are given below. Describe the general pattern using one variable. *(Previous course)*

$2 \cdot 5 + 5 = 3 \cdot 5$
$2 \cdot 7 + 7 = 3 \cdot 7$
$2 \cdot 81 + 81 = 3 \cdot 81$
$2 \cdot 90 + 90 = 3 \cdot 90$

22. The Goldens celebrated their 50th (golden) wedding anniversary in year y. In what year were they married? *(Lesson 3-2)*

23. The Valases will celebrate their nth wedding anniversary in 2000. In what year were they married? *(Lesson 3-2)*

24. Simplify $\frac{50(x + 3)}{2(x + 3)}$ where $x \neq \text{-}3$. *(Lesson 1-8)*

25. Elaine White and Cynthia Wong invested a total of $10,000 to purchase an office supplies business. Ms. White put up $5500 of the investment. What is the difference in investments between White and Wong? *(Lesson 2-6)*

In 26 and 27, use the number line.

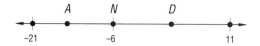

26. Find the coordinate of D if D is 9 units from N. *(Lesson 2-2)*

27. Find the coordinate of A if A is 8 units from N. *(Lesson 2-2)*

28. Jan's Sports sells camping equipment and clothing. In 1985, 71% of sales were of clothing. What percent of the sales were of equipment? *(Previous course)*

29. Estimate the value of x. Do *not* use a calculator.
   a. $999,999 - x = 800,000$
   b. $\frac{1}{2}x = 4.98$
   c. $x + 10^{\text{-}20} = 5$

30. The population of the United States is about $2.35 \cdot 10^8$ and the population of Los Angeles Metropolitan area is about 1.27×10^7. If a person living in the United States is selected randomly, what is the probability that the person lives in the Los Angeles area? *(Lesson 1-8)*

Exploration

31. During gym class, the students formed a circle and counted off. (The first student counted "1," the second student counted "2," etc.)
   a. Student number 7 was directly across the circle from student number 28. How many students are in this class?
   b. Suppose student number 7 was directly across from student number n. Then how many students are in this class?

3-4

Intersection of Sets

Oak Street

Main Street

A police report of an accident stated that "the pedestrian was in the intersection of Main and Oak Streets when he was struck by the car." The intersection of the streets is colored in the map at the left. It is the area where the two streets overlap. Since the pedestrian was in the intersection, he was both in Main Street and in Oak Street.

The term intersection has a similar meaning when used with sets. The **intersection** of two sets is the set of elements in both.

Example 1 Let $A = \{1, 3, 5, 7, 9\}$ and $B = \{1, 4, 7, 10\}$. Give the intersection of A and B.

Solution The elements that are in both A and B are 1 and 7. The intersection is $\{1, 7\}$.

The symbol for intersection is $\cap$. So in Example 1,

$$A \cap B = \{1, 3, 5, 7, 9\} \cap \{1, 4, 7, 10\} = \{1, 7\}.$$

This notation lets us write the following definition.

Definition:

The intersection of sets A and B, written $A \cap B$, is the set of elements that are in both A *and* B.

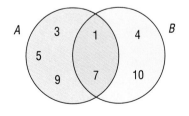

The sets A, B, and $A \cap B$ can be illustrated by the diagram at left, called a **Venn diagram.** The elements in A are in the circle labeled A. The elements in B are in the circle labeled B. The circles overlap because both contain 1 and 7. The overlap represents $A \cap B$.

If sets are infinite, the elements in their intersection cannot be all listed. But they may be described algebraically or graphed.

Example 2   Kim is cooking dinner in a rush and wants to bake a casserole and muffins at the same time. The casserole recipe calls for an oven temperature from 325° to 375°, depending on how long it is cooked. The muffins can bake at any temperature from 350° to 450°.
a. Describe each of the two intervals with an inequality.
b. Describe the intersection of the two intervals (the temperature settings that are right for both the casserole and muffins) algebraically.
c. Graph the intersection.

Solution
a. Let t be an appropriate oven temperature. Then for the casserole, $325 \leq t \leq 375$ and for the muffins, $350 \leq t \leq 450$.
b. The temperature settings all right for both are from 350° to 375°, described by the interval $350 \leq t \leq 375$.
c. First graph the solution to each sentence separately. Be careful to line up the scales on the two number lines.

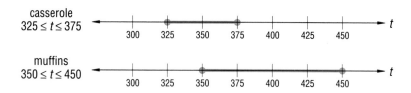

The intersection is that part of the number line where the two graphs overlap.

Many probability problems deal with the intersection of two events. To save space, when E is an event, we often write **P(E)** for "the probability of E."

Example 3   Consider tossing an unbiased die. What is the probability of tossing an even number greater than two?

Solution   Tossing an even number greater than two is the intersection of two events:
A = tossing an even number; B = tossing a number greater than two.
First list the outcomes of each event.

A = tossing an even number = {2, 4, 6}.
B = tossing a number greater than two = {3, 4, 5, 6}.

The question asks for P (even *and* greater than two). $A \cap B = \{4, 6\}$. There are two faces of the die with a number that is both even and greater than two. Since there are six outcomes in all, the probability is $\frac{2}{6}$ or $\frac{1}{3}$. That is, P($A \cap B$) = $\frac{1}{3}$.

Intersections arise whenever there are two or more conditions to be satisfied. When you applied the Triangle Inequality in the last chapter, you found the intersection of three conditions. Example 4 is a type of question you have seen before in Lesson 2-8, but the solution is worded in the language of the intersection of sets.

Example 4 Two sides of a triangle have lengths 30 and 40. What are the possible lengths for the third side?

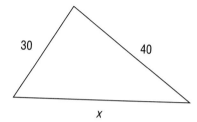

Solution Draw a picture, calling the third length x. Using the Triangle Inequality,

$30 + 40 > x$ *and* $30 + x > 40$ *and* $40 + x > 30$.

So the intersection of the solution sets to these inequalities is desired. Solving them,

$x < 70$ *and* $x > 10$ *and* $x > -10$.

The intersection is the set of numbers between 10 and 70. That set is described as $10 < x < 70$.

Questions

Covering the Reading

1. Define: intersection of two sets.

2. If $A = \{2, 4, 6, 8, 10\}$ and $B = \{3, 6, 9, 12, 15\}$, give the intersection of A and B.

3. Find $\{3, 6, 9, 12, 15, 18\} \cap \{2, 4, 6, 8, 10, 12\}$.

4. One food requires an oven temperature between 275° and 325°. A second food requires a temperature over 300°.
   a. Algebraically describe the temperatures t at which both foods can bake.
   b. Graph the temperatures t at which both foods can bake.

5. Oxygen is a liquid or solid below -183 °C. Nitrogen is liquid or solid below -196 °C. A scientist needs both oxygen and nitrogen in liquid or solid form. Graph the possible temperatures for obtaining both.

In 6 and 7, imagine tossing a fair die with six possible outcomes $\{1, 2, 3, 4, 5, 6\}$.

6. Find the probability of tossing an even prime number.

7. Find the probability of tossing an odd prime number.

8. Draw a Venn diagram to illustrate R, S, and $R \cap S$ if $R = \{2, 5, 6, 11, 13\}$ and $S = \{3, 4, 5, 6, 7\}$.

9. Intersections arise whenever there are __?__ conditions to be satisfied.

10. Two sides of a triangle have lengths 6 and 10.
    a. According to the Triangle Inequality, which three inequalities must be satisfied by the length L of the third side?
    b. Solve the inequalities.
    c. Algebraically describe the possible values of L.

11. Graph the set of all real numbers y, so that
    a. $y \geq -1$ and $y \leq 2$
    b. $-1 < y < 2$

Applying the Mathematics

12. Find $\{1, 3, 7, 10\} \cap \{2, 4, 8, 9\}$.

In 13 and 14, suppose a letter from the alphabet is chosen at random.

13. Find P (the letter is in the word "antic" and in the word "antique").

14. Find P (the letter is a consonant and in the first half of the alphabet).

In 15 and 16, write answers as fractions. If one element x is chosen at random from $\{1, 2, 3, \ldots, 30\}$, find the probability for each of the following:

15. $x > 3$ and $x > 15$ **16.** $x < 8$ and $x > 21$

17. $A = \{1, 3, 5, 7, 9, 11\}$; $B = \{1, 4, 7, 10, 13, 16\}$; $C = \{1, 5, 10, 15, 20, 25\}$
    a. Find $A \cap B$.   **b.** Find $(A \cap B) \cap C$.

18. Remember that N(P) means "the number of elements in set P." If $P = \{20, 22, 28, 29, 35\}$ and $Q = \{25, 26, 27, 28, 29, 30\}$, find
    a. N(P)   **b.** N(Q)   **c.** N($P \cap Q$)

19. T girls are on the track team. S girls are on the swim team. B are on both. How many are on at least one of the teams?

20. Thirty boys are on the school's track team. Twenty-one boys are on the swim team. Six are on both.
   a. Copy and complete the Venn diagram of this situation.
   b. How many are on at least one of the teams?

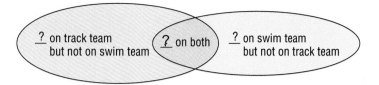

Review

In 21–23, solve. *(Lesson 3-3)*

21. $t - 153 = -12$

22. $4.2 \geq z - 18.3 + 5$

23. $-3b + 4b - 1 = 3$

24. *Multiple choice*   Suppose $j < n$. Then $j - n$
   (a) is always positive
   (b) is always negative
   (c) can be either positive or negative. *(Lesson 3-3)*

25. Simplify: $-10 - p + p - \, -6$. *(Lesson 3-1)*

26. You purchase P granola bars. Here are four expressions.
   i. $P + 3$ **ii.** $P - 3$
   iii. $0.5 \, (P - 1)$ **iv.** $P - 1$

   Match each expression with the description of how many bars you have after
   a. you eat 3 bars,
   b. then you buy 6 more,
   c. then you give 5 to your brother but take 1 back,
   d. then your dog eats half your bars. *(Lessons 2-2, 3-2)*

27. *Skill sequence*   Evaluate. *(Lessons 2-3, 3-1)*
   a. $4 - 7$ **b.** $\dfrac{4}{5} - \dfrac{7}{5}$ **c.** $\dfrac{4}{x} - \dfrac{7}{x}$

28. Evaluate in your head when $x = -3$ and $y = 4$. *(Lessons 2-2, 3-1)*
   a. $x + 2y$ **b.** $x - 2y$ **c.** $2y - x$

Exploration

29. A, B, and C are sets of integers between 1 and 10. What might be the elements of these sets if $N(A) = 4$, $N(B) = 6$, and $N(A \cap B) = 3$?

3-5

Union of Sets

With one game to go in the season the Blues lead the Reds by one game. The Blues are playing the Purples; the Reds are playing the Greens. From the standings at the right, you can see that the Blues will win the league title if
(a) the Blues beat the Purples   *or*
(b) the Greens beat the Reds.

Teams	Wins	Losses
Blues	6	1
Reds	5	2
.	.	.
.	.	.
.	.	.
Greens	2	5
Purples	2	5

The key word in this situation is "or." Unlike the intersection of events, in which two conditions must all be satisfied, here *either* condition *or both* will cause the Blues to win. This is the idea behind the **union** of two sets. The symbol for union looks like the letter U.

Definition:

The union of sets A and B, written $A \cup B$, is the set of elements in either A *or* B (or in both).

Contrast the definition of union with that of intersection. The key word for intersection is "and"; the key word for union is "or." Notice that the union of two sets is not just the result of "putting them together." Elements are not repeated if they are in both sets.

Example 1   **a.** Give the union of {1,3,5,7,9,11} and {1,4,7,10,13,16}.
b. Make a Venn diagram of the two sets and shade the union.

Solution
a. The union is the set of elements in one set or the other (or both).
{1,3,5,7,9,11} ∪ {1,4,7,10,13,16} = {1,3,5,7,9,11,4,10,13,16} Notice that the elements 1 and 7 are in both sets, but they are only written once.
b.

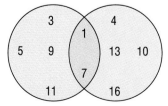

As in the case of intersections, unions can often be represented on the number line.

Example 2 Graph the set of all numbers s such that $s > -2$ or $s \le -10$.

Solution Include all points that satisfy either or both conditions.

Example 3 George remembered that yesterday's temperature was either below 40°F or below freezing. What might have been the temperature?

Solution Let T be the temperature in °F. Either $T < 40$ or $T < 32$. Draw graphs for each inequality. (You may do this in your head.)

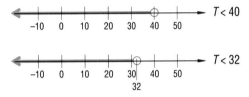

Include all points that are on either graph.

$T < 40$ or $T < 32$ is equivalent to $T < 40$ alone. The temperature might have had any value less than 40 °F.

Example 4 shows how the idea of union can be applied to a relative frequency problem.

Example 4 To determine how many late buses were needed, some of the students in Hillcrest High were surveyed. They were asked, "Do you sometimes stay after school for an activity or sport?" and "How do you get home—bus, car, or walk?" The data are summarized in the table at the right.

		Travel home		
		bus	car	walk
Stay after	yes	20	40	30
	no	85	65	70

a. What is the relative frequency of students walking home?
b. What is the relative frequency of students staying after?
c. What is the relative frequency of students walking home or staying after?

Solution There are a total of 310 students in the survey.

a. 100 walk home; relative frequency $= \dfrac{100}{310} \approx 32\%$

b. 90 stay after; relative frequency $= \dfrac{90}{310} \approx 29\%$

c. There are $20 + 40 + 30 + 70 = 160$ who walk or stay after.
Relative frequency $= \dfrac{160}{310} \approx 52\%$.

Notice that the answers to parts a and b do not add up to the answer in part c. Some students do both.

Questions

Covering the Reading

1. How can the Greens wind up the season ahead of the Purples?

2. Define: union of two sets.

3. Give the union of {2,4,6,8,10} with {3,6,9,12,15}.

4. Find {1,3,7,10} ∪ {2,4,8,9}.

5. Match each sentence below with its graph at the left.
  a. $y < -8$ or $y > 2$
  b. $y < -8$ or $y < 2$
  c. $y > -8$ and $y < 2$

I (number line) -8 ⊕ ⊕ 2 → y

II (number line) -8 ⊕ ⊕ 2 → y

III (number line) -8 ⊕ 2 → y

6. Graph the set of all numbers z such that $z < -2$ or $z \le 4$.

7. Refer to Example 4.
  a. What is the relative frequency of students taking the bus and staying after?
  b. What is the relative frequency of students taking the bus or staying after?

Applying the Mathematics

In 8 and 9, let E = the odd numbers from 1 to 10.
Let F = all multiples of 3 between 1 and 10.

8. Find **a.** $E \cap F$; **b.** $E \cup F$; **c.** $N(E \cup F)$.

9. Now let G = the prime numbers between 1 and 10.
Find **a.** $(E \cap F) \cap G$; **b.** $(E \cup F) \cup G$.

10. Graph the ages that have free admission under the following rule: You will get in free if you are younger than 4 or a senior citizen (62 or older).

11. Copy the Venn diagram at the left and shade in $(A \cap B) \cup C$.

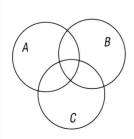

12. $N(P) = 16$. $N(P \cap Q) = 9$. $N(P \cup Q) = 28$. Find $N(Q)$. (Hint: Make a Venn diagram.)

13. Graph the numbers that are solutions to $k + 7 > 6$ or $k + 12 > {-17}$.

14. A group of students who played baseball were asked ''How do you bat—right-handed, left-handed, or both?'' and ''Do you belong to an organized league—yes or no?'' Here are the results.

		Batter's stance		
		right	left	both
Belong	yes	10	2	5
to league	no	15	20	0

What is the relative frequency of students who
a. belong to an organized league?
b. belong to an organized league or bat left-handed?

15. Graph the union of the solution set to ${-10} < y < 1$ and the solution set to $0 < y < 16$.

16. Members of the Jordan family have the following ages: Mr. Jordan, 44; Mrs. Jordan, 43; Michael, 20; Barbara, 16; Louis, 13. If a person from the family is selected at random, what is the probability that
a. it is a teenager or a female?
b. it is a teenager and a female?

Review

17. Solve in your head. *(Lessons 2-6, 3-3)*
  a. $2 = x - 5.4$ **b.** $1\frac{1}{2} + y = 4$ **c.** $z + 1.5 = 0$

18. Solve for v: $a^2 + v = 10$. *(Lesson 2-7)*

19. If $\frac{7}{5}x + 3 = {-2}$, find $\frac{7}{5}x$. *(Lesson 2-6)*

20. Solve for q: ${-2r} < q + 8r$. *(Lesson 2-7)*

21. *Skill sequence*   *(Lesson 2-3)*
  a. $\frac{1}{2} - \frac{2}{3}$ **b.** $\frac{x}{2} - \frac{2x}{3}$ **c.** $\frac{x}{2a} + \frac{2x}{3a}$

22. Write in simplest terms: $\frac{15ab}{12bc} + \frac{3a}{4c}$. *(Lesson 2-3)*

23. Two sides of a triangle both have length 12. What are the possible lengths of the third side? *(Lessons 2-8, 3-3)*

24. A baseball player made 30 hits out of the last 100 times at bat. Estimate the probability that the player will get a hit the next time at bat. *(Lesson (1-8)*

25. *Multiple choice* Which of the following real world situations is represented by the graph below? *(Lesson 2-4)*

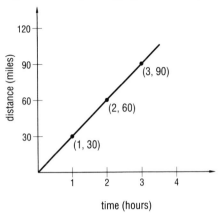

distance (miles)

(3, 90)

(2, 60)

(1, 30)

time (hours)

(a) The distance traveled in h hours at 30 mph
(b) The distance traveled in h hours at 50 mph
(c) The distance you are from home if you started 90 miles from home and traveled home at 30 mph

26. The volume of a cone is $V = \frac{1}{3}\pi r^2 h$ where r is the radius and h the height. Find the volume of a cone with a radius 3 cm and height 15 cm. Round your answer to the nearest cubic centimeter. *(Lesson 1-5)*

27. *Multiple choice* $2y^2 \cdot y^3 = ?$ *(Lesson 1-4)*
(a) $2y^5$ (b) $2y^6$ (c) $(2y)^5$ (d) $4y^6$

28. Camisha is buying a spiral notebook. She has a choice of four colors (red, blue, green, orange) and 2 line widths (narrow, medium).
a. List all the possible types of notebooks.
b. If Camisha picks a type at random, what is the probability that she chooses one that is blue and has narrow lines? *(Lesson 1-7)*

Exploration

29. a. Give an example of two sets A and B with $N(A \cap B) = 10$ and $N(A \cup B) = 12$.
b. Is it possible to have sets A and B with $N(A \cap B)$ larger than $N(A \cup B)$? Explain why or why not.
c. Is it possible to have $N(A \cap B) = N(A \cup B)$? Explain why or why not.

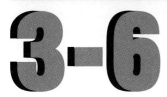

LESSON
3-6

Addition, Subtraction, and Counting

Mr. Drummond's 4th period science class has 23 students. Ms. Whitefeather's 4th period math class has 25 students. If D and W are the sets of students in these classes,

$$N(D) = 23 \text{ and } N(W) = 25.$$

Because the classes meet at the same time, they have no students in common. They are called **mutually exclusive.** It is easy to calculate the number of students in one or the other of the classes.

$$\begin{aligned} N(D \cup W) &= N(D) + N(W) \\ &= 23 + 25 \\ &= 48 \end{aligned}$$

The Venn diagram for mutually exclusive sets, like D and W, is two circles that do not overlap.

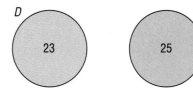

Now consider sets with overlap. Let L be the set of students in Ms. Lee's 5th period science class. Set L has 30 students, 6 of whom are in Ms. Whitefeather's class. Here is a Venn diagram for W and L.

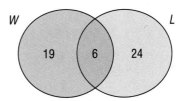

$N(W) = 25$ and $N(L) = 30$, but the total number of students either in Ms. Whitefeather's math class or in Ms. Lee's science class is not 55. If we added the numbers, we would be counting 6 students twice. The six students are in the intersection of the classes; they are the elements of $W \cap L$. The total of 49 students can be found by adding $N(W)$ and $N(L)$, then subtracting $N(W \cap L)$.

$$\begin{aligned} N(W \cup L) &= N(W) + N(L) - N(W \cap L) \\ 49 \quad &= 25 \quad + 30 \quad - 6 \end{aligned}$$

This example illustrates the Fundamental Counting Principle.

Fundamental Counting Principle:

Let A and B be finite sets.
Then $N(A \cup B) = N(A) + N(B) - N(A \cap B)$.

Example 1 A clothing company makes a line of blouses and skirts to be worn as mix-and-match outfits. Of all possible outfits, the company recommends 76 of them as stylish. The orange blouse is used in 5 outfits and the blue skirt in 9 outfits. The orange blouse and blue skirt are used together in 3 outfits. How many outfits include the orange blouse or the blue skirt?

Solution Let G = the set of outfits with the orange blouse.
Let L = the set of outfits with the blue skirt.
Then N (outfits with the orange blouse or blue skirt) =

$$N(G \cup L) = N(G) + N(L) - N(G \cap L)$$
$$= 5 + 9 - 3$$
$$= 11$$

If two sets A and B are mutually exclusive, the situation is simpler. $N(A \cap B) = 0$ and there is nothing to subtract. That is why $N(D \cup W) = N(D) + N(W)$ on page 133.

The Fundamental Counting Principle leads to a property of probabilities. Here is a Venn diagram of the same three classes considered on page 133. There are 72 students in all.

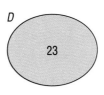 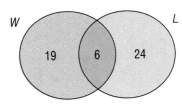

Let a student be chosen randomly from these classes. It's easy to calculate these probabilities.

P(student is in Mr. Drummond's class) $= P(D)$ $= \frac{23}{72}$

P(student is in Ms. Whitefeather's class) $= P(W)$ $= \frac{25}{72}$

P(student is in Ms. Lee's class) $= P(L)$ $= \frac{30}{72}$

Probabilities of unions and intersections are also easy to find. For instance,

$$P(W \cup L) = \frac{49}{72}$$
$$P(W \cap L) = \frac{6}{72}.$$

All these probabilities are the same as the numbers in the sets divided by 72. So it is not surprising that

P(student is in W or L) = P(in W) + P(in L) − P(in both W and L).
That is, $P(W \cup L) = P(W) + P(L) - P(W \cap L)$.
To verify: $\frac{49}{72} = \frac{25}{72} + \frac{30}{72} - \frac{6}{72}$

The general principle is a formula for the probability of a union of events.

Probability of a Union of Events:

$$P(A \cup B) = P(A) + P(B) - P(A \cap B)$$

Written with the words "and" and "or" this formula says that

$$P(A \text{ or } B) = P(A) + P(B) - P(A \text{ and } B).$$

Example 2 At Cardioid Hospital, 65% of the patients have high blood pressure, 70% have clogged arteries, and 2 out of 5 have both conditions. If a patient has either condition, he or she has high risk of a heart attack. What is the probability that a Cardioid Hospital patient has either one of the conditions?

Solution Let H = high blood pressure. Let C = clogged arteries. We want $P(H \cup C)$. The following information is given.

$$P(H) = .65$$
$$P(C) = .70$$
$$P(H \cap C) = \tfrac{2}{5}$$

Use the Probability of a Union of Events.

$$
\begin{aligned}
P(H \cup C) &= P(H) + P(C) - P(H \cap C) \\
&= .65 \ +.70 \ \ - \tfrac{2}{5} \\
&= .65 \ +.70 \ \ -.40 \\
&= .95
\end{aligned}
$$

As a percent, the probability is 95%.

Questions

Covering the Reading

In 1–4, refer to the science and math classes discussed in this lesson.

1. Give the numerical value.
   a. N(D)   b. N(W)   c. N(L)
   d. N($D \cap W$)   e. N($W \cap L$)   f. N($D \cap L$)
   g. N($D \cup W$)   h. N($W \cup L$)   i. N($D \cup L$)

2. Complete the formula: N($W \cup L$) = N(W) + N(L) − __?__.

3. Let a student be randomly selected from these classes. Calculate:
   a. P(W)   b. P(L)   c. P($W \cup L$)

4. Complete the formula: P($W \cup L$) = P(W) + P(L) − __?__.

5. Find N(A or B) if N(A and B) = 6, N(A) = 20 and N(B) = 11.

6. Refer to Example 1. A fuchsia blouse is used in 2 outfits. No outfits use both a fuchsia and an orange blouse.
   a. How many outfits include an orange or a fuchsia blouse?
   b. What is the probability of choosing at random an outfit including an orange or a fuchsia blouse?

7. When are sets mutually exclusive?

8. When is P($A \cup B$) = P(A) + P(B)?

9. Refer to Example 2. At Cardioid Hospital, if P(H) = .59, P(C) = .62, and P($H \cap C$) = .34, what is P($H \cup C$)?

Applying the Mathematics

10. At Central High School, 15% of the students received an A in English, 20% received an A in Math, and 6% received an A in both subjects.
    a. Find the percent of students who got an A in English or an A in Math.
    b. Find N(A in English or A in Math) if there are 100 students at Central.
    c. Find N(A in English or A in Math) if there are 500 students at Central.

11. If N($A \cup B$) = 30, N(A) = 24 and N(B) = 12, calculate N($A \cap B$).

12. A class has s students. On a test 10 students received A's and 4 students received B's. A student is selected at random. What is the probability the student got an A or a B?

13. N(X) = 10; N(Y) = 14; N($X \cup Y$) = 18
    a. Find N($X \cap Y$).
    b. Copy the diagram and fill in the correct numbers in each region.

14. If A = {0, 1, 2, 3, 4, 5}, B = {-2, -1, 0, 1, 2}, and C = {-4, -2, 0, 2, 4}
    a. draw a Venn diagram showing A, B, and C;
    b. find N($A \cup B \cup C$).

15. $P(A) = \frac{1}{2}$, $P(B) = \frac{1}{3}$, and $P(A \cap B) = \frac{1}{4}$. Find $P(A \cup B)$.

16. A company produces 41 different frozen dinners. 14 dinners have beef, 21 have chicken, 5 have broccoli, and 3 have both beef and broccoli. If a dinner is picked at random, what is the probability that
  a. it has beef or broccoli?
  b. it has beef or chicken? (No dinner has both.)

Review

17. *Skill sequence*   Solve. *(Lesson 2-6)*
  a. -2 + x = 7   **b.** -2 + (3 + y) = 7
  c. -2 + (3 + (z + 4)) = 7

18. Solve ((m − 2) − 3) − 4 = -5. *(Lesson 3-3)*

19. If you celebrated your 12th birthday in year *y*:
  a. In what year were you born?
  b. In what year will you celebrate your 50th birthday? *(Lesson 3-2)*

20. Alicia is *d* years older than her brother Jim.   **a.** Five years from now, how many years older than Jim will she be?   **b.** Write an equation relating their ages *A* and *J*. *(Lesson 3-2)*

21. Evaluate $\frac{3x^2}{2y}$ when *x* = 1.1 and *y* = 3.4. Round to nearest tenth. *(Lesson 1-4)*

22. You eat a third of a pizza. A friend eats a fifth of a pizza twice the size. Let *p* stand for the size of the smaller pizza. Simplify $\frac{p}{3} - \frac{2p}{5}$ to determine who has eaten more pizza, and by how much. *(Lesson 3-1)*

23. Christopher played golf on several different courses last summer. A positive result means above par, a negative result means below par. If his results were 5, -3, 8, -4, -2, 0, -1, -2, 3, find his average compared to par. *(Lesson 1-2)*

24. Match each % with its equivalent fraction. *(Previous course)*
  a. 60%   **i.** $\frac{3}{5}$
  b. 62.5%   **ii.** $\frac{2}{3}$
  c. 66.$\overline{6}$%   **iii.** $\frac{5}{8}$

Exploration

25. Do relative frequencies follow formulas like those for counting and probability in this lesson? Try this experiment. Toss two dice 50 times. Let *E* = tossing an even number. Let *L* = tossing a number larger than 9. Let R(*X*) be the relative frequency of an event X.
  a. Find R(*E*), R(*L*), R(*E* ∪ *L*), and R(*E* ∩ *L*).
  b. Do these numbers add to or subtract from each other in some way? If so, how? If not, are they close?

26. Find out how many sides a die would have if it were in the shape of
  a. a tetrahedron; **b.** an icosahedron.

Addition, Subtraction, and Graphing

Suppose Tim wants to buy a dozen muffins. He likes raisin muffins and bran muffins. The ways he can purchase a dozen muffins of the two kinds can be expressed as thirteen ordered pairs, listed in the table at the left.

Using variables makes the pattern easier to write. If Tim buys R raisin muffins and B bran muffins, then Tim's choices can be represented by ordered pairs (R, B) where $R + B = 12$. We could solve for B and write $B = 12 - R$.

R Raisin	B Bran	(R, B) Ordered Pairs
0	12	(0, 12)
1	11	(1, 11)
2	10	(2, 10)
3	9	(3, 9)
.	.	.
.	.	.
.	.	.
10	2	(10, 2)
11	1	(11, 1)
12	0	(12, 0)

Here is a graph of Tim's possible choices for a dozen muffins.

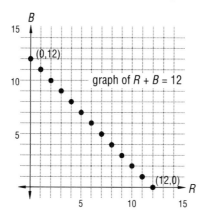

The points are all on the same line. There are 13 points on the graph because there are 13 combinations of muffins Tim can buy. He cannot have a fraction of a muffin and he certainly can't have a negative number of muffins. This graph is discrete and the set of points is finite.

However, suppose you wanted to graph *all the pairs* of numbers whose sum is 12. You would have pairs with negative numbers and fractions, like (15, -3) and (4.5, 7.5). In fact, there are an infinite number of pairs whose sum is 12. But they all still lie in a straight line when they are graphed. This graph is continuous and is the entire line. Its algebraic representation is $x + y = 12$, where x and y can be any real numbers.

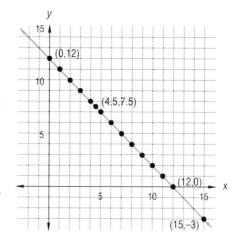

The graph of any constant sum situation will look much like this one.

To graph all pairs of numbers that satisfy a condition, first make a chart showing some sample pairs. Then plot the points on a graph. If the graph is continuous, draw a line through the points.

Example 1 Make a graph of all ordered pairs of real numbers (x, y) for which $y = x - 2$.

x	y	(x, y)
-3	-5	(-3, -5)
-2	-4	(-2, -4)
-1	-3	(-1, -3)
0	-2	(0, -2)
1	-1	(1, -1)
2	0	(2, 0)
3	1	(3, 1)

Solution Choose some numbers for x, say the integers from -3 to 3. For each value of x, find the corresponding value of y and fill in the chart.

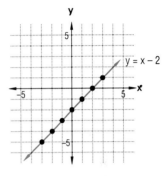

Now plot the ordered pairs on the graph. Since x and y can be any real numbers, the graph should be continuous. Connect the points to form a line. This line has equation $y = x - 2$.

Notice that in Example 1, $y - x = -2$. So the difference of y and x is constant. The graph of any constant difference situation is a line.

Questions

Covering the Reading

In 1–3, consider the muffin example in this lesson.

1. If Tim buys 10 raisin muffins, how many bran muffins did he buy?

2. The table of values in the muffin problem is missing some ordered pairs. Copy and complete.
 a. (_?_ , 4)   **b.** (6, _?_)

3. When a graph is made of all pairs of real numbers whose sum is 12, a _?_ is formed.

4. a. Extend the chart from Example 1 by copying the chart at the right and filling in the empty columns.
 b. Are these points on the line in Example 1?

x	y	(x, y)
4		
5		
6		

5. Xandra and Yvonne each have earned medals in track and field events. Yvonne has 5 more medals than Xandra.

    a. Copy and fill in the chart to show some possible numbers of medals for the girls.

    b. Graph the ordered pairs from the chart.

x Xandra	y Yvonne	(x, y) Ordered Pairs
1	6	
2		
3		(3, 8)
4		
5	10	
6		

Applying the Mathematics

6. In the muffin example, suppose Tim has twice as many bran muffins as raisin muffins. How many bran muffins does he have?

7. Sally and Alvin each have tropical fish. Together they have eight fish.

    a. Find all possible number pairs of Sal's fish and Al's fish.

    b. If Sal has two more fish than Al, how many fish does Al have?

8. a. Copy and fill in the chart at the right to show ordered pairs where the y-coordinate is the opposite of the x-coordinate.

    b. On a coordinate plane, graph all ordered pairs of real numbers for which $y = -x$.

x	y	(x, y)
-2	2	(-2, 2)
-1		
0		
1		
2		

In 9 and 10, make a list of the four ordered pairs graphed. Then choose the equation that describes the points.

9.
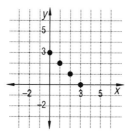

    a. $y = 3x$
    b. $y = x + 3$
    c. $y = 3 - x$

10.
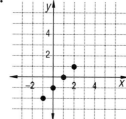

    a. $y = 1x$
    b. $y = x + 1$
    c. $y = x - 1$

11. Write an equation that describes the points in Question 5.

12. Calvin is giving away 10 records to Jorge and Maria.

    a. Make a chart showing the possible ways to divide the records.

    b. Graph the possibilities with the number of Jorge's records on the horizontal axis and the number of Maria's records on the vertical axis.

    c. *True or false* This graph is a line.

Review

13. Solve $10.4 = y - 12.2$. *(Lesson 3-3)*

14. Solve $x - k = 50$ for x. *(Lesson 3-3)*

15. Last weekend, 2250 students went to the football game Friday night. On Saturday night, 276 attended the dance. If 200 students were at both events, find N(attended game or attended dance). *(Lesson 3-6)*

16. If N(A) = 15, N(A ∩ B) = 7 and N(A ∪ B) = 20, find N(B). *(Lesson 3-6)*

17. In the fall election, 80% of the students voted for Millie for president, 60% voted for Moe for vice-president, and 45% voted for both.
   a. What percent voted for Millie or Moe?
   b. If 453 students voted, how many voted for Millie or Moe? *(Lesson 3-6)*

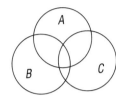

18. Copy the diagram at the left for each question.
   a. Shade A ∪ B.
   b. Shade (A ∪ B) ∩ C.
   c. Shade A ∩ B ∩ C. *(Lessons 3-4, 3-5)*

19. A = {instruments in a symphony orchestra} and B = {instruments in a marching band}. Name an instrument in A but not in A ∩ B. *(Lesson 3-4)*

20. The 1984 population of Illinois was about $1.14 \cdot 10^7$. The population of Chicago was about $7.10 \cdot 10^6$. A person living in Illinois is randomly selected. Approximate the probability that he or she lives in Chicago. *(Lesson 1-7)*

21. *Multiple choice*   The measure of this angle is closest to
   (a) 45°   (b) 90°
   (c) 135°   (c) 180° *(Previous course)*

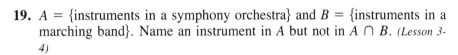

22. At 3:00, what is the measure of the angle between the hands of a clock? *(Previous course)*

23. A factory packs 5-pound packages of mixed peanuts and cashews. If c pounds of cashews are in a package, the price p of the package is found by the formula

$$p = 2.39c + 1.69(5 - c)$$

If a package contains 1.5 pounds of cashews, find its price. *(Lesson 1-5)*

24. Graph: **a.** $y \leq 11$; **b.** $y \leq 11$ or $y \geq 15$; **c.** $y \leq 11$ and $y \leq 7$. *(Lessons 3-4, 3-5)*

25. *Skill sequence* Evaluate in your head when $x = 10$. *(Lesson 1-4)*
  a. x^3 **b.** $2x^3$ **c.** $(\tfrac{1}{2}x)^3$

26. *Multiple choice* Which fraction is not equivalent to the others?
  (Lesson 2-3)
  (a) $\dfrac{-3}{2}$ (b) $\dfrac{3}{-2}$ (c) $-\left(\dfrac{-3}{-2}\right)$ (d) $-\left(\dfrac{-3}{2}\right)$

Exploration

27. a. The year 1986 roughly corresponds to the year 2735 in the Babylonian calendar and the year 5746 in the Jewish calendar. The Norman conquest of England occurred in 1066. What year is 1066 in the Babylonian and the Jewish calendars?
  b. If you graph ordered pairs (J, G), where J is the year in the Jewish calendar and G the year in the Gregorian calendar (the official one in the United States), the graph is a line. Does the line slant up or down as you go to the right?
  c. Look in an almanac or other reference book. What other calendars are there?

28. This graph is roughly symmetric, that is, the graph of the Democrats and that of the Republicans are nearly reflection images of each other. What causes this?

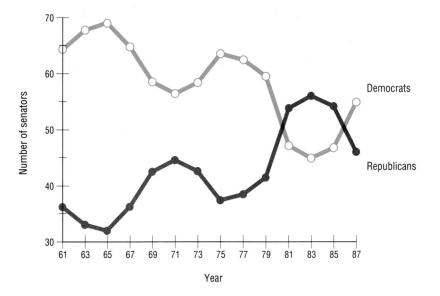

Party Membership of United States Senators

LESSON

3-8

Addition, Subtraction, and Geometry

Constant sums occur in many places in geometry.

In each picture at the right, the pendulum makes two angles with the crossbar of the clock. As the pendulum swings, the angle measures x and y vary. But since the crossbar of the clock is a straight line, the sum of the measures of the angles is always 180°.

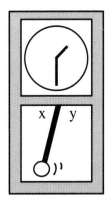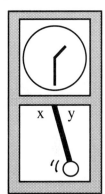

$$x + y = 180$$
Solving for y, $\quad -x + x + y = 180 + -x$
$$0 + y = 180 + -x$$
$$y = 180 - x$$

So if x is known, y can be found by subtracting x from 180°.

■ ■ ■ ■ ■ ■ ■ ■

Example 1   In the angles shown above, find y if x is 128°.

Solution 1 $y = 180 - x$ (from preceding explanation)
$$= 180 - 128$$
$$= 52$$

Solution 2 $x + y = 180$
Substitute 128 for x. $128 + y = 180$
$$y = 52$$

Two angles whose sum is 180° occur often in geometry, so they are given a name. These angles are called **supplementary** (no matter what the position of the angles). The angles are **supplements.**

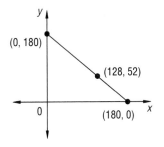

If the pairs of possible measures of supplementary angles are graphed, they lie on the part of the line $x + y = 180$ that is in the first quadrant. At left is a graph of these pairs. The point (128, 52) from Example 1 is identified.

You may already know that in a triangle, regardless of its shape, the sum of the measures of the three angles is 180.

Triangle-Sum Property:

In any triangle with angle measures a, b, and c,
$a + b + c = 180$.

  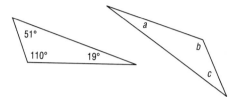

So, if you know the measures of two angles of a triangle, you can find the measure of the third angle.

■ ■ ■ ■ ■ ■ ■ ■
Example 2   In the triangle at the right, find n.

Solution 1   In any triangle with angle measures a, b, and c,
$$a + b + c = 180.$$

Substitute for a, b, and c. $\quad 71 + 86 + n = 180$
Simplify. $\qquad\qquad\qquad\qquad\quad 157 + n = 180$
Solve. $\qquad\qquad\qquad\qquad\qquad\quad n = 23$
So the third angle is 23°.

Solution 2   Add the given measures. Then subtract from 180.

$180 - (71 + 86) = 180 - 157 = 23$

In a right triangle, one of the angles measures 90°. In drawings, a 90° angle is often marked with the symbol ⌐.

■ ■ ■ ■ ■ ■ ■ ■
Example 3   Write an expression for p in the triangle at the right.

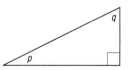

Solution
$$a + b + c = 180$$
Substitute for a, b, and c. $\quad p + q + 90 = 180$
Solve for p and simplify. $\qquad\qquad p = 180 - 90 - q$
$$p = 90 - q$$

In Example 3, angles p and q are called **complements.** Two angles are **complementary** if the sum of their measures is 90°.

Questions

In 1–3, use the drawing at the right.

1. If $x = 42°$, what is y?

2. Find x if y is 137.5°.

3. Represent y in terms of x.

4. If $m\angle F = 58°$ and $m\angle G = 132°$, are $\angle F$ and $\angle G$ supplementary?

5. Find the measure of a supplement of $\angle J$.

In 6–8, find the value of the variable.

6.   7.   8.

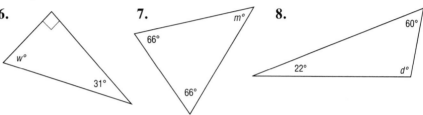

9. *True or false*   65° and 25° are measures of complementary angles.

10. If $m\angle Q$ is 29°, what is the measure of a complement?

11. **a.** Give the measures of five pairs of complementary angles, then write the measures as ordered pairs.
    b. Graph your pairs on the coordinate plane.
    c. Show the possible measures of all pairs of complementary angles on your graph.

12. Angles A and B are complementary. If $m\angle A = W$, write an expression to represent the measure of $\angle B$.

13. $\angle C$ is the supplement of $\angle A$. $\angle T$ is the complement of $\angle C$. $m\angle A = 155°$. Find the measures of $\angle C$ and $\angle T$.

14. This triangle has two angles the same size. The third angle is 102°. Find e.

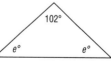

In 15 and 16, use this fact. The sum of the angles of a quadrilateral is always 360°.

15. Find p.   16. Find q.

17. Write an expression to represent $m\angle C$ in terms of $m\angle A$ and $m\angle B$.

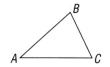

Review

18. Assume for points (x, y) that $x - y = 3$. *(Lessons 1-5, 3-7)*
   a. Make a table of some points (x, y).
   b. Graph $x - y = 3$. The domain for both variables is the set of real numbers.

In 19–21, solve.

19. Solve: $z - 43 < 65$. *(Lesson 3-3)*

20. Solve: $3.1 = w - 7.48$. *(Lesson 3-3)*

21. Solve: $3 + x + \text{-}8 = 5$. *(Lesson 2-6)*

22. The total cost of a hamburger and a milk shake is $1.96. If h is the cost of the hamburger, what is the cost of the milk shake? *(Lesson 3-2)*

23. In a town, 24% of the residents voted for a tax increase. 86% reduced their water consumption. 17% voted for the increase and reduced their water consumption. Of 3526 residents, how many voted for the increase or reduced their water consumption? *(Lesson 3-6)*

24. The only even prime number is 2. There are 25 even numbers and 15 prime numbers less than or equal to 50. If an integer is chosen at random from the interval $1 \le x \le 50$, what is the probability it is even or a prime? *(Lesson 3-6)*

25. The variable y stands for the same number in both figures below. The perimeter of the triangle is 73. What is the area of the square? *(Lesson 2-6)*

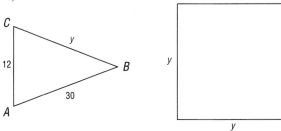

26. Graph: **a.** $y < 2$ or $y > 5$; **b.** $y < 2$ and $y > 5$. *(Lessons 3-4, 3-5)*

27. When this program is run on the computer, what will be printed if 13.94, 6.06 is the input? *(Lesson 1-4)*

```
10 PRINT "EVALUATE A * B"
20 PRINT "GIVE A AND B"
30 INPUT A,B
40 PRINT A * B
50 END
```

28. Here is a list of National Football League scoring leaders over a 10-year period.

Year	Leader	Points
1986	Kevin Butler, Chicago	120
1985	Kevin Butler, Chicago	144
1984	Ray Wersching, San Francisco	131
1983	Mark Mosely, Washington	161
1982	Wendell Tyler, Los Angeles	78
1981	Ed Murray, Detroit	121
1980	Ed Murray, Detroit	116
1979	Mark Mosely, Washington	114
1978	Frank Corrall, Los Angeles	118
1977	Walter Payton, Chicago	96

a. What is the median of points for the leaders?
b. What is the mean of points?
c. What is the range of points? *(Lesson 1-2)*

29. *Multiple choice*   Which of the following is *not* equal to $w + -k$?
(Lesson 3-1)
(a) $w - k$ (b) $k - w$ (c) $-k + w$

30. Simplify. *(Lessons 2-2, 3-1)*
a. $2b + -3b + 5b$ **b.** $15p^5 - 10p^5 + 5p^5$ **c.** $.2n + .3m + .8n - .2m$

In 31 and 32, simplify. *(Lessons 2-3, 3-1)*

31. $\frac{1}{2}x - \frac{1}{3}x - \frac{3}{4}x$ **32.** $\frac{y}{2} - \frac{y}{3} - \frac{3y}{4}$

Exploration

33. Below is a map of Centerville. John lives at the corner marked J. His girl friend Mary lives by the corner marked M. John and Mary each walk directly to the theater along streets and meet there. The sum of the distances they walk is 6 blocks. Copy the diagram. On your copy show all possible locations of the theater.

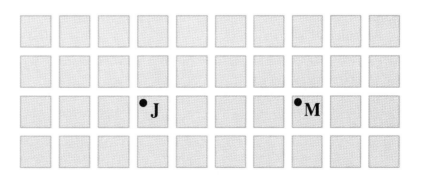

Summary

You can think of this chapter as having three parts. The first part concerns the algebraic definition and uses of subtraction, and solving subtraction sentences. The second part deals with the relationships of addition and subtraction to the union and intersection of sets and to counting and probability. The third part applies these ideas to the graphing of lines and to properties of geometric figures.

The algebraic definition of subtraction is in terms of addition: $a - b = a + -b$. This definition enables subtraction sentences of the form $x - a = b$ or $x - a < b$ to be solved by converting them to addition. Such sentences arise from the uses of subtraction. The two major models for subtraction are take-away and comparison.

Given two sets A and B, their union $A \cup B$ consists of those elements in A *or* in B (or in both). Their intersection $A \cap B$ consists of those elements in A *and* in B. The number of elements in $A \cup B$ is denoted by $N(A \cup B)$.

$$N(A \cup B) = N(A) + N(B) - N(A \cap B)$$

If A and B are events, then $A \cup B$ consists of the event where either A or B occurs, and

$$P(A \cup B) = P(A) + P(B) - P(A \cap B)$$

In both these formulas, subtraction compensates for counting elements in the intersection twice.

When an equation has two variables, its solutions may be graphed on a coordinate plane. In this chapter, ordered pairs (x, y) satisfying $x + y = a$ and $x - y = a$ were graphed. In each case, the graph is a line. Constant sums like these occur in geometry. Two angles whose measures add to $180°$ are supplementary; two angles whose measures add to $90°$ are complementary. The three angles of a triangle give the constant sum $180°$.

Vocabulary

Below are the most important terms and phrases for this chapter. You should be able to give a general description and a specific example of each.

Lesson 3-1
algebraic definition of subtraction

Lesson 3-2
take-away model for subtraction
comparison model for subtraction

Lesson 3-4
intersection of sets, ∩
"and"
Venn diagram

Lesson 3-5
union of sets, ∪
"or"

Lesson 3-6
mutually exclusive
Fundamental Counting Principle
Probability of a Union of Events

Lesson 3-8
supplementary angles, supplements
complementary angles, complements
⌐ symbol for 90° angle
Triangle Sum Property

Progress Self-Test

Take this test as you would take a test in class. You will need graph paper. Then check your work with the solutions in the Selected Answers section in the back of the book.

1. According to the algebraic definition of subtraction, adding -7 is the same as doing what else?

In 2–5, simplify.

2. $n - 16 - 2n - (-12)$

3. $-8x - (2x - x)$

4. $\frac{3}{4} - \frac{7}{8}$

5. $\frac{m}{2a} - \frac{3m}{4a}$

In 6 and 7, solve.

6. $y - 13 = -7$ **7.** $m - 2 < 6$

8. Solve for b: $b - a = 100$.

9. Solve for m: $m - 7n = -22n$.

In 10 and 11, $A = \{-3, -1, 1, 3, 5, 7\}$ and $B = \{-6, -3, 0, 3, 6, 9\}$.

10. Find $A \cap B$.   **11.** $A \cup B$.

In 12 and 13, graph on a number line.

12. $x > 6.7$ or $x \le -2.8$.

13. $p > -5$ and $p \le 2$, where p is an integer

14. Park's Pet Shop mailing list has 480 customers, each of whom has cats or dogs. Of these, 321 customers have dogs and 215 have cats. How many customers have both cats and dogs?

15. If the 1990 population of Vermont is V and the 1990 population of New Hampshire is H, how many more people are in Vermont than in New Hampshire in 1990?

16. A store had B boxes of paper clips at the beginning of the week, sold S of them, used 3 boxes itself, and received a shipment of N new boxes. How many boxes were there after all this?

17. Last summer 30 students in the school spent some time in England. Of these, 15 visited France; 11 visited Germany; and 20 visited either France or Germany. If one of these students is chosen at random to be interviewed by the school newspaper, what is the probability that student visited all three countries?

18. *Multiple choice*   In which pair are angles X and Y complementary?
(a) $m\angle X = 67°$ (b) $m\angle X = 38°$
$m\angle Y = 113°$ $m\angle Y = 38°$
(c) $m\angle X = 5°$ (d) none of
$m\angle Y = 85°$ (a)–(c)

19. Today it is 10 degrees cooler than yesterday. If x = today's temperature and y = yesterday's temperature, graph all ordered pairs (x, y) that fit this description.

20. On a coordinate plane, graph all ordered pairs of numbers whose sum is 7.

21. Which Venn diagram represents $(X \cap Y) \cup Z$?

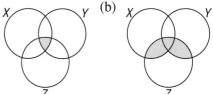

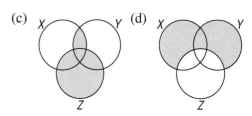

Chapter Review

Questions on SPUR Objectives

SPUR stands for **S**kills, **P**roperties, **U**ses, and **R**epresentations.
The Chapter Review questions are grouped according to the
SPUR Objectives for this chapter.

SKILLS deal with the procedures used to get answers.

■ **Objective A.** *Find the union and intersections of sets. (Lessons 3-4, 3-5)*

In 1–2, $A = \{11, 15, 19, 23, 25\}$, $B = \{10, 15, 20, 25, 30\}$.

   1. Find $A \cap B$.

   2. Find $A \cup B$.

In 3–4, $C = \{2, 8, 9\}$, $D = \{4, 8, 12\}$, $E = \{6, 8, 9\}$.

   3. Find $(C \cup D) \cap E$.

   4. Find $C \cup (D \cap E)$.

■ **Objective B:** *Solve and check sentences involving subtraction. (Lesson 3-3)*

   5. Solve and check: $x - 47 = -2$.

   6. Solve: $2.5 = t - 3.34$.

   7. Solve: $\frac{3}{2} + y - \frac{1}{4} = \frac{3}{4}$.

   8. Solve for x: $x - z = 4$.

   9. Solve: $z - 12 < 11$.

   10. Solve: $-3 < m + 2$.

   11. For the inequality $x - 30 \geq 40$, Brian got the answer $x \geq 10$. Check whether Brian is correct.

   12. Solve for r: $r - 45 \geq d$.

   13. Solve for y: $-8z = y - 7z$

■ **Objective C:** *Find the measure of the supplement or complement of an angle. (Lesson 3-8)*

   14. If $m\angle Q = 17°$, find the measure of the supplement of $\angle Q$.

   15. $\angle A$ and $\angle B$ are complementary. If $m\angle B = 62.5°$, find $m\angle A$.

   16. $m\angle L = a°$. What is the measure of the supplement of $\angle L$?

   17. $\angle R$ and $\angle S$ are complements. $m\angle R = x°$ and $m\angle S = z°$. Write an equation relating x and z.

■ **Objective D:** *Simplify expressions involving subtraction. (Lessons 3-1, 3-3)*

In 18–25, simplify.

   18. $3x - 4x + 5x$

   19. $-\frac{2}{3} - \frac{4}{5}$

   20. $\frac{3a}{2} - \frac{9a}{2}$

   21. $-\frac{28}{3} - \frac{48}{3}$

   22. $c - \frac{c}{3} - 2c - 2$

   23. $y - (2y - y)$

   24. $-41 - 233 + 30 - -52 - 6 - 5$

   25. $z^3 - 7 + 8 - 4z^3$

PROPERTIES deal with the principles behind the mathematics.

■ **Objective E:** *Apply the algebraic definition of subtraction. (Lesson 3-1)*

In 26 and 27, rewrite each subtraction as addition.

   26. $x - y + z$

   27. $-8 - v = 42$

   28. *True or false*   The sum of $m - k$ and $k - m$ is zero.

   29. *Multiple choice*   Which does *not* equal the others?
   (a) $a - b$ (b) $a + -b$
   (c) $-b + a$ (d) $b + -a$

■ **Objective F:** *Use models of subtraction to form expressions and solve sentences involving subtraction.* *(Lessons 3-2, 3-3)*

30. Last week Carla earned E dollars, saved S dollars, and spent P dollars. Relate E, S, and P in a subtraction sentence.

31. An elevator won't run if it holds more than L kilograms. A person weighing 80 kg gets on a crowded elevator and an "overload" light goes on. How much did the other passengers weigh?

32. The total floor area of a three-story house is advertised as 3500 sq ft. If the first floor's area is F sq ft and the third floor's area is 1000 sq ft, what is the area of the second floor?

33. After spending $40, Mort has less than $3 left. If he started with S dollars, write an inequality to describe the possible values of S.

34. Donna is 5 years older than Eileen. If Donna's age is D, how old is Eileen?

35. A plane 30,000 feet above sea level is radioing a submarine 1500 feet below sea level. What is the difference in their altitudes?

36. General Computer Co. had a profit of $4 million last year and a loss of $0.3 million this year. By how much do these amounts differ?

37. There were 5000 people at the game. Herb guessed there were H people. Herb's guess was too low. How far off was the guess?

■ **Objective G:** *Apply the formulas for $N(A \cup B)$ and $P(A \cup B)$.* *(Lesson 3-6)*

38. Of Mrs. Schubert's 20 piano students, 13 can play either a Beethoven sonata or a Chopin mazurka, 6 can play only the Beethoven, and 5 can play only the Chopin. How many can play both?

39. Of the 25 students in Mr. Young's algebra class, 16 studied some on Saturday, 15 studied some on Sunday, and 9 studied both Saturday and Sunday. What is the probability that a student in this class studied on either of the days?

40. $P(A) = \frac{1}{4}$, $P(B) = \frac{1}{13}$, $P(A \cap B) = \frac{1}{52}$. What is $P(A \cup B)$?

41. Fifty-six people signed the Declaration of Independence. Forty people signed the U.S. Constitution. Five people signed both documents. How many people signed the Constitution or the Declaration of Independence?

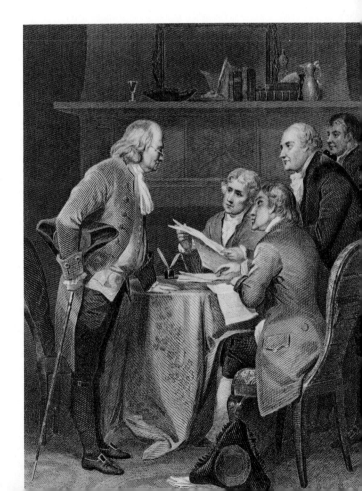

Drawing up the Declaration of Independence are (from left to right) Benjamin Franklin, Thomas Jefferson, Robert R. Livingston, John Adams, and Roger Sherman

REPRESENTATIONS deal with pictures, graphs, or objects that illustrate concepts.

Objective H: *Graph solution sets to inequalities.* (Lessons 3-4, 3-5)

In 42–45, graph on a number line. The domain for all variables is the real numbers.

42. $x > 10$ and $x \geq -\frac{1}{2}$

43. $-2 \leq y \leq 3$ and $y < 0$

44. $d \leq -15$ or $d < 5$

45. $z < \frac{1}{2}$ or $z \geq 1\frac{1}{4}$

Objective I: *Use Venn diagrams to describe union and intersection.* (Lessons 3-4, 3-5)

In 46 and 47, $A = \{-11, -1, 0, 2, 10\}$ and $B = \{-20, -10, 0, 10, 20\}$.

46. Write the set that is represented by the colored area in the Venn diagram.

47. Write the set that is represented by the striped area. $\{-11, -1, 2\}$

In 48 and 49, use the figure below.

48. Copy the figure. Shade $A \cup (B \cap C)$.

49. Copy the figure. Shade $(A \cap B) \cup (C \cap B)$.

Objective J: *Graph sets of ordered pairs (x, y) where x + y = k or x − y = k, where k is a real number.* (Lesson 3-7)

50. Xavier is four years older than his sister Yvonne. Graph all possible ordered pairs that represent their ages. Let x represent Xavier's age and y represent Yvonne's age.

51. The sum of two numbers is 0. Graph all possible pairs of numbers.

52. Mary and Peter have a total of 5 pets. Graph all possible ways the pets may be divided between them.

53. Graph all ordered pairs (x, y) with $x - y = 30$.

REFRESHER

Chapter 4, which discusses multiplication in algebra, assumes that you have mastered certain objectives in your previous mathematics courses. Use these optional questions to check your mastery.

A. Multiply any positive numbers or quantities.

1. $4.7 \cdot 3.21$
2. $0.04 \cdot 312$
3. $666 \cdot 0.00001$
4. $.17 \cdot .02$
5. $\frac{2}{3} \cdot 30$
6. $\frac{5}{2} \cdot 11$
7. $\frac{2}{9} \cdot \frac{3}{4}$
8. $\frac{1}{4} \cdot \frac{1}{3} \cdot \frac{1}{2}$
9. $1\frac{1}{4} \cdot 2\frac{1}{8}$
10. $5 \cdot 6\frac{2}{3}$
11. $17 \cdot \$2.31$
12. $\frac{3}{4} \cdot \$125$
13. $\frac{1}{2} \cdot 10\frac{1}{2}$ inches
14. $30\% \cdot 120$
15. $3\% \cdot \$6000$
16. $5.25\% \cdot 1500$

B. Multiply positive and negative integers.

17. $3 \cdot -2$
18. $11 \cdot -11$
19. $-6 \cdot 4$
20. $-5 \cdot -5$
21. $0 \cdot -1$
22. $-14 \cdot 130$
23. $-60 \cdot -59$
24. $-3 \cdot -2 \cdot -1 \cdot -1$

C. Solve equations of the form $ax = b$, when a and b are positive integers.

25. $3x = 12$
26. $5y = 110$

27. $10z = 5$
28. $1 = 9w$
29. $6 = 50a$
30. $b \cdot 21 = 14$
31. $8c = 4$
32. $7 = 2d$

D. Determine the area of a rectangle below left given its dimensions.

33. length 15″, width 12″
34. length 4.5 cm, width 3.2 cm
35. length 2 m, width 4 m
36. length 100′, width 82′

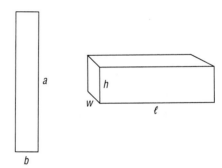

E. Determine the volume of a rectangular solid above right given its dimensions.

37. length 8 cm, width 4 cm, height 3 cm
38. length 12 in., width 9 in., height $1\frac{1}{2}$ in.
39. length $\frac{1}{2}$ ft, width $\frac{1}{2}$ ft, height $\frac{1}{2}$ ft
40. length 60 m, width 50 m, height 10 m

Multiplication in Algebra

Our galaxy looks somewhat like the Andromeda Galaxy pictured here. Is there life on some other planet in our galaxy?

A formula for computing the number of planets N currently supporting intelligent life involves the product of seven numbers.

$$N = T \cdot P \cdot E \cdot L \cdot I \cdot C \cdot A$$

Of course you need to know what the letters in the formula represent. Here are their meanings and an estimated value for each. No one knows these values exactly.

T = *Total* number of stars in our galaxy.
P = fraction of stars with a *Planetary* system.
E = fraction of planetary systems with a planet that has an *Ecology* able to sustain life.
L = fraction of planets able to sustain life on which *Life* actually develops.
I = fraction of planets with life on which that life is *Intelligent*.
C = fraction of planets with intelligent life on which that life could *Communicate* outwardly.
A = fraction of planets with communicating life which is *Alive* now.

In *Space*, a novel by James Michener, these variables are given the values $T = 400,000,000,000$, $P = \frac{1}{4}$, $E = \frac{1}{2}$, $L = \frac{9}{10}$, $I = \frac{1}{10}$, $C = \frac{1}{3}$, and $A = \frac{1}{100,000,000}$. This yields a value of 15 for N. But people disagree on the values. Changing the values can change the value of N by a great deal. For instance if $P = \frac{1}{10}$ with the other values above, then $N = 6$ instead of 15. In this chapter, we discuss many other situations involving multiplication.

4-1

Areas, Arrays, and Volumes

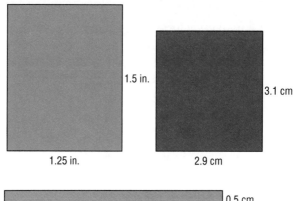

The operation of multiplication can be pictured using area. The area of any rectangle is the product of its two dimensions. The rectangles above are actual size.

Area = 1.25 in. · 1.5 in.	Area = 2.9 cm · 3.1 cm	Area = 6 cm · 0.5 cm
= 1.875 square in.	= 8.99 square cm	= 3 square cm
= 1.875 in.^{2}	= 8.99 cm^2	= 3 cm^2

The area is in square units because both the numbers and units are multiplied.

These examples are instances of a general pattern, the **area model for multiplication.** This model applies only to positive real numbers since the dimensions of a rectangle are lengths.

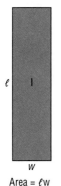

ℓ I

w

Area = ℓw

Area Model for Multiplication:

> The area of a rectangle with length ℓ and width w is ℓw.

Rectangle I at the left has been rotated to give rectangle II. The two rectangles have the same dimensions. They must have the same area. So the area formula could be written $A = wl$ as well as $A = lw$.

In this way, the area model pictures the **Commutative Property of Multiplication,** which holds for all real numbers.

w   II

ℓ

Area = wℓ

Commutative Property of Multiplication:

> For any real numbers a and b,
> $$ab = ba.$$

By combining the area model with models for addition and subtraction, areas of some regions which are not rectangles can be determined.

Example 1 A driveway to a house has the shape at the right. All angles are right angles. Find the area of the driveway.

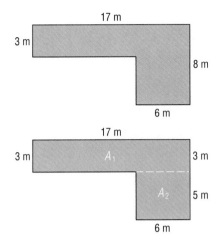

Solution There is no simple formula for an irregular shape like this. But the shape can be split into two rectangles. One way to split the figure is shown at right. This splits the right side into lengths of 3 m and 5 m.

Call the smaller areas A_1 and A_2. The top rectangle has dimensions 3 m and 17 m, so $A_1 = 51$ m^2. Also, $A_2 = 5$ m $\cdot$ 6 m or 30 m^2. Putting the areas together,

$$\begin{aligned} \text{Area of driveway} &= A_1 + A_2 \\ &= 51 \text{ m}^2 + 30 \text{ m}^2 \\ &= 81 \text{ m}^2 \end{aligned}$$

The stars below form a **rectangular array** with 5 rows and 9 columns. It is called a 5-by-9 array. The numbers 5 and 9 are the **dimensions** of the array. The area model of multiplication applies to rectangular arrays. The total number of stars is 45, the product of the dimensions. The graph is discrete.

Area Model (discrete version):

The number of elements in a rectangular array with x rows and y columns is xy.

The area model can be extended to three-dimensional figures. The **volume** of a **rectangular solid** or **box** is the product of its three dimensions.

$$\text{Volume} = \text{length} \cdot \text{width} \cdot \text{height}$$
$$V = lwh$$

Example 2 Find the volume of the rectangular solid at the left.

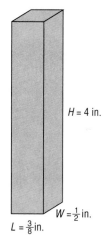

$H = 4$ in.

$W = \frac{1}{2}$ in.

$L = \frac{3}{8}$ in.

Solution $V = \ell wh$
$= \frac{3}{8}$ in. $\cdot \frac{1}{2}$ in. $\cdot 4$ in.

Recall that the product of fractions is the product of the numerators divided by the product of the denominators. So

$$V = \frac{3}{16} \text{ in.}^2 \cdot 4 \text{ in.}$$
$$= \frac{12}{16} \text{ in.}^3$$
$$= \frac{3}{4} \text{ in.}^3.$$

(Volume is measured in cubic units; in.^{3} means cubic inches.)

Check Rewrite the dimensions as decimals.

$$\frac{3}{8} \cdot \frac{1}{2} \cdot 4 = .375 \cdot 0.5 \cdot 4$$
$$= .1875 \cdot 4$$
$$= .75$$

Since $\frac{3}{4} = .75$, the answer checks.

In Example 2, you may have noticed that you did not have to follow order of operations and multiply from left to right. You could first multiply $\frac{1}{2}$ in. by 4 in. This gives 2 in.^{2}, which is easy to multiply by $\frac{3}{8}$ in. This illustrates that

$$(l \cdot w) \cdot h = l \cdot (w \cdot h).$$
Doing left multiplication first = Doing right multiplication first

The general property, true for all numbers, is called the **Associative Property of Multiplication.**

Associative Property of Multiplication:

For any real numbers a, b, and c,
$$(ab)c = a(bc).$$

Since lw is the area of the base, you may see the volume formula expressed as B, the area of the base, times the height.

$$V = Bh$$

Example 3 What is the volume of a box in which the height is $4x$ and the area of the base is $5x^2$?

Solution

$$\text{Volume} = \text{Area of base} \cdot \text{height}$$

$$
\begin{aligned}
V &= (5x^2)4x \\
&= 4(5x^2)x &&\text{Commutative Property of Multiplication} \\
&= (4 \cdot 5)(x^2 \cdot x) &&\text{Associative Property} \\
&= 20x^3 &&x^2 \cdot x = x^3
\end{aligned}
$$

The volume of the box is $20x^3$.

Check Let $x = 2$. Use the volume formula, $V = Bh$. If $x = 2$, the height is 8 and the area of the base is 20, so the volume is 160. Now evaluate the answer $20x^3$.

$$20x^3 = 20 \cdot 2^3 = 20 \cdot 8 = 160, \text{ which checks.}$$

When units are not given, as in Example 3, you can assume they are the same—all inches, all centimeters, or all something else. Otherwise, you should be careful to make sure units are the same before multiplying.

Questions

Covering the Reading

1. State the area model for multiplication.

2. $lw = wl$ is an instance of what property of multiplication?

3. If the length and width of a rectangle are measured in inches, the area is measured in what unit?

In 4 and 5, find the area of a rectangle with the given dimensions.

4. length 7.2 cm and width 4.3 cm

5. length 8 in. and width y in.

6. In the figure at the right, all angles are right angles. Find its area.

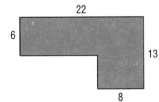

7. Compute the number of dots in a rectangular array containing 48 rows and 83 columns.

In 8–10, find the volume of the rectangular solid in which:

8. the dimensions are $\frac{1}{4}$ ft by $\frac{2}{3}$ ft by $\frac{1}{8}$ ft.

9. the length is 3, width is $7x$, height is y.

10. the area of the base is $9p^2$ and the height is $12p$.

11. How many square tiles, each one foot on a side, are needed to cover an area $9a$ feet by $5b$ feet?

12. The O'Leary's garden is a rectangle 45 ft by 60 ft.
   a. What is the area of their garden?
   b. If they put a fence around the garden, how long will the fence be?

13.

   Write an expression for the area of the rectangle at the left.
   Write an expression for its perimeter.

14. How many cubic inches are in a box 5 inches wide, 1 foot long, and 3 inches high?

15. What is the volume of a rectangular solid in which the length is $6x$ cm, the width is $2x$ cm, and the height is x cm?

16. What is the volume of a rectangular solid in which the width and height have the same measurement z and the length is twice that measurement?

In 17 and 18, use the following information. Area can be used to picture multiplication with fractions. The sides of the square below are 1 unit long. To picture $\frac{3}{4} \cdot \frac{2}{3}$, one side is divided into fourths, an adjacent side into thirds. Each of the small rectangles is $\frac{1}{12}$ the area of the square.

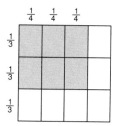

At the left, the rectangle with dimensions $\frac{3}{4}$ and $\frac{2}{3}$ is colored. The area of the colored part is $\frac{6}{12}$ of the total area since 6 small rectangles are colored. Since $A = lw$, $A = \frac{3}{4} \cdot \frac{2}{3}$ = colored area $= \frac{6}{12}$. The fraction $\frac{6}{12}$ can be reduced to $\frac{1}{2}$.

17. a. What multiplication problem is pictured?
    b. What fraction of the square is colored?
    c. The picture shows $\frac{1}{2} \cdot \frac{3}{4} = ?$

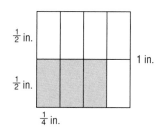

18. Draw a one-inch square. Divide one side into fifths and an adjoining side into thirds. Color the drawing to represent $\frac{1}{3} \cdot \frac{3}{5}$.

19. Find the area of the figure below. All angles are right angles. The unit is meters.

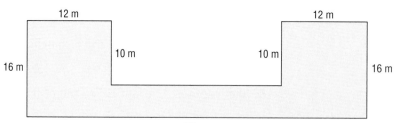

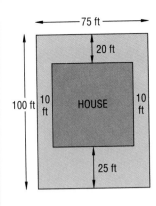

— 75 ft —

20 ft

100 ft | 10 ft | HOUSE | 10 ft

25 ft

20. A single-story house is to be built on a lot 75 feet wide by 100 feet deep. The shorter side of the lot faces the street. The house must be set back from the street 25 feet. It must be 20 feet from the back lot line, and 10 feet from each side lot line. What is the maximum square footage (area) the house can have?

21. The volume of the rectangular solid at the right is 84. Give four different sets of dimensions the box might have.

In 22 and 23, use the associative and commutative properties to do the multiplication in your head.

22. $25 \cdot x \cdot 4 \cdot 34$ **23.** $(2 \cdot 3x)(8x \cdot 5)$

Review

24. Which property tells you that $(x + {-3}) + 3 = x + ({-3} + 3)$? *(Lesson 2-1)*

In 25–27, evaluate each expression when $a = 7$, $b = 5$, $x = 4.7$, and $y = 2.3$.

25. $3(x + y) - 2(a - b)$ *(Lesson 1-4)*

26. $\dfrac{5(4b - 2a)}{b^2}$ *(Lesson 1-4)*

27. ab^2 *(Lesson 1-4)*

28. Carl runs more than three miles every morning.
  a. Write an inequality to describe this situation.
  b. Graph on a number line. *(Lesson 1-1)*

29. A caterer estimates that 20 lb of meat will be eaten. She is within 3 lb of the actual amount used.
  a. Write an inequality showing the actual amount possibly eaten.
  b. Graph on a number line. *(Lessons 1-1, 3-9)*

30. Compute in your head. *(Previous course)*
  a. $-2 \cdot -\frac{1}{2}$ **b.** $3 \cdot -5 \cdot -0.1$ **c.** $-2 \cdot 8.5$

Exploration

31. You can make a box by cutting out square corners from a rectangular piece of paper.
  a. Begin with a piece of $8\frac{1}{2}''$ by $11''$ notebook paper. Cut a $1''$ square from each corner. Fold to make a box and use tape to hold it.
  b. What is the volume of the box you have made?
  c. Estimate the size of the square you should cut out from each corner to make the box with largest volume. (Use trial and error.)

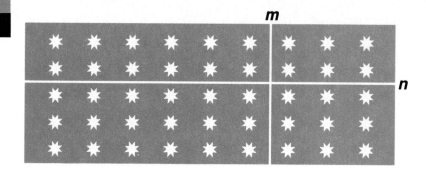

Multiplying Fractions and Rates

In the array above, $\frac{6}{9}$ of the stars are to the left of line m. Also, $\frac{2}{5}$ of the stars are above the horizontal line n. What fraction of the stars are both to the left of m and above n?

There are $6 \cdot 2$ or 12 stars both left and above. There are $9 \cdot 5$ or 45 stars in all. So $\frac{12}{45}$ of the stars are to the left and above. Another way of answering this question is by multiplying $\frac{6}{9} \cdot \frac{2}{5}$ in the familiar way.

$$\frac{6}{9} \cdot \frac{2}{5} = \frac{6 \cdot 2}{9 \cdot 5} = \frac{12}{45}$$

The general pattern is very easy to describe with variables.

Rule for Multiplication of Fractions:

For all real numbers a, b, c, and d, with b and d not zero,

$$\frac{a}{b} \cdot \frac{c}{d} = \frac{ac}{bd}.$$

Example 1   Multiply $\frac{3}{4}$ by $\frac{14x}{15}$.

Solution 1 $\dfrac{3}{4} \cdot \dfrac{14x}{15} = \dfrac{3 \cdot 14x}{4 \cdot 15} = \dfrac{42x}{60}$

Since 6 is a common factor of the numerator and denominator, the answer can be simplified using the Equal Fractions Property.

$$= \frac{6 \cdot 7x}{6 \cdot 10}$$

$$= \frac{7x}{10}$$

Solution 2   Take out the common factors 2 and 3 before multiplying.

$$\frac{\overset{1}{\cancel{3}}}{\underset{2}{\cancel{4}}} \cdot \frac{\overset{7}{\cancel{14}x}}{\underset{5}{\cancel{15}}} = \frac{7x}{10}$$

Example 2 Simplify $\dfrac{2}{5y} \cdot 10y$.

Solution

$$\frac{2}{5y} \cdot \frac{10y}{1} = \frac{2}{\underset{1\cdot 1}{\cancel{5}y}} \cdot \frac{\overset{2\cdot 1}{\cancel{10}y}}{1}$$

$$= 4$$

Fraction forms often appear in situations involving rates. Rate units can be expressed using a slash "/" or horizontal bar "—." The slash or bar is read "per."

rate	with a slash	with a bar
99 cents per pound	99 cents/lb	$99 \dfrac{\text{cents}}{\text{lb}}$
80 kilometers per hour	80 km/hr	$80 \dfrac{\text{km}}{\text{hr}}$
W words per minute	W words/min	$W \dfrac{\text{words}}{\text{min}}$

Rates can be multiplied by other quantities. The units are treated as if they were factors in a multiplication.

Example 3 An airplane flies 380 miles each hour. How far will it have flown in 12 hours?

Solution Write the rate of travel as

$$380 \frac{\text{miles}}{\text{hour}}.$$

Multiply by the number of hours. The "hours" unit on the left cancels the "hour" in the denominator.

$$12 \;\cancel{\text{hours}} \cdot 380 \frac{\text{miles}}{\cancel{\text{hour}}} = 4560 \text{ miles}$$

Two or more rates may be multiplied. The units are treated like numbers in fractions.

Example 4 Each week a student must study 4 grammar lessons. There are 3 pages per lesson. The student spends 20 minutes per page. How many minutes does she spend studying grammar each week?

Solution Think rates: 3 pages per lesson, 20 minutes per page.

$$4 \;\cancel{\text{lessons}} \cdot 3 \frac{\cancel{\text{pages}}}{\cancel{\text{lessons}}} \cdot 20 \frac{\text{min}}{\cancel{\text{page}}} = 240 \text{ minutes}$$

The general idea behind these situations is a second model for multiplication, the **rate factor model.**

Rate Factor Model for Multiplication:

When a rate is multiplied by another quantity, the unit of the product is the product of units. Units are multiplied as though they were fractions. The product has meaning when the units have meaning.

The rate factor model can be applied to both positive and negative rates. Suppose a farmer's topsoil is eroding at a rate of 0.5 inch per year. The rate of change is

$$-0.5 \, \frac{\text{inch}}{\text{year}}.$$

If erosion continues at this rate for 20 years, multiplying gives the total loss.

$$20 \, \cancel{\text{years}} \cdot -0.5 \, \frac{\text{inch}}{\cancel{\text{year}}} = -10 \text{ inches}$$

The final answer is negative, which means that 10 inches of topsoil will be *lost* over the twenty years.

This instance confirms the rules for multiplication of positive and negative numbers.

If two numbers have the same + or − sign, their product is positive. If two numbers have different signs, their product is negative.

Sometimes a given rate does not fit into the rate factor model, but its reciprocal does. For example, if a computer printer prints 12 pages in 1 minute, then it takes $\frac{1}{12}$ minute to print 1 page. The two rates $\frac{1}{12} \, \frac{\text{min}}{\text{page}}$ and $12 \, \frac{\text{pages}}{\text{min}}$ are *reciprocal rates*, each describing the same situation from a different point of view. Reciprocals do not usually equal each other, but reciprocal *rates* are equal quantities.

Example 5 A computer prints at the rate of 12 pages/min. How long will it take to print 2400 documents with 3 pages per document?

Solution There are two rates: 12 pages/min and 3 pages/document. Write the product.

$$2400 \text{ doc} \cdot 3 \frac{\text{pages}}{\text{doc}} \cdot 12 \frac{\text{pages}}{\text{min}}$$

The document units cancel but pages and min do not cancel. So use the reciprocal rate to rewrite $12 \frac{\text{pages}}{\text{min}}$ as $\frac{1}{12} \frac{\text{min}}{\text{page}}$.

$$2400 \text{ doc} \cdot 3 \frac{\text{pages}}{\text{doc}} \cdot \frac{1}{12} \frac{\text{min}}{\text{pages}} = 600 \text{ minutes}$$

It will take 600 minutes to print the documents.

Questions

Covering the Reading

1. What fraction of the dots are above the horizontal line ℓ and to the right of the vertical line m?

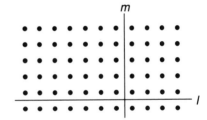

2. Multiply. **a.** $\frac{5}{x} \cdot \frac{3}{2}$ **b.** $\frac{a}{b} \cdot \frac{m}{n}$

In 3–5, compute.

3. $\frac{3m}{n} \cdot \frac{7m}{9}$ 4. $\frac{3}{8} \cdot 32z$ 5. $81 \cdot \frac{2k}{27}$

6. *Multiple choice*   Which is not equal to the others?
   (a) $\frac{20}{11} \cdot r$ (b) $\frac{20r}{11}$ (c) $\frac{20}{11r}$ (d) $\frac{r}{11} \cdot 20$

In 7 and 8, **a.** copy the sentence and underline the rate; **b.** use a slash to write the rate unit; **c.** use a fraction bar to write the rate unit.

7. A typist types 70 words per minute.

8. Roast beef costs $2.99 a pound.

9. Joanne runs 5.85 miles each hour. At the same rate, how far will she run in 3 hours?

10. Each week Myron must study k grammar lessons. There are 4 pages per lesson. He spends 30 minutes per page. How many minutes does he spend studying grammar each week?

11. a. A farmer's topsoil is eroding at a rate of 0.3 inches per year. What is the total change after 5 years ?

b. $5 \cdot -0.3 =$ _____

12. *Multiple choice*   What is the reciprocal rate of $40 \frac{\text{miles}}{\text{gallon}}$?

(a) $40 \frac{\text{gallons}}{\text{mile}}$ (b) $\frac{1}{40} \frac{\text{gallon}}{\text{mile}}$ (c) $\frac{1}{40} \frac{\text{mile}}{\text{gallon}}$

13. Which choice of Question 12 is a quantity equal to $40 \frac{\text{miles}}{\text{gallon}}$?

In 14 and 15, **a.** use the reciprocal of the rates to do the multiplication; **b.** invent a situation for which the multiplication would be appropriate.

14. $6 \frac{\text{dollars}}{\text{lb}} \cdot 30 \frac{\text{shrimp}}{\text{lb}} =$ _____ $\frac{\text{shrimp}}{\text{dollar}}$

15. $15 \frac{\text{pages}}{\text{doc}} \cdot \frac{1}{2} \frac{\text{page}}{\text{min}} =$ _____ $\frac{\text{min}}{\text{doc}}$

16. Suppose a laser printer prints 5 pages per minute. How long will it take to print 2400 documents with 3 pages per document?

Applying the Mathematics

17. Multiply $\frac{8b}{7c} \cdot \frac{21a}{2x} \cdot 5c$.

18. While Phyllis exercises, her heart rate is 150 beats per minute. If she exercises for m minutes at this rate, how many times will her heart beat?

19. There are 16 bottles per case. There are 8 ounces of liquid per bottle.
a. How many ounces are in 12 cases?
b. How many ounces are in c cases?

20. a. Find the rent for two years on an apartment which rents for $595 per month.
b. Find the rent for y years on this apartment.

21. In 1983, the U.S. birth rate was 15.5 babies per 1000 population. The population was 226,000,000. Use this multiplication,

$$15.5 \frac{\text{babies}}{1000 \text{ people}} \cdot 226{,}000{,}000 \text{ people},$$

to determine how many babies were born in 1983.

22. Marty can wash k dishes per minute. His sister Sue is twice as fast.
a. How many dishes does Sue wash per minute?
b. How many minutes does Sue spend per dish?

23. As a magician at a birthday party, Kelly charges D dollars per party. If he performed at s parties, how much did he earn altogether?

24. Pat adds K dollars to her bank account each week.
a. How much has she added after 20 weeks?
b. If she started with $150 in her account, write an expression for the amount she has after 20 weeks.

25. A sky diver "free falls" at about 120 mi/hr which is 176 ft/sec. How many seconds does it take the diver to "free fall" 2000 ft?

Review

26. Write an expression for the volume of a cube having edges of length $5e$. *(Lesson 4-1)*

5e

27. What is the area of the colored portion of the figure at the left? Write your answer as a reduced fraction. *(Lesson 4-1)*

$\frac{1}{4}''$

$\frac{1}{6}''\{$

1"

28. Thirty-seven people attended the Spanish Club picnic. Twenty-one people ate hamburgers and twenty-eight ate hot dogs. If everyone ate either hamburgers or hot dogs, how many people ate both? *(Lesson 3-6)*

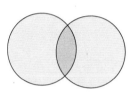

29. What is the output from line 60 of the program below if 5, 12 is input in line 30? *(Lesson 1-4)*

```
10 PRINT "TOTAL SURFACE AREA OF A CYLINDER"
20 PRINT "GIVE RADIUS AND HEIGHT"
30 INPUT R, H
40 LET A = 2 * 3.1416 * R * R + 2 * 3.1416 * R * H
50 PRINT "RADIUS", "HEIGHT", "AREA"
60 PRINT R, H, A
70 END
```

30. Use the triangle at the right. Find x.
(Lessons 2-6, 3-8)

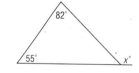

31. *Skill sequence   True or false (Lesson 1-6)*
   a. $4x = 18$ if $x = 4.5$
   b. $-9y = 42$ if $y = -\frac{14}{3}$
   c. $\frac{4}{5}z = -96$ if $z = 120$

32. Compute in your head using the associative and commutative properties of multiplication. *(Lesson 4-1)*
   a. $2 \cdot 7 \cdot 4 \cdot 5$ **b.** $\frac{1}{2} \cdot \frac{3}{4} \cdot \frac{2}{3}$ **c.** $2.5 \cdot 6 \cdot 2$

Exploration

33. The abbreviation mph stands for the rate unit "miles per hour." For what rate does each of these abbreviations stand?
   a. mpg **b.** rpm **c.** psi

LESSON

4-3

Special Numbers in Multiplication

The numbers 1, 0, and -1 have special roles in multiplication. The number 1 is the **multiplicative identity.**

Multiplicative Identity Property of 1:

For any real number a,
$$a \cdot 1 = 1 \cdot a = a.$$

This simple property is used in ways that are not always obvious. For instance, 2.54 cm = 1 inch. Since the quantities are equal, dividing one by the other gives reciprocal rates

$$\frac{2.54 \text{ cm}}{1 \text{ inch}} \text{ and } \frac{1 \text{ inch}}{2.54 \text{ cm}}$$

each equal to the number one.

Multiplying by such a rate does not change a quantity. It only converts the quantity to different units.

▪ ▪ ▪ ▪ ▪ ▪ ▪ ▪

Example 1   Elise needs to buy 7 inches of ribbon. It is sold in the store by the centimeter. How many centimeters does she need?

Solution   Multiply 7 inches by the rate factor $2.54 \frac{\text{cm}}{\text{inch}}$.

$$7 \cancel{\text{ inches}} \cdot 2.54 \frac{\text{cm}}{\cancel{\text{inch}}} = 17.78 \text{ cm}$$

In Example 1, because 7 inches is multiplied by 1, the actual length of ribbon Elise needs is not being changed. Some other useful rates equal to 1 are $\frac{5280 \text{ ft}}{1 \text{ mi}}$, $\frac{16 \text{ ounces}}{1 \text{ pound}}$, and $\frac{1 \text{ min}}{60 \text{ sec}}$.

168

Just as multiplication by 1 has special uses, multiplying a number by -1 yields an important result. Think about the farmer in Lesson 4-2 whose topsoil is eroding by 0.5 inch per year. So one year ago there was 0.5 inch *more* soil than there is now. Going back is a negative direction in time. The time is -1 year. Here is the multiplication.

$$-1 \text{ year} \cdot -0.5 \frac{\text{inch}}{\text{year}} = 0.5 \text{ inch}$$

$$\text{a year ago} \qquad \text{rate} \qquad \text{0.5 inch more soil}$$

This situation illustrates that multiplication by -1 changes a number to its opposite.

Multiplication Property of -1:

For any real number a,
$$a \cdot -1 = -1 \cdot a = -a.$$

Example 2 Simplify $-x \cdot -y$.

Solution Since $-x = -1x$ and $-y = -1y$
then $-x \cdot -y \qquad = -1x \cdot -1y$
$\qquad\qquad\qquad = (-1 \cdot -1) \, xy$
$\qquad\qquad\qquad = 1xy$
$\qquad\qquad\qquad = xy$

When x and y are positive, $-x$ and $-y$ are negative. The result of Example 2 shows that the product of two negative numbers is positive. $-4 \cdot -3 = 4 \cdot 3 = 12$.

Remember that 8 and -8 are opposites (additive inverses) because their sum is the additive identity, 0. Similarly, the numbers 8 and $\frac{1}{8}$ are **reciprocals** or **multiplicative inverses** because their product is the multiplicative identity, 1.

Property of Reciprocals:

For any *nonzero* real number a,

$$a \cdot \frac{1}{a} = \frac{1}{a} \cdot a = 1.$$

The reciprocal of n can always be written $\frac{1}{n}$ or $1/n$. The bar or slash indicates division. You can calculate the reciprocal of a number by dividing 1 by the number.

■ ■ ■ ■ ■ ■ ■ ■

Example 3   Give the reciprocal of 1.25.

Solution   The reciprocal of 1.25 is $\dfrac{1}{1.25}$. A calculator shows the quotient to be 0.8.

Check $1.25 \cdot 0.8 = 1$. So 1.25 and 0.8 are reciprocals because their product is 1.

If you know that the reciprocal of 1.25 is 0.8, you also know that the reciprocal of -1.25 is -0.8, since $-1.25 \cdot -0.8 = 1$.

Most scientific calculators have a reciprocal key $\boxed{1/x}$. To find the reciprocal of n, key in n $\boxed{1/x}$. The answer will be displayed as a decimal. You must apply a different process to write the reciprocal as a fraction. Use the fact that the reciprocal of a/b is b/a (unless a or b is zero).

■ ■ ■ ■ ■ ■ ■ ■

Example 4   Give the reciprocal of $-\dfrac{3}{7}$.

Solution   The reciprocal of $-\dfrac{3}{7}$ is $-\dfrac{7}{3}$.

Check $-\dfrac{3}{7} \cdot -\dfrac{7}{3} = \dfrac{21}{21} = 1$

Look back at the property of reciprocals. The term *nonzero* is emphasized because the property is not true for $a = 0$. Zero does not have a reciprocal because there is no number such that $0 \cdot x = 1$. In fact, you know that 0 times any real number is 0.

Multiplication Property of Zero:

For any real number a,
$$a \cdot 0 = 0 \cdot a = 0.$$

■ ■ ■ ■ ■ ■ ■ ■

Example 5   Evaluate $(w + 4.7)(2.6 - w)(w + 7.1)$ when $w = -4.7$.

Solution   Substituting gives
$$(-4.7 + 4.7)(2.6 - -4.7)(-4.7 + 7.1).$$

Since -4.7 and 4.7 are opposites, their sum is 0.
$$= (0)(2.6 - -4.7)(4.7 + 7.1)$$
$$= 0$$

The product is 0. There is no need to do the computation in the second or third parentheses.

Questions

Covering the Reading

1. **a.** What number is the multiplicative identity?
   b. What number is the additive identity?

2. State the property of reciprocals.

3. Reciprocals are also called __?__.

4. Zero has the same role in __?__ that 1 has in __?__.

5. Copy and complete the following:
   a. -18 and 18 are __?__ because their sum is __?__.
   b. 18 and $\frac{1}{18}$ are __?__ because their product is __?__.

6. To what real number does the property of reciprocals *not* apply?

In 7–11, give the reciprocal and check your answer.

7. 10   **8.** 1/9   **9.** $4\frac{3}{5}$ **10.** 2.5   **11.** $-\frac{6}{7}$

12. Evaluate $(x + 1)(x + 2)(x + 3)(x + 4)$ when $x = -3$.

13. Refer to Example 1.
    a. Suppose Elise needs 10 more inches of ribbon. How many more centimeters must she buy?
    b. Suppose Elise needed x inches of ribbon. How many centimeters is this?

14. **a.** Simplify: $z \cdot -w$.
    b. If z and w are positive numbers, then part **a** indicates that the product of a __?__ number and a __?__ number is __?__.

Applying the Mathematics

15. Simplify: $a \cdot -b \cdot c \cdot -d \cdot e$.

16. *Multiple choice*   Which does *not* equal the multiplicative identity?
    (a) c/c when $c \neq 0$ (b) $0.8 \cdot 5/4$ (c) $4.1 - 4.1$ (d) $.9 + .1$

17. Which pairs of numbers are reciprocals?
    a. 200 and 0.005   **b.** 7 and .7   **c.** $\frac{1}{4}$ and 0.25
    d. 1.5 and $\frac{2}{3}$ **e.** $\frac{3}{5}$ and $-\frac{3}{5}$

18. How many feet are in x miles if 5280 ft = 1 mile?

19. Don bought a 34-inch belt. His waist measures 75 cm. Is the belt long enough for him to wear?

20. To attract people to a grocery store, the manager decides to sell milk at her cost and make 0¢ profit for each bottle sold. How much profit is made when b bottles of milk are sold?

21. Find the particular values of a and b such that $x = ax + b$ is true for all values of x.

22. Suppose $D = (w + 2)(w - 3)(w + 6)$. For what three values of w will D have a value of 0?

In 23 and 24, simplify.

23. $\dfrac{4}{3} \cdot 3a \cdot \dfrac{3}{4} \cdot \dfrac{1}{3a} \cdot 10x$

24. Compute in your head.

    a. $3 \cdot \dfrac{5}{3}$ **b.** $\dfrac{9}{x} \cdot x$ **c.** $a \cdot \dfrac{b}{a}$

25. *Skill sequence*   Multiply. *(Lesson 4-2)*

    a. $\dfrac{4}{3} \cdot 24p$ **b.** $\dfrac{-5}{y} \cdot 25y$ **c.** $\dfrac{x}{3} \cdot 3x$

26. *Multiple choice*   Which is not equal to the others? *(Lesson 4-2)*

    (a) $\dfrac{7}{4} \cdot 6x$ (b) $\dfrac{42}{4}x$ (c) $\dfrac{42x}{4}$ (d) $\dfrac{7}{24x}$

27. Write the rate "33 math problems per night" using a slash. *(Lesson 4-2)*

28. Suppose a seven-month-old baby weighing w kg has been gaining 0.01 kg per day. At this rate,
    a. How much will the baby gain in 60 days?
    b. How much will the baby weigh in 60 days? *(Lesson 4-2)*

29. An auditorium has dimensions $40'$ by $80'$. It contains a stage that is $8'$ by $15'$.
    a. What is the area of the remaining floor space?
    b. If each audience member needs 6 square feet of floor space, how many people will the auditorium hold? *(Lesson 4-1)*

30. Solve for y: $3x + y = 2$. *(Lesson 2-6)*

31. John spent T total hours on homework. One hour was spent on science and $\frac{1}{2}$ hour on math. How much time did John spend on other subjects? *(Lesson 3-2)*

32. Let x and y be any real numbers. Graph all solutions to $x = y - 6$. *(Lesson 3-7)*

33. The expression *numero uno* is Spanish for "number one." Translate 1 into three other languages.

Solving $ax = b$

Sara earns $400/week as a zoo attendant. This is equal to a salary rate of $20,800/year.

Original salary: $400/week = $20,800/year

To reward Sara for excellent work the zoo decides to multiply her salary by 1.2. (This is a 20% raise.)

After raise: 1.2 · $400/week = 1.2 · $20,800/year
                         $480/week = $24,960/year

Multiplied by 1.2, her salary is $480/week or $24,960/year. These rates are still equal. This situation is an instance of **Multiplication Property of Equality.**

Multiplication Property of Equality:

For all real numbers a, b, and c: if $a = b$,
                                             then $ca = cb$.

The most important use of this property is in solving equations.

Example 1   An auditorium can seat 40 people in each row. For a talk, the audience should sit as close to the stage as possible. How many rows are needed if 600 people are expected to attend the talk?

Solution   It helps to draw a picture.

r rows $\{$ •••
                    ••
                    ⋮   40 seats/row
                    ••

The total number of seats is $40r$. The equation to solve is $40r = 600$. The solution for this equation is on page 174.

To solve, multiply both sides by $\frac{1}{40}$, the reciprocal of 40.

$$40r = 600$$
$$\frac{1}{40} \cdot 40r = \frac{1}{40} \cdot 600 \qquad \text{Multiplication Property of Equality}$$
$$\left(\frac{1}{40} \cdot 40\right)r = \frac{600}{40} \qquad \text{Associative Property of Multiplication}$$
$$1 \cdot r = \frac{600}{40} \qquad \text{Property of Reciprocals}$$
$$r = 15 \qquad \text{Multiplication Property of 1}$$

So 15 rows are needed.

Check   Will 15 rows hold 600 seats? Yes. $15 \cdot 40 = 600$.

To solve $ax = b$ for x (when a is not zero) multiply both sides of the equation by the reciprocal of a.

In the term ax, a is called the **coefficient** of x. So to solve $ax = b$, multiply both sides by the reciprocal of the coefficient of x.

Example 2   Solve $-6x = 117$.

Solution   The reciprocal of -6 is $-\frac{1}{6}$.
Multiply both sides of the equation by $-\frac{1}{6}$.

$$-6x = 117$$
$$-\frac{1}{6} \cdot -6x = -\frac{1}{6} \cdot 117$$
$$x = -\frac{117}{6}$$
$$x = -19.5$$

Check $\qquad\qquad -6x = -6 \cdot -19.5 = 117$

Example 3   Solve $\frac{4}{5}y = 56$.

Solution   The reciprocal of $\frac{4}{5}$ is $\frac{5}{4}$.
Multiply both sides of the equation by $\frac{5}{4}$.

$$\frac{4}{5}y = 56$$
$$\frac{5}{4} \cdot \frac{4}{5}y = \frac{5}{4} \cdot 56$$
$$y = \frac{5}{\overset{}{\underset{1}{\cancel{4}}}} \cdot \overset{14}{\cancel{56}}$$
$$y = 70$$

Check $\qquad\qquad \frac{4}{5} \cdot y = \frac{4}{5} \cdot 70 = 56$

Example 4 Solve $3.4w = -85$.

Solution The reciprocal of 3.4 can be written as $\frac{1}{3.4}$. Multiply both sides of the equation by $\frac{1}{3.4}$.

$$3.4w = -85$$
$$\frac{1}{3.4} \cdot 3.4w = \frac{1}{3.4} \cdot -85$$
$$w = -\frac{85}{3.4}$$
$$w = -25$$

Check $\qquad\qquad\qquad\qquad 3.4w = 3.4 \cdot -25 = -85$

A rate like $\dfrac{\text{miles}}{\text{hr}}$ is calculated as follows: divide miles representing distance by hours representing time. This gives the formula:

$$\frac{\text{distance}}{\text{time}} = \text{rate}.$$

In short, $\qquad\qquad\qquad\qquad \dfrac{d}{t} = r.$

Solving this equation for d gives a related formula.

Multiply both sides by t. $\quad t \cdot \dfrac{d}{t} = r \cdot t$

$$d = rt$$

That is, $\qquad\qquad \text{distance} = \text{rate} \cdot \text{time}.$

Example 5 How long would it take to drive from Detroit to Indianapolis, a distance of 290 miles, if you travel at 55 mph?

Solution Let t be the length of time, in hours.

$$d = r \cdot t$$
$$290 \text{ miles} = 55 \frac{\text{miles}}{\text{hour}} \cdot t \text{ hours}$$
$$290 = 55t$$

Multiply both sides of the equation by $\frac{1}{55}$.

$$\frac{1}{55} \cdot 290 = \frac{1}{55} \cdot 55t$$
$$5.27 \approx t$$

It will take about 5.27 hours.

Check If you travel at 55 mph for 5.27 hours, will you travel about 290 miles? Yes, because

$$55 \frac{\text{mi}}{\text{hr}} \cdot 5.27 \,\cancel{\text{hr}} \approx 290 \text{ mi}$$

Questions

Covering the Reading

1. **a.** If $a = b$, then $6a = $ _?_ .
   b. What property is used to answer part a?

2. If 920 people were expected at the talk of Example 1, how many rows of seats would be needed?

In 3–5, **a.** What is the coefficient of the unknown? **b.** By what number should you multiply both sides to solve the equation? **c.** Solve.

3. $-32x = 416$ 4. $-210 = 4.2y$ 5. $\frac{3}{32}A = \frac{3}{4}$

6. To solve $ax = b$ for x, multiply both sides of the equation by _?_ .

7. Solve $-240x = 12$.

8. Solve $\frac{1}{5} = \frac{1}{80}y$.

9. In the formula $d = rt$, what do d, r, and t represent?

10. Refer to Example 5. If you travel from Detroit to Indianapolis at 65 mph, how long will it take to get there?

11. Julie thinks $\frac{1}{4}$ is the solution to the equation $\frac{1}{3} \cdot m = \frac{4}{3}$. Is she correct?

Applying the Mathematics

12. Solve for k: $-3k = 4m$.

13. Jose and Maria want to use the seesaw at the left. Jose weighs 40 kg and sits 150 cm from the turning point. Maria weighs 30 kg. Where should she sit to balance Jose? (Note: To balance, weight times distance on one side must equal weight times distance on the other.)

14. On a day in the U.S., a mean of about 10,205 people each give a pint of blood. About how many days will it take to get one million pints of blood?

15. The circumference of a circle is 39 cm. Find its diameter to the nearest hundredth. (Remember: $C = \pi d$.)

16. The volume of a box needs to be 500 cubic centimeters. The base of the box has dimensions 12.5 cm and 5 cm. How high must the box be?

17. Density D of a material is defined as the rate M/V, where M is the weight of material and V is the volume.
    a. What equation results when both sides of $D = M/V$ are multiplied by V?
    b. If the density of water is 62.4 pounds per cubic foot, find the weight of 10.4 cubic feet of water.

18. *Multiple choice*   Which results are equal to the multiplicative identity? *(Lesson 4-3)*

(a) $\dfrac{2y}{2y}$ (b) $.4 \cdot \dfrac{2}{5}$ (c) $-6.4 + 6.4$ (d) $.8 + .2$

19. *Multiple choice*   Which pair of numbers are reciprocals? *(Lesson 4-3)*

(a) $2y$ and $\dfrac{1}{2y}$ (b) 0.4 and $\dfrac{2}{5}$ (c) -6.4 and 6.4 (d) 0.8 and 0.2

20. About how many quarts are in x liters, if 1 liter ≈ 1.06 quarts? *(Lesson 4-3)*

21. In this lesson, the answer to Example 5 is 5.27 hr. How many minutes is 0.27 hr? *(Lesson 4-2)*

22. If Irma dribbles a basketball 2 times per second, and moves 4.5 ft per second, how many dribbles will she make moving 60 ft down court? *(Lesson 4-2)*

23. The formula for the area of a circle is $A = \pi r^2$, where π is about 3.14 and r is the radius. Find the area of the colored region if the radius of the circle is 8 cm and the rectangle is 3 cm by 10 cm *(Lesson 3-2, 4-1)*

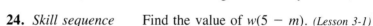

24. *Skill sequence*   Find the value of $w(5 - m)$. *(Lesson 3-1)*
a. $w = 20$, $m = -7$
b. $w = 20$, $m = 7$
c. $w = -20$, $m = 7$

25. Solve for a. *(Lessons 2-6, 2-7)*
a. $a + 43 = 36$
b. $a - 31 \le 17$
c. $a + c > 166$

26. Graph the point $(2, -3)$ and its image after a slide of 7 units to the left and 6 units up. *(Lesson 2-5)*

27. Write as a percent: **a.** 0.24 **b.** 0.3 **c.** 1.4 *(Previous course)*

28. A number of well-known equations are of the same form as $d = rt$. Some of them are listed below. For each equation, find out what the variables represent.

a. $A = \dfrac{1}{2}bh$ **b.** $P = 2\ell + 2w$ **c.** $C = 2\pi r$
d. $I = prt$ **e.** $F = ma$

177

4-5

0%
financing
limited time only
BUY NOW !

Special Numbers in Equations

When a and b are real numbers, the equation $a + x = b$ has exactly one solution. This may lead you to think that all simple equations have exactly one solution. That is not so, even with some equations of the form $ax = b$.

Example 1   Solve $0x = 4$.

Solution   By the Multiplication Property of 0, for any value of x, $0x = 0$. So $0x$ cannot equal 4. There is no solution.
Write: There is no solution.

Example 2   Solve $0y = 0$.

Solution   This is the Multiplication Property of 0. It is always true.
Write: All real numbers are solutions.

Notice that in Examples 1 and 2, you cannot multiply both sides by the reciprocal of 0. Why?—because 0 has no reciprocal! That signals you to be careful when 0 is the coefficient of the unknown. Zero is the only number that causes such problems.

In the next example, the coefficient of t has a reciprocal. The result is that there is exactly one solution.

Example 3   Solve $13t = 0$.

Solution 1   This equation can be solved by multiplying both sides by $\frac{1}{13}$.

$$\frac{1}{13} \cdot 13t = \frac{1}{13} \cdot 0$$
$$t = 0$$

Solution 2   Do it in your head. The only number that 13 can be multiplied by to get 0 is 0 itself. So $t = 0$.
Write: The solution is 0.

To avoid writing sentences as explanations of solutions, solution sets can be written. For Examples 1, 2, and 3, the solutions are as follows.

Equation	Sentence	Solution set
$0x = 4$	There is no solution.	$\{\}$ or $\emptyset$
$0y = 0$	All real numbers are solutions.	R
$13t = 0$	The solution is 0.	$\{0\}$

Notice that $\{0\}$ and $\{\}$ are different. The set $\{\}$ or $\emptyset$ has no elements and indicates that no number works in $0x = 4$. The set $\{0\}$ contains the element 0, and indicates that 0 works in $13t = 0$.

The Multiplication Property of -1, $-1 \cdot a = -a$, is useful in solving equations where the opposite of the unknown has been found.

Example 4   Solve $-x = 3.824$.

Solution 1   Use the Multiplication Property of -1.
$$-1 \cdot x = 3.824$$

Multiply both sides by -1 (which happens to be the reciprocal of -1).
$$-1 \cdot -1 \cdot x = -1 \cdot 3.824$$
$$x = -3.824$$

Solution 2   Translate the equation into words and find the solution in your head. The opposite of what number is 3.824? Answer: -3.824

Equations like the one above arise when a variable has been subtracted. The next example leads to a situation of the form $a - x = b$.

Example 5   A rod was 1.2 cm too long. It was shortened and found to be 0.03 cm too short. How much was cut off?

Solution   Let the amount cut off be c. "Too short" means -0.03. From the Take-away Model for Subtraction, the answer is the solution to
$$1.2 - c = -0.03.$$

Use the definition of subtraction. Change to addition.
$$1.2 + -c = -0.03$$

Apply the Addition Property of Equality. Add -1.2 to both sides.
$$-1.2 + 1.2 + -c = -1.2 + -0.03$$
$$-c = -1.23$$

Apply the Multiplication Property of Equality. Multiply both sides by -1.
$$c = 1.23$$

The amount cut off was 1.23 cm.

Questions

Covering the Reading

1. Why can't both sides of $3 = 0x$ be multiplied by the reciprocal of 0?

In 2–4, describe the solutions **a.** with a sentence; **b.** with the solution set.

2. $7y = 0$ **3.** $0 \cdot w = 14$ **4.** $0 = a \cdot 0$

5. *Multiple choice*   Which set is the same as ø?
(a) {ø}   (b) {0}   (c) 0   (d) { }

6. Write the Multiplication Property of -1, beginning with "For any real number c, …"

In 7–12, solve.

7. $-1 \cdot x = 40$ **8.** $-y = -3$ **9.** $-z = 0$

10. $16 - w = 102$ **11.** $1.74 - v = -2$ **12.** $-6 - x = 8$

13. A TV program is found to be 1 minute 14 seconds too long for its time slot. It is shortened and a second version is 15 seconds too short. By how much was it shortened?

Applying the Mathematics

14. If $90 - x = 11$, what is x?

15. a. Solve for y: $m - y = 25$.
b. Check your answer to part a by letting $m = 27$ and $y = 2$.

16. Solve for z: $300z - 299z - z = 0$.

17. Refer to the formula on page 155. Some people believe that, other than the Earth, the value of $L = 0$. What effect does this have on the value of N?

Review

In 18–20, solve. *(Lesson 4-4)*

18. $\frac{7}{9}q = 140$ **19.** $-4p = 12$ **20.** $12r = -4$

21. The circumference C of a circle can be found from the formula $C = 2\pi r$, where r is the radius of the circle. If $C = 60\pi$, what is r? *(Lesson 4-4)*

22. The volume of a box is half a cubic meter. The length of the box is 1.5 meters and the width is 0.8 meter. **a.** Is this possible? **b.** If so, find the height of the box. If not, explain why not. *(Lessons 4-1, 4-4)*

23. Simplify: $-1 \cdot 4 \cdot -8 + 2 \cdot -5 \cdot 9 - -3 \cdot 6 \cdot -10$. *(Lessons 2-2, 3-1, 4-3)*

24. Graph all solutions to $n > -1$. *(Lesson 1-1)*

25. a. Evaluate $-1 \cdot -1 \cdot -1 \cdot -1 \cdot -1 \cdot -1 \cdot -1 \cdot -1$.
b. Evaluate $-1 \cdot -1 \cdot \ldots \cdot -1$, where there are 25 factors.
c. Give a general rule for answering questions like part b. *(Previous course)*

26. A crate contains 12 cases. Each case has 24 boxes. Each box has 60 packages of batteries. Each package has 2 batteries. How many crates will it take to ship 100,000 batteries? *(Lessons 4-2, 4-4)*

In 27–29, which is larger? *(Previous course)*

27. 3 + -4 or 4 + -3?

28. 1 · -8 or 2 · -8?

29. -432 · -175 or -346 · 811?

Exploration **30.** Consider the following pattern.

row 1	$x = 10$
row 2	$\frac{1}{2}x = 10$
row 3	$\frac{1}{3}x = 10$
⋮	⋮
row 100	

    a. What will be written in row n?
    b. What is the solution to the equation in row n?
    c. What will be written in row 100?
    d. What is the solution to the equation in row 100?
    e. To what equation are the equations getting closer and closer?

4-6

Solving $ax < b$

Here are some numbers in increasing order.

$$-10 \quad -6 \quad 5 \quad 30 \quad 30.32 \quad 870$$

Because the numbers are in order, you could put the inequality sign $<$ between any two of them. Now multiply these numbers by some fixed *positive* number, say 11. Here are the products.

$$-110 \quad -66 \quad 55 \quad 330 \quad 333.52 \quad 9570$$

The order is the same. You could still put an $<$ sign between any of the numbers. This illustrates that if $x < y$, then $11x < 11y$. In general, multiplication by a positive number keeps order.

Multiplication Property of Inequality (part 1):

If $x < y$ and a is positive, then $ax < ay$.

Each of the signs $>$, $\leq$, or $\geq$ between numbers or expressions also indicates order. The Multiplication Property of Inequality works with any of those signs. Many inequalities can be solved using this property.

Example 1 Solve $4x \leq 20$.

Solution Multiply both sides by $\frac{1}{4}$. Since $\frac{1}{4}$ is positive, the inequality sign remains the same.

$$\frac{1}{4} \cdot 4x \leq \frac{1}{4} \cdot 20$$
$$x \leq 5$$

Check As with other inequalities you have solved, there are two steps to the check.
Step 1: Substitute 5 in the original inequality. It should make the two sides equal. It does. $4 \cdot 5 = 20$.
Step 2: Check some number that works in $x \leq 5$. We pick 3.
Is $4 \cdot 3 \leq 20$? Yes, $12 \leq 20$.
Since both steps worked, the solution $x \leq 5$ is correct.

Example 2 The length of a rectangle is 50 cm. Its area is greater than 175 cm². Find the width of the rectangle.

length = 50 cm

width = w cm

Solution It helps to draw a picture. $A = \ell w$, and $\ell = 50$, so the area of the rectangle is $50w$. The situation is described by the inequality $50w > 175$.

To solve, multiply both sides by $\frac{1}{50}$, the reciprocal of 50.

$$\frac{1}{50} \cdot 50w > \frac{1}{50} \cdot 175$$

It's just like solving equations. Simplifying yields

$$w > 3.5.$$

The width is more than 3.5 cm.

Check Step 1: Does $50 \cdot 3.5 = 175$? Yes.
  Step 2: Pick some value that works in $w > 3.5$ cm, such as 10. Is $50 \cdot 10 > 175$? Yes.

Multiplying both sides of an inequality by a *positive* number is straightforward. Multiplying by a *negative* number requires one more step. Here are the numbers from the beginning of this lesson.

$$-10 \quad -6 \quad 5 \quad 30 \quad 30.32 \quad 870$$

Multiplying these numbers by -3,

$$30 \quad 18 \quad -15 \quad -90 \quad -90.96 \quad -2610$$

The first row of numbers is in *increasing* order. The second row is in *decreasing* order. The order has been reversed. Multiplication by a negative number changes order. So, if you multiply both sides of an inequality by a negative number, you must **change the direction of the inequality.**

Multiplication Property of Inequality (part 2):

If $x < y$ and a is negative then $ax > ay$.

■ ■ ■ ■ ■ ■ ■ ■

Example 3 Solve $-7x \geq 126$.

Solution Multiply both sides by $-\frac{1}{7}$, the reciprocal of -7.
Since $-\frac{1}{7}$ is a negative number, remember to *change the inequality sign* from $\geq$ to $\leq$.
$$-\frac{1}{7} \cdot -7x \leq -\frac{1}{7} \cdot 126$$

Now simplify.
$$x \leq -\frac{126}{7}$$
$$x \leq -18$$

Check Step 1: Does $-7 \cdot -18 = 126$? Yes.
  Step 2: Try a number that works in $x \leq -18$. We use -20.
  Is $-7 \cdot -20 \geq 126$? Yes, $140 \geq 126$.

Changing from $<$ to $>$, or from $\leq$ to $\geq$, or vice-versa, is called **changing the sense** of the inequality. Note this is the same as changing the direction of the inequality. The only time you have to change the sense is when you are multiplying both sides by a negative number. Otherwise, solving $ax < b$ is just like solving $ax = b$.

You can see why the two-step check of an inequality is important. The first step checks the number in the solution. The second step checks the sense of the inequality.

Questions

Covering the Reading

In 1–3, consider the inequality $20 < 30$. What inequality results if:

1. you multiply both sides of this inequality by 6?

2. you multiply both sides of this inequality by $\frac{2}{3}$?

3. you multiply both sides of this inequality by -4?

4. Give both parts of the multiplication property of inequality.

In 5 and 6, change the sense of each inequality.

5. $<$ 6. $\geq$

7. Tell whether or not these numbers are solutions to $-9x < -18$.
   a. 2   b. -2   c. 3   d. -3   e. -1   f. 0

In 8–13, solve and check each sentence.

8. $5x \geq 10$ 9. $-3y < 300$ 10. $-4A < -124$

11. $13 > 2z$ 12. $\frac{2}{3}P \leq \frac{1}{4}$ 13. $0.09 > -9c$

14. The length of a rectangle is 20 cm. Its area is less than 154 cm². Find the width of the rectangle.

Applying the Mathematics

In 15–17, multiply both sides by -1 to solve the sentence.

15. $-m < 8$ 16. $-2 \leq -n$ 17. $-t < 0$

18. The area of the foundation of a rectangular building is not to exceed 20,000 square feet. The width of the foundation is to be 125 feet. How long can the foundation be?

19. An auditorium has at least 1500 seats. There are 50 seats in each row. There must be at least how many rows?

20. Parents of the bride have budgeted $2500 for the dinner after the wedding. Each person's dinner will cost $27.50. At most how many people can attend the dinner?

21. Three fourths of a number is less than two hundred four. What are the possible values of the number?

22. You are to travel more than 100 km in 3 hours. Write an inequality to describe your rate.

23. Use the clues to find x.
Clue 1: x is an integer.
Clue 2: $2x < 10$.
Clue 3: $-3x < -9$.

Review

In 24–29, solve. *(Lessons 2-2, 4-4, 4-5)*

24. $8 = -2a$

25. $\frac{1}{2}b = 5$

26. $-200 - c = -144$

27. $3d + 3d = 42$

28. $8 + g = 0$

29. $30\pi = \pi d$

30. One mile is 1760 yards. How many yards are in m miles? *(Lesson 4-3)*

31. A box is made by folding the pattern at the left along the dotted lines and taping the edges. What is the volume of the box? *(Lesson 4-1)*

32. *Skill sequence*   Subtract. *(Lesson 3-1)*
a. $143 - 256$ **b.** $143 - {-256}$ **c.** $-143 - 256$ **d.** $1.43 - 2.56$

33. A number n is randomly selected from $\{10, 11, 12, 13, 14, 15\}$
(Lessons 1-7, 3-6)
a. Find $P(n > 12)$.   **b.** $P(n$ is odd$)$
c. $P(n > 12$ and n is odd$)$ **d.** $P(n > 12$ or n is odd$)$

34. a. Find a nonzero value of x for which $\dfrac{6 + x}{2 + x}$ does not equal 3.

b. Find a nonzero value of x for which $\dfrac{6 \cdot x}{2 \cdot x}$ does not equal 3.
(Lessons 1-4, 1-8)

35. You can see that the formula $A = \frac{1}{2}bh$ for the area of a triangle is true for a right triangle because a right triangle is half of a rectangle with sides b and h. Write a formula for the area of a triangle that is half of a square with sides of length s. *(Lesson 1-5)*

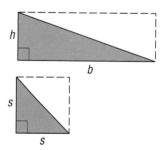

36. Graph all solutions to $x + y = 8$. Start with letting x and y be whole numbers. *(Lesson 3-7)*

37. *Skill sequence*   Divide. *(Previous course)*
a. $\frac{2}{5} \div \frac{1}{3}$ **b.** $2 \div \frac{1}{3}$ **c.** $2.5 \div \frac{1}{3}$ **d.** $\frac{1}{2} \div 3$

Exploration

38. Using your calculator, find a number $x < 0$ such that $.05 < x^2 < .06$.

2.5
2   2
2   2

11.4

2   2
2   2
2.5

Multiplication Counting Principle

In mathematics, a procedure is said to be *elegant* if it is clever and simple at the same time. For instance, area and multiplication are helpful in solving many types of counting problems. The area model of multiplication allows you to organize the items being counted into rectangular arrays, and then multiply to get totals. Here is an example of such a situation.

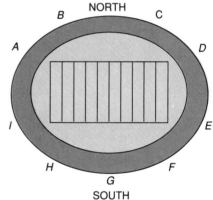

Example 1   A stadium has 9 gates. Gates *A*, *B*, *C*, and *D* are on the north side. Gates *E*, *F*, *G*, *H* and *I* are on the south side. In how many ways can you enter the stadium through a north gate and leave through a south gate?

Solution   Create a rectangular array where each entry in the array is an ordered pair. The first letter represents a gate you enter. The second letter stands for a gate you exit. The ordered pair (*C*, *H*) means you enter at gate *C* and leave through gate *H*.

		South Gate				
		E	*F*	*G*	*H*	*I*
	A	(*A*,*E*)	(*A*,*F*)	(*A*,*G*)	(*A*,*H*)	(*A*,*I*)
North	B	(*B*,*E*)	(*B*,*F*)	(*B*,*G*)	(*B*,*H*)	(*B*,*I*)
Gate	C	(*C*,*E*)	(*C*,*F*)	(*C*,*G*)	(*C*,*H*)	(*C*,*I*)
	D	(*D*,*E*)	(*D*,*F*)	(*D*,*G*)	(*D*,*H*)	(*D*,*I*)

Since the array has 4 rows and 5 columns, there are $4 \cdot 5 = 20$ pairs in the table. There are 20 ways of entering through a north gate and leaving through a south gate.

This elegant use of multiplication occurs often enough that it is given a special name, the **Multiplication Counting Principle.**

Multiplication Counting Principle:

If one choice can be made in *m* ways and a second choice can be made in *n* ways, then there are *mn* ways of making the first choice followed by the second choice.

The Multiplication Counting Principle can be extended to situations where more than two choices must be made.

Example 2   A high school student wants to take a foreign language class in period 1, a music course in period 2, and an art course in period 3. The language classes available are French, Spanish, and German. The music classes available are chorus and band. The art classes available are drawing and painting. In how many ways can the student choose the three classes?

Solution   Make a blank for each decision to be made.

———————————— · ———————————— · ————————————
ways to choose language   ways to choose music   ways to choose art

There are 3 choices in foreign language, 2 choices in music, and 2 choices in art. Use the multiplication counting principle.

————— 3 ————— · ————— 2 ————— · ————— 2 —————
ways to choose language   ways to choose music   ways to choose art

There are $3 \cdot 2 \cdot 2 = 12$ choices.

Check   Organize the possibilities using a **tree diagram.**

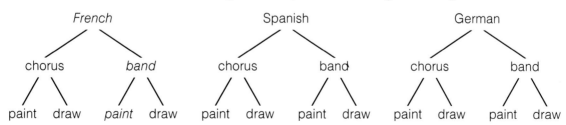

Each choice can be found by following a path along the diagram. One possible choice is shown in italics: *French—band—paint.* Counting shows there are 12 paths. Twelve different choices are possible.

Example 3 Mr. Graff gave his algebra class a quiz with five questions. Since Angie had not done her homework, she had to guess the answers. The quiz had two multiple-choice questions with choices A, B, C, and D, and three true-false questions. What is the probability that Angie gets all the questions correct?

Solution There are 4 ways to answer the multiple-choice question and 2 ways to answer each true-false question. Use the Multiplication Counting Principle.

$$\underset{\substack{\text{ways to}\\\text{answer \#1}}}{4} \cdot \underset{\substack{\text{ways to}\\\text{answer \#2}}}{4} \cdot \underset{\substack{\text{ways to}\\\text{answer \#3}}}{2} \cdot \underset{\substack{\text{ways to}\\\text{answer \#4}}}{2} \cdot \underset{\substack{\text{ways to}\\\text{answer \#5}}}{2}$$

There are $4 \cdot 4 \cdot 2 \cdot 2 \cdot 2 = 128$ different ways of answering the five questions, so there are 128 outcomes. Since only one of those outcomes is "all correct answers," with random guessing, Angie only has 1 chance out of 128 of getting all the questions correct.

P(all answers correct) $= \frac{1}{128}$

Example 4 Mr. Graff has already written the chapter test. It has three multiple-choice questions each with m possible answers, two multiple-choice questions each with n possible answers and five true-false questions. How many ways are there to answer the questions?

Solution Make a blank for each of the 10 questions. Fill it with the number of possible answers for that question.

Question #	1	2	3	4	5	6	7	8	9	10
Choices	m	m	m	n	n	2	2	2	2	2

Apply the Multiplication Counting Principle and multiply these to get $m \cdot m \cdot m \cdot n \cdot n \cdot 2 \cdot 2 \cdot 2 \cdot 2 \cdot 2$ sets of answers. Using exponents, this can be expressed as $m^3 n^2 2^5$, or $32 m^3 n^2$.

Questions

1. In mathematics, a procedure is said to be elegant if __?__.

2. State the Multiplication Counting Principle.

3. Can the Multiplication Counting Principle be applied in problems involving more than 2 choices?

In 4 and 5, refer to Example 1.

4. In how many ways can a person enter through a north gate and leave through gate *G*?

5. Suppose the stadium closed gates *D*, *G*, and *H*. Using an array, list the ways a person could enter the stadium through a north gate and leave through a south gate.

6. Suppose the school in Example 2 offered Russian as a fourth language choice. How many choices could a student make?

7. Using exponents, simplify $5 \cdot m \cdot m \cdot n \cdot n \cdot n \cdot n$.

8. Suppose Mr. Graff's quiz had two questions with *x* choices, and three true-false questions. Give the number of ways to answer the items on the test with an expression:
a. not using exponents **b.** using exponents.

9. All the questions on a quiz can be answered with true or false. There are 20 questions. If a student guesses randomly,
a. how many ways are there of answering the test?
b. what are the chances of getting all the answers correct?

10. In Example 4, if all the multiple choice questions had *q* possible answers, how many different ways are there for answering the test?

Applying the Mathematics

11. Satoshi, Izumi, Mitsuo, and Takeshi are candidates for Winter Carnival King. Reiko, Akiko, and Kimiko are the nominees for Winter Carnival Queen. Name all the possible "Royal Couples."

12. Radio station call letters must start with W or K, like WNEW or KYRZ.
a. How many choices are there for the first letter?
b. How many choices are there for the second letter?
c. How many different 4-letter station names are possible?

13. Telephone area codes consist of 3 digits. The first digit must be chosen from 2 through 9. The second digit must be 0 or 1. The third digit cannot be 0. How many area codes are possible?

14. Suppose in Example 1 that a person could enter through any gate and leave through any gate. In how many ways can this be done?

15. Suppose in Example 1 that a person could enter through any gate and leave through any *other* gate. In how many ways can this be done?

16. At the Fulton College cafeteria, the main dish last Thursday was chicken. The vegetables served were carrots and beans. There were three dessert choices: an apple, pudding, and cake.
a. Organize the possible meals consisting of a main dish, vegetable, and a dessert using a tree diagram.
b. How many different such meals are possible?

17. A quiz has *Q* true-false questions. How many sets of answers are possible? (Hint: Make a list. Express the answer when $Q = 1$, $Q = 2$, $Q = 3$, and $Q = 4$. Then find the general pattern.)

In 18–23, solve. *(Lessons 4-4, 4-5, 4-6)*

18. $6j = 11$ **19.** $-20k = \frac{2}{5}$ **20.** $-\ell = 3 - 5$

21. $4m < \frac{1}{10}$ **22.** $-90 \geq -6n$ **23.** $d + 2d < 39$

24. If apples cost 49¢ a pound and 5 apples weigh 3 pounds, about how much should 1 apple cost? *(Lesson 4-2)*

25. A farmer harvests 40 bales of hay per acre. How many acres would produce a harvest of 236 bales? *(Lessons 4-2, 4-4)*

26. All angles in this figure are right angles. Write an expression for the area of the figure. *(Lesson 4-1)*

27. The length of a rectangle is 5 units longer than twice its width. If the width is w units long, write an expression for its area. *(Lesson 4-1)*

28. David weighs D pounds. His brother, Carl, weighs 17 pounds more.
a. Write an expression for Carl's weight.
b. Write an expression for their total weight. *(Lesson 2-2)*

In 29–32, name the property illustrated. *(Lessons 2-2, 2-3)*

29. $a(3 + b) = (3 + b)a$ **30.** $a(3 + b) = a(b + 3)$

31. $a + (3 + -3) = a + 0$ **32.** $(a + 3) + 0 = a + 3$

33. Evaluate $\dfrac{4\pi r^3}{3}$ on your calculator when $r = 2.1$. Round your answer to the nearest tenth. *(Lesson 1-4)*

34. Evaluate $6(3x - y)$ when $x = -\dfrac{7}{3}$ and $y = -\dfrac{32}{5}$. Write your answer as a fraction. *(Lessons 1-4, 2-3)*

35. Evaluate when $d = \frac{1}{2}$. *(Lesson 1-4)*
a. d^2 **b.** $3d + 4$ **c.** $-8d^3$

36. At Harwood High, $\frac{3}{8}$ of the students take French and $\frac{1}{4}$ of the French students are in Ms. Walker's French class. What fraction of Harwood High Students are in Ms. Walker's French class? *(Previous course)*

37. If the probability of rain tomorrow is 30%, what is the probability of no rain tomorrow? *(Lesson 1-7)*

In 38 and 39, use the fact that an *acronym* is a name made from first letters of words or parts of words.

38. Here are some famous acronyms. Tell what the letters stand for.
a. NASA **b.** UNICEF **c.** CIA
d. IBM **e.** ICBM **f.** AFL-CIO
g. NFL **h.** NATO **i.** UNESCO

39. *True or false* Over a half million 4-letter acronyms are possible in English. (Hint: Use the Multiplication Counting Principle.)

LESSON

4-8

Multiplying Probabilities

Many TV watchers seldom change channels. So TV networks often put a new show after a popular one to boost the ratings of the new show. Suppose 25% of viewers watched show A and 80% of these stayed to watch show B. Then, since .80 · .25 = .20, 20% of the viewers watched both shows.

If a viewer is called at random, then each of the percents can be interpreted as a probability.

$$25\% = .25 = \text{P(watched show A)} = P(A)$$
$$80\% = .80 = \text{P(watched show B having}$$
$$\text{already watched show A)} = P(B \text{ given } A)$$
$$20\% = .20 = \text{P(watched shows A and B)} = P(A \text{ and } B)$$

You have seen $P(A)$ and $P(A \text{ and } B)$ before. $P(B \text{ given } A)$ is the probability that B occurs *given that A occurs*. $P(B \text{ given } A)$ is called a **conditional probability.** This situation illustrates the **Conditional Probability Formula.**

Conditional Probability Formula:

$$P(A \text{ and } B) = P(A) \cdot P(B \text{ given } A).$$

You may see the Conditional Probability Formula with the intersection symbol.

$$P(A \cap B) = P(A) \cdot P(B \text{ given } A).$$

Here is an example that makes use of this formula.

Example 1   A department store knows that $\frac{4}{5}$ of its customers are female. The probability that a U.S. female wears contact lenses is about $\frac{3}{50}$. If a customer is chosen at random, what is the probability that the customer is a female and wearing contact lenses?

Solution You are asked to find P(A ∩ B), where A = customer is a female and B = customer is wearing contacts. You are given P(A) = $\frac{4}{5}$.

P(B given A) means the probability that a person wears contacts given that the person is a female. Here

$$P(B \text{ given } A) = \tfrac{3}{50}$$
$$P(A \text{ and } B) = P(A) \cdot P(B \text{ given } A)$$
$$= \tfrac{4}{5} \cdot \tfrac{3}{50}$$
$$= \tfrac{12}{250}$$
$$= 0.048$$

Since 0.048 = 4.8%, there is about a 5% chance that a randomly chosen customer is a female who wears contact lenses.

The situation of Example 1 can be pictured in a tree diagram.

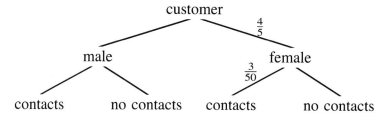

The tree diagram helps to see what other probabilities could be calculated. For instance, $\frac{47}{50}$ of females do not wear contacts. Can you find the probability that the customer is a female who doesn't wear contact lenses?

Example 2 Suppose that two cards are drawn from a well-shuffled deck. The first card is *not* put back before the second is drawn. What is the probability that both cards are hearts?

Solution
Let A = drawing a heart as the first card.
Let B = drawing a heart as the second card.
The problem is to compute P(A and B).
There are 13 hearts in the deck of 52, so P(A) = $\frac{13}{52}$.
After the first heart has been drawn there are 12 hearts left in a deck of 51. So P(B given A) = $\frac{12}{51}$.

$$P(A \text{ and } B) = P(A) \cdot P(B \text{ given } A)$$
$$= \tfrac{13}{52} \cdot \tfrac{12}{51}$$
$$= \tfrac{1}{4} \cdot \tfrac{4}{17}$$
$$= \tfrac{1}{17}$$

In the following example, the occurrence of the first event does not affect the probability of the second event. However, the formula for multiplying probabilities still works.

Example 3 Ken plays a game at a carnival in which he draws a marble from each of two jars. He wins a prize if both marbles are green.

Look at the jars at the left. Find the probability that both marbles Ken draws are green.

Solution Looking at the jar, P(1st is green) = $\frac{1}{10}$. Notice that the second color drawn does not depend on what color the first marble is. So P(2nd is green given 1st is green) = $\frac{3}{7}$.

P(1st is green and 2nd is green) = P(1st is green) · P(2nd is also green)

$$= \frac{1}{10} \cdot \frac{3}{7}$$

$$= \frac{3}{70} \approx .04$$

Probability situations like these can lead to equations of the form $ax = b$, where both a and b are fractions.

Example 4 In a survey, $\frac{2}{15}$ of the people surveyed were men who liked Brand X Toothpaste. If $\frac{3}{5}$ of the people surveyed were men, what fraction of men like Brand X?

Solution
Let A = person surveyed is a man.
Let B = person liked Brand X.
Then we know that P(A and B) = $\frac{2}{15}$ and P(A) = $\frac{3}{5}$.
We want P(B given A), the probability that a man liked Brand X. Call this t.

$$\text{Now P(A and B)} = \text{P(A)} \cdot \text{P(B given A)}$$
$$\frac{2}{15} = \frac{3}{5} \cdot t$$

Solve the equations as usual. Multiply both sides by $\frac{5}{3}$.

$$\frac{5}{3} \cdot \frac{2}{15} = \frac{5}{3} \cdot \frac{3}{5} \cdot t$$
$$\frac{2}{9} = t$$

Thus the probability that a man likes Brand X is $\frac{2}{9}$.

Check Draw a tree diagram with the given information and the solution found.

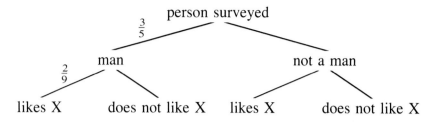

person surveyed

$\frac{3}{5}$

man

$\frac{2}{9}$

not a man

likes X   does not like X   likes X   does not like X

From the diagram, $\frac{3}{5} \cdot \frac{2}{9} = \frac{2}{15}$, the given probability that a person surveyed is a man who likes Brand X.

Questions

Covering the Reading

1. a. Give the Conditional Probability Formula for P(A and B).
   b. If $P(A) = \frac{1}{10}$ and $P(B$ given $A) = \frac{2}{3}$, what is P(A and B)?

2. a. If 28% of TV viewers watched Channel 5 from 7:00 to 7:30 and 75% of these stayed to watch the Channel 5 program at 7:30, what percent of viewers watched both?
   b. Part **a** can be done using the Conditional Probability Formula. Describe the events A and B, B given A, and A.

In 3 and 4, refer to Example 1.

3. a. What is the probability that a customer in the store is female?
   b. What is the probability that a customer in the store is female and wears contact lenses?

4. a. What is the probability that a female in the U.S. does not wear contacts?
   b. Use the tree diagram to calculate the probability that a customer in the store is female and does not wear contacts.

5. Refer to Example 1. At Horst's Shoes, 5 out of 7 customers are females. If a customer is chosen at random, find the probability that the person is a female and is wearing contact lenses. Write your answer as a percent.

6. Here is a tree diagram for Example 2. Copy and compute the probabilities.

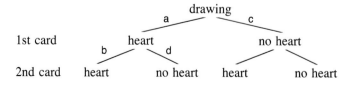

drawing

a   c

1st card   heart   no heart

b   d

2nd card   heart   no heart   heart   no heart

7. In the situation of Example 2, what is the probability that neither card drawn is a heart?

8. Suppose that two cards are drawn from a deck. The first card is not put back before the second is drawn. What is the probability that both cards are kings?

9. Ken draws a marble from each of two jars. The probability that the first marble is red is $\frac{8}{15}$. The probability that the second one is red is $\frac{1}{3}$. Find the probability that both marbles are red.

10. Two thirds of the people surveyed were male. Three tenths of those surveyed were males who had watched television the previous night. What fraction of males watched television last night?

11. *Multiple choice* Suppose $X =$ is over 20 years old and $Y =$ likes opera. Which describes P(Y given X)?
(a) the probability that a person is over 20 and likes opera
(b) the probability that a person who is over 20 also likes opera
(c) the probability that a person who likes opera is over 20

12. In a certain town it was determined that 90% of the blocks have fire hydrants. Of those blocks with fire hydrants, 72% have more than one. What percent of blocks in the town have more than one fire hydrant?

13. There are two traffic lights on the way to the store. One is red 60% of the time. The other light is red 50% of the time no matter what color the first light is. What is the probability of being stopped by a red light at both traffic signals?

14. There are r red balls out of a total of t balls in a jar. A ball is drawn and kept out of the jar. Then a second ball is drawn. Write an expression for each of the following:
a. P(first ball is red and second ball is red)
b. P(first ball is *not* red and second ball is red)

15. There are 20 books in a pile: 3 history books, 7 biographies, and 10 novels. You close your eyes and choose a book at random without putting it back into the pile. Your friend then closes his eyes and chooses a book at random. Find each of the following:
a. P(you choose a novel)
b. P(your friend then chooses a biography)
c. P(you choose a novel and your friend chooses a biography)

In 16 and 17, find P(A and B).

16. P(A) = $\frac{x}{z}$, P(B given A) = $\frac{y}{x}$ **17.** P(A) = $\frac{m}{3n}$, P(B given A) = $\frac{5m}{3n}$

18. Before an oil company decides to drill a well, geologists test the site to determine if there is a good chance of striking oil. In 1986, the Oliff Oil Company geologists found that 23% of the sites were favorable. Wells were drilled on those sites and 8% of the drilled wells struck oil.
a. Estimate the probability that a site will seem favorable and produce oil.
b. Next year the Oliff Oil Company plans to test 50 sites. How many sites can be expected to produce oil?

19. If the probability of a newborn being a girl is $\frac{1}{2}$, find the probability of the birth of
a. two girls in a row.   **b.** three girls in a row.

Review

20. How many different outfits can be made from 3 shirts, 4 pairs of slacks and 2 pairs of shoes, all of which go together? *(Lesson 4-7)*

21. A certain state makes car license plates with 3 letters and then 3 numbers. For example, WEZ 123 is a car license number in this state. So is WEW 002. How many license plates are possible? *(Lesson 4-7)*

22. *Skill sequence*   Solve. *(Lessons 3-3, 4-4, 4-5, 4-6)*
a. $x - \frac{3}{8} = -\frac{1}{4}$ **b.** $\frac{3}{8} - y = -\frac{1}{4}$
c. $\frac{3}{8}z = -\frac{1}{4}$ **d.** $\frac{3}{8}z < -\frac{1}{4}$

23. A rectangular array of d dots has c columns and r rows. How are c, r, and d related? *(Lesson 4-1)*

24. *Multiple choice* $3x^2 \cdot x^4 =$
(a) $3x^8$ (b) $3x^6$ (c) $9x^8$ (d) $24x$ *(Lesson 1-4)*

In 25–27, multiply. *(Lessons 1-8, 4-2)*

25. $\frac{3 \cdot 2 \cdot 1}{5 \cdot 4 \cdot 3 \cdot 2 \cdot 1} \cdot 4 \cdot 3 \cdot 2 \cdot 1$ **26.** $\frac{5ab}{8m} \cdot 12am$ **27.** $\frac{6x^2}{4a} \cdot \frac{2}{3x^2}$

28. Five instances in a general pattern are given below. **a.** Give the next instance. **b.** Calculate the values of each instance. *(Previous course)*

$$32 \div 4 \qquad 32 \div 2 \qquad 32 \div 1 \qquad 32 \div \frac{1}{2} \qquad 32 \div \frac{1}{4}$$

29. Solve for x: $mx = {}^-12$. *(Lesson 4-4)*

30. In your head, estimate each result to the nearest whole number. *(Previous course)*
a. $8.93 \cdot 2.06$ **b.** $\frac{1}{2} + \frac{5}{8} + \frac{3}{4}$ **c.** $75.36 \div 2.0000965$

31. Find the image of the point $(^-7, 2)$ after a slide of:
a. 8 units to the right and 5 units down
b. r units up
c. m units to the right and n units down *(Lesson 2-5)*

32. A restaurant found that out of 1750 people who made reservations last month, 1714 actually showed up. What is the relative frequency of people who did not show up? *(Lesson 1-8)*

Exploration

33. Repeat the experiment of Example 2, one hundred times.
a. How often do you pick two hearts?
b. How often do you pick two cards of the same suit?
c. Is the probability calculated in Example 2 verified in your experiment?

4-9

The Factorial Symbol

A special case of the Multiplication Counting Principle occurs when a list of things is to be ranked or ordered.

Some students were asked to rank these famous basketball stars in order of preference: Magic Johnson, Michael Jordan, Larry Bird, Dominique Wilkins. How many possible rankings can there be in the poll?

To answer this question, you might try to list all the possible outcomes. Here are just three of the possible rankings.

Larry Bird	Dominique Wilkins	Michael Jordan
Magic Johnson	Michael Jordan	Magic Johnson
Dominique Wilkins	Larry Bird	Dominique Wilkins
Michael Jordan	Magic Johnson	Larry Bird

It seems like it would take a long time to list all rankings. But the number of rankings can be found using the Multiplication Counting Principle. There are 4 people who could finish first. After choosing one for first place there are only 3 people left who could finish second. Then, after 1st and 2nd place have been chosen, there are only 2 people left who could finish third, and the remaining person will finish last.

$$\underset{\substack{\text{ways to choose} \\ \text{1st place}}}{\underline{\quad 4 \quad}} \cdot \underset{\substack{\text{ways to choose} \\ \text{2nd place}}}{\underline{\quad 3 \quad}} \cdot \underset{\substack{\text{ways to choose} \\ \text{3rd place}}}{\underline{\quad 2 \quad}} \cdot \underset{\substack{\text{ways to choose} \\ \text{4th place}}}{\underline{\quad 1 \quad}}$$

The answer is $4 \cdot 3 \cdot 2 \cdot 1 = 24$.

A shortcut way to write $4 \cdot 3 \cdot 2 \cdot 1$ is 4! This is read "four factorial."

Definition:

The symbol ***n!*** (read ***n* factorial**) means the product of the integers from *n* down to 1.

Example 1   Evaluate 5!

> **Solution** $5! = 5 \cdot 4 \cdot 3 \cdot 2 \cdot 1$
> $= 120$

An arrangement of letters, names, or objects is called a **permutation.** We have found that there are 4! permutations of 4 names. In general:

Permutation Theorem:

There are $n!$ possible permutations of n objects when each object is used exactly once.

Factorials are used when you are making arrangements of all the items in a set.

Scientific calculators usually have a factorial key $\boxed{n!}$. To evaluate $n!$, key in n $\boxed{x!}$. On some calculators you may have to use a second function key, $\boxed{inv}$ or $\boxed{2nd}$. Then key in n $\boxed{inv}$ $\boxed{x!}$.

Example 2   A baseball manager is setting a batting order for his 9 starting players. How many batting orders are possible?

> **Solution**   The batting order is an arrangement of the starting players. Each player is used only once so the permutation theorem applies.
>   There are 9 starting players,
>   so there are 9! possible batting orders.
> Using the calculator, key in 9 $\boxed{x!}$
>
> $$9! = 362,880$$

Sometimes you can do a problem more quickly with pencil and paper than with a calculator. When working with fractions, see whether you can simplify first. Large numbers can often be managed using this method.

Example 3   Evaluate $\dfrac{12!}{10!}$.

Solution 1   Express the factorials as multiplications and simplify.

$$\frac{12!}{10!} = \frac{12 \cdot 11 \cdot 10 \cdot 9 \cdot 8 \cdot 7 \cdot 6 \cdot 5 \cdot 4 \cdot 3 \cdot 2 \cdot 1}{10 \cdot 9 \cdot 8 \cdot 7 \cdot 6 \cdot 5 \cdot 4 \cdot 3 \cdot 2 \cdot 1}$$
$$= 12 \cdot 11$$
$$= 132$$

Solution 2   Only write out as many of the factors of the factorials as necessary.

$$\frac{12!}{10!} = \frac{12 \cdot 11 \cdot 10!}{10!}$$
$$= 12 \cdot 11$$
$$= 132$$

Check   Use a calculator to evaluate the factorials. Key in 12 $\boxed{x!}$ $\boxed{\div}$ 10 $\boxed{x!}$ $\boxed{=}$.

$$\frac{12!}{10!} = \frac{4.79 \cdot 10^8}{3{,}628{,}800}$$
$$= 132$$

12! is so large that the calculator must express it in scientific notation.

Questions

Covering the Reading

1. Make a list of all possible rankings in a poll with Michael Jordan, Magic Johnson, and Larry Bird.

2. If you take a poll ranking 4 basketball stars, how many outcomes can there be in the poll?

3. A short way to write $4 \cdot 3 \cdot 2 \cdot 1$ is __?__.

4. The symbol $n!$ means __?__.

5. Evaluate $n!$ when n equals:
   a. 1   **b.** 2   **c.** 3   **d.** 4
   e. 5   **f.** 6   **g.** 7   **h.** 8

In 6 and 7, evaluate with a calculator.

6. 15!   **7.** 30!

In 8 and 9, evaluate.

8. $\dfrac{6!}{3!}$

9. $\dfrac{100!}{99!}$

10. An arrangement of objects is called a __?__.

11. The number of permutations of n objects is __?__.

12. In softball, there are 10 people who can bat. In how many ways can the manager of a softball team arrange the batting order?

Applying the Mathematics

In 13 and 14, evaluate.

13. $(3!)!$

14. $\dfrac{7!}{4!\ 3!}$

15. Suppose 8 horses are in a race.
  a. In how many ways can first and second place be awarded?
  b. In how many different orders can all eight horses finish?

16. Twenty pictures are placed in a line on a wall. In how many ways can they be arranged? Write your answer in scientific notation.

17. a. Evaluate $\dfrac{n!}{(n-1)!}$ when $n = 10$.   **b.** Generalize part a.

Review

18. Michelle has three outfits she can wear to school tomorrow. One is blue, one is green, and the third is purple. Let $A =$ she wears the blue outfit in tomorrow's math class. Let $B =$ she wears the blue outfit in tomorrow's English class. Estimate: *(Lesson 4-8)*
  a. P(A)   **b.** P(B)   **c.** P(B given A)   **d.** P($A \cap B$)

19. Six students in a class of 25 have the flu. Two of these six are girls. Thirteen of the 25 students in the class are boys. Draw a tree diagram.
  a. What is the probability that a randomly chosen student is a girl?
  b. What is the probability that a randomly chosen student with the flu is a girl?
  c. What is the probability that a randomly chosen student is a girl with the flu? *(Lesson 4-8)*

20. On three questions of a five question multiple-choice test there are four choices. On two questions there are five choices. If Mary guesses on all the questions, what are her chances of getting them all correct? *(Lesson 4-7)*

In 21–26, solve. *(Lessons 4-4, 4-5, 4-6)*

21. $10x = 723$

22. $7a \geq -2$

23. $\frac{3}{8}y < \frac{5}{4}$

24. $1.6 = 2.5x$

25. $(2 - 3)t \leq 8$

26. $8v - 8v = 0$

27. Simplify: $\frac{7}{5} \cdot 5x + 0 \cdot x + -7x$. *(Lesson 4-3)*

28. Write the reciprocal of $\frac{50n^3}{233m^5}$. *(Lesson 4-3)*

29. There are about 16,000 grains of sand per cubic inch and 1728 cubic inches per cubic foot. About how many grains of sand are in a 25 cubic foot sandbox? *(Lesson 4-2)*

30. A rectangular solid is twice as long as it is wide. It is three units higher than it is wide. If w is the width, write an expression for its volume. *(Lesson 4-1)*

In 31–33, simplify. *(Lesson 4-1)*

31. $3.3(2a)$

32. $(6b)(3b)$

33. $(8c)^2$

Exploration

In 34–36, use a calculator.

34. What is the smallest value of n for which $n!$ is divisible by 100?

35. What is the largest value of n for which the calculator can calculate or estimate $n!$?

36. Key in 2.5 $\boxed{x!}$. Explain what happens.

Summary

Multiplication has many uses. The product xy of two numbers x and y may stand for:

Area Model   the area of a rectangle with length x and width y

Area Model (discrete version)   the number of elements in a rectangular array with x rows and y columns

Rate Factor Model   the result of multiplying a rate x by a secondary quantity y

Multiplication Counting Principle   the number of ways of making a first choice followed by a second choice if the first choice can be made in x ways and the second choice can be made in y ways

Conditional Probability Formula $P(A \cap B)$ if $P(A) = x$ and $P(B$ given $A) = y)$

The product of three numbers xyz is the volume of a box with dimensions, x, y, and z. The product of the integers from 1 to n, written $n!$, is the number of ways of arranging n objects.

The numbers 0, 1, and -1 are special in multiplication. Multiplying any number by zero gives the same result ... 0. For this reason, equations of the form $0x = b$ are either true for all real numbers or for none. Multiplying any number by 1 yields that number. A conversion factor is the number 1 written using different units, so multiplying by it does not change the value of a quantity. Multiplying any number by -1 changes it to its opposite.

These many applications mean that equations of the form $ax = b$ and inequalities of the form $ax < b$ are quite common. When $a \neq 0$, such sentences can be solved by multiplying both sides by the number $\frac{1}{a}$, the reciprocal of a. The only caution is to remember to change the sense of the inequality if a is negative.

Vocabulary

Below are the most important terms and phrases for this chapter. You should be able to give a general description and a specific example of each.

Lesson 4-1
Area Model for Multiplication
Commutative Property of Multiplication
Associative Property of Multiplication
rectangular array, dimensions
rectangular solid, box
volume

Lesson 4-2
Rule for Multiplication of Fractions
Rate Factor Model for Multiplication
reciprocal rates

Lesson 4-3
Multiplicative Identity Property of 1
Multiplication Property of -1
reciprocal, multiplicative inverse, $\boxed{1/x}$
Property of Reciprocals
Multiplication Property of Zero

Lesson 4-4
Multiplication Property of Equality

Lesson 4-6
Multiplication Property of Inequality
changing sense of an inequality, changing direction of an inequality

Lesson 4-7
Multiplication Counting Principle
tree diagram

Lesson 4-8
$P(A$ and $B)$
$P(B$ given $A)$, conditional probability
Conditional Probability Formula

Lesson 4-9
n factorial, $n!$
permutation
Permutation Theorem

Progress Self-Test

Take this test as you would take a test in class. You will need graph paper. Then check your work with the solutions in the Selected Answers section in the back of the book.

1. Evaluate $\frac{5!}{3!2!}$.

2. Evaluate $7(v + 2.9)(2v + 3.1)(2.4 - v)$ when $v = 2.4$.

3. Simplify $1\frac{2}{3} \cdot \frac{x}{5}$.

4. **a.** Give an instance of the Associative Property of Multiplication.
   b. Give an instance of the Property of Reciprocals.

5. Use a square to show that $\frac{1}{2} \times \frac{3}{5} = \frac{3}{10}$.

In 6–10, solve.

6. $30x = 10$ 7. $\frac{1}{4}k = -24$

8. $15 \leq 3m$ 9. $-y \leq -2$

10. $1.46 = 2.7 - t$

11. What is the reciprocal of $3m$?

12. Write using symbols: The product of a number and the multiplicative identity is that number.

13. Find the area of the figure below. (All angles are right angles.)

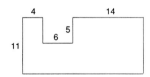

14. A sports stadium has 4 north gates and 5 south gates. In how many ways can a person enter through a north gate and leave through a south gate?

15. Dianne, Juan, Myisha, Sari, and Ted ran for class representative. No two of them received the same number of votes. If they are ranked from most votes to least votes, how many possible rankings are there?

16. What is the price of s shirts that sell for $15.50 a shirt?

17. How much driving time is there from Chicago to St. Louis, a distance of 300 miles, if you travel at an average of 55 mph?

18. There are 2.54 cm/in. How many centimeters are there in 8 inches?

19. A drawer contains 4 black socks and 3 red socks. If one sock is picked at random from the drawer and not put back, and then another is picked, what is the probability that both are red?

20. A box holds 1000 cm³. Its length is 16 cm and its depth is 5 cm.
   a. Write an equation to find the width w of the box.
   b. Solve for w.

21. Fourteen of the first forty U.S. Presidents served before the Civil War. Of the Presidents who served before that war, half were born in Virginia. Only one of the Presidents who served after that war was born in Virginia. What is the probability that a randomly chosen U.S. President served before the Civil War and was born in Virginia?

Chapter Review

Questions on **SPUR** Objectives

SPUR stands for **S**kills, **P**roperties, **U**ses, and **R**epresentations.
The Chapter Review questions are grouped according to the
SPUR Objectives for this chapter.

SKILLS deal with the procedures used to get answers.

■ **Objective A.** *Multiply fractions. (Lesson 4-2)*

1. $\dfrac{9x}{10} \cdot \dfrac{3}{4x}$

2. $1\frac{5}{8} \cdot 2\frac{2}{3}$

3. $\dfrac{a}{b} \cdot \dfrac{c}{d} \cdot \dfrac{b}{a}$

4. $\dfrac{ax}{3} \cdot \dfrac{6}{a}$

■ **Objective B.** *Solve and check equations of the form $ax = b$. (Lessons 4-4, 4-5)*

In 5–10, solve and check.

5. $2.4m = 360$

6. $-\frac{1}{2}k = -10$

7. $-10f = 23$

8. $-2 = 0.4h$

9. $\frac{4}{25}A = \frac{6}{5}$

10. $\frac{2}{9} = 4c - 2c - 2c$

■ **Objective C.** *Solve and check equations of the form $a - x = b$. (Lesson 4-5)*

In 11–14, solve and check.

11. $31 - x = 43$

12. $7.6 = -5.2 - y$

13. $\frac{1}{5} - z = \frac{2}{5}$

14. $0 - x = 1$

■ **Objective D.** *Solve and check inequalities of the form $ax < b$. (Lesson 4-6)*

In 15–20, solve and check.

15. $8m \le 16$

16. $-250 < 5y$

17. $6u > -12$

18. $-x \ge -1$

19. $\frac{1}{2}g \le 5$

20. $3.6h < 720$

■ **Objective E.** *Evaluate expressions containing a factorial symbol. (Lesson 4-9)*

In 21–23, write as a decimal.

21. $4! + 3!$

22. $\dfrac{6!}{4!2!}$

23. $\dfrac{16!}{14!}$

In 24–26, use a calculator to estimate.

24. $15!$

25. $\dfrac{20!}{15! \ 5!}$

26. $(4!)!$

PROPERTIES deal with the principles behind the mathematics.

■ **Objective F.** *Identify and apply properties of multiplication.* *(Lessons 4-1, 4-3, 4-4, 4-6)*

      Commutative Property
      Associative Property
      Property of Reciprocals
      Multiplication Property of Zero
      Multiplication Property of Equality
      Multiplication Property of Inequality
      Multiplicative Identity Property of 1
      Multiplication Property of -1.

27. **a.** Simplify in your head: $4 \cdot x \cdot 25 \cdot 22$.
    b. What properties aid in the simplification?

28. $3 \cdot a = a \cdot 3$ is an instance of what property?

29. Write in symbols: The product of a number and its reciprocal is the multiplicative identity.

In 30–32, write the reciprocal of the given number.

30. -2   **31.** 0.6   **32.** $\frac{3}{4x}$

33. **a.** Evaluate $(k + 3.8)(4.3 - k)(k + 6.2)$ when $k = -6.2$.
    b. What property aids in this evaluation?

34. Of what property is this an instance? If $m = n$, then $12m = 12n$.

35. Multiplication by -1 changes a number to its __?__.

36. If $-12x < 4$, what inequality results from multiplying both sides by $-\frac{1}{12}$?

In 37 and 38, give an example of an equation of the form $ax = b$ that has:

37. no solution.

38. all real numbers as solutions.

USES deal with applications of mathematics in real situations.

■ **Objective G.** *Apply the area and rate factor models for multiplication.* *(Lessons 4-1, 4-2)*

39. Consider the sketch of the 9′ × 12′ area rug at the right. What is the area of the colored part?

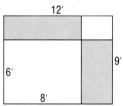

40. What is the volume of a box that is 12 cm long, 15 cm high, and 8 cm wide?

41. Find the rent for k months on an apartment that rents for $350 per month.

42. At 30 miles per gallon of gas and $1.00 per gallon of gas, what is the gas cost per mile?

43. There are 43,560 sq ft/acre. How many square feet are there in 24 acres?

44. A hairdresser charges 15 dollars per cut. How much will he earn in 5 hours if he does 3 cuts per hour?

45. On average, B books fit on 1 foot of shelf space. A bookcase has 24 feet of shelf. How many books can fit on C bookcases?

■ **Objective H.** *Apply the Multiplication Counting Principle.* *(Lesson 4-7)*

ΑΒΓΔΕΖΗΘΙΚΛΜ
ΝΞΟΠΡΣΤΥΦΧΨΩ

46. The Greek alphabet has 24 letters. How many 3-letter monograms are possible? (3-letter Greek monograms often are used to name fraternities and sororities.)

47. All 10 questions on a quiz are multiple choice, each with 5 possible choices. How many different sets of answers are possible on the test?

48. A sports stadium has n north gates and s south gates. In how many ways can a person enter through a north gate and leave through a south gate?

49. Claire has 3 skirts. One is red and two are black. She has 8 blouses: 2 are white, 2 are yellow, 3 are floral, and 1 is plaid. In a rush one morning she pulls out a skirt and a blouse at random. What is the probability that she takes a red skirt and a white blouse?

■ Objective I. *Apply the Conditional Probability Formula. (Lesson 4-8)*

50. Bill is a streak hitter in baseball. He gets hits 25% of the time he is at bat. But when he gets a hit his first time up, the probability he will get a hit the next time up is 32%. What is the probability Bill will get hits twice in a row at the beginning of a game?

51. Ten of the 30 students in a class are boys. Three students were absent yesterday; two of those were boys. A student from the class is to be randomly selected to get a special test on this chapter.
    a. What is the probability that if a boy is selected, the boy was absent yesterday?
    b. What is the probability that a boy who was absent yesterday will be selected?

52. The estimated probability of being able to roll your tongue is $\frac{1}{8}$. The estimated probability of having attached earlobes is $\frac{1}{16}$. What is the probability of a person being able to roll his or her tongue and having attached earlobes?

53. Two cards are drawn from a standard deck. What is the probability of drawing:
    a. two clubs? b. two kings?

■ Objective J. *Apply the Permutation Theorem. (Lesson 4-9)*

54. The number of permutations of n objects is ___?___.

55. Nine students are lining up outside a classroom. a. In how many ways can they arrange themselves? b. Write the answer in scientific notation.

56. Jesse, Simon, Arlene, and Ed ran for class president. No two received the same number of votes. How many possible orders, by votes received, are there?

■ Objective K. *Solve sentences of the form $ax = b$ and $ax < b$ to answer questions from real situations. (Lessons 4-1, 4-2, 4-3, 4-4, 4-6)*

57. How long does it take to drive from Chicago to Minneapolis, a distance of 411 miles, if a person can average 50 mph?

58. The volume of a rectangular storage area needs to be 10,000 cubic feet. The floor has dimensions 40 feet and 80 feet. What should the height be?

59. Daniel budgeted $550 for accommodations. His hotel costs $45 a day. At most how many days can he stay at the hotel?

60. At least how many 16 oz bottles are needed to hold 300 oz of mineral water?

REPRESENTATIONS deal with pictures, graphs, or objects that illustrate concepts.

■ Objective L. *Use area, arrays, and volume to picture multiplication. (Lesson 4-1)*

61. What property of multiplication is illustrated below, where $A_1 = A_2$?

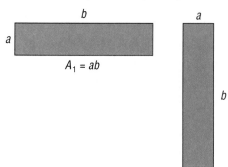

62. The large square at the right has length 1. What multiplication of fractions does the drawing represent?

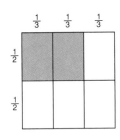

63. Use a square to show that $\frac{1}{2} \times \frac{3}{4} = \frac{3}{8}$.

64. Use two arrays to show that $2 \times 3 = 3 \times 2$.

65. If a dot is picked at random, what is the probability that it is to the right of line *l* and below line *m*?

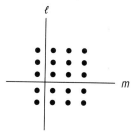

66. How many dots are in the array below?

67. How many square tiles, each one foot on a side, are needed to cover an area 4*q* feet by 5 feet?

68. How many cubes, one centimeter on each edge, can be placed in a box with base *p* cm² and height 3 cm?

69. What is the volume of a box with dimensions 3*k*, 2*k*, and 4*k*?

70. What is the volume of a box with dimensions 0.3 meter, 0.45 meter, and *x* meters?

REFRESHER

Chapter 5, which discusses division in algebra, assumes that you have mastered certain objectives in your previous mathematics work. Use these optional questions to check your mastery.

A. Divide.

1. $40 \div 100$ **2.** $100 \div 40$

3. $7.2 \div 3$ **4.** $80 \div .05$

5. $.06 \div .3$ **6.** $6.8 \div 34$

7. $1 \div 625$ **8.** $11.27 \div 2.35$

9. $\frac{12}{7} \div \frac{2}{7}$ **10.** $\frac{3}{5} \div \frac{3}{4}$

11. $\frac{2}{9} \div \frac{1}{3}$ **12.** $\frac{3}{2} \div \frac{4}{5}$

13. $6 \text{ ft} \div 2$ **14.** $10m \div 4$

15. $100 \text{ kg} \div 7$ **16.** $6 \text{ lb} \div 25$

B. Convert any simple fraction to a decimal or percent.

In 17–20, give the decimal and percent equivalent.

17. $\frac{1}{2}$ **18.** $\frac{3}{4}$

19. $\frac{1}{40}$ **20.** $\frac{73}{100}$

In 21–24, round to the nearest hundredth.

21. $\frac{1}{7}$ **22.** $\frac{20}{3}$

23. $\frac{110}{17}$ **24.** $\frac{3467}{1103}$

In 25–28, give to the nearest percent.

25. $\frac{11}{5}$ **26.** $\frac{27}{101}$

27. $\frac{8}{9}$ **28.** $\frac{11}{60}$

C. Convert percents to decimals and fractions.

29. 30% **30.** 1%

31. 300% **32.** 2.46%

33. .03% **34.** $\frac{1}{4}$%

D. Give a percent of a number or quantity.

35. 32% of $750

36. 94% of 72 questions

37. 7.3% of 40,296 voters

38. 100% of 12,000 square miles

39. 0% of 60 students

E. Divide.

40. $-8 \div -4$ **41.** $-40 \div 5$

42. $60 \div -120$ **43.** $2 \div -80$

44. $\frac{-3}{-6}$ **45.** $\frac{400}{-4}$

F. Identify the divisor, dividend, and quotient in a division situation.

46. $21 \div 3 = 7$

47. $0.2 = 20 \div 100$

48. $\frac{56}{7} = 8$

CHAPTER 5

Division in Algebra

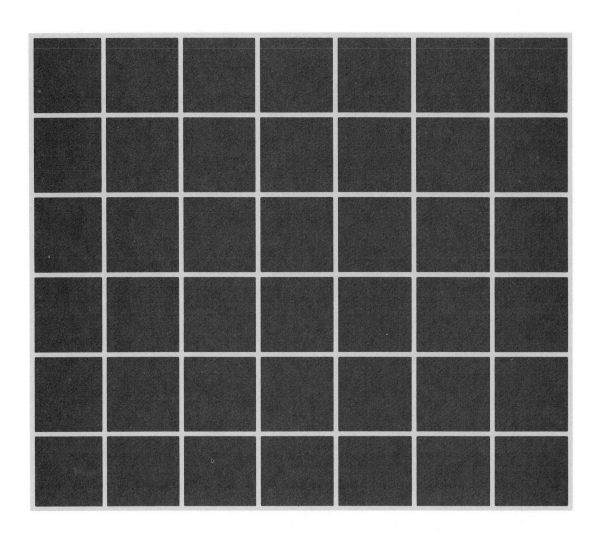

Can the big 6-by-7 rectangle be split into little regions of the T-shape at the left below? A person could start solving this problem by using trial and error.

Below is a start that leads to a situation which does not work. But maybe the start was wrong.

There's another way to examine the problem. The rectangle's area is 42 square units. The T-shaped region has area 4 units. If the rectangle can be split up, there will be $\frac{42}{4}$, or 10.5 T-shaped regions. Since the number of regions must be an integer, no way will work.

Division starts with questions of splitting up things. In this chapter, you will study this and other uses of division.

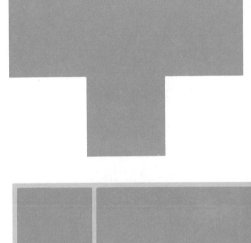

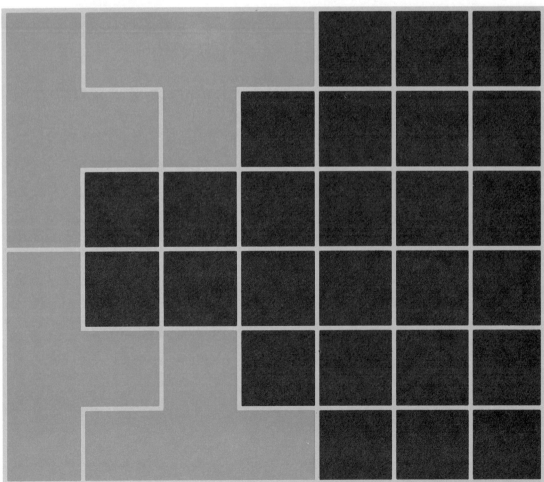

The Definition of Division

Abe divided a quart of orange juice equally among his five children. How many ounces of orange juice did each child receive?

This question can be answered either by multiplication or division. Either way you must change 1 quart to 32 ounces. You can view this problem as 32 ounces divided among 5 children,

$$32 \div 5 = 6.4$$

or each child gets $\frac{1}{5}$ of the orange juice,

$$32 \cdot \frac{1}{5} = 6.4$$

Each way the answer is 6.4 ounces. Dividing by 5 is the same as multiplying by $\frac{1}{5}$.

In general, dividing by b is the same as multiplying by its reciprocal, $\frac{1}{b}$.

Algebraic Definition of Division:

For any real numbers a and b, $b \neq 0$,

$$a \div b = a \cdot \frac{1}{b}.$$

Recall that $a \div b$ can also be written as a/b or $\frac{a}{b}$. So $a \div b = \frac{a}{b} = a/b = a \cdot \frac{1}{b}$.

The definition of division allows any division situation to be converted to multiplication. For instance, consider $\frac{a}{b} \div \frac{c}{d}$. According to the definition of division, dividing by $\frac{c}{d}$ is the same as multiplying by $\frac{d}{c}$. So,

$$\frac{a}{b} \div \frac{c}{d} = \frac{a}{b} \cdot \frac{d}{c}.$$

For example $\frac{5}{2} \div \frac{9}{7} = \frac{5}{2} \cdot \frac{7}{9} = \frac{35}{18}$. Variables are dealt with in the same way.

Example 1 Simplify $\dfrac{x}{5} \div \dfrac{3}{4}$.

Solution $\dfrac{x}{5} \div \dfrac{3}{4} = \dfrac{x}{5} \cdot \dfrac{4}{3} = \dfrac{4x}{15}$

Check Let $x = 2$. Does $\dfrac{2}{5} \div \dfrac{3}{4} = \dfrac{4 \cdot 2}{15}$? To determine this, change each fraction to a decimal. Does $0.4 \div 0.75 = \frac{8}{15}$? Yes, each side is equal to $0.\overline{53}$.

Dividing fractions with negative numbers can be dealt with in the same way. In Example 2, the fraction is converted to the equivalent division problem.

Example 2 Simplify $\dfrac{\frac{1}{6}}{-\frac{2}{3}}$.

Solution $\dfrac{\frac{1}{6}}{-\frac{2}{3}}$ means $\dfrac{1}{6} \div -\dfrac{2}{3}$

Use the definition of division to convert the division to a multiplication. The reciprocal of $-\frac{2}{3}$ is $-\frac{3}{2}$. Thus

$$\frac{1}{6} \div -\frac{2}{3} = \frac{1}{\overset{}{\underset{2}{\cancel{6}}}} \cdot -\frac{\overset{1}{\cancel{3}}}{2}$$

$$= -\frac{1}{4}$$

In Example 2, a positive number is divided by a negative number. This is converted into multiplication of a positive by a negative. So the quotient is negative. Generally, the signs of answers to division problems follow the same rules as in multiplication.

If two numbers have the same + or − sign, their quotient is positive. If two numbers have different signs, their quotient is negative.

Remember that π is a number and can be multiplied and divided like any other number. Example 3 on page 212 uses π in a division problem.

Example 3 Simplify $\dfrac{7\pi}{\dfrac{\pi}{21}}$

Solution Use the definition of division to rewrite as a multiplication.

$$7\pi \div \frac{\pi}{21} = 7\pi \cdot \frac{21}{\pi}$$

Follow the properties of multiplication.

$$= \frac{7\cancel{\pi}}{1} \cdot \frac{21}{\cancel{\pi}}$$

$$= 147$$

The definition of division leads to another method for solving equations of the form $ax = b$. In the next example, one method of solution uses multiplication. However, multiplying by $-\frac{1}{31}$ is the same as dividing by -31. So a second method uses division.

Example 4 Solve $-31m = 527$.

Solution 1 Multiply both sides by $-\frac{1}{31}$.

$$-\frac{1}{31} \cdot -31m = -\frac{1}{31} \cdot 527$$

$$m = -\frac{527}{31}$$

$$m = -17$$

Solution 2 Divide both sides by -31.

$$-31m = 527$$

$$\frac{-31m}{-31} = \frac{527}{-31}$$

$$m = -17$$

Both methods of Example 4 are acceptable. You may prefer one or the other.

Questions

Covering the Reading

1. Suppose Abe had divided a quart of orange juice equally among his five children *and himself*. Show two ways to determine how many ounces of orange juice each person received.

2. State the algebraic definition of division.

In 3–5, fill in the blanks.

3. a. $\dfrac{m}{n} = m \div \underline{}$.

  b. $\dfrac{m}{n} = m \cdot \underline{}$

4. a. $\dfrac{\frac{a}{b}}{\frac{c}{d}} = \dfrac{a}{b} \div \underline{}$.   **b.** $\dfrac{a}{b} \div \dfrac{c}{d} = \dfrac{a}{b} \cdot \underline{}$.

5. a. $50 \div \frac{1}{2} = 50 \cdot \underline{}$.

  b. $\dfrac{50}{\frac{1}{2}} = 50 \cdot \underline{}$.

  c. $\dfrac{50}{2} = 50 \cdot \underline{}$.

  d. If 50 boxes of flyers are split so each campaign worker gets $\frac{1}{2}$ of a box, how many campaign workers are needed to distribute all the flyers?

6. Consider $\frac{3}{4} \div \frac{3}{16}$. **a.** Rewrite as a multiplication. **b.** Evaluate.

In 7 and 8, simplify.

7. $\dfrac{5}{4} \div \dfrac{n}{10}$

8. $\dfrac{12\pi}{5} \div \dfrac{\pi}{4}$

In 9 and 10, **a.** rewrite with the $\div$ sign; **b.** rewrite as a multiplication; **c.** simplify.

9. $\dfrac{\frac{6}{25}}{\frac{10}{7}}$

10. $\dfrac{x}{-\frac{1}{2}}$

11. When solving $-3j = -48$,
  a. by what would you multiply both sides?
  b. by what would you divide both sides?

In 12–14, solve.

12. $143 = -13x$ **13.** $-1.5q = -75$ **14.** $2.5k = -0.7$

Applying the Mathematics

15. Simplify: $1\frac{2}{3} \div 3\frac{1}{3}$.

16. Ruth divided a gallon of milk equally among x people. How many ounces did each person receive?

17. Half of a pizza was divided equally among 3 people. How much of the original pizza did each person receive?

18. Le Parfum Company produces perfume in 200-ounce batches and bottles it in quarter-ounce bottles. Suppose you want to know how many bottles are needed per batch.
  a. Write a division problem that will tell you.
  b. Find the answer.

19. **a.** Write as a decimal: $\frac{-7}{2}$, $\frac{7}{-2}$, and $-\frac{7}{2}$.

    b. Write as a decimal: $\frac{-3}{11}$, $\frac{3}{-11}$, and $-\frac{3}{11}$.

    c. Generalize the pattern of parts a and b using variables.

20. Simplify $\frac{xy}{21} \div \frac{x}{4y}$.

21. In 1577, Guillaume Gosselin published a book titled *De Arte Magna*. It contained some of the first work on positive and negative numbers. It was written in Latin. Here are some rules Gosselin wrote.

    1. P in P diviso quotus est P.   2. M in M diviso quotus est P.
    3. M in P diviso quotus est M.   4. P in M diviso quotus est M.

    Translate these rules into English.

Review

22. Suppose $x = -.0046928146$ and $y = -.0046928146$. What is $\frac{x}{y}$? *(Previous course)*

23. Your uncle is having a party and he asked you to buy 5 pounds of meat. You can buy sliced turkey or roast beef. Graph all the possible ways you could buy the two different meats so they total 5 pounds. Let $t =$ amount of turkey and $b =$ amount of beef. *(Lesson 3-7)*

24. Simplify: **a.** $7 + \frac{3}{7}$ **b.** $7 \cdot \frac{3}{7}$ **c.** $x \cdot \frac{y}{x}$ *(Lesson 2-3, 4-2)*

25. You bought 5 new shirts today. How many different ways can you choose the order in which to wear them? *(Lesson 4-9)*

26. After paying $213.00 tax, Cliff still had more than $4700 in his account. How much did he have before the tax? *(Lesson 3-3)*

27. The perimeter of quadrilateral *URSA* at the left is 25.
    a. Find x.   **b.** Find *UR*. *(Lesson 2-6)*

28. *Skill sequence*   Solve and check. *(Lessons 2-6, 4-4)*
    a. $\frac{3}{5} + r = 8$ **b.** $\frac{3}{5}r = 8$

    c. $\frac{48}{6} = 0 \cdot r + r + .6$ **d.** $\frac{3rt}{5t} = 2^3$

29. A circle has a radius of 1.2 m. Find its area to the nearest tenth of a square meter. (Hint: $A = \pi r^2$.) *(Lesson 1-5)*

30. Evaluate in your head. *(Previous course, Lessons 4-2, 4-3)*
    a. $(2)^3$ **b.** $(-1)^3$ **c.** $\left(\frac{2}{3}\right)^3$

Exploration

31. Congruent figures are figures with the same size and shape. Split this region into 6 congruent pieces.

5-2

Rates

In Lesson 4-2, you multiplied with rates. For example, if you travel at 45 miles per hour, then in 5 hours, you will travel 225 miles because

$$5 \text{ hours} \cdot 45 \, \frac{\text{miles}}{\text{hour}} = 225 \text{ miles}.$$

If you are not given a rate, you may be able to calculate it using division.

Example 1   Joe took a trip of 400 miles. The trip took him 8 hours. What was his average rate?

Solution 1   Divide the distance in miles by the time in hours.

$$\frac{400 \text{ miles}}{8 \text{ hours}}$$

Separate the numerical parts from the measurement units.

$$\frac{400 \text{ miles}}{8 \text{ hours}} = \frac{400}{8} \, \frac{\text{miles}}{\text{hours}}$$
$$= 50 \text{ miles per hour}$$

He was traveling at a rate or speed of 50 miles per hour.

Solution 2   You could also divide the time by the distance.

$$\frac{8 \text{ hours}}{400 \text{ miles}} = \frac{8}{400} \, \frac{\text{hours}}{\text{miles}}$$
$$= \frac{1}{50} \text{ hour per mile}$$

This means that, on the average, it took him $\frac{1}{50}$ of an hour to travel each mile.

The first solution gives Joe's rate in *miles per hour*. You are familiar with the meaning of this rate from speed limit signs or watching the speedometer in a car. The second solution gives the rate in *hours per mile*, or how long it takes to travel one mile. Notice that $\frac{1}{50}$ of an hour is $\frac{1}{50} \cdot 60$ min or 1.2 minutes. In other words, it takes a little over a minute to go one mile. Either rate is correct. The one you use depends on the situation in which you use it.

This situation is an instance of the **Rate Model for Division.**

Rate Model for Division:

If a and b are quantities with different units, then $\frac{a}{b}$ is the amount of quantity a per quantity b.

Example 2 In 1986, total medical costs in the United States were about $458,200,000,000. That year the population was about 240,500,000. What did a person spend for medical costs on the average?

Solution This situation should prompt you to find dollars per person. To get this rate, divide the total spent by the number of people.

$$\frac{458,200,000,000 \text{ dollars}}{240,500,000 \text{ people}} = \frac{458,200,0\!\!\!/0\!\!\!/0,0\!\!\!/0\!\!\!/0 \text{ dollars}}{240,5\!\!\!/0\!\!\!/0,0\!\!\!/0\!\!\!/0 \text{ people}}$$
$$\approx 1905 \text{ dollars/person}$$

Rates can be negative quantities.

Example 3 The temperature goes down 12 degrees in 5 hours. What is the rate of temperature change?

Solution The rate here is degrees per hour. To find it, divide the number of degrees by the number of hours.

$$\frac{\text{drop of 12 degrees}}{5 \text{ hours}} = \text{drop of 2.4 degrees per hour}$$

This can be written as

$$\frac{\text{-12 degrees}}{5 \text{ hours}} = \text{-2.4 degrees per hour.}$$

The temperature drops at a rate of 2.4 degrees per hour.

Consider the rate $\dfrac{10 \text{ meters}}{0 \text{ seconds}}$. It means you are traveling 10 meters in 0 seconds. To be true, you would have to be in two places at the same time. Since that is impossible, this model illustrates that it is meaningless to think of $\dfrac{10}{0}$. *Division by zero is impossible.* The denominator of a fraction cannot be zero. When variables appear in divisions, you sometimes must take precautions so that you don't attempt to divide by zero.

Example 4 What value can k *not* have in $\dfrac{7}{k-5}$?

Solution The denominator can't be 0, so $k-5$ cannot be 0. Thus k cannot be 5.

Check If $k = 5$, then $\dfrac{7}{k-5} = \dfrac{7}{5-5} = \dfrac{7}{0}$.

If a computer is instructed to divide by zero, it won't do it! The program will stop running and an error message will appear on the screen. To avoid this, you can have the computer decide in advance if an expression will involve division by zero. An **IF-THEN** command has the computer make a decision. When the sentence between the words IF and THEN is true, the computer executes the statement following the THEN. When the sentence is false, the statement following the THEN is completely ignored by the computer.

In Example 4 you saw that if k is 5, the expression $\dfrac{7}{k-5}$ has no meaning. The following program evaluates $\dfrac{7}{k-5}$ for different values of k, as long as $k \neq 5$. Otherwise, it prints a message about division by zero.

```
10 PRINT "EVALUATE 7/(K − 5)"
20 PRINT "ENTER VALUE OF K"
30 INPUT K
40 PRINT "VALUE OF 7/(K − 5)"
50 IF K = 5 THEN PRINT "DIVISION BY ZERO IS IMPOSSIBLE"
55 REM      THE COMPUTER SYMBOL FOR "NOT EQUAL" IS < >.
60 IF K <> 5 THEN PRINT 7/(K - 5)
70 END
```

Questions

Covering the Reading

1. Suppose Joe took a trip of 300 miles and it took him 8 hours.
  a. What was his average speed in miles per hour?
  b. On the average, how long did it take him to travel a mile?

2. State the rate model for division.

3. a. To calculate degrees per hour, you should divide __?__ by __?__.
  b. The temperature drops 11 degrees in 5 hours. What is the rate of temperature change?

4. Give a rate suggested by each situation.
  a. A family drove 24 miles in $\frac{2}{3}$ hours.
  b. A family drove m miles in $\frac{2}{3}$ hours.
  c. A family drove 24 miles in h hours.
  d. A family drove m miles in h hours.

In 5–7, calculate a rate suggested by each situation.

5. In 9 days Julian earned $495.

6. c cans of natural lemonade cost $2.10.

7. You travel 270 miles and use 7.8 gallons of gasoline.

In 8 and 9, what value can the variable *not* have in each expression?

8. $\dfrac{18}{k-4}$

9. $\dfrac{x-6}{x+1}$

In 10 and 11, refer to the computer program in this lesson.

10. Tell what the computer will print if the value entered for K is:
  a. 19 **b.** 0 **c.** 5.

11. Why are parentheses needed around $K - 5$?

Applying the Mathematics

12. In one store a 20-ounce can of pineapple costs 89¢ and a 6-ounce can of the same kind of pineapple costs 39¢.
  a. Calculate the unit cost of the 20-ounce can.
  b. Calculate the unit cost of the 6-ounce can.
  c. Based on the unit cost, which is the better buy?

A street in Bangladesh

13. For each place, find the number of people per square mile. (This is called the *population density*.)
    a. Bangladesh: population 85,122,000; area 55,126 sq mi
    b. Greenland: population 50,000; area 840,000 sq mi

14. A fast runner can run a half mile in 2 minutes. What is the average rate, then, in miles per *hour*?

15. *Multiple choice* In m minutes, a copy machine made c copies. At this rate, how many copies per *hour* can be made?
    (a) $\dfrac{60m}{c}$ (b) $\dfrac{60c}{m}$ (c) $\dfrac{c}{60m}$ (d) $\dfrac{m}{60c}$

16. In $\dfrac{5n + 1}{2n}$, what value can n not have?

17. What happens on your calculator when you divide by 0?

18. Write a computer program to calculate rate of speed in miles per hour. It should ask for the number of miles and number of hours. If the user enters zero hours, it should print IMPOSSIBLE.

Review

19. $\dfrac{x}{3} \div 5$ is the same as what multiplication problem? *(Lesson 5-1)*

In 20–24, simplify. *(Lesson 5-1)*

20. $\dfrac{5}{\frac{2}{3}}$ **21.** $\dfrac{^-3}{4} \div \dfrac{^-3}{2}$ **22.** $\dfrac{x}{2y} \div \dfrac{11y}{3}$

23. $6 \div \dfrac{1}{2}$ **24.** $\dfrac{x}{2} \div \dfrac{x}{4}$

In 25 and 26, solve. *(Lessons 4-4, 5-1)*

25. $.45x = {}^-135$

26. $\frac{5}{8}m = \frac{10}{3}$

27. Solve $^-19 = 1.9tn$ for n. *(Lesson 5-1)*

28. If the area of square S is 9 and the perimeter of square T is 20, find the area of square R. *(Previous course)*

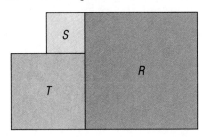

29. In the summer of 1987, an American dollar was worth about .62 English pound and one dollar would buy about 6.1 French francs. About how many francs could be bought for 75 pounds? *(Lesson 4-3)*

30. There are 3 teachers of freshman English, 4 of freshman math, and 2 of freshman science. Lionel wants Mr. Novel for English, Mrs. Euclid for math, and Mr. Bunson for science. If students are randomly assigned to classes, what is the probability Lionel will get all his favorite teachers? *(Lesson 4-8)*

In 31–33, solve and check. *(Lessons 2-3, 3-3)*

31. $8 = x - 5$

32. $^-7 = y - 2.5$

33. $13 - 15 + z = 2\frac{1}{2} + \frac{^-3}{4}$

Exploration

34. Use an almanac or some other source to find the national debt. Find the U.S. population. Then calculate the average debt per capita (the phrase "per capita" means "per person").

35. See Question 13. What is the population density of the community or city where you live?

Ratios

Being overweight increases the probability that a person will suffer from heart disease. The chart below left shows a way to test whether an adult has an increased risk. Dividing the waist measure by the hip measure results in a **ratio** which can be used to compare them.

Example 1 Ms. Mott's waist measure is 26". Her hip measure is 35". According to the chart at left, does she run an increased risk of heart disease?

Solution $\dfrac{\text{waist measure}}{\text{hip measure}} = \dfrac{26''}{35''} \approx 0.74$

Since $0.74 < 0.8$, her risk (according to this test) is not high.

1.   Measure waist and hips. Call these w and h.
2.   For women, risk of heart disease increases if $\frac{w}{h} > 0.8$.
3.   For men, risk of heart disease increases if $\frac{w}{h} > 1.0$.

The direction of the comparison is important. If Ms. Mott compared her hip measure to her waist measure, the result would be $\frac{35''}{26''} \approx 1.35$, which is greater than 0.8. She would be using the wrong number and would misinterpret the test.

Subtraction provides one way to compare quantities. Division is another way.

Ratio Model for Division:

Let a and b be quantities with the same units. Then the ratio $\dfrac{a}{b}$ compares a to b.

The ratio $\dfrac{b}{a}$ compares b to a.

Notice the difference between a rate and a ratio. In a rate, the units for a and b are different. When forming a ratio, be sure that the quantities are measured in the same units.

Example 2 It takes Mr. Garcia $\frac{3}{4}$ hour to go to work and it takes Ms. Harper 10 minutes to go to work. Find a ratio comparing Ms. Harper's time to Mr. Garcia's time.

Solution The units of measure for 10 minutes and $\frac{3}{4}$ hour are not the same. We change hours to minutes. Since Ms. Harper is first in the word description of the ratios, her time must be in the numerator. The ratio of Ms. Harper's time to Mr. Garcia's time is

$$\frac{\text{Ms. Harper's time}}{\text{Mr. Garcia's time}} = \frac{10 \text{ minutes}}{\frac{3}{4} \text{ hour}} = \frac{10 \text{ minutes}}{\frac{3}{4} \cdot 60 \text{ minutes}} = \frac{10 \text{ minutes}}{45 \text{ minutes}} = \frac{2}{9}$$

This means that, on the average, Ms. Harper travels 2 minutes for every 9 minutes that Mr. Garcia travels. You could also say that it takes Ms. Harper $\frac{2}{9}$ of the time it takes Mr. Garcia to go to work.

Often ratios are written as percentages. In Example 2, $\frac{2}{9} = .\overline{22} \approx$ 22%, and you could say that it takes Ms. Harper about 22% of the time it takes Mr. Garcia. A percentage can always be interpreted as a ratio, in this case about $\frac{22}{100}$.

Example 3 The Illinois Department of Public Health reported that for the years 1971 to 1982, 4865 skunks were examined for rabies. Of that number, 2162 actually had rabies. What percent of the skunks tested actually had rabies?

Solution You are asked to compare the skunks with rabies to the entire group of skunks.

$$\frac{2162 \text{ skunks}}{4865 \text{ skunks}} = \frac{2162}{4865} \approx .444 \approx 44.4\%$$

Thus about 44% of the tested skunks actually had rabies.

If asked to compare quantities without being given any order, you can make the comparison either way.

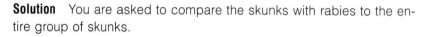

Example 4 Life expectancy has increased from an estimated 18 years in 3000 B.C. to 73.8 years in 1980. Using a ratio, compare these life expectancies.

Solution 1 $\dfrac{1980 \text{ expectancy}}{3000 \text{ B.C. expectancy}} \approx \dfrac{73.8 \text{ years}}{18 \text{ years}} = 4.1$

This means that the life expectancy in 1980 was about 4 times what it was in 3000 B.C.

Solution 2 $\dfrac{3000 \text{ B.C. expectancy}}{1980 \text{ expectancy}} \approx \dfrac{18 \text{ years}}{73.8 \text{ years}} \approx .244$

Since $.244 \approx .25$, the answer indicates that life expectancy in 3000 B.C. was about $\frac{1}{4}$ of what it was in 1980.

In a problem involving discount, the *percent of discount* is the ratio of the discount to the selling price.

■ ■ ■ ■ ■ ■ ■ ■

Example 5 The selling price of an item originally costing $30 is reduced $6. What is the percent of discount?

Solution $\dfrac{\text{amount of discount}}{\text{original price}} = \dfrac{6 \text{ dollars}}{30 \text{ dollars}}$

$= \dfrac{1}{5}$

$= .2$

$= 20\%$

Questions

Covering the Reading

In 1 and 2, refer to the method for testing heart disease risk.

1. Does a woman run an increased risk of heart disease if her waist and hip measurements are 32″ and 37″, respectively?

2. Does a man run an increased risk of heart disease if his waist is 34″ and his hips are 36″?

3. Let x and y be two quantities with the same units. Write two ratios comparing x and y.

4. Suppose it takes Ms. Lopez 25 minutes to complete a particular job and it takes Mr. Sampson half an hour to complete the same job. Write a ratio which:
   a. compares Ms. Lopez's time to Mr. Sampson's time;
   b. compares Mr. Sampson's time to Ms. Lopez's time.

5. The Illinois Department of Health reported that for the years 1971 to 1982, of 230 horses tested for rabies only 16 actually had the disease. What percentage of horses tested had rabies?

6. An item costing $40 is reduced by $5. What is the percent of discount?

7. What is the difference between a rate and a ratio?

8. *Multiple choice*   Which is not a ratio?

(a) $\dfrac{14 \text{ seconds}}{23 \text{ seconds}}$ (b) $\dfrac{150 \text{ miles}}{3 \text{ hours}}$ (c) $\dfrac{27 \text{ cookies}}{13 \text{ cookies}}$

9. *Multiple choice*   A girl born today in the United States has a life expectancy of about 79 years. A boy has a life expectancy of about 72 years. Compare a girl's life expectancy to a boy's.

(a) 1.09   (b) 1.1   (c) .91   (d) 9

10. The ratio of x to y is the __?__ of the ratio of y to x.

11. If w is a man's waist measure and h is his hip measure, write a formula describing when a man's risk of heart disease increases.

12. Give an example of waist and hip measurements which would not run a risk of heart disease for a man, but would for a woman.

13. According to a teacher group, the average teacher salary for 1987–88 was $28,031. In 1870 it was $189. The 1987–88 salary is about how many times the 1870 salary?

14. An item in a store sells for $15.00. The store's profit on that item is $6.00.
  a. Write a ratio of profit to selling price.
  b. What is the percent of profit?

15. A store charges 64¢ tax on a $16.00 purchase.
  a. Write a ratio of tax to purchase price.
  b. What is the percent of tax?

16. Banner High School has won 36 of its last 40 football games.
  a. Write a ratio of games won to games played.
  b. What percent of these 40 games has it won?
  c. Winning percentages are often written as a three-place decimal. Write the answer from part a as a three-place decimal.

In 17 and 18, refer to the circles. Lengths of their radii are given.

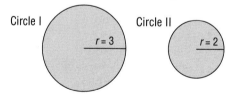

17. The diameter of a circle is twice its radius. Find the ratio of the diameter of Circle I to the diameter of Circle II.

18. In a circle, Area $= \pi r^2$.
  a. Give the ratio of the area of Circle I to the area of Circle II, in lowest terms.
  b. Write this ratio as a decimal.

19. Frieda owns x sweaters while her brother-in-law Fred owns y sweaters. How many times as many sweaters does Fred own as Frieda?

20. It took Patrick 6 hours and 30 minutes to travel 314 miles. What was his average rate for the trip? *(Lesson 5-2)*

21. Simplify: $-\frac{8}{3} \div \frac{7}{2}$. *(Lesson 5-1)*

22. *Multiple choice* Which of the following is *not* equal to $-\frac{a}{b}$? *(Lesson 5-1)*

(a) $\frac{a}{-b}$ (b) $\frac{-a}{b}$ (c) $\frac{-a}{-b}$ (d) $-\frac{-a}{-b}$

23. To use a mainframe computer, a school pays $500 per hour. What is this in minutes per dollar? *(Lesson 5-2)*

24. In Seattle, Washington, it rained or snowed an average of 150 days per year over a 30-year period. What was the relative frequency of precipitation? *(Lesson 1-8)*

25. Remember the area of a rectangle is $\ell \cdot w$. Find the area of the rectangle. *(Lesson 4-1)*

a.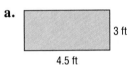
3 ft
4.5 ft

b.
$\frac{2}{3}$ m
$\frac{1}{2}$ m

c.
$2x$
$3x$

26. *Skill sequence* Solve. *(Lessons 3-3,4-5)*
a. $b - 12 = -4$ **b.** $12 - c = -4$ **c.** $(5 + d) - 12 = -4$

27. In 1986, salaries for state governors were as high as $100,000 (New York) and as low as $35,000 (Arkansas and Maine).
a. What is the range of salaries?
b. What is the mean of the two extreme salaries? *(Lesson 1-2)*

28. In a slide, the preimage (3, 4) has image (6, 1). Under the same slide
a. what is the image of (0, 8)?
b. what is the preimage of (-2, -7)? *(Lesson 2-5)*

29. Evaluate $\frac{y_2 - y_1}{x_2 - x_1}$ when $x_1 = 6$, $y_1 = 1$, $x_2 = 8$ and $y_2 = 23$. *(Lesson 1-5)*

30. Consider this right triangle with an angle of about 37°. The three sides can form six different ratios.
a. Write the values of all six of these ratios.
b. Some of these ratios have special names. One of these is called the sine. For the triangle to the left, it is written sin 37°. Compute sin 37° on your calculator using the key sequence

$$37 \quad \boxed{\sin}$$

and determine which of the six ratios in part a is the sine.
c. Compute cos 37° on your calculator and determine which of the six ratios in part a is the cosine.
d. Compute tan 37° on your calculator and determine which of the six ratios in part a is the tangent.

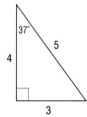

5-4

Solving Percent Problems Using Equations

The word **percent** (often written as the two words per cent) comes from the Latin words *per centum,* meaning "per 100." So 7% literally means 7 per 100, or the ratio $\frac{7}{100}$, or 0.07. The symbol % for percent is only about 100 years old.

In many situations, a percent of an original quantity is a known quantity.

$$50\% \text{ of } 120,000 \text{ is } 60,000.$$
$$5\tfrac{1}{4}\% \text{ of } \$3000 \text{ is } \$157.50.$$

Often, however, you only know two of the three numbers. One method for solving such percent problems is to translate the words into an equation of the form $ab = c$.

Example 1   **a.** 112% of 650 is what number?
b. 7% of what number is 31.5?
c. What percent of 3.5 is .84?

Solution

a. 112% of 650 is what number?

$$a \quad \cdot \quad b \quad = \quad c$$
$$1.12 \cdot 650 \quad = c$$
$$728 \quad = c$$

The answer is 728.

Change 112% to 1.12.

b. 7% of what number is 31.5?

$$a \quad \cdot \quad b \quad = \quad c$$
$$.07 \cdot b \quad = 31.5$$
$$\frac{.07b}{.07} \quad = \frac{31.5}{.07}$$
$$b \quad = 450$$

The answer is 450.

7% is .07.

Divide both sides by .07.

Check   7% of 450 = .07 · 450 = 31.5

c. What % of 3.5 is .84?

$$a \quad \cdot \quad b = c$$
$$a \cdot 3.5 \quad = .84$$
$$\frac{a \cdot 3.5}{3.5} \quad = \frac{.84}{3.5}$$
$$a \quad = .24$$
$$\quad = 24\%$$

Divide both sides by 3.5.

Check   24% · 3.5 = .24 · 3.5 = .84

Example 2

It was reported in 1987 that 55% of the 51.5 million married couples in the U.S.A. had two incomes. Approximately how many couples had two incomes?

Solution Let c be the number of couples with two incomes.

$$.55 \cdot 51.5 = c$$
$$28.325 = c$$

Approximately 28.3 million couples had two incomes.

Example 3

In 1983, only 43,200 or 2.4% of America's enlisted military personnel were college graduates. How many enlisted personnel were there in all?

Solution Let x be the number of enlisted personnel then,

$$2.4\% \cdot x = 43,200$$
$$.024 \cdot x = 43,200$$
$$x = \frac{43,200}{.024}$$
$$x \approx 1,800,000$$

There were approximately 1,800,000 enlisted personnel.

Problems related to business often involve percents.

Example 4   A camera is on sale for $60. Its original cost was $80. What is the percent of discount?

Solution   The discount is $20. So we ask:

$$\underbrace{\text{What \%}}_{a} \; \cdot \; \underbrace{\text{of \$80}}_{b} \; = \; \underbrace{\text{is \$20?}}_{c}$$

$$a \cdot 80 = 20$$
$$a = \frac{20}{80}$$
$$a = .25$$

The percent of discount is 25%.

Check   75% of 80 = .75 · 80 = 60

Percent problems are so common that almost all calculators have a %️ key. If you press 5 %️ you will see .05, which equals 5%. This key is not needed for percent problems like those in this chapter.

Questions

Covering the Reading

1. In 1987, how many married couples were there in the U.S.?

2. In 1983, what percent of America's enlisted military personnel were college graduates?

3. One method of solving percent problems is to use an equation of the form __?__ .

In 4–6, use an equation to solve.

4. 123% of 780 is what number?

5. 40% of what number is 440?

6. What % of 4.7 is .94?

7. A camera is on sale for $49. Its original cost was $59. What is the percent of discount?

8. Refer to Example 2. How many married couples in 1987 did not have two incomes?

9. About 70,000 or 12.4% of America's navy personnel are officers.
   a. Write an equation that you can use to determine how many navy personnel there are in all.
   b. Solve the equation.

Applying the Mathematics

10. If 2.4% of America's enlisted military personnel are college graduates, what percent are *not* college graduates?

11. Convert .325823224 to the nearest whole percent.

In 12 and 13, use this information. In 1985, 57.7 million people in the U.S. attended at least one game of baseball, football, basketball, or ice hockey.

12. Suppose 36% of all spectators attended a baseball game. How many people attended a baseball game? (Round to the nearest tenth of a million.)

13. Suppose 18.8 million spectators attended a football game. What % of the people went to a football game? (Round your answer to the nearest whole number percent.)

14. Clearwater High School expects a 14% decrease in enrollment next year. There are 1850 students enrolled this year. How many students will the school lose?

15. On a mathematics test there were 8 As, 12 Bs, 10 Cs, 2 Ds and 0 Fs. What percent of the students earned As?

16. It takes Angela 60 minutes to do her homework. What percent of her homework time is 45 minutes?

17. Mr. and Mrs. Thompson insured their house for $74,800, which is 85% of its value. What is the value of their house?

18. A TV originally cost $320. It is on sale for $208. What is the percent of discount?

19. A bicycle costs $256 wholesale. If Better Bikes sells it for $425, write this price compared to the wholesale cost as a percent.

20. $1.72 = \underline{\ ?\ }\%$ *(Previous course)*

21. In 1987, the Los Angeles Lakers defeated the Boston Celtics in the NBA Championship series 4 games to 2. What percent of the series games were won by the Lakers? *(Lesson 5-3)*

22. Stanley took $1\frac{1}{4}$ hours to do his homework; Jenile took 35 minutes.
  a. What is the ratio of Jenile's homework time to Stanley's?
  b. What is the ratio of Stanley's homework time to Jenile's?
  (Lesson 5-3)

23. Ben charges $2.25 per hour to babysit. How much will he earn if he starts a job at 7:00 p.m. and leaves at 10:45 p.m.? Round to the nearest 50 cents. *(Lesson 4-2)*

In 24 and 25, simplify. *(Lesson 5-1)*

24. $\frac{5}{6} \div \frac{1}{3}$

25. $-\dfrac{4x}{y} \div \dfrac{2x}{3y}$

26. What is the area of a circle with radius r? *(Previous course)*

27. $\frac{2}{3}$ of a number is 87. Find the number. *(Lesson 4-4)*

28. Use the pentagon at the right. Find x if
  a. its perimeter is 56 cm. *(Lesson 2-2)*
  b. its perimeter is at least 56 cm. *(Lesson 2-7)*

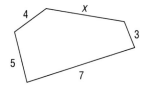

29. Solve for y: $-4x + 3y = 10x$. *(Lesson 2-6)*

In 30 and 31, solve. *(Lessons 4-4, 4-6)*

30. $\frac{2}{3}a = 6$

31. $5c > 2$

32. On a coordinate plane, graph the following points. Use the letters to label each point. $A = (-6, 2)$ $B = (1, 3)$ $C = (-2, 4\frac{1}{2})$ $D = (7, 2)$ $E = (-1, -8)$ $F = (0, 5)$ *(Lesson 2-4)*

33. Take a number and increase it by 30%. Decrease this result by 30%. You shouldn't end up with the number you started with.
  a. What number should you end up with?
  b. Explain why this happened.

5-5

Probability Without Counting

In Chapter 1, we defined the probability of an event as the ratio of the number of successes to the total number of equally-likely possibilities. Since probabilities are ratios, they are often found by dividing. Division can help find probabilities even when it is not possible to find an answer by counting.

Example 1 The picture at the left represents a square dart board with dimensions as shown. Assume a person throws a dart which sticks to the board at a random place. What is the probability that it is a "bull's eye"? (The bull's eye circle is the smaller circle.)

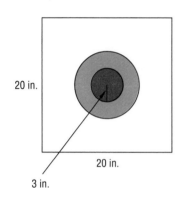

20 in.

20 in.

3 in.

Solution Recall that the area of a circle with radius r is πr^2. Compare the area of the bull's eye circle to the area of the whole board.

$$\frac{\text{area of bull's eye circle}}{\text{area of whole board}} = \frac{(\pi \cdot 3^2) \text{ sq in.}}{(20 \cdot 20) \text{ sq in.}}$$

$$= \frac{9\pi}{400}$$

$$\approx 0.071 \text{ or about } 7\%$$

In Example 1, the probability is the ratio of two areas. In Example 2, the probability is the ratio of two lengths.

Example 2 A, B, C, D, and E are exits on an interstate highway. If accidents occur at random between points A and E, what is the probability that an accident occurs between B and C?

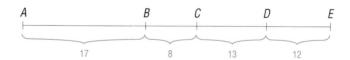

A   B   C   D   E

17   8   13   12

Solution Using the Putting-together Model of Addition, the length of $\overline{AE}$ is 50 miles.

$$\text{probability the accident is in } \overline{BC} = \frac{\text{length of } \overline{BC}}{\text{length of } \overline{AE}}$$

$$= \frac{8 \text{ miles}}{50 \text{ miles}}$$

$$= .16 \text{ or } 16\%$$

Here is an example which involves angle measure.

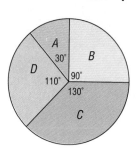

Example 3   At the left is a picture of a spinner like that used in many games. Suppose the spinner is equally likely to point in any direction. The spinner is spun once. What is the probability that it lands in region *A*?

Solution   The sum of the measures of all angles around the center of the circle is 360°. Probability (spinner lands in A) $= \frac{30°}{360°} = \frac{1}{12}$.

(Notice that in this example it is convenient to write the probability as a fraction.)

All of these examples have illustrated the following **Probability Formula for Geometric Regions.**

Probability Formula for Geometric Regions

If all points occur randomly in a region, then the probability P of an event is given by

$$P = \frac{\text{measure (area, length, etc.) of region for event}}{\text{measure of entire region}}$$

Example 4   A circle is inscribed in a square as shown here. If a point is selected at random from inside the square, what is the probability that it lies in the colored region (outside the circle)?

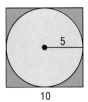

Solution   Probability of a point in shaded region =
$$\frac{\text{area of shaded region}}{\text{area of square}}$$

Using subtraction,

the area of the shaded region = area of square − area of circle

$$= \quad 10^2 \quad - \quad \pi 5^2$$

$$= \quad 100 \quad - \quad 25\pi$$

Thus the probability is $\dfrac{100 - 25\pi}{100}$

$$\approx \frac{100 - 25(3.14)}{100}$$

$$\approx \frac{21.5}{100}$$

$$\approx 21.5\%$$

Questions

Covering the Reading

1. Consider the dart board at the right. A person throws a dart. What is the probability that it is a bull's eye?
   a. Write your answer as a decimal.
   b. Write your answer as a percent.

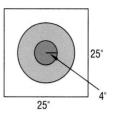

In 2 and 3, refer to Example 2. What is the probability an accident occurs

2. between exits *B* and *E*?   **3.** between exits *C* and *E*?

In 4 and 5, refer to Example 3.

4. What is the probability that the spinner will land in region *B*?

5. What is the probability that the spinner will land in region *C*?

6. In general, how can you calculate the probability of an event involving a geometric region?

7. Refer to Example 4.
   a. What is the area of the square?
   b. What is the area of the shaded region?
   c. What is the probability that a randomly selected point in the square lies inside the circle?
   d. If *p* is the probability that the point lies in the colored region and *q* is the probability that it lies in the circle, what is $p + q$? Why?

Applying the Mathematics

8. An electric clock with a second hand is stopped by a power failure. What is the probability that the second hand stopped between
   a. 12 and 3?
   b. 1 and 5?
   c. 11 and 1?

9. Suppose the larger circle of the dart board of Example 1 has a radius of 6 in.
   a. What is the probability that the dart lands inside the larger circle?
   b. What is the probability that the dart lands inside the larger circle but outside the bull's eye?

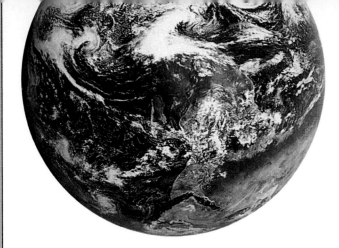

10. The land area of the earth is about 57,510,000 square miles and the water surface area is about 139,440,000 square miles. Give the probability that a meteor hitting the surface of the earth will:
a. fall on land;   **b.** fall on water.

11. In a rectangular yard of dimensions q ft by p ft, there is a garden of dimensions b ft by a ft. If a newspaper is thrown randomly into the yard, what is the probability that a point on it lands in the garden?

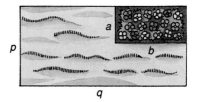

Review

12. Answer in your head. *(Lesson 5-4)*
    a. What is 25% of 60?   **b.** 50% of what number is 13?
    c. 8 is what percent of 24?

13. According to the census, in 1980 about 167 million people lived in urban areas in the U.S. while 59.5 million lived in rural areas. What percent of the U.S. population lives in a rural area? *(Lesson 5-4)*

14. A world record in the 100-meter dash was set by American Calvin Smith in 1983, at 9.93 seconds. The world record in the 200-meter dash, set by Italian Pietro Mennea in 1979, is 19.72 seconds.
    a. Find the average number of meters per second in each record run.
    b. By this measure, which runner is faster? Can you account for this?
    (Lesson 5-2)

In 15–17, perform the operation and simplify. *(Lessons 4-2, 5-1)*

15. $1\frac{5}{9} \div 2\frac{1}{7}$ **16.** $\dfrac{2x}{s} \div \dfrac{x}{10}$ **17.** $bd \cdot \dfrac{a}{b}$

18. In the U.S. army, a *squad* is usually 10 enlisted men. A *platoon* is 4 squads. A *company* is 4 platoons. A *battalion* is 4 companies. A *brigade* is 3 battalions. A *division* is 3 brigades. A *corps* is 2 divisions. A *field army* is 2 corps. How many enlisted men are there in a field army? *(Lesson 4-2)*

19. The volume of a box is to be more than 1700 cm³. The base has dimensions 8 cm by 15 cm. What are the possible heights of the box? *(Lesson 4-6)*

20. If $12(n - 7) = 48$, **a.** find the value of $n - 7$ and **b.** find the value of n. *(Lesson 4-4)*

21. *Skill sequence*   Evaluate. *(Lesson 4-9)*

    a. $10!$ **b.** $\dfrac{12!}{10!}$ **c.** $\dfrac{8!}{3!\,5!}$

22. *Skill sequence*   Solve. *(Lessons 2-6, 2-7, 4-4, 4-6)*
    a. $x + .7 = 1$ **b.** $x - .7 = 1$
    c. $.7x = 1$ **d.** $-.7x > -1$

23. Let x and y be whole numbers. Graph all solutions to $x + y = 11$. *(Lesson 3-7)*

Exploration

24. In 1760, the French mathematician Buffon discovered the following amazing property. Draw a set of parallel lines as close to exactly ℓ units apart as you can, where ℓ is the length of a needle. Drop the needle onto the lines and count how often it touches a line and how often it doesn't.

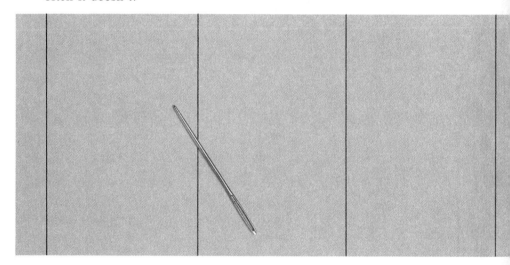

The probability the needle touches a line is $\dfrac{2}{\pi}$. That is,

$$\frac{\text{number of times needle touches line}}{\text{number of times needle is dropped}} = \frac{2}{\pi}.$$

Try Buffon's experiment, drawing the lines and dropping the needle at least 100 times. How close do you get to $\dfrac{2}{\pi}$?

Size Changes

Some photocopy machines can reduce a picture to 75% or 64% of its original size. They can also enlarge it to 120% of its original size. In these examples the numbers 75%, 64%, and 120% are called **size change factors**. Notice below that lengths and widths (not the area) of the top figure are multiplied by each size change factor.

Original figure

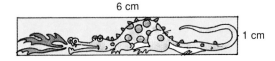

6 cm

1 cm

A. Size change factor of 75%:

   .75 · 6 = 4.5
   .75 · 1 = .75

4.5 cm

.75 cm

B. Size change factor of 64%:

   .64 · 6 = 3.84
   .64 · 1 = .64

3.84 cm

.64 cm

C. Size change factor of 120%:

   1.20 · 6 = 7.2
   1.20 · 1 = 1.20

7.2 cm

1.2 cm

These and similar situations lead to a **Size Change Model for Multiplication.**

Size Change Model for Multiplication:

If a quantity x is multiplied by a size change factor k, $k \neq 0$, then the resulting quantity is kx.

When the size change factor k is greater than 1 or less than -1, the size change is an **expansion**. Since 120% = 1.2, the size change pictured in C above illustrates an expansion. If k is between -1 and 1, the size change is a **contraction**. Since 75% = 0.75 and 64% = 0.64, the size changes pictured in A and B above are contractions.

Expansions and contractions also occur with figures in a coordinate plane. In geometric situations, the size change factor is called the **magnitude of the size change.**

Example 1 A quadrilateral in a coordinate plane has vertices (3, 0), (0, 3), (6, 6), and (6, 3). What is its image under a size change of magnitude $\frac{1}{3}$?

Solution Multiply the coordinates of each vertex by $\frac{1}{3}$ and draw the new quadrilateral.

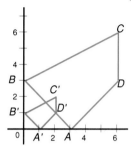

Preimage point	Image point
$A = (3, 0)$	$A' = (\frac{1}{3} \cdot 3, \frac{1}{3} \cdot 0) = (1, 0)$
$B = (0, 3)$	$B' = (\frac{1}{3} \cdot 0, \frac{1}{3} \cdot 3) = (0, 1)$
$C = (6, 6)$	$C' = (\frac{1}{3} \cdot 6, \frac{1}{3} \cdot 6) = (2, 2)$
$D = (6, 3)$	$D' = (\frac{1}{3} \cdot 6, \frac{1}{3} \cdot 3) = (2, 1)$

In this example, notice that under a size change of magnitude $\frac{1}{3}$, the image of (x, y) is $(\frac{1}{3}x, \frac{1}{3}y)$. In general, on the coordinate plane,

---

Multiplying coordinates of all points of a figure by k, $k \neq 0$, performs a size change of magnitude k.

In Example 1, the magnitude k is $\frac{1}{3}$. Since $-1 < \frac{1}{3} < 1$, the size change is a contraction. The image is smaller than the preimage. Negative size change factors are possible. In Example 2, the coordinates of the points are multiplied by -2. Notice that multiplying by a negative number changes the size and rotates the figure 180° about the point (0, 0). It turns the figure upside down.

Example 2 A figure in a coordinate plane has vertices (-2, 0), (-1, -4), (-2, -2), and (-4, -2). What is its image under a size change of magnitude -2?

Solution Multiply the coordinates of each vertex by -2 and draw the new quadrilateral.

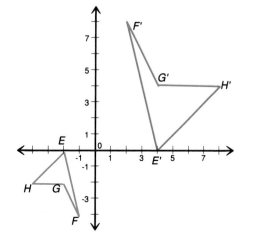

Preimage points
$E = (-2, 0)$
$F = (-1, -4)$
$G = (-2, -2)$
$H = (-4, -2)$

Image points
$E' = (4, 0)$
$F' = (2, 8)$
$G' = (4, 4)$
$H' = (8, 4)$

The size change in Example 2 is an expansion since -2 < -1. In an expansion the image is larger than the preimage. Because the magnitude is negative, the figure has been rotated 180°. Turn the page upside down to see that it looks like the original, only it is larger.

Examples 1 and 2 show that size changes do not change shape. Each angle and its image have the same measure. Also, the corresponding sides of the preimage and image are parallel. (However, the corresponding sides need not be parallel.) When two figures have the same shape, they are called **similar figures**. In a size change, the preimage and image are always similar figures.

Example 3   After a size change of magnitude 75%, a figure is 10 cm long. What was its original length?

Solution   75% · original length = 10 cm
Let L be the original length. Change 75% to 0.75. The equation becomes

$$0.75L = 10$$
$$\frac{0.75L}{0.75} = \frac{10}{0.75} \qquad \text{Divide both sides by 0.75.}$$
$$L = 13.\overline{3}$$

The original length was $13\frac{1}{3}$ cm.

Check   75% of 13.3 is just about 10. So the solution checks.

The idea of size change can involve quantities other than lengths.

Example 4   A window washer receives time-and-a-half for overtime. If he gets $14.10 for an overtime hour, what is his normal hourly wage?

Solution   Time-and-a-half means the overtime wage is $1\frac{1}{2}$ times the normal wage. Let W be the normal hourly wage.

$$1\frac{1}{2} \cdot W = \$14.10$$

That is, $1.5W = 14.10$

Divide both sides by 1.5. $W = \dfrac{14.10}{1.5}$

                                                       $W = 9.40$

The normal hourly wage is $9.40.

Check   Half of $9.40 is $4.70, and that added to $9.40 gives $14.10.

Questions

Covering the Reading

1. In the photocopy machine example at the beginning of this lesson, the magnitude of the size change factor in A is .75. What is the magnitude of the size change factor in B?

2. A(n) __?__ results from a size change of magnitude k where $-1 < k < 1$.

3. A(n) __?__ results from a size change of magnitude k where $k > 1$ or $k < -1$.

4. In a size change a preimage figure and its image are always __?__.

5. A picture with dimensions 10 inches by 15 inches is reduced on a photocopy machine with a factor of 64%. What are the dimensions of the reduced picture?

6. The image at the right is to be enlarged on a photocopy machine by a factor of 120%. How tall will the image be in the new diagram?

5.5 cm

7. State the size change model for multiplication.

8. What is the image of (3, -9) under a size change of magnitude -7?

9. What is the image of (x, y) under a size change of magnitude 8?

10. **a.** Graph the quadrilateral *ABCD* in Example 1.
    b. Graph its image under a size change of magnitude 2.
    c. Graph its image under a size change of magnitude -2.

11. $J'K'L'M'N'$ is the image of pentagon *JKLMN* under a size change of magnitude $\frac{3}{2}$.
    a. Find the coordinates of J', K', L', M' and N'.
    b. Graph pentagon $J'K'L'M'N'$.

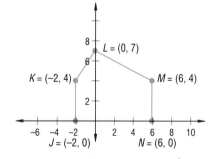

12. After a size change of 120%, a figure is 18 cm long.
    a. Write an equation using *L* for the original length.
    b. What was its original length?

13. A worker receives time-and-a-half for overtime. If the worker gets $11.70 for an overtime hour, what is the worker's normal hourly wage?

14. Lorenzo is one year old and weighs 21 lb. This is $3\frac{1}{2}$ times his birth weight. What did he weigh at birth?

15. A human hair is 0.1 mm thick. Under a microscope it appears 15 mm thick. How many times is it magnified?

16. Under a size change of magnitude 6, the image P' of a point P is (9, -42). What are the coordinates of P?
$(\frac{3}{2}, -7)$

17. a. Describe the graph of the image of the quadrilateral at the right under a size change of magnitude 1.
  b. What property of multiplication does this size change represent?

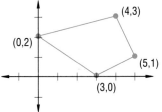

(4,3)

(0,2)

(5,1)

(3,0)

18. Refer to the poster below.
  a. What would it cost for a party of five people to fly to Hawaii?
  b. What would it cost for a party of p people? (Assume $p \geq 3$.)

19. A football field has measurements given at the right. If a balloon floats down to a random spot on the field, what is the probability it will land in the darker area of play? *(Lesson 5-5)*

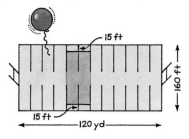

20. In 1985, 93.5 million of the 110 million workers in the U.S. commuted by car. What percent is this? *(Lesson 5-5)*

21. A scarf normally sells for $23.95. It is on sale for 30% off.
  a. What percent of the price does the customer pay?
  b. Write an equation to find the sale price.
  c. How much does the scarf cost on sale? *(Lesson 5-5)*

In 22–24, solve. *(Lessons 4-4, 4-6)*

22. $-6.5x = -117$ **23.** $-117 \geq -6.5x$

24. $19 = \frac{x}{4}$ (Hint: rewrite $\frac{x}{4}$ as a multiplication.)

25. Solve for y: $xy = 53$.

In 26–28, **a.** write an equation, and **b.** solve.

26. Five-eighths of what number is 80? *(Lesson 4-4)*

27. Five-eighths more than what number is 80? *(Lesson 2-6)*

28. Fifty-eight more than what number is 80? *(Lesson 2-6)*

In 29–31, estimate without a calculator. *(Previous course)*

29. $\frac{1}{4}$ of 398   **30.** $987,925 \div 991$ **31.** 35% of 24

32. Find the value of $-9a(b - 2)$ when $a = 0$ and $b = -20$. *(Lesson 4-3)*

33. Suppose $x < 0$ and $y < 0$.
  a. Is $-x \cdot -y$ positive or negative?
  b. Is $-x + -y$ positive or negative? *(Lessons 2-2, 4-3)*

Hawaiian alphabet

34. A manufacturer of copy machines advertises that its new machine will produce 3 times as many copies per minute as brand A's machine. Brand A's machine produces c copies per minute. In 4 minutes, how many copies will the new machine produce? *(Lesson 4-2)*

35. How many different two-letter "words" are possible using the 12 letters of the Hawaiian alphabet? *(Lesson 4-7)*

36. Draw a rectangle. Now draw a new rectangle whose sides are $\frac{1}{4}$ the length of sides of the first rectangle. How many of the small rectangles can fit in the large rectangle?

5-7

Proportions

A **proportion** is a statement that two fractions are equal. In general, any equation of the form

$$\frac{a}{b} = \frac{c}{d}$$

is called a proportion. The numbers b and c are the **means** of the proportion. The numbers a and d are the **extremes** of the proportion. For example, $\frac{30}{x} = \frac{6}{7}$ is a proportion. Its means are 6 and x. Its extremes are 30 and 7. The equation $\frac{x}{2} + \frac{5}{4} = \frac{1}{2}$ is not a proportion because the left side of the equation is not a single fraction.

Suppose two fractions are equal. We show a specific case and the general idea.

$$\frac{9}{12} = \frac{30}{40} \qquad\qquad \frac{a}{b} = \frac{c}{d}$$

Multiply both sides by the product of the denominators. $12 \cdot 40 = 480$, and $b \cdot d = bd$.

$$480 \cdot \frac{9}{12} = 480 \cdot \frac{30}{40} \qquad\qquad bd \cdot \frac{a}{b} = bd \cdot \frac{c}{d}$$

Simplifying leads to another equation using the same numbers as in the original fraction.

$$40 \cdot 9 = 12 \cdot 30 \qquad\qquad da = bc$$

The **Means-Extremes Property** enables you to shorten your work by skipping the middle step.

Means-Extremes Property:

For all real numbers a, b, c, and d (b and d nonzero),

$$\text{if } \frac{a}{b} = \frac{c}{d}, \text{ then } ad = bc.$$

Many questions can be answered by writing a proportion from two equal ratios or two equal rates.

Example 1   A motorist traveled 283.5 miles on 9 gallons of gas. With the same driving conditions, how far could the car go on a full tank of 14 gallons of gas?

Solution   Let x be the number of miles traveled on 14 gallons. Since conditions are the same, $\frac{\text{mi}}{\text{gal}}$ using 9 gallons $= \frac{\text{mi}}{\text{gal}}$ using 14 gallons.

$$\frac{283.5 \text{ miles}}{9 \text{ gallons}} = \frac{x \text{ miles}}{14 \text{ gallons}}$$

$$\frac{283.5}{9} = \frac{x}{14}$$

Use the Means-Extremes property.

$$283.5 \cdot 14 = 9x$$
$$441 = x$$

The car could travel 441 miles on 14 gallons.

The above example was set up by equating *rates*. Proportions can also be set up using equal *ratios*. In Example 2 below, Solution 1 uses rates. Solution 2 uses ratios.

Example 2 In the map below, 1 inch represents 880 miles. On the map the distance between New York and Miami is $1\frac{1}{4}$ inches. How far apart are the cities?

Solution 1 Let x be the actual distance between the cities. Set up a proportion using equal rates in inches per mile.

$$\frac{1 \text{ inch}}{880 \text{ miles}} = \frac{1\frac{1}{4} \text{ inches}}{x \text{ miles}}$$
$$1 \cdot x = 880 \cdot 1\frac{1}{4}$$
$$x = 1100$$

The cities are 1100 miles apart.

0   440   880
└────────┴────────┘
        miles

Scale: 1 inch = 880 miles

Solution 2 Set up a proportion using equal ratios. One ratio compares inches, the other miles.

$$\frac{1 \text{ inch}}{1\frac{1}{4} \text{ inches}} = \frac{880 \text{ miles}}{x \text{ miles}}$$

$$1 \cdot x = 880 \cdot 1\frac{1}{4}$$

$$x = 1100$$

The cities are 1100 miles apart.

Often several correct proportions can be used to solve a problem. There are two other proportions which could have been used to find the distance from New York to Miami.

$$\frac{880 \text{ miles}}{1 \text{ inch}} = \frac{x \text{ miles}}{1\frac{1}{4} \text{ inches}} \quad \text{or} \quad \frac{1\frac{1}{4} \text{ inches}}{1 \text{ inch}} = \frac{x \text{ miles}}{880 \text{ miles}}$$

Each gives the same final answer.

Questions

Covering the Reading

1. *True* or *false*   A proportion results when two fractions are equal.

In 2–5, tell whether the expression or sentence is a proportion.

2. $2 + \frac{x}{3} = \frac{1}{5}$

3. $\frac{a}{d} = \frac{b}{x}$

4. $\frac{a}{b} \cdot \frac{c}{d}$

5. $\frac{2}{3} + \frac{4}{5}$

6. According to the Means-Extremes Property, if $\frac{x}{y} = \frac{z}{w}$, then __?__.

7. A motorist keeps records of his car's gas mileage. The last time he filled the tank, he had gone 216 miles on 13.8 gallons of gas. At this rate, how far can the car go on a full tank of 21 gallons?

8. On the map in Example 2, the distance between Seattle and San Francisco is $\frac{3}{4}$ in.
   a. Write two proportions that can be used to find the distance between the two cities.
   b. What is the distance?

9. For parts a–c, write the equation that results from using the Means-Extremes Property.
   a. $\frac{3}{5} = \frac{n}{7}$
   b. $\frac{3}{7} = \frac{n}{5}$
   c. $\frac{7}{5} = \frac{n}{3}$
   d. Which of the proportions in part a, b, or c has a different solution than the others?

10. If $\frac{3}{4x} = \frac{5}{11}$ then $3 \cdot 11 =$ __?__.

A picture is worth a thousand words.

In 11 and 12, **a.** write the equation that you get from using the Means-Extremes Property; **b.** solve. Write your answer as a fraction.

11. $\dfrac{x}{7} = \dfrac{3}{11}$

12. $\dfrac{5}{9} = \dfrac{3m}{2}$

Applying the Mathematics

13. If a picture is worth a thousand words, what is the worth of 325 words?

14. A basketball team scores 17 points in the first 6 minutes of play. At this rate, how many points will the team score in a 32-minute game?

15. About 0.75 inch of rain fell in 3 hours. At this rate, how many inches of rain will fall in 9 hours?

In 16 and 17, solve using the Means-Extremes Property. Write your answer as a fraction.

16. $\dfrac{2}{a} = \dfrac{-14}{15}$

17. $\dfrac{2}{b} = 7$

The Means-Extremes Property can be used to determine whether fractions are equal. The fractions $\dfrac{a}{b}$ and $\dfrac{c}{d}$ will be equal only if $ad = bc$. In 18 and 19, tell whether the given fractions are equal.

18. $\dfrac{1}{3}, \dfrac{33}{100}$

19. $\dfrac{4.5}{-5}, \dfrac{-153}{170}$

20. a. If $A \neq 0$ and $B \neq 0$, solve $\dfrac{A}{B} = \dfrac{C}{x}$ for x.

b. Complete line 40 so the program solves $\dfrac{A}{B} = \dfrac{C}{X}$, when the values for A, B, and C are entered.

```
10 PRINT "SOLVE PROPORTION A/B = C/X"
20 PRINT "ENTER A, B AND C"
30 INPUT A, B, C
40 LET X =  ?
50 PRINT "X IS "; X
60 END
```

21. **a.** Find the image of (-3, 6) under a size change of magnitude $-\frac{5}{3}$.
    b. Is the size change an expansion or a contraction? *(Lesson 5-6)*

22. **a.** Copy the figure at the right. Draw the image of figure *QRSTU* under a size change of magnitude 3. Call the image $Q'R'S'T'U'$.
    b. What is the length of $\overline{QR}$?
    c. What is the length of $\overline{Q'R'}$?
    (Lesson 5-6)

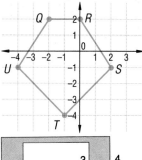

23. What is the probability of a randomly picked point being in the colored region at right? *(Lesson (5-5)*

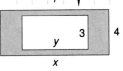

24. *Karat* is a measure of fineness used for gold and other precious materials. Pure gold is 24-karats. Gold of 18-karat fineness is 18 parts pure gold and 6 parts other metals giving 24 parts altogether.
    a. A bracelet is 18-karat gold. What percent gold is this? (Hint: Write a ratio.)
    b. A necklace is 10-karat gold. What percent gold is this? *(Lesson 5-3)*

25. In the 1988 congressional election, Senator Lloyd Bentsen of Texas got 3,120,348 votes out of 5,232,113 cast. Express his part of the votes as **a.** a ratio; **b.** a percent. *(Lesson 5-3)*

26. What value(s) can *m not* have in the expression $\dfrac{m - 7}{m + 5}$? *(Lesson 5-2)*

27. Evaluate $\dfrac{y_2 - y_1}{x_2 - x_1}$ when $x_1 = 4$, $y_1 = 2$, $x_2 = 3$, and $y_2 = 5$. *(Lesson 1-5)*

28. Solve. *(Lessons 2-6, 4-4, 4-5)*
    a. $p + 8 = 23$ **b.** $8p = 23$
    c. $8 - p = 23$ **d.** $\frac{1}{8}p = 23$

29. The statement, "driver is to car as pilot is to airplane" is called an *analogy*. This analogy corresponds to the proportion $\dfrac{a}{b} = \dfrac{c}{d}$, where a = driver, b = car, c = pilot and d = airplane. Here is a list of analogies. Solve for the missing word.
    a. Soup is to bowl as water is to __?__.
    b. Inch is to centimeter as __?__ is to kilogram.
    c. __?__ is to earth as earth is to sun.
    d. Cow is to __?__ as hen is to chick.
    e. Hoop is to __?__ as net is to soccer.
    f. Washington is to first as Reagan is to __?__.
    g. __?__ is to Maryland as Sacramento is to California.
    h. Motorist is to car as __?__ is to bicycle.

30. Make up two analogies similar to those in Question 29.

Similar Figures

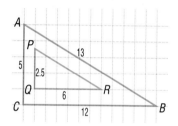

In Lesson 5-6, you learned that with expansions and contractions, the image and the preimage are similar. At the left $\triangle PQR$ is the image of $\triangle ABC$ under a contraction of magnitude $\frac{1}{2}$. So the triangles are similar. In this lesson you will learn more about what happens when two figures are similar.

A segment and its image are corresponding sides. Notice what happens when we write ratios to compare the lengths of two of the three pairs of corresponding sides. PQ is the length of $\overline{PQ}$.

$$\frac{PQ}{AC} = \frac{2.5}{5} = \frac{1}{2} \qquad\qquad \frac{QR}{CB} = \frac{6}{12} = \frac{1}{2}$$

The ratios are equal: $\dfrac{PQ}{AC} = \dfrac{QR}{CB}$. This suggests the following generalization.

In similar figures, ratios of lengths of corresponding sides are equal.

A ratio of corresponding sides for two similar figures is called a **ratio of similitude**. This ratio is the same as the magnitude of a size change between the figures. It can represent an expansion or a contraction. In the figure at the left, the ratio of similitude is $\frac{1}{2}$ or 2. This ratio equals the magnitude of the contraction.

■ ■ ■ ■ ■ ■ ■ ■

Example 1   In the figure at the top left, find PR.

Solution $\overline{PR}$ and $\overline{AB}$ are corresponding sides. $\overline{PR}$ is smaller.

Since the triangles are similar, $\dfrac{PR}{AB} = \dfrac{QR}{CB}$.

Substitute for AB, QR, and CB. $\dfrac{PR}{13} = \dfrac{6}{12}$

Apply the Means-Extremes Property. $12 \cdot PR = 78$
Divide both sides by 12. $PR = 6.5$

Check $\dfrac{PR}{AB} = \dfrac{6.5}{13} = 0.5$, which is $\frac{1}{2}$.

Example 2 The two quadrilaterals below are similar. The figures have been drawn with corresponding sides parallel. (This means that the lines containing the corresponding sides never intersect or that the corresponding sides lie on the same line.) Find *CD*.

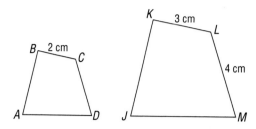

Solution You want to find the length of $\overline{CD}$, so find the side corresponding to it. That is $\overline{LM}$. Now find corresponding sides whose lengths you know. These are $\overline{BC}$ and $\overline{KL}$. Since the figures are similar, the ratios of lengths of these corresponding sides are equal.

$$\frac{CD}{LM} = \frac{BC}{KL}$$

Substitute.

$$\frac{CD}{4} = \frac{2}{3}$$

Use the Means-Extremes Property.

$$3 \cdot CD = 2 \cdot 4$$

$$CD = \frac{8}{3} \text{ cm}$$

Check 1 *CD*, at $\frac{8}{3}$ cm, is bigger than *BC*. This is fine, because *LM* is bigger than *KL*.

Check 2 Find the ratios of similitude first using *BC* and *KL*, then using *CD* and *LM*.

$$\frac{BC}{KL} = \frac{2}{3} \qquad \frac{CD}{LM} = \frac{\frac{8}{3}}{4} = \frac{8}{3} \div 4 = \frac{8}{3} \cdot \frac{1}{4} = \frac{8}{12} = \frac{2}{3}$$

They are equal.

Using similar figures, you can find the height of an inaccessible object.

Harold Hanking wanted to find the height h of the flagpole in front of Hatcher Heights High. Here is how he did it. He held a yardstick parallel to the flagpole and measured the length of its shadow. Then he measured the length of the shadow of the flagpole. He drew the following picture.

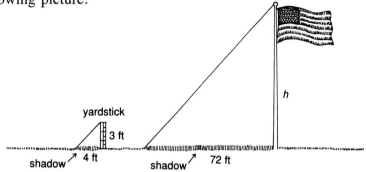

yardstick

3 ft

h

shadow 4 ft   shadow 72 ft

He recognized that the two triangles were similar. Then using ratios of corresponding sides he wrote:

$$\frac{3}{h} = \frac{4}{72}$$
$$4h = 72 \cdot 3$$
$$4h = 216$$
$$h = 54$$

The flagpole is 54 feet tall.

Questions

Covering the Reading

1. In similar figures, __?__ of lengths of corresponding __?__ are equal.

2. Refer to the triangles at the beginning of the lesson. *True or false*
$\frac{CB}{QR} = \frac{AC}{PQ}$.

3. The two triangles below are similar. Corresponding sides are parallel. Which side of $\triangle BIG$ corresponds to: **a.** $\overline{AT}$; **b.** $\overline{CT}$?

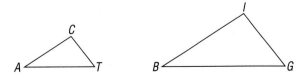

4. How can you calculate a ratio of similitude for two similar figures?

5. Refer to these similar quadrilaterals. Corresponding sides are parallel.

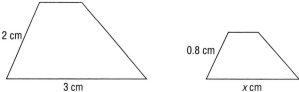

a. Write a proportion that could be used to find x.
b. Solve the proportion you wrote in part a.
c. What are two possible ratios of similitude?

6. A tree casts a shadow that is 14 ft long. A yardstick casts a shadow that is 2.25 ft long.

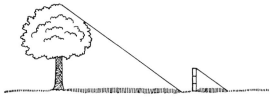

a. Copy this diagram and put in the given lengths.
b. Write a proportion that describes this situation.
c. How tall is the tree?

7. Write four equal ratios for these similar figures. Corresponding sides are parallel.

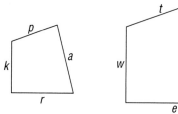

8. The quadrilaterals below are similar. Corresponding sides are shown in the same color. Write a proportion and solve for x.

9. On a sunny day Jim, who is 6 feet tall, casts a shadow that is 10 feet long. A nearby tree, which is t feet tall, casts a shadow that is 25 feet long.
a. Draw a diagram of this situation. Write in the lengths.
b. Write a proportion that describes this situation.
c. How tall is the tree?

10. Triangles *ABC* and *DEF* are similar. Corresponding sides are parallel.

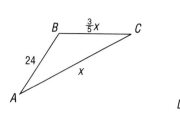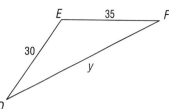

a. *Multiple choice*   Which proportion should you use to find x?

I. $\dfrac{24}{30} = \dfrac{x}{y}$ II. $\dfrac{\frac{3}{5}x}{y} = \dfrac{x}{35}$ III. $\dfrac{24}{30} = \dfrac{\frac{3}{5}x}{35}$

b. Solve the equation you chose for x.

c. Use your solution to part **b** to find y.

d. Based on your answers to parts b and c, what is the ratio of the *perimeter* of *ABC* to the *perimeter* of *DEF*?

11. Use the scale drawing of a house shown at the left. The actual width of the base of the house is 30 ft. Use a ruler and properties of similar figures.

a. Write a fraction comparing the width of the house in the drawing to the actual width of the house.

b. Write a proportion you could use to find the actual height to the peak of the roof.

c. Solve your answer to part b.

12. Pentagon *ABCDE* is given. $\overline{PT}$ is given and corresponds to $\overline{AE}$. Copy the second drawing and draw the complete similar pentagon *PQRST*.

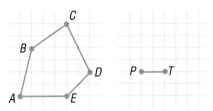

Review

In 13–16, solve. *(Lessons 2-3, 5-7)*

13. $\frac{1}{2} + A = \frac{3}{4}$

14. $\dfrac{3}{B} = \dfrac{9}{12}$

15. $\dfrac{2}{3} = \dfrac{5}{C}$

16. $\dfrac{48}{2x} = \dfrac{9}{5}$

17. This year there are 1250 students in Harwood Junior High School. Five years ago there were only 800 students. Calculate the rate of growth of the school population during this period. *(Lesson 5-2)*

$\left(\text{Hint: rate of growth} = \dfrac{\text{change in population}}{\text{change in time}}\right)$

18. Refer to the sketches in Question 12. What is a ratio of similitude of the two pentagons? *(Lesson 5-6)*

19. Pixley is 38 kilometers by road from Mayberry. If a bicyclist stops at random on a trip between the two towns, what is the probability of her stopping within $\frac{1}{2}$ kilometer of the old barn? *(Lesson 5-5)*

20. *Multiple choice* $150.75, or 35%, of the yearbook committee's income comes from advertising. What equation describes the total income T of the committee? *(Lesson 5-4)*
(a) $.35T = 150.75$ (b) $.35(150.75) = T$
(c) $.35 = 150.75T$ (d) $\dfrac{.35}{150.75} = T$

21. A sweater on sale costs $17. It originally cost $21. What is the percent of discount? *(Lesson 5-3)*

22. Jan walked 0.4 mile to the bus stop. She rode the bus to town and walked 0.7 mile to work. In the evening she took the same route home. If her round trip is 11.4 miles, how many miles does she ride on the bus each day? *(Lesson 2-7)*

23. Simplify: $\dfrac{3\pi}{y^2} \div \dfrac{\pi}{5}$. *(Lesson 5-1)*

24. How much would you pay for 7 records at $6.98 each? *(Lesson 4-5)*

25. A menu includes 4 appetizers, 6 entrees, and 3 desserts. How many different meals are possible with one entree and either an appetizer or a dessert? *(Lesson (4-7)*

26. Carla rode two dolphins for a distance of 100 yd. The trip took S seconds. What was their average rate of travel in ft/sec? *(Lesson 5-2)*

In 27–29, solve. *(Lessons 2-7, 3-3, 4-6)*

27. $3a \le 45$

28. $k + 7 > -25$

29. $2b - b - 11 \le 58 + 6$

30. Find the highest point of a tree, or a building, or some other object, using the shadow method described in this lesson.

Summary

Division is closely related to multiplication. The definition of division states that to divide by a number is the same as multiplying by its reciprocal. This definition is directly applied to divide fractions. Because zero has no reciprocal, division by zero is impossible.

Rates and ratios are models for division. A rate compares quantities with different units; ratios compare quantities with the same units. An equation with two equal rates or ratios is called a proportion. The Means-Extremes Property can be used to find missing values in a proportion. One important use of proportions is in similar figures.

Percent, probability in geometry, and size changes are applications involving rates and ratios. It is possible to translate percent problems to equations of the form $ab = c$. Solving the equation then gives an answer to the problem. Since it is impossible to count the infinite number of points in a geometric region, a ratio of lengths or areas is used to compute probabilities in geometric situations. Size changes yield similar figures. If the magnitude is k, where $k > 1$ or $k < -1$, then the size change is an expansion. If $-1 < k < 1$, the size change is a contraction.

Vocabulary

Below are the most important terms and phrases for this chapter. You should be able to give a general description and a specific example of each.

Lesson 5-1
Algebraic Definition of Division

Lesson 5-2
Rate Model for Division
IF-THEN command

Lesson 5-3
ratio
Ratio Model for Division
percent of discount

Lesson 5-4
percent, %

Lesson 5-5
Probability Formula for Geometric Regions

Lesson 5-6
size change factor
Size Change Model for Multiplication
expansion, contraction
magnitude of a size change
similar figures

Lesson 5-7
proportion
means, extremes
Means-Extremes Property

Lesson 5-8
ratio of similitude

Progress Self-Test

Take this test as you would take a test in class. You will need a calculator and graph paper. Then check your work with the solutions in the Selected Answers section in the back of the book.

In 1–4, simplify.

1. $15 \div -\frac{3}{2}$

2. $\frac{x}{9} \div \frac{2}{3}$

3. $\frac{2b}{3} \div \frac{b}{3}$

4. $\dfrac{\frac{4}{7}}{\frac{3}{m}}$

In 5 and 6, solve.

5. $\frac{y}{11} = \frac{2}{23}$

6. $\frac{10}{3y} = \frac{5}{6}$

7. If 14% of a number is 60, what is the number?

8. 4.5 is 25% of what number?

9. $\frac{1}{2}$ is what percent of $\frac{4}{5}$?

10. Dividing by 8 is the same as multiplying by what number?

11. In $\frac{2}{3} = \frac{x}{10}$, which terms are the means?

12. What value can v not have in the expression $\frac{10v}{v + 1}$?

13. *True or false*   Using the Means-Extremes Property is an appropriate first step in solving the equation.

$$d + \tfrac{4}{5} = \tfrac{2}{3}.$$

14. Horatio spent 36 minutes on his English homework. Mary Ellen spent a half hour on her English homework.
   a. Horatio's time is what percent of Mary Ellen's time?
   b. Horatio studied __?__ percent longer than Mary Ellen.

15. A cat has a life expectancy of c years. A dog has a life expectancy of d years. If $c < d$, then the life expectancy for a dog is how many times that of a cat?

16. Which is faster, reading p pages in $7y$ minutes or p pages in $8y$ minutes?

17. The profit on an item with a selling price of $40 is $16. What percent of the selling price is profit?

18. If the electricity goes out and a clock stops, what is the probability that the second hand stops between 2 and 3?

19. Suppose your class covers three chapters of Algebra in 35 school days. At this rate, how many school days will it take to cover 13 chapters?

20. A car travels 250 miles on 12 gallons of gas. At this rate, about how far (to the nearest mile) can the car travel on 14 gallons of gasoline?

21. Here is a computer program. For what input value will the computer print 2?

```
10 INPUT B
11 IF B=0, PRINT "IMPOSSIBLE"
12 IF B <> 0, PRINT 10/B
13 END
```

22. a. Graph the triangle with vertices (2, 3), (-1, -3), and (3, 1).
   b. Graph its image under a size change of -2.

23. The two triangles below are similar. Corresponding sides are parallel. Find z.

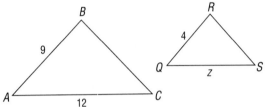

24. The pentagons below are similar. Corresponding sides are parallel. Find x.

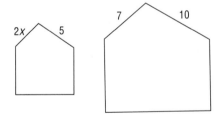

25. When Sasha stands in the back center of a tennis court she can cover (return the ball from) a rectangle 20 feet by 25 feet. Each side of the court is 27 feet by 39 feet. Find P(Sasha returns the ball), assuming her opponent's shot randomly bounces anywhere inside Sasha's side of the court.

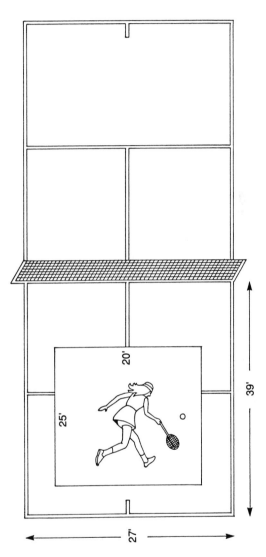

Chapter Review

Questions on **SPUR** Objectives

SPUR stands for **S**kills, **P**roperties, **U**ses, and **R**epresentations.
The Chapter Review questions are grouped according to the
SPUR Objectives for this chapter.

SKILLS deal with the procedures used to get answers.

Objective A. *Divide real numbers and simplify division expressions. (Lesson 5-1)*

In 1–10, simplify.

1. $25 \div \frac{1}{5}$

2. $\dfrac{60}{\frac{1}{4}}$

3. $\frac{3}{4} \div \frac{7}{8}$

4. $\dfrac{a}{14} \div \dfrac{7}{2}$

5. $\dfrac{2\pi}{9} \div \dfrac{\pi}{6}$

6. $\dfrac{\frac{3}{2}}{-\frac{15}{16}}$

7. $\dfrac{\frac{a}{b}}{\frac{c}{d}}$

8. $\dfrac{x}{-\frac{1}{4}}$

9. $\frac{4}{5} \div \frac{2}{5}$

10. $\dfrac{x}{z} \div \dfrac{x}{y}$

Objective B. *Solve percent problems using equations or in your head. (Lesson 5-4)*

11. What is 75% of 32?

12. 10 is what percent of 5?

13. What is 20% of 18?

14. 85% of what number is 170?

15. 12 is what percent of 36?

16. 105% of 64 is what number?

17. 30% of a number is $\frac{3}{4}$. What is the number?

18. 1.2 is what percent of 0.8?

Objective C. *Solve proportions. (Lesson 5-7)*

In 19-24, solve.

19. $\dfrac{x}{130} = \dfrac{6}{5}$

20. $\dfrac{6}{25} = \dfrac{-10}{m}$

21. $\dfrac{2w}{-5} = 3$

22. $\dfrac{4}{x} = 7.5$

23. $\dfrac{-1.1}{y} = \dfrac{2.3}{0.4}$

24. $\dfrac{24}{12} = \dfrac{3b}{10}$

PROPERTIES deal with the principles behind the mathematics.

Objective D. *Identify restrictions on a variable in a division situation. (Lesson 5-2)*

In 25–28, what value(s) can the variable not have?

25. $\dfrac{3}{2y}$

26. $\dfrac{x+1}{x+4}$

27. $\dfrac{15x}{x-.2}$

28. $\dfrac{y}{y+\frac{1}{2}}$

Objective E. *Use the language of proportions and the Means-Extremes Property. (Lessons 5-7, 5-8)*

29. In $\frac{5}{8} = \frac{15}{24}$,
 a. Which numbers are the means?
 b. Which numbers are the extremes?

30. If $\dfrac{m}{n} = \dfrac{p}{q}$, then by the Means-Extremes Property $\underline{\ ?\ } = \underline{\ ?\ }$.

31. If $\dfrac{2}{3} = \dfrac{x}{5}$, then $\dfrac{5}{x} = \underline{\ ?\ }$.

32. If $\dfrac{a}{b} = \dfrac{c}{d}$, then $\dfrac{b}{a} = \underline{\ ?\ }$.

■ **Objective F.** *Use the rate model for division.* (*Lesson 5-2*)

In 33–35, ask a question and calculate a rate to answer the question.

33. The Johnsons drove 30 miles in $\frac{3}{4}$ hours.

34. In *d* days Tony lost $400.

35. Four weeks ago the puppy weighed 3 kilograms less.

In 36 and 37, assume a 22-mile bike trip took 2 hours.

36. What was the rate in miles per hour?

37. What was the rate in hours per mile?

38. A train travels from Newark to Trenton, a distance of 48.1 miles, in 30 minutes. At what average speed, in miles per hour, does the train travel?

39. Marlene worked $3\frac{1}{2}$ hours and earned $14. How long did it take her to earn a dollar?

40. In one store a 46-ounce can of tomato juice costs $1.03, and a 6-ounce can costs 23 cents.
   a. Calculate the unit cost of each can.
   b. Based on the unit cost, which is the better buy?

41. Which is faster, reading *w* words in *m* minutes or 6*w* words in 2*m* minutes?

42. *Multiple choice* Suppose *s* sweaters cost *d* dollars. At this rate how many sweaters can be bought for $75?

   (a) $\dfrac{s}{75d}$ (b) $\dfrac{75s}{d}$

   (c) $\dfrac{75d}{s}$ (d) $\dfrac{d}{75s}$

■ **Objective G.** *Use a ratio to compare two quantities.* (*Lesson 5-3*)

43. In the first period algebra class, there are 10 girls in a class of 27 students. Write the ratio of boys to girls.

44. An item selling for $36 is reduced by $6. What is the percent of the discount?

45. The profit on an item selling for $20 is $8. What percent of the item's selling price is profit?

46. David paid *k* dollars for a tennis racket. Joe paid *j* dollars for the same racket on sale. How many times as much did David pay as Joe paid?

47. Sue charges $1.50 an hour to babysit. Lou charges $2.00 an hour.
   a. Lou charges what percent of Sue's price?
   b. Lou charges ___?___ percent more than Sue.
   c. Sue charges ___?___ less than Lou.

■ **Objective H.** *Solve percent and size change problems from real situations.* (*Lessons 5-4, 5-6*)

48. A sofa is on sale for $450. It originally cost $562.52. What percent of the original price is the sale price?

49. In 1983, 49% of all accidental deaths occurred in motor vehicle accidents. If 91,000 people died accidentally, how many were killed in motor accidents?

50. In 1987, Kevin McHale from Boston had the highest field goal shooting percentage in professional basketball, 60.4%. If he attempted 1307 shots, how many did he make?

51. A $15.99 tape is on sale for $11.99. To the nearest percent, what is the percent of discount?

52. A sales clerk at Clark's Department Store receives time-and-a-half for overtime. If normal time is $9.00/hr, what is the overtime pay?

53. Model trains of HO-gauge are $\frac{1}{87}$ actual size (no fooling!). If a model locomotive is 30 cm long, how long is the real locomotive it models?

■ **Objective I.** *Solve problems involving proportions in real situations.* *(Lesson 5-7)*

54. Anne was saving for a class trip. For every $35 that Anne earned her mother added an extra $15. If Anne earned $245, how much would her mother add?

55. A $\frac{3}{4}$ cup of sugar is equivalent to 12 tablespoons of sugar. How many tablespoons are there in 3 cups of sugar?

56. A family decided to keep track of the number of phone calls made. In the first seven days of November, 45 calls were made. At this rate, about how many calls will be made for the month?

57. For every dollar, you could get 2.03 deutsche marks (the currency in West Germany) in 1986. If a crystal vase cost 120 deutsche marks in 1986, what would the cost be in dollars?

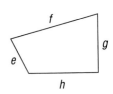

REPRESENTATIONS deal with pictures, graphs, or objects that illustrate concepts.

■ **Objective J.** *Find probabilities involving geometric regions.* *(Lesson 5-5)*

58. A 3-cm square inside a 4-cm square is drawn at the right. If a point is selected at random from the figure, what is the probability it lies in the shaded region?

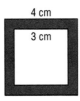

4 cm
3 cm

59. At the right is a picture of a spinner. Suppose any position of the spinner is equally likely. What is the probability that the spinner lands in region *A* or *B*?

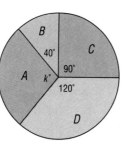

60. In a storm, the electricity went out. A clock stopped. What is the probability that its second hand stopped between the 5 and the 7?

61. If accidents occur randomly on the roads from town *A* to town *B* as shown below, what is the probability that an accident on these roads occurs on the 8 km stretch of road shown in color?

5 km
2 km
8 km
1 km
B
A

■ **Objective K.** *Find missing lengths in similar figures.* *(Lesson 5-8)*

In 62 and 63, refer to the similar figures below. Corresponding sides are parallel.

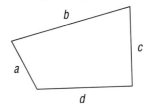

b
c
a
d
f
g
e
h

62. *Multiple choice* Which proportion is *incorrect*?

(a) $\frac{d}{h} = \frac{c}{g}$ (b) $\frac{e}{a} = \frac{f}{b}$

(c) $\frac{g}{c} = \frac{a}{e}$ (d) $\frac{h}{d} = \frac{f}{b}$

63. If $a = 12$, $b = 15$, and $e = 10$, what is f?

64. Refer to the similar triangles below. Corresponding sides are shown in the same color.

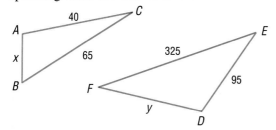

a. What is the ratio of similitude?
b. Find the length of $\overline{AB}$.
c. Find the length of $\overline{DF}$.

65. The quadrilaterals below are similar. Corresponding sides are parallel.
a. Solve for y.
b. Solve for x.

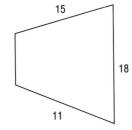

66. A man 6 feet tall casts a shadow 15 feet long. At the same time the shadow of a tree is 140 feet long. How tall is the tree?

67. A tree casts a shadow that is 9 feet long. A yardstick casts a shadow n feet long. How tall is the tree?

◼ **Objective L.** *Use the IF-THEN command in computer programs.* *(Lesson 5-2)*

68. What will be printed if the input for the program below is 8?

```
10 PRINT "EVALUATE AN EXPRESSION"
20 PRINT "ENTER VALUE OF A"
30 INPUT A
40 IF A = 0 THEN PRINT "IMPOSSIBLE"
50 IF A <> 0 THEN PRINT (40 + A)/A
60 END
```

69. Write a program to evaluate $\dfrac{k + 1}{k - 2}$ for any value of k that is entered. If the input makes the expression meaningless, the computer should print IMPOSSIBLE.

◼ **Objective M:** *Apply the Size Change Model for Multiplication on the coordinate plane.* *(Lesson 5-6)*

70. Give the image of (2, 4) under a size change of magnitude 3.

71. Give the image of (-8, -12) under a size change of magnitude $-\frac{1}{4}$.

In 72 and 73, copy the figure below. Then draw its image under the size change with the given magnitude.

72. $\frac{1}{2}$

73. -3

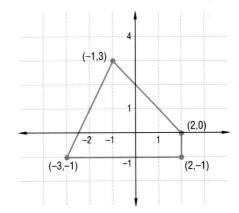

CHAPTER 6

Linear Sentences

A **linear sentence** is one in which the variable or variables are all to the first power. Examples are

$$3x + 1000 = 2800$$
$$9a - 2(b - a) = 15$$
$$-\tfrac{2}{3}t \leq 12$$
$$m + n > 43.61$$

Most of the sentences you solved in preceding chapters are linear sentences.

An enormous variety of situations can lead to linear sentences. For instance, answers to all four questions below can be found by solving

$$3x + 1000 = 2800.$$

1. You want to go to Africa to see the wild animals. You can get $1000 from your savings and your family, but you need $2800. If you save $3 a week, how many weeks will it take you to save up for the trip?

2. There is $1000 in a school fund for a big dance. Tickets will be sold for $3 apiece. How many tickets will need to be sold to cover the anticipated cost of $2800 for the band, the food, the decorations, and the publicity?

3. There are 1000 students in a school in an area that is growing quickly. Each day it seems that about 3 more students are entering the school. The school building has a capacity of 2800. In about how many days will the school be at capacity?

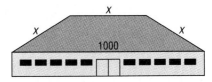

4. The perimeter of the roof line of the warehouse above is 2800 feet. If the front edge is 1000 feet and the other three edges have the same length, what is the length of each of the other edges?

In this chapter, you will learn methods which enable the solving of $3x + 1000 = 2800$ and many other linear sentences. This will give you the power to answer many questions about real situations.

6-1

Solving $ax + b = c$

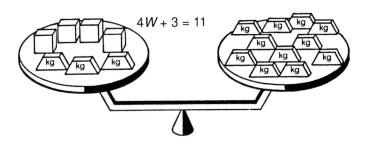

An equation is like a balance scale. Here is a picture of the equation $4W + 3 = 11$. On the left side of the scale are 4 boxes and 3 one-kilogram weights. They balance with the 11 kilograms on the right.

You can find the weight of one box in two steps. Each step keeps the scale balanced.

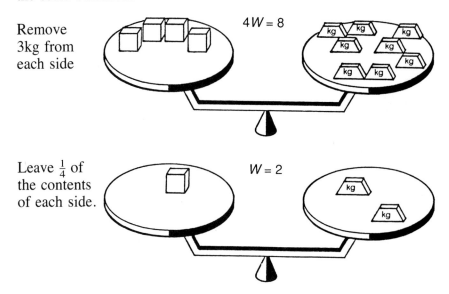

Remove 3kg from each side

Leave $\frac{1}{4}$ of the contents of each side.

Example 1 shows the same steps without the balance scale.

Example 1 Solve: $4W + 3 = 11$

Solution $4W = 8$ Addition Property of Equality (added -3 to each side)

            $W = 2$ Multiplication Property of Equality (multiplied both sides by $\frac{1}{4}$)

Check   Substitute 2 for W in the original equation. Does $4 \cdot 2 + 3 = 11$? Yes.

Any equation like the one in Example 1 can be solved in two steps. First add a number to both sides. Then multiply both sides by a number. The only tasks are to determine the numbers and to do the arithmetic correctly. To help with these tasks, you may put in work steps showing what you did. Work steps are shown in Example 2, which is the equation for the situations on page 261.

Example 2   Solve: $3x + 1000 = 1800$

Solution   Here is a work step, showing that -1000 is added to each side.

$$3x + 1000 + \text{-}1000 = 1800 + \text{-}1000$$
$$3x = 800$$

Another work step shows that both sides are multiplied by $\frac{1}{3}$.

$$\frac{1}{3} \cdot 3x = \frac{1}{3} \cdot 800$$
$$x = \frac{800}{3}$$
$$= 266\frac{2}{3}$$

Check   Substitute $266\frac{2}{3}$ for x in the original equation.

Does $3 \cdot 266\frac{2}{3} + 1000 = 1800$? Yes, it does.

Using Example 2, you can answer the questions on page 261. For Question 2, 267 tickets will need to be sold. You are asked to answer the other questions yourself.

Equations involving subtraction are solved using the same two steps. Again we show work.

Example 3   Solve: $7x - 2 = -23$

Solution   Add 2 to both sides. $7x - 2 + 2 = -23 + 2$
Simplify. $7x = -21$
Multiply both sides by $\frac{1}{7}$, or do $\frac{7x}{7} = \frac{-21}{7}$
the equivalent, divide by seven.
                                         $x = -3$

Check   Substitute -3 for x in the original equation.
Does $7(-3) - 2 = -23$?
Yes, since $21 - 2 = -23$.

Covering the Reading

1. The boxes are of equal weight. Each weight marked kg is 1 kilogram.
   a. What equation is pictured by this balance scale?

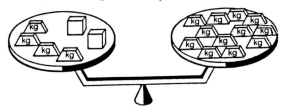

   b. What two steps can be done with the weights on the scale to find the weight of a single box?

2. a. When solving $4n + 8 = 60$, first add __?__ to both sides. Then __?__ both sides by __?__.
   b. Solve and check: $4n + 8 = 60$

3. Here is an equation solved, but the work steps are not put in.

   Given: $55v - 61 = 434$
   Step 1: $55v = 495$
   Step 2: $v = 9$

   a. What was done to get Step 1?
   b. What was done to get Step 2?

In 4–7, solve and check.

4. $8x + 15 = 47$ **5.** $7y - 11 = 52$

6. $n + 5 = -6$ **7.** $2z + 32 = 288$

8. *Multiple choice*   Which equation does *not* have the same solution as $3x + 5 = 9$?
   (a) $5 + 3x = 9$ (b) $9 = 5 + 3x$
   (c) $9 = 3 + 5x$ (d) All have same solutions.

In 9–12, solve and check.

9. $312 = 36w + 60$ **10.** $22 + 11a = 66$

11. $7 + 8t = 207$ **12.** $-81 = 2x + 5$

13. *Multiple choice*   Which is the solution to $3a - 11 = -40$?
   (a) $-\frac{51}{3}$ (b) $-\frac{29}{3}$ (c) $\frac{51}{3}$ (d) $\frac{29}{3}$

In 14 and 15, solve and check.

14. $5y - 6 = 2$ **15.** $7x + 11 = -4$

16. Answer Question 1 on page 261.

17. Answer Question 3 on page 261.

18. Answer Question 4 on page 261.

Answer Question 3 on page 261.

Answer Question 4 on page 261.

In 19 and 20, use these formulas relating shoe size S and foot length L in inches.

Applying the Mathematics

In 19 and 20, use these formulas relating shoe size S and foot length L in inches.

$$\text{for men:} \quad S = 3L - 26$$
$$\text{for women:} \quad S = 3L - 22$$

19. If Sam wears a size 9 shoe, about how long are his feet?

20. If Bernice wears a size 6 shoe, about how long are her feet?

21. a. Simplify: $-2x + 5x + 16$.
  b. Solve: $-2x + 5x + 16 = 46$.

Review

22. To the nearest integer, what is the solution to $1.34m = 13.5$? *(Lesson 4-4)*

23. Evaluate $(-7)^3 + (-6)^2 + (-5)^1$. *(Lesson 1-4)*

24. If a discount is 10%, then the sale price is $100\% - 10\% = 90\%$ of the regular price.
  a. Copy and complete the chart. *(Lesson 3-2)*

discount	sale price (as % of regular price)
10%	90%
25%	?
50%	?
?	30%
$n\%$	?

  b. Graph the ordered pairs from the first four rows of part a. Label the horizontal axis "discount" and the vertical axis "sale price." *(Lesson 3-7)*

In 25 and 26, when $x = -4$ and $y = -5$, find the value of each expression. *(Lesson 1-4)*

25. $-5 - 4xy$

26. $-4x + 5y^2$

27. Solve in your head. *(Lesson 2-6)*
  a. $20 = 35 + x$
  b. $20 + y = 35$
  c. $20 + (z - 30) = 35$

28. There are 162 games in the major league baseball season. In 1987, Eric Davis of the Cincinnati Reds had 20 home runs after 55 games. If he continued at this rate, would he have broken the all-time record of 61 home runs set by Roger Maris of the New York Yankees in 1961? *(Lesson 5-7)*

29. Joe is 5 feet tall. He stands so that the top of his shadow hits exactly the same place as the end of the flagpole's shadow. If his shadow is 8 feet long and the flagpole's shadow is 18 feet long, how high is the flagpole? *(Lesson 5-7)*

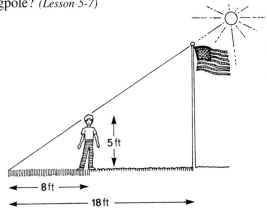

Exploration

30. A person usually wears shoes of the most comfortable size. Referring to the formula of Question 19, describe *all* the lengths of feet that would be best fit by a shoe of size 9.

266

The three examples of the last lesson are all of the same form.

$$4w + 3 = 11$$
$$3x + 1000 = 1800$$
$$7x - 2 = -23$$

The general form is $ax + b = c$

where the equation is to be solved for x. All equations of this type can be solved with the two major steps you used in the last lesson, even when a, b, and c are fractions or decimals.

More Solving $ax + b = c$

Example 1   Solve: $\frac{2}{3}x - 16 = 11$

Solution   Begin as you did in Lesson 6-1. Add 16 to both sides.

$$\frac{2}{3}x - 16 + 16 = 11 + 16$$

$$\frac{2}{3}x = 27$$

Multiply both sides by $\frac{3}{2}$. $\frac{3}{2} \cdot \frac{2}{3}x = \frac{3}{2} \cdot 27$

$$x = \frac{81}{2}$$
$$= 40\frac{1}{2}$$

Check   Substitute 40.5 for x. Does $\frac{2}{3} \cdot 40\frac{1}{2} - 16 = 11$?

$$\frac{81}{3} - 16 = 11?$$
$$27 - 16 = 11? \text{ Yes.}$$

Example 1 and all the examples of the last lesson use the same two steps to solve equations of the form $ax + b = c$.

To solve $ax + b = c$ for x, add $-b$ to both sides. Then multiply both sides by $\dfrac{1}{a}$.

Sometimes fractions and decimals occur in the same problem. Then it is convenient to convert all to either decimals or fractions.

Example 2   A formula relating Fahrenheit and Celsius temperatures is

$$F = \frac{9}{5}C + 32.$$

What is the Celsius equivalent of normal body temperature, 98.6°F?

Solution Substitute 98.6 for F and solve for C.

$$98.6 = \tfrac{9}{5}C + 32$$

We change $\tfrac{9}{5}$ to a decimal because the decimal terminates. $\tfrac{9}{5} = 1.8$. So the problem becomes

$$98.6 = 1.8C + 32.$$

Add -32 to each side.

$$66.6 = 1.8C$$

Multiply both sides by $\tfrac{1}{1.8}$.

$$\frac{66.6}{1.8} = C$$
$$37° = C$$

Check Substitute 98.6 for F and 37 for C in the original formula. Does $98.6 = \tfrac{9}{5} \cdot 37 + 32$?

$$= \tfrac{333}{5} + 32?$$
$$= 66.6 + 32? \text{ Yes.}$$

In fact, in almost the entire world, 37° is the value used for normal body temperature, not 98.6°, because the U.S. is the only major country that still measures temperature using the Fahrenheit scale.

Often equations are complicated but can be simplified into ones that you can solve. Simplifying each side is an important strategy for all equation solving.

Example 3 When Val works overtime at the zoo on Saturday she earns $9.80 per hour. She is also paid $8.00 for meals and $3.00 for transportation. Last Saturday she was paid $77.15. How many hours did she work?

Solution Let $h =$ the number of hours Val worked. She earned 9.80h dollars in those hours. So

$$9.80h + 8.00 + 3.00 = 77.15.$$

Simplify the left side and the equation looks more familiar.

$$9.80h + 11 = 77.15$$
$$9.80h + 11 + \text{-}11 = 77.15 + \text{-}11$$
$$9.80h = 66.15$$
$$\frac{9.80h}{9.80} = \frac{66.15}{9.80}$$
$$h = 6.75$$

Val worked $6\tfrac{3}{4}$ hours.

Check If she worked 6.75 hours at $9.80 per hour, she earned
6.75 · 9.80 dollars. That comes to $66.15. Now add $8 for meals and
$3 for transportation. The total is $77.15, as desired.

The coefficient of x in $ax + b = c$ can be negative. Still the equation is solved in the same way.

Example 4 Solve: $-5x + 1025 = 685$

> **Solution** Add -1025 to each side. $-5x + 1025 + -1025 = 685 + -1025$
> $$-5x = -340$$
>
> Divide each side by -5.
> $$\frac{-5x}{-5} = \frac{-340}{-5}$$
> $$x = 68$$

Check Does $-5 \cdot 68 + 1025 = 685$? Yes, $-340 + 1025 = 685$.

Questions

Covering the Reading

1. To solve an equation of the form $ax + b = c$, first add __?__ to both sides. Then __?__ both sides by __?__.

2. **a.** To solve $4 + 2.5x = 21.5$, what is an appropriate first step?
   b. Solve this equation.

In 3 and 4, solve and check.

3. $\frac{2}{3}x + 15 = 27$ 4. $32 = 0.2y - 3$

5. What is the Celsius equivalent of a room temperature of 68°F?

6. If an equation is complicated, what important strategy of equation solving should you follow to solve it?

7. **a.** What should be the first step in solving $3.5 + 2x + 5.6 = 10$?
   b. Solve and check this sentence.

In 8 and 9, solve and check.

8. $8 + 27 + 4p + 1 = -44$ 9. $6 - 30n - 18 = 69$

10. Refer to Example 3. If Val's pay two Saturdays ago was $89.40, how many hours did she work?

In 11 and 12, solve and check.

11. $-4x + 12 = -100$ 12. $2 = -8y - 1$

FEMUR

FEMUR

In 13 and 14, use this information. Lengths of human bones are related by linear sentences. These can be used to estimate heights of individuals if certain bones are found. Archaeologists, paleontologists, and forensic scientists all use these techniques. Here are formulas relating the length f of the femur (thigh bone) to the height h of a person, both in centimeters.

$$\text{for men:} \quad h \approx 69.089 + 2.238f$$
$$\text{for women:} \quad h \approx 61.412 + 2.317f$$

13. A 160-cm tall female has a femur of about what length?

14. Clues from footprints show that a man was 180 cm tall. A partial skeleton is found in which the femur is 50 cm long. Could this be a matchup?

15. Solve for x: $ax + b = c$.

16. A copy machine begins work with 1025 sheets of paper. It uses $5t$ sheets each second. After how many seconds will 685 sheets be left?

In 17–20, solve. Do at least one of these in your head. *(Lesson 6-1)*

17. $9x + 40 = 40$

18. $2 + 5z = 3$

19. $-2y - 11 = 63$

20. $100 = 20h + 160$

21. When the singing group The Three Blungets goes on tour, they earn $1260 per concert. The group also gets $10 per mile for travel expenses. Their last tour covered 3500 miles.
  a. How much did the group get for travel expenses?
  b. If their earnings and money for travel expenses totaled $100,000, how many concerts did they give? *(Lesson 6-1)*

22. Simplify: **a.** $2y + y + 16 + y$ **b.** $(2y)(y)(16)(y)$ *(Lessons (2-2, 4-1))*

23. Find the area of the right-angled figure at the left. *(Lesson 4-1)*

24. $\angle A$ and $\angle B$ are supplementary. $m\angle A = 20°$. Find $m\angle B$. *(Lesson 3-8)*

2 in.

0.75 in.

0.75 in.

1.5 in.

25. Three vocations are mentioned in Questions 13 and 14. What do these people do?

26. Give four linear sentences of the form $ax + b = c$ whose solution is $\frac{2}{3}$.

LESSON

6-3

The Distributive Property

You know that $6x + 8x$ equals $14x$. This property, an instance of adding like terms, is one example of a more general property called the **Distributive Property.**

Distributive Property:

For any real numbers a, b, and c:

$$ac + bc = (a + b)c$$
$$ac - bc = (a - b)c.$$

The Distributive Property enables you to solve difficult-looking equations.

■ ■ ■ ■ ■ ■ ■ ■ ■

Example 1   Solve: $-3x + 4 + 5x = -6$

Solution   First use the Commutative Property of Addition to put like terms next to each other.

$$-3x + 5x + 4 = -6$$

Now apply the Distributive Property.

$$(-3 + 5) x + 4 = -6$$
$$2x + 4 = -6$$

Now solve as usual.

$$2x + 4 + -4 = -6 + -4$$
$$2x = -10$$
$$x = -5$$

Recall the Multiplicative Identity Property of 1. For any real number n,

$$n = 1n.$$

When this property is combined with the Distributive Property, many expressions can be simplified. For instance, $k + 8k = 1k + 8k = 9k$. Here are some problems involving 1 as a coefficient.

Example 2 A $140,000 estate is to be split among three children and a grand-child. Each child gets the same amount and the grandchild gets one half as much. How much should each child receive?

Solution Let c represent each child's portion. Then $\frac{1}{2}c$ is the grand-child's portion.

$$c + c + c + \tfrac{1}{2}c = 140{,}000$$

Use the Multiplication Property of 1.

$$1c + 1c + 1c + \tfrac{1}{2}c = 140{,}000$$

Use the Distributive Property and change $\frac{1}{2}$ to .5.

$$3.5c = 140{,}000$$

Divide each side by 3.5.

$$c = 40{,}000$$

Each child should receive $40,000.

Check The grandchild receives half as much as each child does, and so gets $20,000.
Does 40,000 + 40,000 + 40,000 + 20,000 = 140,000? Yes.

The arithmetic in Example 2 could have been made easier by thinking of all amounts "in thousands." The equation would become $3\frac{1}{2}c = 140$. It would still be solved in the same way.

In Examples 3 and 4, the original price of an item is raised or lowered by a given percent. The idea of percent change can be generalized. If an item is 15% off, you pay 85%, which is 100% − 15%. The 15% is called a **discount.** If the item were 33% off, you would pay 100% − 33% or 67% of the price. If there is a 4% **tax,** you pay 104% of the price. If there is an 8% tax, you pay 108% of the price. The 4% or 8% is an example of a **markup.**

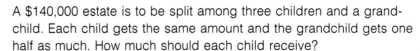

Example 3 A microwave oven is on sale for 15% off the regular price. The sale price is $279.65. What is the regular price?

Solution Let m be the regular price of the microwave.

$$\text{sale price} = \text{regular price} - 15\% \text{ of regular price}$$
$$279.65 = m - .15m$$

Remember $m = 1m$. $279.65 = (1 - .15)m$

$$279.65 = .85m$$
$$\frac{279.65}{.85} = \frac{.85m}{.85}$$
$$329 = m$$

The regular price is $329.

Check   If the regular price is $329, the discount is $0.15 \cdot \$329$, or $49.35.
The sale price is $329 − $49.35 = $279.65. This checks.

■ ■ ■ ■ ■ ■ ■ ■

Example 4   The Richardsons bought a new van. The total amount they were charged was $11,864. Included in this amount was 4% sales tax and $60 for licenses. What was the cost of the van before the tax?

Solution   Let V be the cost of the van. Then

$$\text{cost} + 4\% \text{ of cost} + \$60 = \$11864$$
$$V + .04V + 60 = 11864$$
$$1.04V + 60 = 11864$$
$$1.04V + 60 - 60 = 11864 - 60$$
$$1.04V = 11804$$
$$\frac{1.04V}{1.04} = \frac{11804}{1.04}$$
$$V = 11350$$

The cost of the van before tax and licenses was $11,350.

Check   4% of $11,350 is $454; $11,350 + 454 + 60 = 11,864$

In general:

If an item is **discounted** $x\%$, you pay $(100 - x)\%$ of the price.

If an item is **marked up** or **taxed** $y\%$, you pay $(100 + y)\%$ of the price.

Questions

Covering the Reading

In 1–4, simplify one side of the equation. Then solve.

1. $h + h = 41$

2. $.75y + y + 10 = 45$

3. $g + .04g + 60 = 8796$

4. $x - .20x = \$35.00$

5. \$90,000 is to be divided among three heirs. Two of the heirs each receive equal amounts. The third heir receives one fourth that amount. Write an equation and find how much money each heir will receive.

6. In Question 5, suppose that \$10,000 is given to charity before the heirs get any money. Write an equation and solve it to determine how much each heir will receive.

7. In Example 3, suppose the sale price of another microwave is \$350. What is the regular price?

8. *Multiple choice* The price of a bicycle is B. The total paid T includes 6% sales tax. Which equation describes this situation?
 (a) $B + .06 = T$
 (b) $.06B = T$
 (c) $1.06B = T$

9. Julian went out for dinner. The bill came to \$11.55. Included in this amount was a 5% sales tax. What was the cost of the meal?

10. If a coat is discounted 30%, you pay __?__ of the original price.

Applying the Mathematics

11. Some taxicab companies allow their drivers to keep $\frac{3}{10}$ of all fares collected. The rest goes to the company. If a driver collects F dollars in fares, write an expression for how much the company gets.

In 12–14, **a.** write an equation that can help answer the question and **b.** use your equation to answer the question.

12. The population of Arlington, Texas was about 214,000 in 1984 after a growth of 33.6% from 1980 to 1984. What was Arlington's population in 1980?

13. Customers at Flo's Meat Market must pay 6% state sales tax. One day Flo takes in \$2650, including tax. How much of the \$2650 does Flo have left after sending the tax money to the state?

14. In a boxing match, \$250,000 goes to the promoter and, of the rest, the loser will receive one fifth of what the winner will get. The total purse is \$750,000. How much will each boxer receive?

15. A suit has a regular price of \$189.00. The final discounted price is \$90.00. What percent *discount* is this?

Review

In 16–21, solve. Do at least one of these in your head. *(Lessons 6-1, 6-2)*

16. $3a - 2 = 13$

17. $7 + 5x = 24$

18. $\frac{1}{2}t + 4 = -2$ **19.** $-4y + 30 = 12$

20. $6B - \frac{2}{3} = \frac{3}{4}$ **21.** $-c + 3.4 = 6.21$

22. You estimate that a trip to South America to see relatives will cost $1000 for air fare and $60 a day for living expenses.
  a. What will it cost to stay n days?
  b. How long can you stay for $1500? *(Lesson 6-1)*

23. Use the formula $F = \frac{9}{5}C + 32$ to find the Celsius equivalent of 32°F. *(Lesson 6-2)*

24. Simplify $-q \cdot -r \cdot -r$. *(Lesson 4-3)*

25. The average mass of air molecules is $30 \cdot 1.66 \cdot 10^{-24}$ grams. Write this number in scientific notation. *(Appendix)*

26. Calculate in your head using the division fact $1625 \div 25 = 65$. *(Previous course)*
  a. $162.5 \div 25$
  b. $16.25 \div 25$
  c. $1.625 \div 25$

27. What are the coordinates of the point 2 units to the right and 4 units down from (x, y)? *(Lesson 2-5)*

Exploration

28. In 1985, when IBM announced that it would no longer market the PC Jr. computer, *USA Today* reported:
> "Omni Computer Inc. of Des Moines, Iowa, cleared out its last 12 PC Jrs. in one day by announcing it would cut the price 10% (of the previous price) every hour, starting at the list of $1269, until the store closed at 5 P.M."

Suppose the store opened at 9 A.M. and the first price cut occurred at 10 A.M.
  a. What price would be charged at 1:15 P.M.?
  b. What price would be charged at 4:49 P.M.?
  c. Write an expression that gives the cost after n hours.

Repeated Addition and Subtraction

PARKING CHARGES

1 HOUR	$1.50
2 HOURS	2.30
3 HOURS	3.10
4 HOURS	3.90
5 HOURS	4.70

each additional
hour $ .80

Equations of the form $ax + b = c$ often arise from situations of repeated addition or subtraction. For instance, consider the rates charged by a parking garage in a big city, as shown in the sign at the left. If a person left his car in the garage for a week, what would the parking charge be?

With 24 hours per day for 7 days, the person would be charged for 168 hours. Surely we don't want to add 80 cents each hour all these times.

A good strategy is to make a table to find the pattern. In this situation, we work back from the sign to figure out what 0 hours would cost. Then

Hours	Charges	Cost Pattern
0	.70 =	$.70 + .80 \cdot 0$
1	.70 + .80 =	$.70 + .80 \cdot 1$
2	.70 + .80 + .80 =	$.70 + .80 \cdot 2$
3	.70 + .80 + .80 + .80 =	$.70 + .80 \cdot 3$
4	.70 + .80 + .80 + .80 + .80 =	$.70 + .80 \cdot 4$

Notice how the hours and cost pattern are related. The number on the far right is the number of hours. This signals an expression for the cost of parking t hours

$$t \qquad .70 + .80t$$

To find the charge for 168 hours, substitute 168 for t in the cost pattern.

$$168 \qquad .70 + .80 \cdot 168 = \$135.10$$

The pattern of costs can be used to answer questions about the parking situation.

Example 1 An uncle of yours parks in this garage and is charged $11.10 for parking. How many hours did he park?

Solution The cost is .70 + .80t for parking t hours. So solve

$$.70 + .80t = 11.10.$$

You have solved equations like this in previous lessons. We omit the work steps.

Add -.70 to both sides.	$.80t = 10.40$
Divide both sides by .80.	$t = 13$

Your uncle parked there for 13 hours.

Check You should convince yourself that the answer is reasonable.

The method in Example 1 can be applied to find a formula for any situation involving repeated addition or subtraction of a constant quantity.

Example 2 Find the 88th term in the sequence below where each term is 6 more than the previous one.

$$7, 13, 19, 25, 31, 37, \ldots$$

Solution Make a table. We need a term before the first term, a row 0.

term number	term
0	1
1	$7 = 1 + 6$
2	$13 = 1 + 6 \cdot 2$
3	$19 = 1 + 6 \cdot 3$
4	$25 = 1 + 6 \cdot 4$
5	$31 = 1 + 6 \cdot 5$
6	$37 = 1 + 6 \cdot 6$
$\vdots$	$\vdots$

In the table, the number 6 is added n times to get the nth term. So

$$n \qquad\qquad 1 + 6n$$

To find the 88th term in the sequence, substitute 88 for n.

$$88 \qquad\qquad 1 + 6 \cdot 88 = 529$$

The 88th term is 529.

In Example 2, $1 + 6n$ is an expression for the nth term in the sequence. This expression enables you to answer some questions easily.

Example 3   In the sequence 7, 13, 19, 25, 31, 37, ..., of Example 2, what term is 595?

Solution   The nth term is $1 + 6n$. Here,

$$1 + 6n = 595$$
$$6n = 594$$
$$n = 99.$$

595 is the 99th term.

In the next example, each term decreases by a certain amount. Since we know the beginning amount, no row 0 is needed.

Example 4   A newspaper publishing company buys paper by the boxcar load. It now has 8000 tons of paper. Each week the company uses 475 tons of paper. How much paper will be left n weeks from now?

Solution

End of week	Paper left (tons)
1	$8000 - 475 \quad\; = 7525$
2	$8000 - 475 \cdot 2 = 7050$
3	$8000 - 475 \cdot 3 = 6575$
4	$8000 - 475 \cdot 4 = 6100$

The pattern is now easy to see.

n	$8000 - 475n$

After n weeks there will be $8000 - 475n$ tons of paper unless a new shipment comes in.

The examples in this lesson involve sequences in which a constant amount is repeatedly added or subtracted. Computers are most useful when jobs require speed or a lot of repetition. The following computer program will make a list of the first 200 terms of the sequence in Example 2. The program uses a **FOR/NEXT loop.** The FOR statement tells the computer the number of times to execute the loop. The first time through the loop, N is 1. Each time through, N increases by one. The NEXT statement sends the computer back to the FOR statement.

```
10 PRINT "TERMS OF A SEQUENCE"
20 FOR N = 1 TO 200    The loop will be executed 200 times.
30  LET T = 1 + 6 * N   The formula from Example 2 is evaluated.
40  PRINT T             Print the value of the term.
50 NEXT N               Go back to line 20 with a new value for N.
60 END
```

When the program is run, the computer will print 200 terms. The first three terms printed are: The last three terms are:

```
    7
   13
   19
----------
 1189
 1195
 1201
```

Questions

Covering the Reading

In 1–3, use the parking lot cost pattern .70 + .80t.

1. a. What does the t in the formula represent?
   b. Where does .80 in the formula come from?
   c. What would it cost to park 50 hours in this garage?

2. a. Does the formula work for parking one hour?
   b. Does it work for $2\frac{1}{2}$ hours?
   c. What is the domain for t in the formula?

3. A person is charged $35.10 for parking. How long did the person park?

In 4 and 5, refer to Example 2.

4. Find the 72nd term in the sequence.

5. Which term is 493?

In 6 and 7, refer to Example 4.

6. How many tons of paper are left at the end of the 10th week?

7. The company orders new paper when down to 4000 tons. In what week will that occur?

In 8 and 9, refer to the computer program in this lesson.

8. Suppose line 20 read FOR N = 1 TO 99. How many times would the loop be executed?

9. Find the value of T in line 30 for N = 86.

Applying the Mathematics

10.

term number	term
1	5
2	14
3	23
4	32
5	41
6	50

a. What is the first term?
b. How much larger is each term than the previous one?
c. What is the 9th term?
d. Write an expression for the nth term.

11. 1000, 997, 994, 991, 988, 985,
   a. Write an expression for the nth term of the sequence. (Hint: Begin with a row 0.)
   b. Use your expression to find the 56th term of the sequence.
   c. 400 is which term of the sequence?

12. One large city school system with 140 school buses uses 7500 gallons of gas per day.
   a. Make a chart showing gas consumption for 1, 2, 3, 4, 5 and 6 days.
   b. Find a formula for gas consumed in n days.
   c. Calculate the number of gallons of gas consumed in 90 days.
   d. Write a computer program that will print a list of the gallons of gas consumed for 1 through 20 days.

13. These graphs picture situations discussed in the lesson.

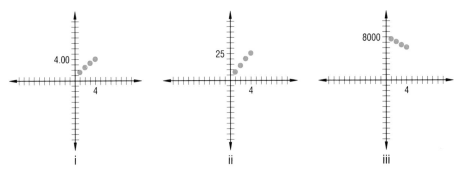

a. Which of these graphs represents the situation in Example 1?
b. Which of these graphs represents the situation in Example 2?
c. Which of these graphs represents the situation in Example 4?
d. What do these graphs have in common?

In 14 and 15, simplify. *(Lessons 2-2, 6-3)*

14. $3 + \text{-}2p + \text{-}2q - 7 - 3$ **15.** $\frac{1}{5} - \frac{2}{3}p + \frac{5}{6}p$

In 16 and 17, solve. *(Lessons 2-2, 6-3)*

16. $3x - 5x + 12 + \text{-}15 = \text{-}4$ **17.** $x - .1x = 1.8$

18. After a 15% discount, a suit costs $144.46. What was the original price? *(Lesson 6-3)*

19. In $\triangle ABC$, $m\angle A = x$, $m\angle B = 2x$, and $\angle C$ is a right angle. What are the measures of the three angles of the triangle? *(Lesson 3-8)*

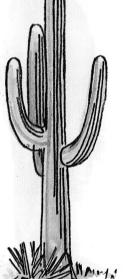

In 20–23, solve. *(Lessons 2-7, 4-6, 6-1)*

20. $6 + 4t = \text{-}6$ **21.** $\frac{5}{2}x - 1 = \frac{1}{2}$

22. $3 + y < 50$ **23.** $3z < 231$

24. Roberto traveled 53 mph for $2\frac{1}{2}$ hours. How many miles did he travel? *(Lesson 4-2)*

25. Refer to the graph below. Douglas, Arizona is typical of a monsoon area, an area in which a wind system causes yearly rain to be concentrated in a few months. *(Lesson 2-4)*
   a. During what months is the average rainfall in Douglas less than 1 in.?
   b. What months have the highest average rainfall?
   c. Give the average rainfall for April in Douglas.
   d. Between which two consecutive months is there the greatest change in rainfall?

Average monthly rainfall in Douglas, Arizona (inches)

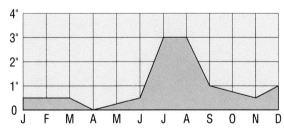

26. A picture frame is enlarged from a width of 4 inches to a width of 6 inches. Find the magnitude of the size change factor. *(Lesson 5-6)*

27. Calculate in your head. *(Lesson 4-2)*
   a. $4 \cdot \frac{3}{4}$ **b.** $12 \cdot \frac{3}{4}$
   c. $12 \cdot 1\frac{3}{4}$ **d.** $12 \cdot \frac{4}{3}$

28. Write a computer program to ask for the values of A, B, and C and then print the solution to $Ax + B = C$.

Mid-Chapter Review

You need to be able to solve equations of the form $ax + b = c$ in order to solve the equations of the next lessons. You also need to be able to translate situations into expressions of the form $ax + b$. This lesson gives you practice should you need it. There are no new ideas.

To solve equations of the form $ax + b = c$, add $-b$ to both sides. Then multiply both sides by $\frac{1}{a}$. Try these.

1. $7a + 40 = 33$ **2.** $46 + 32b = 1230$

3. $-3c - 2 = -4$ **4.** $-100 = 6D + 500$

5. $0 = 7e + 3$ **6.** $-f + 36 = 50$

The same algorithm works if a, b, and c are decimals or fractions. Solve these.

7. $\frac{2}{3}g + \frac{1}{3} = \frac{7}{3}$ **8.** $\frac{3}{5}h - 2 = 13$

9. $8.8 + 3.1i = 42.9$ **10.** $1024 = 1000 + .06j$

11. $-\frac{1}{2}k + 3 = 4$ **12.** $8L - 0.4 = 9.24$

Sometimes one or both sides of an equation need to be simplified to get the equation in the form $ax + b = c$.

13. $11m + 12m + 13m + 14 = 15$

14. $31 = 2n - 3n + 4$

15. $p + .2p = 120$

16. $7.46 = q - .25q$

17. $1000 + r + 3r + 3r = 9400$

18. $5s + 55 = 555 + 5555$

Many situations lead to equations of this form. For these, first write an equation. Then solve it.

19. With a 6% tax, a sofa cost $720.80. How much did the sofa cost without tax?

20. After a 30% discount, a compact disc player cost $139.30. What was the original cost of the player?

21. There is $150 in a fund for a school dance. Tickets will be $4.50 apiece. How many persons must attend in order to cover the estimated expenses of $400?

22. The formula relating shoe size S and foot length L in women is $S = 3L - 22$. What is the approximate length of a foot of a woman who wears a size 7 shoe?

23. Connie babysat last weekend and earned $8.50 for sitting for 4 hours. This included an extra $1.50 because she took some long phone messages. What does she charge for sitting?

24. If the height h of a man and the length f of his femur (both in centimeters) are approximately related by the formula $h = 69.089 + 2.238f$, what should be the length of a femur of a 6-ft man? (Note: Remember that 1 inch = 2.54 cm. You will have to first calculate the number of cm in 6 feet.)

25. The Judson family used 160 New Year's cards this year. Six cards were spoiled; the rest were mailed. Mrs. Judson addressed twice as many as her husband and three times as many as her son. How many cards did her son address?

26. If you work 40 hours a week for $5.25 an hour and make time-and-a-half for overtime, how long must you work in order to earn $250 in a week? (Hint: Let h be the number of hours of overtime.)

Repeated addition and subtraction situations also lead to equations of the form $ax + b = c$. In these situations, first write a formula for a general term. Then use the formula to answer the question.

27. A sequence begins 3, 11, 19, 27, 35, ..., and every term is 8 more than the preceding. Which term of the sequence is 331?

28. A parking garage charges $.30 for the first hour and $.20 each succeeding hour. How long was a car parked if the cost of parking is $1.90?

29. If you began the school year with 1000 sheets of notebook paper and are using 6 sheets a day, in how many days will you have only 100 sheets left?

30. The number of court cases in a city was 729 in 1985 and has been increasing by about 15 a year since then. In what year, if this trend continues, will the number of cases hit 1000? (Hint: Let the unknown n equal the number of years after 1985.)

Solving $ax + b = cx + d$

This diagram of a balance scale is similar to the one in Lesson 6-1. The circles represent one-kilogram weights and the weight of each box is unknown. If B is the weight of one box, then the situation is described by

$$5B + 6 = 2B + 18.$$

Notice that the variable B is on each side of the equation.

To solve this equation pictorially, remove 2 boxes from each pan.

$$3B + 6 = 18$$

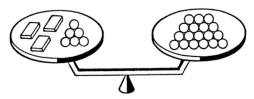

Remove 6 weights from each pan.

$$3B = 12$$

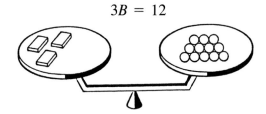

Then leave one third of the contents of each scale.

$$B = 4$$

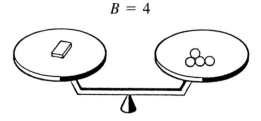

One box weighs 4 kilograms.

Example 1 shows this process algebraically. Work steps are put in. Reasons are given for the key steps.

Example 1 Solve: $5B + 6 = 2B + 18$

Solution $-2B + 5B + 6 = -2B + 2B + 18$ Addition Property of Equality
(Remove 2 boxes.)

$$3B + 6 = 18$$
$$3B + 6 + -6 = 18 - 6$$ Addition Property of Equality
(Remove 6 kg.)

$$3B = 12$$
$$\tfrac{1}{3}(3B) = \tfrac{1}{3}(12)$$ Multiplication Property of Equality
$$B = 4$$ (Leave $\tfrac{1}{3}$ of each side.)

Check Substitute 6 for B in the original equation.
Does $5 \cdot 4 + 6 = 2 \cdot 4 + 18$?
$20 + 6 = 8 + 18$? Yes.

The equation in Example 1 has the unknown variable on each side. It is of the form

$$ax + b = cx + d$$

and is called the **general linear equation.** To solve such equations, you need three key steps. In the first step, add either $-cx$ or $-ax$ to both sides. This removes a variable from one side and leaves an equation of the kind you have solved in previous lessons. Here is Example 1, written as an expert might, without the work steps.

Given	$5B + 6 = 2B + 18$
Step 1	$3B + 6 = 18$
Step 2	$3B = 12$
Step 3	$B = 4$

Example 2 Solve $\qquad -7y + 54 = 2y - 36$

Solution We could add either $7y$ or $-2y$ to both sides. We choose to add $-2y$.

$$-2y + -7y + 54 = -2y + 2y - 36$$

Adding like terms gives a simpler equation.

$$-9y + 54 = -36$$

Now add -54 to each side and simplify.

$$-9y + 54 + -54 = -36 + -54$$
$$-9y = -90$$

Finally, multiply both sides by $-\tfrac{1}{9}$ (or divide by -9).

$$y = 10$$

Check   Substitute 10 for y wherever y appears in the original equation.
Does $-7 \cdot 10 + 54 = 2 \cdot 10 - 36$?
$$-70 + 54 = 20 - 36?$$
Yes, both sides equal -16.

Many situations lead to the solving of equations with variables on both sides.

Example 3   The 1980 population of Columbus, Ohio was 565,000 people. In the 1970s it had been increasing at an average rate of 2500 people per year. In 1980, the population of Milwaukee, Wisconsin was 636,000. In the 70s it was decreasing at the rate of 8000 people a year. If these rates continued, in what year would the populations in Columbus and Milwaukee be the same?

Solution   After n years the population in Columbus will be 565,000 + 2500n. After n years the population in Milwaukee will be 636,000 − 8000n. The populations would be the same when

$$565{,}000 + 2500n = 636{,}000 - 8000n.$$

Solve this equation for n. We add 8000n to each side.

$$565{,}000 + 2500n + 8000n = 636{,}000 - 8000n + 8000n$$
$$565{,}000 + 10{,}500n = 636{,}000$$
$$-565{,}000 + 565{,}000 + 10{,}500n = -565{,}000 + 636{,}000$$
$$10{,}500n = 71{,}000$$
$$n \approx 6.76$$

At these rates, in 1986 or 1987 (that is, after 6.76 years), the populations would be the same.

Check   The population of Columbus would be $565{,}000 + 6.76 \cdot 2500$, or 581,900. The population of Milwaukee would be $636{,}000 - 6.76 \cdot 8000$, or 581,920. These are close enough given the accuracy of the information.

Questions

1. The boxes are of equal weight. Each circle weighs 1 kg.
   a. What equation is pictured by this balance scale?

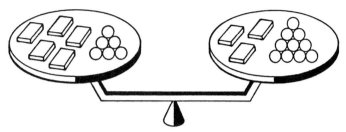

   b. What three steps can be done with the scale to find the weight of a box?

2. **a.** To solve $8x + 7 = 2x + 9$, what could you add to both sides to collect all of the variables on one side of the equation?
   b. Solve the equation in **a**.

3. Solve the equation of Example 2 by adding $7y$ to both sides.

In 4–7, solve.

4. $12x + 1 = 3x - 8$

5. $43 - 8w = 25 + w$

6. $7y = 5y - 3$

7. $2 - z = 3 - 4z$

8. Alaska has a population of 480,000 which has been increasing at a rate of 6000 people a year. Delaware's population of 606,000 has been increasing at a rate of 4000 people a year. If the rates of increase do not change, in how many years will the populations be equal?

In 9–12, solve.

9. $\frac{1}{2}a + 3 = \frac{1}{3}a + 4$

10. $3 - 8b + 2b = 4 + 2b - b$

11. $1.5c + 17 = 0.8c - 32$

12. $3d + 4d + 5 = 6d + 7d + 8$

13. Kim has $20 and is saving at a rate of $6 per week. Jenny has $150 but is spending at a rate of $4 per week. **a.** After how many weeks will each have the same amount of money? **b.** After how many *days* will each have the same amount of money?

14. Five more than twice a number is three more than four times the number. What is the number?

15. In 1988, the women's world record for the 100-m freestyle in swimming was 54.73 seconds. It had been decreasing at a rate of 0.33 seconds a year. The men's record was 49.36 seconds and had been decreasing at 0.18 seconds a year. Assume that these rates continued.
a. What would the women's 100-meter record be x years after 1988?
b. What would the men's 100-meter record be x years after 1988?
c. After how many years would the records be the same?

16. Stuart is offered two payment plans when he buys his new stereo. Under Plan A, he pays $100 down and x dollars a month for 4 months. In Plan B, he pays nothing down and $10 more per month than in Plan A for 6 months.
a. Write an equation to find x.
b. What is the monthly payment under each plan?

Review

17. Evaluate these expressions in your head for $r = 4$. *(Lesson 1-4)*
   a. r^2 **b.** $r^2 + r$ **c.** $-10r^2$

In 18–20, use the property named to write another expression equivalent to the one given.

18. $(5x + 5y) + 6$; Associative Property of Addition. *(Lesson 2-1)*

19. xyz; Commutative Property of Multiplication. *(Lesson 4-1)*

20. $13x - x$; Distributive Property. *(Lesson 6-3)*

21. A retailer wants to sell an item for exactly $10 including sales tax. If the sales tax is 5%, write an equation to find how much he or she should charge for the item without tax. *(Lesson 6-3)*

22. During one week in Chicago in the spring, it rained on r days and was sunny the other days.
a. Write an expression in terms of r to tell how many days were sunny.
b. What is the probability that a randomly selected day from that week was sunny? *(Lessons 3-6, 1-7)*

23. In a rectangular room, $\frac{2}{3}$ of the floor area is covered by a 9 ft by 12 ft rug.
    a. Write an equation to find the area A of the floor.
    b. Solve the equation. *(Lesson 2-6)*

In 24–26, solve. *(Lesson 4-4)*

24. $\frac{4}{3}y = -2$ **25.** $\frac{4}{3}z - \frac{11}{3}z = -2$

26. $\frac{4}{3}w - \frac{11}{6}w = -2$

Exploration

27. Here is a puzzle from the book *Cyclopedia of Puzzles*, written by Sam Loyd in 1914. If a bottle and a glass balance with a pitcher, a bottle balances with a glass and a plate, and two pitchers balance with three plates, can you figure out how many glasses will balance with a bottle?

6-7

Solving
$ax + b < cx + d$

The same steps for solving linear equations that you learned in Lesson 6-6 also work for solving inequalities of the form $ax + b < cx + d$. These are called **linear inequalities.**

Example 1   Graph all solutions to the inequality

$$13x + 18 > 10x + 12.$$

Solution   The choice is to add either $-13x$ or $-10x$ to both sides. We add $-10x$.

$$-10x + 13x + 18 > -10x + 10x + 12$$
$$3x + 18 > 12$$

Now add -18. $3x > -6$

Multiply both sides by $\frac{1}{3}$. $\frac{1}{3} \cdot 3x > \frac{1}{3} \cdot -6$

$$x > -2$$

Here is the graph.

Check   Recall that checking an inequality requires two steps.

Step 1: Does $x = -2$ make both sides of the original sentence equal?
Does $13 \cdot -2 + 18 = 10 \cdot -2 + 12$?

$-26 + 18 = -20 + 12$? Yes, each equals -8.

Step 2: Try a number that works in $x > -2$. We try 0. Does it work in the original sentence? You should verify that it does.

The reason that linear inequalities and linear equations can be solved with the same steps is that the Addition and Multiplication Properties of Inequality are just like the Addition and Multiplication Properties of Equality. Recall the only exception: multiplying by a negative number reverses the sense of the inequality. This is an important thing to remember, because it can happen in any linear inequality.

For instance, suppose you choose to add $-13x$ to both sides in Example 1. You would get

$$-13x + 13x + 18 > -13x + 10x + 12$$
$$18 > -3x + 12$$

Now add -12 to each side. The inequality remains $>$.

$$18 + -12 > -3x + 12 + -12$$
$$6 > -3x$$

To find x, multiply both sides by $-\frac{1}{3}$. This multiplication reverses the sense of the inequality.

$$-\frac{1}{3} \cdot 6 < -\frac{1}{3} \cdot -3x$$
$$-2 < x$$

This inequality is equivalent to that found in Example 1. Any number bigger than -2 works.

As always, many situations lead to these kinds of sentences. Example 2 gives an instance of such a situation.

Example 2   A crate weighs 6 kg when empty. A lemon weighs about 0.2 kg. For economical shipping the crate with lemons must weigh at least 45 kg. How many lemons should be put in the crate?

Solution   First, let n be the number of lemons. Then the weight of n lemons is $0.2n$. The weight of the crate with n lemons is $0.2n + 6$, which must be at least 45 kg, so the question can be answered by solving the inequality

$$0.2n + 6 \geq 45.$$

This is of the form $ax + b \geq c$ and is solved like $ax + b = c$. First, add -6 to both sides and simplify.

$$0.2n + 6 + -6 \geq 45 + -6$$
$$0.2n \geq 39$$
$$\frac{0.2n}{0.2} \geq \frac{39}{0.2} \qquad \text{Multiply both sides by } \frac{1}{0.2} \text{ or divide by 0.2.}$$
$$n \geq 195$$

There must be at least 195 lemons in the crate.

Check   Step 1: Does $0.2(195) + 6 = 45$?
$$39 + 6 = 45? \text{ Yes.}$$
Step 2: Pick some value that works for $n \geq 195$. We choose 200.
Is $0.2(200) + 6 \geq 45$?
$$40 + 6 \qquad \geq 45? \text{ Yes.}$$

The lemon crate question of Example 2 could be answered by solving $0.2n + 6 = 45$ and avoiding the inequality. This is true in many situations. But sometimes the inequality can help you determine if an answer is reasonable.

■ ■ ■ ■ ■ ■ ■ ■

Example 3 Three times a number is less than two times the same number. Find the number.

Solution It may seem that there is no such number. But work it out to see. Let n be such a number. n must be a solution to

$$3n < 2n.$$

Solve this as you would any other linear inequality. Add $-2n$ to each side.

$$3n + -2n < 2n + -2n$$
$$n < 0$$

So if n is any negative number, 3 times it will be less than 2 times it.

Check Let $n = -5$; $3 \cdot -5$ is -15, $2 \cdot -5$ is -10, and -15 is less than -10.

Questions

Covering the Reading

1. The method for solving $ax + b < cx + d$ is like the method for solving the equation __?__.

2. **a.** Solve $4k + 3 > 9k + 18$ by first adding $-4k$ to each side.
   b. Solve $4k + 3 > 9k + 18$ by first adding $-9k$ to each side.
   c. Should you get the same solutions to parts a and b?

In 3–6, solve and graph the solutions.

3. $3x + 4 < 19$

4. $-48 + 10a < -8 + 20a$

5. $6 \leq 4b + 10$

6. $0.12 - 0.03y > 0.27 + 0.07y$

7. A crate weighs 10 kg when empty. A grapefruit weighs about 0.5 kg. How many grapefruit can be put in the crate and still keep the total weight under 40 kg?

8. Five times a number is less than three times the same number. Find the number.

9. Suppose admission to a carnival is $4.00. You allow $3.00 for lunch and $1.00 for a snack. Each ride is $.80. You have $15 to spend. How many rides can you go on?

10. A card printer charges $5.00 to set up each job and an additional $4.00 per box of 100 cards printed.
    a. How much would it cost to print n boxes of cards?
    b. What is the greatest number of boxes you could have printed for under $100?

11. Sending a package by Fast Fellows Shipping costs $3.50 plus 10 cents per ounce. Speedy Service charges $4.75 plus 6 cents per ounce. What weight packages are cheaper at Speedy Service?

12. Find the two integers that satisfy both these properties.
    (i) If you add six to it, the sum is greater than twice it.
    (ii) If you add six to it, the sum is less than three times it.

In 13–16, solve. *(Lesson 6-6)*

13. $4x + 12 = -2x - 6$

14. $60t - 1 = 48t$

15. $109 - m = 18m - 5$

16. $3n - n + 5 = 4n - n + 20$

17. Three instances of a general pattern are given below. Write the pattern using two variables. *(Previous course)*

$$-(7 - 4) = 4 - 7$$
$$-(\tfrac{8}{3} - \tfrac{5}{6}) = \tfrac{5}{6} - \tfrac{8}{3}$$
$$-(1.2 - 2.1) = 2.1 - 1.2$$

18. Ten club members celebrate at a restaurant. Because it is a large group, the restaurant automatically adds on a 15% tip. If the bill is $103.90 with the tip, what was the bill without the tip? *(Lesson 6-3)*

19. In 1986, according to the Federal Communications Commission, 7.8% of all households in the U.S. were without a telephone. Of 100,000 households, how many would you expect to be without a phone? *(Lesson 5-4)*

20. A car normally priced at $6990 is on sale for $6291. What percent discount is this? *(Lesson 5-3)*

In 21–23, try to answer the question in your head. If you cannot, then use an equation. *(Lesson 5-4)*

21. What is 150% of 6?

22. 10% of what number is 12?

23. $\frac{1}{2}$ is what percent of 2?

24. Mr. Roberts sells appliances. He earns $230 per month plus 5% commission on all he sells. He earns E dollars in a month when he sells S dollars worth of appliances. Write an equation that describes how E and S are related. *(Lesson 6-1)*

25. The school lunch counter will make up sandwiches in three kinds of bread (white, rye, and whole wheat) and three fillings (ham, chicken, or tuna salad). How many different kinds of sandwiches are possible? *(Lesson 4-7)*

26. What will a computer print on the screen when this program is run? *(Lesson 1-4)*

```
10 PRINT "SOLVE 5X + 3.4 = 10.18"
20 LET A = 5
30 LET B = 3.4
40 LET C = 10.18
50 LET X = (C − B)/A
60 PRINT "X = "; X
70 END
```

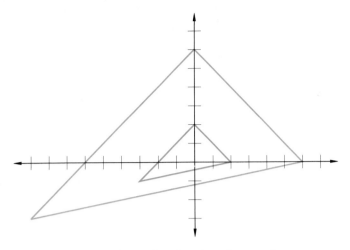

27. Give the magnitude for the size change shown. *(Lesson 5-6)*

Preimage point	Image point
(-9, -3)	(-3, -1)
(0, 6)	(0, 2)
(6, 0)	(2, 0)

28. Use this diagram of mileage along Interstate 10. The distance from Los Angeles to New Orleans is 1946 miles. How far is it from San Antonio to Houston? *(Lessons 2-1, 2-6)*

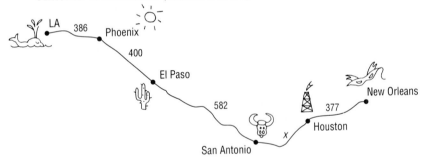

29. Write 725 billion in scientific notation. *(Appendix)*

In 30–31, simplify. *(Lessons 2-3, 5-1)*

30. $\dfrac{x}{4} + \dfrac{x}{3}$

31. $\dfrac{\frac{m}{4}}{\frac{12}{n}}$

32. There are certain numbers which are less than their squares.
   a. Find one such number.
   b. Find all such numbers.

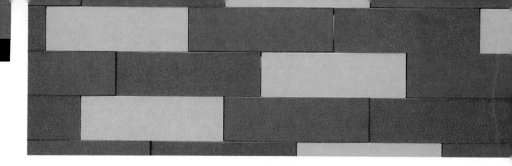

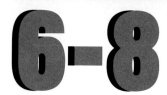

Why the Distributive Property Is So Named

Two rectangles, each having width c, are placed end to end. What is the total area?

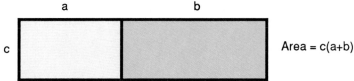

Since the length of this rectangle is $a + b$ and the width is c, the total area is $c(a + b)$. However, the area can also be expressed in a different way, by finding the areas of each smaller rectangle and adding.

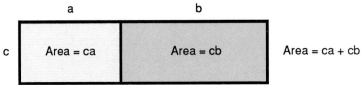

The area is the same, no matter which way it is calculated. This illustrates the pattern

$$c\,(a + b) = ca + cb.$$

This is not a new property. If the sides of this equation are switched and then each side is rearranged using the Commutative Property, the result is the familiar Distributive Property.

The property $c(a + b) = ca + cb$ is another form of the Distributive Property. It tells why the Distributive Property is so named. Multiplying by c is *distributed* over the terms in the sum. This form of the Distributive Property is used to **eliminate parentheses.** Here are some examples.

Examples

1. $6(y + 7) = 6y + 6 \cdot 7$
$\qquad\qquad = 6y + 42$

2. $-60\left(\dfrac{x}{6} - \dfrac{y}{5}\right) = -60 \cdot \dfrac{x}{6} - -60 \cdot \dfrac{y}{5}$
$\qquad\qquad\qquad = -10x + 12y$

3. $4a(3 + 2b - 5c) = 4a \cdot 3 + 4a \cdot 2b - 4a \cdot 5c$
$\qquad\qquad\qquad = 12a + 8ab - 20ac$

This form of the Distributive Property shows how to perform some calculations mentally.

Example 4 Calculate in your head how much 5 records cost if they sell for $8.97 each.

> **Solution** Think of $8.97 as $9 − 3¢.
> So 5 · $8.97 = 5($9 − 3¢)
> = 5 · $9 − 5 · 3¢ } Do all this in your head.
> = $45 − 15¢
> = $44.85
>
> **Check** Multiply 5(8.97) with pencil and paper or with your calculator. You should get 44.85.

This form of the Distributive Property also enables multiplication involving mixed numbers like $1\frac{1}{2}$.

Example 5 Monica's hourly wage as a carpenter is $12.00. If she receives time-and-a-half for overtime, what is her overtime hourly wage?

> **Solution** Time-and-a-half means $1\frac{1}{2}$ times the regular hourly wage. Since $1\frac{1}{2} = 1 + \frac{1}{2}$, use the Distributive Property to mentally calculate the wage.
>
> $$12 \cdot 1\frac{1}{2} = 12 \cdot (1 + \frac{1}{2}) = 12 \cdot 1 + 12 \cdot \frac{1}{2}$$
> $$= 12 + 6$$
> $$= 18$$
>
> Monica receives $18 as an hourly wage for overtime.
>
> **Check** $18 is halfway between her $12 salary and double that, $24.

The Distributive Property is *very important* in algebra. It may be used more than once in the same problem.

Example 6 Solve $-5(x + 2) + 3x = 8$.

Solution 1 We show reasons at the right.

$-5x + -5 \cdot 2 + 3x = 8$	Distributive Property (to eliminate parentheses)
$-5x + -10 + 3x = 8$	arithmetic
$-5x + 3x + -10 = 8$	Commutative Property of Addition
$-2x + -10 = 8$	Distributive Property (to add like terms)
$-2x + -10 + 10 = 8 + 10$	Addition Property of Equality
$-2x = 18$	arithmetic
$x = -9$	Multiplication Property of Equality

Solution 2 Experts do arithmetic and work steps mentally and may write down only a few steps:

$$-5x - 10 + 3x = 8$$
$$-2x - 10 = 8$$
$$-2x = 18$$
$$x = -9$$

Check Substitute -9 for x in the original sentence. Follow the order of operations.

Does

$$-5(-9 + 2) + 3 \cdot -9 = 8?$$
$$-5 \cdot -7 + 3 \cdot -9 = 8?$$
$$35 + -27 = 8? \text{ Yes.}$$

Questions

Covering the Reading

1. Fill in the blanks to express the area in two different ways.

$$\underline{\ ?\ } (\underline{\ ?\ } + \underline{\ ?\ }) = \underline{\ ?\ } + \underline{\ ?\ }$$

2. Which two of these true sentences are forms of the Distributive Property?

a. $n(x + y) = nx + ny$ **b.** $n(x + y) = (x + y)n$

c. $nx + ny = ny + nx$ **d.** $nx + ny = n(x + y)$

In 3–8, use the Distributive Property to eliminate parentheses.

3. $12(k + 5)$ **4.** $2(y - 1.5)$

5. $7a(3 - 5b)$ **6.** $10bc(-a + b + c)$

7. $\frac{1}{2}x(4y + 6)$ **8.** $-3(-2 + 2x)$

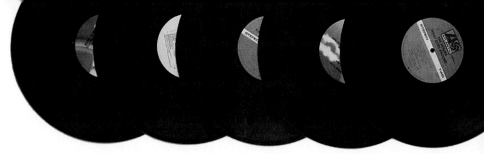

9. Show how the Distributive Property can help you mentally compute the price of 5 records if each record costs $7.96.

10. Mentally compute the total cost of four gallons of milk at $2.07 each.

In Questions 11 and 12, mentally compute the overtime hourly wage (at time-and-a-half for overtime) if the normal hourly wage is the given amount.

11. $14.00   12. $6.50

In Questions 13–16, use the Distributive Property to simplify, then solve and check.

13. $4(m + 7) = 320$

14. $11(3 - n) = 6(5 - 2n)$

15. $2(x + 3.1) = 9.8$

16. $6(4y - 1) - 2y = 82$

Applying the Mathematics

17. Mentally compute 6 times 999,999.

18. For each hour of television, there is an average of $8\frac{1}{2}$ minutes of commercials. If you watch 6 hours of television in a week, compute in your head how many commercial minutes you will see.

19. Solve the equation $\frac{1}{2}x + \frac{2}{3} = \frac{11}{15}$ in two ways:
   a. by first adding $-\frac{2}{3}$ to both sides and proceeding from that;
   b. by multiplying both sides by 30 and proceeding from that.

20. Solve $.05x - 1.03 = 2.92$ by first multiplying both sides by 100 to get an equation with no decimals.

21. The area A of a trapezoid with parallel bases b_1 and b_2 and height h is given by the formula

$$A = \frac{1}{2}h(b_1 + b_2).$$

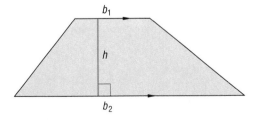

If a trapezoid has base $b_1 = 5$ cm, height 6 cm, and area 60 cm^2, what is the length of its other base?

In Questions 22 and 23, solve.

22. $\dfrac{3x}{4} = \dfrac{3x + 1}{6}$

23. $2n + 3(5 + 2n) \le 14$.

Review

24. The perimeter of triangle *TRI* is 120 units. Find the length of $\overline{RT}$. *(Lesson 6-2)*

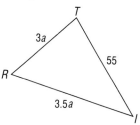

In Questions 25–27, solve. *(Lessons 6-1, 6-6, 6-7)*

25. $4x + 3 = 12$ **26.** $3 - 4x = 12$ **27.** $5x + 2 < -x - 1$

In 28–30, simplify. *(Lessons 2-2, 4-1, 4-2)*

28. $-3(-\frac{2}{3}a)$ **29.** $4b - 6b + 8b - 2b$ **30.** $(3c)(2c)(-c)$

31. Here is a scale. On the left side, a $\frac{1}{8}$-lb weight together with 2 boxes balances with a $\frac{3}{4}$-lb weight and $\frac{1}{24}$-lb weight on the right. What is the weight of a box? *(Lessons 6-1,6-2)*

32. 6 is what percent of 45? *(Lesson 5-4)*

33. In 1980, the population of Mississippi was 2,520,638 and was increasing at a rate of 27,400 people per year. If this rate continued, what was the population in 1985? *(Lesson 6-4)*

Exploration

34. Write the area of the large rectangle in three different ways as the sum of the areas of smaller rectangles.

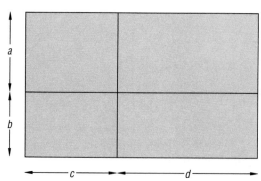

LESSON

6-9

Subtracting Quantities

The Distributive Property allows us to write

$$c(a + b) = ca + cb.$$

When $c = -1$ this means $-1(a + b) = -1a + -1b$.

Recall that the Multiplication Property of -1 states that for any real number n, $-1n = -n$. So the above property can be rewritten as

$$-(a + b) = -a + -b.$$

We say that *the opposite of the sum is the sum of the opposites of its terms*.

An algebraic expression in parentheses, like $a + b$ above, is called a **quantity.** You can think of the above discussion as showing that the opposite sign distributes over the terms of a quantity. The same idea holds if the quantity in parentheses has subtractions.

Example 1   Simplify $-(4a - 7)$.

> **Solution** $-(4a - 7) = -1(4a - 7)$ Multiplication Property of -1
> $= -4a + 7$ Distributive Property
>
> **Check**   Substitute some number for a. We substitute 5.
>
> Does $-(4 \cdot 5 - 7) = -4 \cdot 5 + 7$?
> $-(20 - 7) = -20 + 7$?
> $-13 = -13$? Yes.

Remember that the definition of subtraction says that $x - y = x + -y$. In Example 2, y is the quantity $4a - 7$ found in Example 1. Example 2 is read "Subtract the quantity $4a - 7$ from the quantity $10a + 6$."

Example 2   Simplify $(10a + 6) - (4a - 7)$.

> **Solution 1**   Change the subtracting of $(4a - 7)$ to addition.
>
> $$\begin{array}{rcl} x \quad - \quad y & = & x \quad + -y \\ (10a + 6) - (4a - 7) & = & (10a + 6) + -(4a - 7) \\ & = & 10a + 6 + -4a + 7 \\ & = & 10a + -4a + 6 + 7 \\ & = & 6a + 13 \end{array}$$

Solution 2 An expert might write the following.

$$(10a + 6) - (4a - 7) = 10a + 6 - 4a + 7$$
$$= 6a + 13$$

These ideas can be verified by considering a real situation in which a quantity is subtracted.

Example 3 Anton had $500 in his savings account. He went to the bank and withdrew x dollars. Deciding that this was not enough, he went back and withdrew y more dollars. Express the amount of money left in his savings account in two different ways.

Solution Anton can subtract each amount from 500. This is written as

$$500 - x - y.$$

Anton could also add the two withdrawals and subtract the total. This is written as

$$500 - (x + y).$$

Check Pick some values for x and y. We pick 300 for x and 20 for y. Does $500 - 300 - 20 = 500 - (300 + 20)$? Always follow the order of operations. You should verify that each side equals 180.

CAUTION: Remember to distribute the opposite over each term of the sum or difference. For example, $-(a - b - c) = -a + b + c$. Notice the first step in Example 4.

Example 4 Solve: $3x - (2 + 4x) = 25$

Solution

$13x - 2 - 4x = 25$	Distributive Property (opposite of a sum)
$9x - 2 = 25$	Distributive Property (like terms)
$9x = 27$	Addition Property of Equality
$x = 3$	Multiplication Property of Equality

Check Does $13 \cdot 3 - (2 + 4 \cdot 3) = 25$?
$$39 - (14) = 25? \text{ Yes.}$$

Example 5 Solve: $8t = 3t - 2(t + 14)$

Solution Simplify the right side. Because of order of operations, the multiplications come first, so the Distributive Property should be applied.

$$8t = 3t - (2t + 28)$$
$$8t = 3t - 2t - 28$$
$$8t = t - 28$$
$$7t = -28$$
$$t = -4$$

Check Does $8 \cdot -4 = 3 \cdot -4 - 2(-4 + 14)$?
Yes, $-32 = -12 - 2 \cdot 10$.

Questions

Covering the Reading

1. What property does this sentence illustrate? $-n = -1 \cdot n$

2. *Multiple choice* $-(x + 4) =$
   (a) $-x - 4$ (b) $x + -4$ (c) $-x + 4$

3. *Multiple choice* Which does *not* equal $-(x + y)$?
   (a) $-x + y$ (b) $-1x + -1y$ (c) $-x - y$

In 4–7, simplify.

4. $-(x + 15)$ 5. $-(4n - 3m)$

6. $x - (x + 2)$ 7. $3y - 5(y + 1)$

In 8 and 9, simplify and check.

8. $(3k + 4) - (7k - 9)$ 9. $-(5 + k) + (k - 18)$

10. A clerk has a 20-yard bolt of cloth. She first cuts r yards of cloth from the bolt and then cuts t yards from the same bolt.
    a. Express the amount of cloth left in two different ways.
    b. Let $r = 3$ yards and $t = 7$ yards to check that the two expressions in part a are equal.

303

11. You begin the day with D dollars. You spend L dollars for lunch and $10 for a book. Write two expressions for the amount of money you have left.

Applying the Mathematics

12. Rewrite without parentheses: $-(a + 2b - c)$.

13. a. Simplify the left side of $12 - (2y - 4) = 18$.
  b. Solve.

In 14–17, solve.

14. $-(A - 9) = 11$ **15.** $4x = 3x - 2(x + 6.5)$

16. $3(x + 9) - (9 + x) + 6(x + 9) = 80$

17. $\dfrac{3x}{2} - \dfrac{x + 1}{2} = 7$

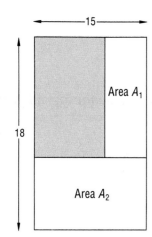

18. Write an expression for the area of the colored part of this 15 by 18 rectangle at the left.

19. A carpenter has a board 96 inches long. He measures R inches, then another S inches and makes a cut. What is the length of the board after $R + S$ inches have been cut off?

20. Rayette had $400 in her savings account. She withdrew $37 Monday, and withdrew some more on Tuesday, but forgot to record the amount. The teller said she currently has $318 in her account.
  a. Write an equation to describe the situation.
  b. Solve the equation to find out how much she withdrew on Tuesday.

21. Isaac thought himself an expert, so he wrote

$$13 - x + y - (x - 2 + y) = 11 - 2x + 2y.$$

Is Isaac right?

22. a. Solve: $-(y - 7) - (-2) < 12$.
  b. Graph the answer on a number line.

23. *Multiple choice*   Which is *not* the opposite of $a + b$?
  (a) $-a + -b$ (b) $-(a + b)$ (c) $-a + b$ (d) $-1 \cdot a + -1 \cdot b$

Review

24. After a 10% raise, Myretta's salary was $8.80 an hour. What was her salary before the raise? *(Lesson 6-3)*

In 25–28, solve. *(Lessons 6-2, 6-7, 6-8)*

25. $\frac{1}{2}x - \frac{1}{3} = -\frac{5}{6}$ **26.** $30 < 3n + 3$

27. $8z + 2(z + 1) = 106$ **28.** $\dfrac{x}{4} = \dfrac{x + 1}{6}$

29. If the dots continue on a line in the graph below, how much would you pay for 20 tapes? *(Lesson 2-4)*

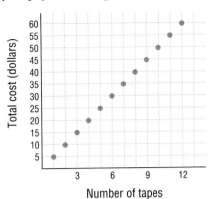

Number of tapes

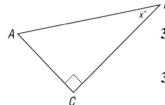

30. Triangle *ADC* is a right triangle. Write an expression for the measure of ∠*A* in terms of *x*. *(Lesson 3-8)*

31. *Skill sequence*   Evaluate. *(Lesson 4-9)*

    a. 6!
            b. $\dfrac{6!}{3!}$
            c. $\dfrac{96!}{93!}$

32. Carol Sue has 5 pairs of earrings, one pair of which is purple. If she picks two earrings at random, **a.** what is the probability that the first earring she picks will be purple, and **b.** what is the probability that both earrings are purple? *(Lesson 4-8)*

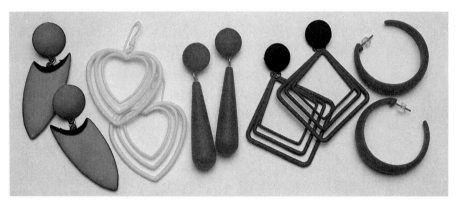

33. Estimate in your head. *(Previous course)*
    a. How much will 4 shirts cost at $19.95 each?
    b. If 3 VCR tapes sell for $17.95, how much does one cost?

Exploration

34. In this lesson, it was noted that $a - b$, $a + {}^-b$, and $a + {}^-1 \cdot b$ are all equal. Write five more expressions that are equal to $a - b$.

35. The difference of two numbers is subtracted from their sum. What can be said about the answer?

Summary

A linear sentence is one equivalent to a sentence of the form $ax + b = cx + d$ or $ax + b < cx + d$. An example is $3x + 4 = 5x - 2$; another example is $y = 4x + 5$. This chapter emphasizes how to solve linear sentences. In the next two chapters you will see why they are called "linear."

To solve linear sentences:
1. Simplify each side of the sentence, if possible.
2. Add a quantity to both sides to get all variables on one side, then simplify.
3. Add a quantity to both sides to get all numerical terms on the other side, then simplify.
4. Multiply both sides by a number that will get the variable alone.

A basic property needed in working with many linear sentences is the Distributive Property. The Distributive Property has many forms. As

$$ax + bx = (a + b)x,$$

it helps to combine like terms. In the form

$$c(a + b) = ca + cb,$$

it is used to remove parentheses. When $c = -1$, there is the special case $-(a+b) = -a + -b$, which is used when sums are subtracted. When $a = 1$ and b is a percent change, a special case arises which is useful in many practical situations such as markup and discounts: for example, $p + .05p = 1.05p$.

You have now studied how to solve virtually all linear equations or inequalities. The fundamental idea in solving is to change the given sentence into a simpler sentence. So, to solve an equation with the unknown on both sides, add something to change it to a sentence with the unknown on just one side. Then add to get an equation of the form $ax = b$. This last equation is one you know how to solve.

Vocabulary

Below are the most important terms and phrases for this chapter. You should be able to give a general description and a specific example for each.

Lesson 6-1
linear sentence

Lesson 6-3
Distributive Property
discount
markup, tax

Lesson 6-4
FOR/NEXT loop

Lesson 6-6
general linear equation

Lesson 6-7
linear inequality

Lesson 6-8
eliminate parentheses

Lesson 6-9
quantity (in parentheses)

Progress Self-Test

Take this test as you would take a test in class. You will need a calculator. Then check your work with the solutions in the Selected Answers in the back of the book.

In 1–4, simplify.

1. $m - 0.23m$

2. $k + 3(k + 3)$

3. $\frac{5}{2}(4v + 100 - w)$

4. $6t - 2(9 - 4t)$

In 5–12, solve.

5. $8r + 14 = 74$

6. $-4q + 3 + 9q = -12$

7. $6w - 37 = 8w + 35$

8. $6(x + 2) = 3(x + 6)$

9. $14m - 7(3 - m) = 21$

10. $13 \le 7 - x$

11. $v - 3v > 2$

12. $\frac{1}{5}(10h + 1) = 4h + \frac{1}{5}$

13. If both sides of $\frac{1}{4}x + \frac{3}{20} = \frac{3}{5}$ are multiplied by 20, what equation results?

14. Irving bought 6 pairs of socks at $2.99 per pair. Show how Irving could calculate the total cost of the socks in his head.

15. A stereo system costs $385 after a 30% discount. What did it cost before the discount?

16. Some books cost a total of $20.41. If an 8% tax was included in this price, how much did the books originally cost?

17. A sequence begins 2, 14, 26, 38, 50, ... , and each term is 12 more than the preceding. Which term is 470?

18. Each minute a computer printer prints 6 sheets of paper. Suppose the printer starts with 1100 sheets of paper and prints continuously. After how many minutes will 350 sheets be left?

19. If $F = \frac{9}{5}C + 32$, find the Celsius equivalent of a Fahrenheit temperature of 50°.

20. A person's savings account has $137.25 and $2.50 is added each week. Disregarding interest, how much will there be after w weeks?

21. Jack and Jill had a lemonade stand. Jack worked twice as long as Jill and so they decided to split their profits so that Jack received twice as much. If the total profits were $19.50, how much should Jill receive?

22. Town A has a population of 25,000 and is growing at a rate of 1200 people per year. Town B has a population of 37,000 and is declining at a rate of 200 people a year. If these rates continue, in how many years will the two towns have the same population?

23. Explain how the picture below represents the Distributive Property.
$c(a + b) = ca + cb$

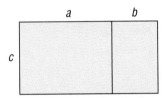

24. In the U.S., the life expectancy L of a 12- to 16-year-old white male is given by the formula
$$L - 60.7 = .95(A - 12).$$
where A is the age of the person. What is the life expectancy of a 14-year-old white male according to this formula?

25. From a roll of wallpaper y yards long, a decorator cuts off c yards for one wall and 15 more yards for another wall. Give two different expressions for how much is left.

Chapter Review

Questions on **SPUR** Objectives

SPUR stands for **S**kills, **P**roperties, **U**ses, and **R**epresentations.
The Chapter Review questions are grouped according to the
SPUR Objectives for this chapter.

SKILLS deal with the procedures used to get answers.

Objective A. *Use the distributive property to remove parentheses and collect like terms.* (*Lessons 6-3, 6-8, 6-9*)

In 1–10, simplify.

1. $c + \frac{1}{2}c$
2. $5w + 8z - 7w$
3. $m - .08m$
4. $.2x + x$
5. $3(a + 2b)$
6. $-\frac{2}{3}(6 - v)$
7. $11(3x - 2) + 8(4 - 2x)$
8. $4(m - 1) - 5(2 + 3m)$
9. $1 - (z - 1)$
10. $3w + 2 + z - 4(w + 8 - z)$

Objective B. *Solve linear equations.* (*Lessons 6-1, 6-2, 6-5, 6-6, 6-8, 6-9*)

In 11–24, solve.

11. $4n + 3 = 15$
12. $470 + 2n = 1100$
13. $\frac{3}{4}x - 12 = 39$
14. $66 = \frac{z}{5} + 2$
15. $m - 3m = 10$

16. $2x + 3(1 + x) = 18$
17. $A + 5 = 9A - 11$
18. $3n = 2n + 4$
19. $216 = 2(w - 30) - 8$
20. $(5x - 8) - (3x + 1) = 36$
21. $2a - 6 = -2a$
22. $10e = 4e - 5(3 + 2e)$
23. $\frac{5 + g}{3} = \frac{g + 2}{2}$
24. $\frac{1}{x - 1} = \frac{3}{x - 2}$

Objective C. *Solve linear inequalities.* (*Lesson 6-7*)

In 25–32, solve.

25. $2x + 11 < 201$
26. $11h + 71 \geq 13h - 219$
27. $4 < 16g - 7g + 5$
28. $4x - 1 < 2x + 1$
29. $32 - y < 45$
30. $\frac{3}{4}t + 21 > 12$
31. $-4(2y - 1) \leq 12$
32. $5(5 + z) < 3(2 + 2z)$

PROPERTIES deal with the principles behind the mathematics.

■ **Objective D.** *Apply and recognize properties associated with linear sentences.*
Addition Property of Equality *(Lessons 6-1, 6-6)*
Addition Property of Inequality *(Lesson 6-7)*
Multiplication Property of Equality *(Lessons 6-1, 6-6)*
Multiplication Property of Inequality *(Lesson 6-7)*
Distributive Property *(Lessons 6-3, 6-8)*
Multiplication Property of 1 *(Lesson 6-3)*
Multiplication Property of -1 *(Lesson 6-9)*
In 33–36, identify the property being applied.

33. $4(x - y) = 4x - 4y$

34. $-(a + b + c) = -1(a + b + c)$

35. If $2x + 3 < 5$, then $2x < 2$.

36. Since $\frac{2m}{3} = 12$, $m = \frac{3}{2} \cdot 12$.

In 37–40, perform the indicated operation.

37. $a + 2 < 3a + 4$; add $-a$.

38. $4000x + 300 = 11,000$; multiply on both sides by $\frac{1}{100}$.

39. $B - 6B + 4(B + 2) \geq B - 5$; apply the Distributive Property

40. $-V \geq 3$; multiply on both sides by -1.

■ **Objective E.** *Use the Distributive Property to perform calculations in your head. (Lesson 6-3)*
In 41–44, show how the Distributive Property can be used to do the calculations mentally.

41. $7 \cdot \$3.04$

42. $101 \cdot 35$

43. the cost of 9 shirts if each one costs $19.99

44. $3 \cdot 118$

USES deal with applications of mathematics in real situations.

■ **Objective F.** *Answer questions involving markups or discounts. (Lesson 6-3)*

45. The Jacobsons went out for dinner. The total amount they were charged was $58.83. Included in this amount was a 6% sales tax. What was the cost of the meal before tax?

46. A stereo is on sale for a 25% discount. The sale price is $325. What is the regular price?

47. If phone rates have gone up 20% and you now pay $10.80 a month, what was the previous rate?

48. On sale at 40% off, a radio costs $39.96. What was the original price?

■ **Objective G.** *Describe patterns and answer questions in repeated addition or repeated subtraction situations. (Lessons 6-4, 6-6)*

49. Liz saves $15 per week. She has $750 in the bank and needs $1500 in the bank before she can afford to go on vacation. For how many weeks must she save before she has the required amount?

50. Mel adds 2 cards to his collection each week. If he has 1406 cards now, how many cards will he have n weeks from now?

51. Kate has $1500 in an account and adds $45 each month. Melissa has $2000 and adds $20 a month. When will they have the same amount of money in their accounts?

52. Len has $25 and is saving at the rate of $9 a week. Basil has $100 and is spending $5 a week. When will Len have more money than Basil?

53. A sequence begins 12, 17, 22, 27, ..., and each term is 5 more than the preceding.
   a. What is the 1100th term?
   b. Which term is 182?

54. A sequence begins 10, 9.8, 9.6, 9.4, 9.2, ..., and each term is 0.2 less than the preceding. Which term is 3.4?

■ **Objective H.** *Answer questions involving linear sentence formulas. (Lesson 6-2)*

55. If $S = 3L - 26$ relates a man's shoe size S and foot length L (in inches), find the foot length of a man with a size 11 shoe.

56. If $F = \frac{9}{5}C + 32$, find the Celsius equivalent of a Fahrenheit temperature of 100°.

57. If the height h of a woman and the length f of her femur (in cm) are related by the formula $h \approx 61.412 + 2.317f$, estimate the length of the femur of a woman 150 cm tall.

58. In a trapezoid, $A = \frac{1}{2}h(b_1 + b_2)$. If the area is 100 square feet, the height is 12.5 feet, and one base has length 7 feet. Find the length of the other base.

■ **Objective I.** *Answer questions about situations combining addition and multiplication. (Lessons 6-3, 6-6)*

59. A hamburger bun has about 200 calories. One ounce of hamburger has about 80 calories. How large a plain hamburger with bun can you eat and still be under 600 calories?

60. Angle 2 has 3 times the measure of angle 1. What are their measures?

61. If you work 35 hours a week for $4.50 an hour and receive time-and-a-half for overtime after that, how long must you work to earn over $200 in a week?

62. A home team is to get 5 times as much from gate sales as the visiting team. If the total receipts are $1572, how much should each team receive?

■ **Objective J.** *Translate balance scale models and rectangle area models into expressions and equations.* *(Lessons 6-1, 6-6, 6-8)*

63. Use the picture of a balance scale below. The boxes are equal in weight and the others are one-kilogram weights.
   a. Write an equation describing the situation with *B* representing the weight of one box.
   b. What is the weight of one box?

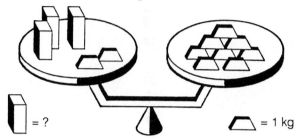

= ?   = 1 kg

64. Use the picture below.
   a. Write an equation to describe this situation with *W* representing the weight of one box.
   b. What is the weight of one box?

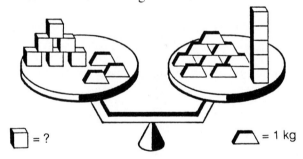

= ?   = 1 kg

In 65 and 66, write two different expressions to describe the total areas of the rectangles.

65.

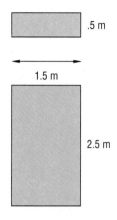

.5 m

1.5 m

2.5 m

66.

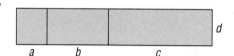

a   b   c   d

CHAPTER 7

Lines and Distance

An *optical illusion* is something that looks one way but is actually another.

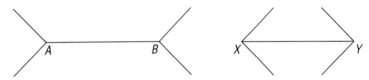

For instance, $\overline{AB}$ appears longer than $\overline{XY}$, but they are the same length. (If you don't believe this, measure them with a ruler!)

Most optical illusions rely on the fact that people have poor perceptions about distance and angle measure. The photograph at the left illustrates a real optical illusion. When the sun is seen near the horizon it appears larger than when it is higher in the sky. It is actually the same size both places since it is the same distance from the earth.

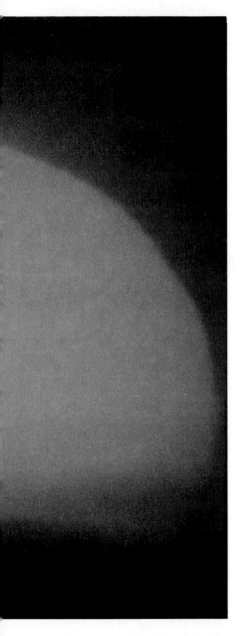

These illusions show that it is important to have more precise ways of computing lengths and distances. In this chapter you will learn more about lines and distance.

Graphing Lines

Suppose Beth begins with $10 in the bank and adds $5 to her account each week. After w weeks she will have $10 + 5w$ dollars. If t represents the total amount in her account at the end of w weeks, Beth's bank balance can be described in three different ways:

(1) with an equation $\qquad$ $t = 10 + 5w$

(2) with a chart

w	t
0	10
1	15
2	20
3	25
4	30
⋮	⋮

(3) with a graph

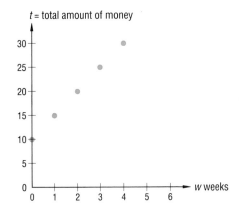

The chart lists the ordered pairs (0, 10), (1, 15), (2, 20), (3, 25) and (4, 30). All these pairs make the equation $t = 10 + 5w$ true. The equation $t = 10 + 5w$ is called a **linear equation** since the points of its graph lie on a line. Because Beth puts money into her account at specific intervals, it does not make sense to connect the points. The graph is *discrete*.

Whereas the equation $t = 10 + 5w$ uses the two variables w and t, the letters x and y are more commonly used to describe graphs. Many situations lead to straight-line graphs.

Example 1 A flooded stream is 14 inches above its normal level. The water level is dropping 2 inches per hour. Its height y above normal after x hours is given by $y = 14 - 2x$. Graph this relationship.

Solution Find the height at 0, 1, 2, 3, and 4 hours and make a table.

hour (x)	height ($y = 14 - 2x$)
0	$14 - 2 \cdot 0 = 14$
1	$14 - 2 \cdot 1 = 12$
2	$14 - 2 \cdot 2 = 10$
3	$14 - 2 \cdot 3 = \ 8$
4	$14 - 2 \cdot 4 = \ 6$

Graph the ordered pairs found in the table and draw the line through them.

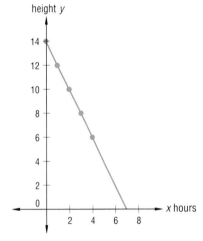

In Example 1, the points found are connected because the height is steadily decreasing. So the graph is *continuous*.

If you are not told the domain for a variable, assume that any number is allowed. The graph will be continuous. In later chapters you will see equations like $y = 3x^2 + 5$ and $y = 2^x$ whose graphs are not lines. They are not linear equations because both variables are not to the first power.

Example 2 Draw the graph of $y = -2x + 1$.

Solution 1 You can choose *any* values for *x*. We chose -1, 0, 1, 2, and 3. Make a table of solutions. Graph the points and connect them.

x	y = -2x + 1
-1	-2 · -1 + 1 = 3
0	-2 · 0 + 1 = 1
1	-2 · 1 + 1 = -1
2	-2 · 2 + 1 = -3
3	-2 · 3 + 1 = -5

Solution 2 The following program will print a table of values similar to the one shown above. Lines 30–60 build the table for *x* = -1, 0, 1, 2, and 3. From each *x* value, *y* is computed. The ordered pairs are printed in a table.

```
10 PRINT "TABLE OF (X,Y) VALUES"
20 PRINT "X VALUE", "Y VALUE"
30 FOR X = -1 TO 3
40    LET Y = -2 * X + 1
50    PRINT X,Y
60 NEXT X
70 END
 -1   3
  0   1
  1  -1
  2  -3
  3  -5
```

You can change line 30 to specify a different set of points for your table. You can change line 40 to specify a different linear equation.

Questions

Covering the Reading

1. How much money will Beth have in her account after 3 weeks?

2. Suppose a person begins with $5 and adds $2 per week.
   a. Write an equation that represents *t*, the sum of money after *w* weeks.
   b. Copy and complete this chart.
   c. Graph the ordered pairs (*w*, *t*).

weeks (*w*)	total (*t*)
0	
1	
2	
3	
4	

3. The graphs in this lesson are graphs of __?__ equations.

In 4–6, refer to Example 1.
4. After how many hours will the stream be 10 inches above normal?

5. How high above normal will the stream be after 6 hours?

6. After how many hours will the stream level be back to normal? (Hint: y will be equal to zero.)

7. **a.** In Example 2, when $x = \frac{1}{2}$, what is y?
   b. Will this point lie on the line that is graphed?

8. Rewrite line 30 in the computer program so that ordered pairs are printed for $x = 0, 1, 2, 3, 4, 5, 6$, and 7 when the program is run.

9. Rewrite line 40 so the computer program will print a table of values for $y = 8x - 3$.

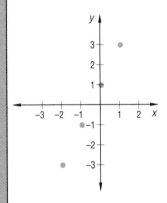

10. **a.** Copy and complete the chart when $y = 4x - 2$.
    b. Graph the equation.

x	y
-1	
0	
1	
2	

11. For what value of x will $(x, 3)$ be on the line: $12x - 5 = y$?

12. Copy and complete the chart for the graph at the left.

x	y
-2	
-1	
0	
1	

13. *Multiple choice* Which equation describes the ordered pairs?

x	y
0	5
1	6
2	7
3	8

   (a) $y = 5x$
   (b) $y = x + 5$
   (c) $y = x - 5$
   (d) $y = \dfrac{x}{5}$

14. When a weight x is attached to a spring, the spring stretches y inches, where $y = 1.5x + 2$. Make a table of solutions and draw the graph. (Choose your own x values for the table.)

15. **a.** Draw the graph of $y = 3x$.
    b. On the same grid as part a, draw the graph of $y = -3x$.
    c. What point(s) do the graphs of parts a and b have in common?

16. What is the minimum number of points needed to graph a straight line?

17. Kristin worked at a hotel restaurant. She earned $3.35 an hour but had to pay 14¢ for each dish she broke. Last week she worked h hours and broke a tray of y dishes. How much did she make last week? *(Lessons 3-2, 4-2)*

18. Calculate in your head. *(Previous course)*
    a. 25% of 120   **b.** 25% of 18   **c.** 25% of 11

In 19–21, solve. *(Lessons 6-3, 6-6, 6-8)*

19. $x + .03x + 50 = 113.86$ **20.** $3(w + 4) = 26 - 5(w + 12)$

21. $4x + 9 = 2x$

22. Six less than twice a number is equal to two more than the number. Find the number. *(Lesson 6-6)*

23. Twenty-four is two fifths of what number? *(Lesson 4-4)*

24. *Skill sequence* Simplify.
    a. $\dfrac{3}{x} + \dfrac{12}{x}$ **b.** $\dfrac{x}{7} - \dfrac{2x}{5}$ **c.** $\dfrac{x}{y} + \dfrac{w}{z}$ *(Lesson 2-3)*

25. Write $\dfrac{12!}{3!\ 4!}$ in scientific notation. *(Lesson 4-9, Appendix B)*

8.4 in.

R 11 in.

26. Recall that a formula for the area of a triangle is $A = \frac{1}{2}bh$. Find the area of $\triangle RST$ at the left. *(Lesson 1-5)*

27. Write as a decimal.
    a. 1 divided by 4
    b. 1 divided by .4
    c. 1 divided by .04
    d. 1 divided by .000004 *(Previous course)*

28. Consider the two equations:
    i. $\dfrac{x + 2}{5} = \dfrac{x}{4}$ **ii.** $\dfrac{x}{3} + 5 = \dfrac{x + 1}{7}$
    a. Which equation is a proportion?
    b. Solve the equation you chose in part a. *(Lesson 5-7)*

29. Which activity will produce a graph like this?

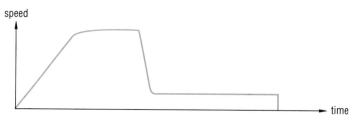

Choose the best answer from the following and explain how it fits the graph.

    fishing   pole vaulting   100-meter sprint
    sky diving   golf   archery
    drag racing   javelin throwing

Horizontal and Vertical Lines

Consider the official temperature readings at the Chicago lakefront on October 27, 1985, as shown in the table below at the left. When the ordered pairs (1, 60), (2, 60), and (3, 60) are graphed, the points lie on a **horizontal** line.

time	temperature
1 A.M.	60°
2 A.M.	60°
3 A.M.	60°

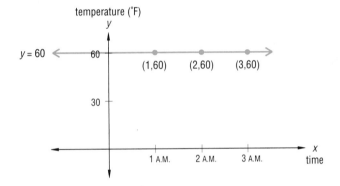

All the points have the same *y-coordinate*. The *y*-coordinate always equals 60. An equation for the line is therefore $y = 60$. Every horizontal line has an equation that is this simple. The equation says that the *y*-coordinate must be 60, but there are no restrictions on *x*. The line is parallel to the *x*-axis.

Every horizontal line has an equation of the form **y = k** where *k* is a fixed real number.

A **vertical** line is drawn below. Notice that each point on the line has the same *x-coordinate*, 2.5. Thus an equation for the line is $x = 2.5$. This means *x* is fixed at 2.5, but *y* can be any number. The line is parallel to the *y*-axis.

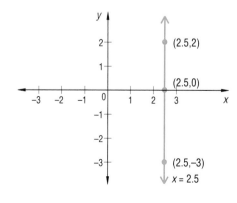

> Every vertical line has an equation of the form **x = h** where *h* is a fixed real number.

When you see an equation having only one variable, such as $x = -5$, you need to decide from the directions or the context of the problem whether it is graphed on a number line (in which case it is a point),

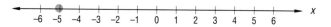

or on a coordinate plane (in which case it is a line).

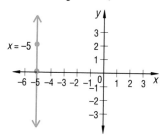

In a similar way, *inequalities* with one variable can be graphed on a number line or on a coordinate plane.

Example 1   Graph $y < 3$:
  a. on a number line;
  b. on a coordinate plane.

Solution
  a.

Recall that the *open circle* at 3 means that y is not equal to 3. The *ray* to the left of 3 is drawn. All points on this ray have a coordinate less than 3.

  b.

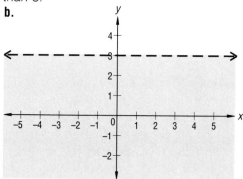

First graph the horizontal line $y = 3$. The line is *dotted* to indicate that the points having a y-coordinate of 3 are *not* to be included. Then the **half-plane** below the line is shaded. All points in the shaded half-plane have a y-coordinate which is less than 3.

When you graph an inequality on a number line, you use open and closed dots to show whether the boundary point is included. In a similar fashion, on a coordinate plane, use solid or dotted lines to show if the **boundary line** is included in the graph.

Example 2 Give a sentence describing all points in each shaded region.

a.

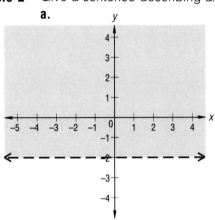

b.

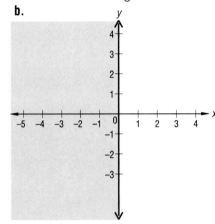

Solution

a. Every point in the shaded region has y-coordinate greater than -2. The dotted line shows that points with y-coordinate equal to -2 are not included. So a sentence describing this shaded region is $y > -2$.

b. Every point to the left of the y-axis has a negative x-coordinate. The solid line indicates that all points with x-coordinate equal to 0 are also to be included. So a sentence describing the shaded region is $x \leq 0$.

Questions

Covering the Reading

1. All points on a horizontal line have the same __?__.

2. All points on a vertical line have the same __?__.

3. *True or false* In the coordinate plane, points on the line $y = 60$ can have any real number for their x-coordinate.

4. *True or false* The graph of all points in the coordinate plane satisfying $x = -.19$ is a vertical line.

In 5–8, **a.** graph the points on a number line satisfying each equation; **b.** graph the points in the coordinate plane satisfying each equation.

5. $y = \frac{1}{2}$

6. $x = -15$

7. $y \geq 0$

8. $x < 5$

In 9–11, write an equation or inequality for each graph.

9.

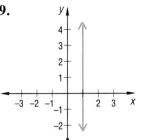

10.

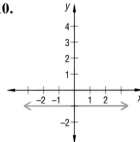

11.

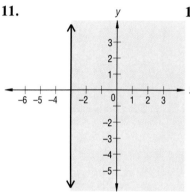

12.

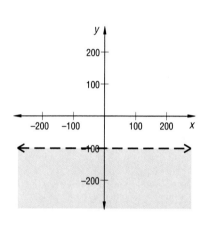

13. a. Graph all points on a number line satisfying $3x > 6$.
 b. Graph all points in the coordinate plane satisfying $3x > 6$.

In 14–16, write an equation for the line containing the points. The equation will be of the form $y = k$ or $x = h$.

14. (-9, 12), (4, 12), (.3, 12) **15.** (-6, -3), (-6, 0), (-6, 200)

16. $(4, m)$, $(0, m)$, $(-2, m)$

17. Graph the lines $y = -4$ and $x = 15$ and find the coordinates of the point of intersection.

18. a. Write an equation of the horizontal line through (7, -13).
 b. Write an equation of the vertical line through (7, -13).

19. a. Write an equation for the x-axis.
 b. Write an equation for the y-axis.

x	y
-1	
0	
1	
2	

20. a. Copy and complete the chart shown at the right for $y = -2x - 1$.
 b. Graph $y = -2x - 1$. *(Lesson 7-1)*

21. Write a computer program to print a table of (x , y) values for the equation $y = -.63 + 19x$ where x is 0, 1, 2, 3, 4, and 5. *(Lesson 7-1)*

22. In your head, find y if $y = -x - 10$ and
 a. $x = 15$; **b.** $x = 2$; **c.** $x = 4$. *(Lesson 1-4)*

23. A glacier has already moved 25 inches and moves at a rate of 2 more inches per week. Let t be the total number of inches the glacier has moved after w weeks. Write a formula for t in terms of w. *(Lesson 6-4)*

24. A contractor estimates the cost of building a house at $85,000. His actual cost C is within $5,000 of his estimate. *(Lesson 1-3)*
   a. What are the endpoints of the interval in which C lies?
   b. What is the length of the interval?

25. Simplify: $3(2x^2 + 15x - 11) + 10x^2$. *(Lessons 2-2, 6-3)*

26. Find the value of x in triangle ABC at left below. *(Lessons 3-8, 6-1)*

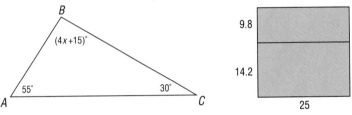

27. Show two ways to find the total area of the rectangles at right above. *(Lesson 6-8)*

In 28–31, solve. *(Lessons 4-4, 6-6)*

28. $-750y = 1350$

29. $-7.5 = 10.7 - 1.35w$

30. $y = 2.5y$

31. $.4 + .7x = -2.1x + 2.5$ *(Lesson 6-6)*

Exploration

32. Marie is between 150 and 160 cm tall and weighs between 50 and 55 kg. Copy the graph on the right. Display all Marie's possible heights and weights on your graph.

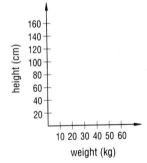

33. The robot below is described by the following inequalities:

Chest: $2 \le x \le 6$ and $4 \le y \le 8$
Middle: $3 \le x \le 5$ and $2 \le y \le 4$
Legs: $3 \le x \le 3.5$ and $0 \le y \le 2$
      $4.5 \le x \le 5$ and $0 \le y \le 2$

Copy the diagram and put a head on the robot. Describe the head with inequalities.

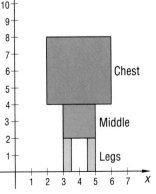

7-3

Distance and Absolute Value

Subtraction can be used to compare two numbers, to find out how much larger one is than another. 2000 is 45 larger than 1955 because $2000 - 1955 = 45$. In Lesson 3-2, we called this the Comparison Model for Subtraction.

Sometimes, when you compare, you do not care which number is larger. Suppose you are to guess the number of marbles in a large jar and the closest guess wins. Here are the guesses of five people: 350, 400, 382, 586, and 290. If the jar contains 487 marbles, then notice how the errors are calculated.

$$487 - 350 = 137$$
$$487 - 400 = 87$$
$$487 - 382 = 105$$
$$487 - 586 = \text{-}99$$
$$487 - 290 = 197$$

If the smallest error wins, then 586 is the winning guess because -99 is the smallest error, $\text{-}99 < 87$. But of course the guess with error 87 should win. If there were other guesses, then there might be other negative errors. What we would like is to change all the errors to positive numbers.

The operation which keeps positive numbers and zero the same and changes negatives to positives is called taking the **absolute value.**

Definition:

If a number is negative, its absolute value is its opposite. If a number is positive or zero, it equals its absolute value.

Taking the absolute value of a negative number changes it to a positive. The absolute value of -116 is 116. Taking the absolute value of a nonnegative number leaves it unchanged. The absolute value of 29 is 29. The absolute value of 0 is 0.

The symbol for absolute value is $|\ |$. So $|\text{-}116| = 116$, $|29| = 29$, and $|0| = 0$. With this symbol and algebra, you can make the definition shorter.

If $n < 0$, then $|n| = -n$.
If $n \geq 0$, then $|n| = n$.

Notice in the definition that $-n$ is positive because $n < 0$. The absolute value of a number cannot be negative.

Example 1 Solve and graph the solution set on a number line: $|x| = 5$.

> **Solution** You can think: The absolute value of a number is 5. What is the number? The number could be 5 or -5. So $x = 5$ or $x = -5$. Here is the graph.

> **Check** $|-5| = 5$; or $|5| = 5$.

The number line enables a geometric interpretation to be given to absolute value. *The absolute value of a number is its distance from the origin.*

This idea can be extended to find a formula for the distance between any two points on a number line. In the marble-guessing contest, the expression $|487 - G|$ gives the error between a guess G and the actual quantity of 487 marbles. On the number line, $|487 - G|$ represents the distance between 487 and G. The general formula for the distance between two points on a line is of the same form.

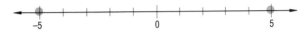

Distance Formula on a Number Line:

> If two points on a line have coordinates x_1 and x_2, the distance between them is $|x_1 - x_2|$.

Instead of $|x_1 - x_2|$ you could use $|x_2 - x_1|$. Since $x_2 - x_1$ and $x_1 - x_2$ are opposites, they have the same absolute value. As an example, the distance between 72 and 92 is either $|72 - 92|$ or $|92 - 72|$. Either way you get 20.

Example 2 Solve and graph on a number line: $|487 - G| = 100$.

Solution 1 The number between the absolute value signs, $487 - G$, represents either 100 or -100.

$$487 - G = 100 \qquad \text{or} \qquad 487 - G = -100$$

Solve each equation.

$$G = 387 \qquad \text{or} \qquad G = 587$$

Solution 2 Think of $|487 - G| = 100$ as the error in the guess. Again the solutions are 387 or 587.

The description $4 \leq x \leq 14$ of an interval is not a simple sentence because there are two $\leq$ signs. You can read this as "x is less than or equal to 14 *and* x is greater than or equal to 4." By using absolute value, intervals can be described using one $\leq$ sign.

Example 3 Write the interval $4 \leq x \leq 14$ using the absolute value symbol and one $\leq$ sign.

Solution First, calculate the midpoint of the interval. This is the mean of 4 and 14, which is 9.

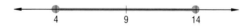

Now calculate the distance between the midpoint 9 and either endpoint using the distance formula (or in your head). That distance is found to be 5. Every other point on the interval is within 5 of the midpoint. Let x be a point on the interval.

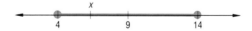

Then $|x - 9|$ is the distance from x to the midpoint. So the answer is

$$|x - 9| \leq 5.$$

You can read this as "the absolute value of $x - 9$ is less than or equal to 5" or "the distance from x to 9 is less than or equal to 5."

Another way to describe an interval is with an error description. If the mean of the interval is 9, then ± 5 describes the amount of possible error.

You now have learned three ways to describe an interval such as the one from 4 to 14.

1. endpoint description: $\qquad$ $4 \le x \le 14$
2. error description or tolerance: $\qquad$ 9 ± 5
3. distance description: $\qquad$ $|x - 9| \le 5$

The advantage of the distance description is that it lends itself to graphing sentences. Solutions to the three related sentences $|x - 9| = 5$, $|x - 9| < 5$, and $|x - 9| > 5$ are all related to the interval from 4 to 14. Think: The distance from x to 9 equals, is less than, or is greater than 5.

Example 4 Graph all solutions:
   a. $|x - 9| = 5$
   b. $|x - 9| < 5$
   c. $|x - 9| > 5$

Solutions
   a. $|x - 9| = 5$
   b. $|x - 9| < 5$
   c. $|x - 9| > 5$

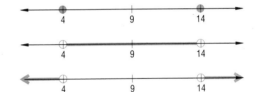

Computer languages have commands that take the absolute value of a number. In BASIC, $|x|$ is written **ABS(X).** The following program asks for the coordinates of two points on a number line and then finds the distance between them.

```
10 PRINT "ENTER THE FIRST COORDINATE"
20 INPUT X1
30 PRINT "ENTER THE SECOND COORDINATE"
40 INPUT X2
50 PRINT "THE DISTANCE IS "; ABS(X1 - X2)
60 END
```

On a coordinate plane, every horizontal or vertical line can be considered as being a number line. Therefore you can use the distance formula to find the distance between two points that are on the same horizontal or vertical line.

Example 5 $P = (-3, 5)$ and $Q = (-3, -2)$. Find the distance between P and Q.

Solution The two points are on the same vertical line. All points on this line of the graph have x-coordinates of -3. Only the y-coordinates are different. Find the distance between the two points by using their y-coordinates and the distance formula. The distance is $|5 - (-2)| = 7$.

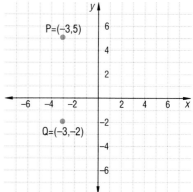

In Lesson 7-7, you will learn to find the distance between two points that are *not* on the same horizontal or vertical line.

Questions

Covering the Reading

1. A jar contains 487 marbles. Determine the error in your guess if you guess it contains:
a. 438 marbles;   **b.** 512 marbles;   **c.** G marbles.

2. Give the definition of absolute value:
a. in words;
b. in symbols.

In 3–5, give the absolute value of each number.
3. 72   **4.** -3.4   **5.** 0

In 6–8, simplify.
6. $|7\frac{1}{2}|$ **7.** $|5 - 6|$ **8.** $|5| - |6|$

In 9–11, find the distance between the two points whose coordinates are given.

9. $\xrightarrow{\quad \bullet \qquad\qquad \bullet \quad}$
     -28   11

10. $\xrightarrow{\quad \bullet \qquad\qquad \bullet \quad}$
     -81   -57

11. $\xrightarrow{\quad \bullet \qquad\qquad \bullet \quad}$
     17.5 x

12. What will the computer print if the following line is entered?
PRINT ABS(-6 − 20)

13. If the input for the program in this lesson is 30 and -8, what will the computer print in line 50?

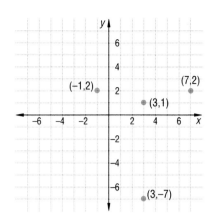

In 14 and 15, use the graph above. Find the distance between

14. (7, 2) and (-1, 2);

15. (3, 1) and (3, -7).

In 16 and 17, each equation has two solutions. Solve and graph on a number line.

16. $|t| = 25$ **17.** $|300 - x| = 10$

18. Graph all solutions to each sentence.
     a. $|x - 10| = 40$
     b. $|x - 10| > 40$
     c. $|x - 10| < 40$

Applying the Mathematics

19. Simplify. **a.** $|2| + |-2|$ **b.** $-3 \cdot |-2|$

In 20–22, make choices from these equations.
(a) $|n| = 0$ (b) $|n| = -6$ (c) $|n| = 31$

20. Which equation has two solutions?

21. Which equation has one solution?

22. Which equation has no solution?

23. Find the perimeter of the rectangle at the left below.

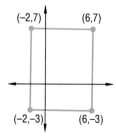
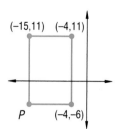

24. Find the coordinates of point P in the rectangle at the right above.

25. M is the midpoint of $\overline{PQ}$ and has coordinate m. The interval is described by $m \pm 4$.
   a. Write an expression for the coordinate of Q.
   b. Write an expression for the coordinate of P.
   c. What is the distance between P and M?
   d. What is the distance between P and Q?

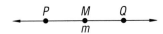

26. The distance between two points on a number line with coordinates $2v - 8$ and $4v + 1$ is 12. Find all possible values of v.

Review

In 27–29, graph in the plane.

27. $y = \frac{1}{2}x - 5$ *(Lesson 7-1)*

28. $y = 3$ *(Lesson 7-2)*

29. $x < 1$ *(Lesson 7-2)*

30. Give the property which justifies each step. *(Lessons 6-1, 6-3)*

$$3(\tfrac{1}{2}x - 4) = 10$$

   a. $\frac{3}{2}x - 12 = 10$
   b. $\frac{3}{2}x = 22$
   c. $x = \frac{44}{3}$

31. How many square centimeters are in a rectangle with width $9x$ cm and length $12x$ cm? *(Lesson 4-1)*

32. Calculate in your head. *(Lesson 1-4)*
   a. 4^2 **b.** $(\tfrac{1}{3})^2$ **c.** $(0.1)^2$

Exploration

33. **a.** Make a table of various values for $y = |x - 3|$ where $-10 \le x \le 10$.
   b. Graph the points you found in part a.
   c. Describe the graph of *all* points for which $y = |x - 3|$.

Square Roots

In Lesson 7-3, you saw that the equation

$$|x| = 5$$

has two solutions, 5 and -5. Another equation with these same two solutions is $x^2 = 25$. We call the numbers 5 and -5 the **square roots** of 25.

Definition:

If $A = s^2$, then s is called a square root of A.

Because $16 = 4^2$, we say that 4 is a square root of 16. Similarly 20.25 is the square of 4.5, and 4.5 is a square root of 20.25. The term *square root* comes from the geometry of squares and their sides.

Pictured are the squares of 4 and of 4.5.

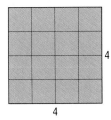

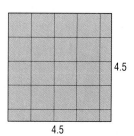

Area = $4 \cdot 4 = 4^2 =$
16 sq units

Area = $4.5 \cdot 4.5 = 4.5^2 =$
20.25 sq units

The symbol for square root is $\sqrt{}$ and is called a **radical sign.** From the above, $\sqrt{16} = 4$ and $\sqrt{20.25} = 4.5$.

You are familiar with squares of whole numbers. These are called **perfect squares.**

$$0^2 = 0 \quad 1^2 = 1 \quad 2^2 = 4 \quad 3^2 = 9 \quad 4^2 = 16 \quad \text{and so on.}$$
$$0 = \sqrt{0} \quad 1 = \sqrt{1} \quad 2 = \sqrt{4} \quad 3 = \sqrt{9} \quad 4 = \sqrt{16} \quad \text{and so on.}$$

Some square roots are not whole numbers. You can get a rough estimate of the square root by placing it between whole numbers. A good approximation can be found using the calculator.

Example 1 Estimate $\sqrt{2}$.

Solution 1 A rough estimate: since $\sqrt{1} = 1$ and $\sqrt{4} = 2$, $\sqrt{2}$ must be somewhere between 1 and 2.

Solution 2 A good approximation: use a calculator.
Key sequence: 2   [√x̄]

Display: ⌐   2.⌐   ⌐ 1.4142136 ⌐

The actual decimal for $\sqrt{2}$ is infinite and does not repeat. The number 1.4142136 is an estimate of $\sqrt{2}$.

Check Multiply 1.4142136 by itself. One calculator shows that
$1.4142136 \cdot 1.4142136 \approx 2.0000001$

Example 2 The area of a square is 198 sq in. Give the length of a side: **a.** exactly, using a radical symbol; **b.** approximated to two decimal places.

Solution
a. Since the area is 198 sq in., the side is $\sqrt{198}$ in.
b. Use a calculator to evaluate $\sqrt{198}$. The side is approximately 14.07 in.

Check Does $14.07 \cdot 14.07 = 198$? $(14.07)^2 = 197.9649$, so it checks.

The check to Example 2 was approximate because 14.07 was used to approximate $\sqrt{198}$. But $\sqrt{198} \cdot \sqrt{198} = 198$ *exactly*, as does $\sqrt{2} \cdot \sqrt{2} = 2$. In general,

---

For any nonnegative number n, $\sqrt{n} \cdot \sqrt{n} = n$ and $(\sqrt{n})^2 = n$.

The equation $W^2 = 49$ has an obvious solution 7, because $7 \cdot 7 = 49$. However, $-7 \cdot -7 = 49$, so -7 is also a solution. Every positive number has *two* square roots, one positive and one negative. The radical sign symbolizes only the *positive* one. $\sqrt{49}$ means "the positive square root of 49" so $\sqrt{49} = 7$. The symbol for the *negative* square root of 49 is $-\sqrt{49}$. So $-\sqrt{49} = -7$.

The two square roots are opposites of each other. For example, the square roots of 97 are $\sqrt{97}$ and $-\sqrt{97}$, which are approximately 9.85 and -9.85, respectively. That is, $\sqrt{97} \cdot \sqrt{97} = 97$ and $-\sqrt{97} \cdot -\sqrt{97} = 97$.

When solving equations like $W^2 = 49$ or $97 = n^2$, you should give all solutions. When the solutions are not integers, your teacher may expect two versions: **a.** the exact solution and **b.** decimal approximations rounded to the nearest hundredth.

Example 3 Solve $\dfrac{2}{d} = \dfrac{d}{11}$. Express the solutions **a.** exactly and **b.** rounded to the nearest hundredth.

Solution Apply the Means-Extremes Property.

$$d^2 = 22$$

a. There is no integer whose square is 22. The solution is $d = \sqrt{22}$ or $d = -\sqrt{22}$.

b. A calculator shows $\sqrt{22} \approx 4.69$. The approximate solutions to the proportion are 4.69 and -4.69.

Check

a. Is $\dfrac{2}{\sqrt{22}} = \dfrac{\sqrt{22}}{11}$? Yes, since $\sqrt{22} \cdot \sqrt{22} = 22$.

b. Is $\dfrac{2}{4.69} \approx \dfrac{4.69}{11}$? Yes, since $\dfrac{2}{4.69} \approx 0.4264$ and $\dfrac{4.69}{11} \approx 0.4264$.

Suppose you were asked to solve the equation $x^2 = -4$. Your first guess might be that $x = -2$. But $-2 \cdot -2$ is 4, so -2 is not a solution. In the real number system, the equation $x^2 = -4$ has *no solution* because no number multiplied by itself gives -4. When n is negative, $\sqrt{n}$ is not a real number. To summarize:

Every positive number has *exactly two* real square roots. Every negative number has *no* real square roots. Zero has *exactly one* square root, itself. $\sqrt{0} = 0$

The equation $x^2 = k$ can be solved by taking square roots of each side. This results in $x = \pm\sqrt{k}$. The equation $\sqrt{y} = k$ can be solved for y by doing the reverse process, *squaring both sides*. This results in $y = k^2$.

Example 4 Solve $\sqrt{t} = 16$.

Solution Multiply the left side by $\sqrt{t}$, the right side by 16. Since they are equal, this can be done. The results are the squares of both sides.

$$(\sqrt{t})^2 = (16)^2$$
$$t = 256$$

Check Does $\sqrt{256} = 16$? Yes, $16^2 = 256$.

You can also use the idea that $(\sqrt{x})^2 = \sqrt{x} \cdot \sqrt{x} = x$ to simplify expressions.

■ ■ ■ ■ ■ ■■■

Example 5 Multiply $4\sqrt{10} \cdot \sqrt{10}$.

Solution Think of $4\sqrt{10}$ as being $4 \cdot \sqrt{10}$.

$$4\sqrt{10} \cdot \sqrt{10} = 4 \cdot \sqrt{10} \cdot \sqrt{10}$$
$$= 4 \cdot (\sqrt{10})^2$$
$$= 4 \cdot 10$$
$$= 40$$

Check Convert to decimals. $\sqrt{10} \approx 3.16$, so $4\sqrt{10} \approx 12.64$. $4\sqrt{10} \cdot \sqrt{10} \approx 12.64(3.16) = 39.9424 \approx 40$. It checks.

A computer will also calculate square roots. In BASIC, **SQR(X)** is $\sqrt{x}$. Consider this program:

```
10 REM PROGRAM TO COMPUTE SQUARE ROOTS
20 INPUT "TYPE IN A NUMBER"; A
30 PRINT "CALCULATE SQUARE ROOTS"
40 IF A > 0 THEN PRINT SQR(A), -SQR(A)
50 IF A = 0 THEN PRINT A
60 IF A < 0 THEN PRINT "NO REAL SQUARE ROOTS"
70 END
```

■ ■ ■ ■ ■ ■■■

Example 6 What will the computer print if $A = 39$?

Solution Since $39 > 0$, the computer will print:

```
TYPE IN A NUMBER ?39
CALCULATE SQUARE ROOTS
6.244997998     -6.244997998
```

Questions

Covering the Reading

1. Because $169 = 13 \cdot 13$, 13 is called a __?__ of 169.

2. $72.25 = 8.5 \cdot 8.5$; so __?__ is a square root of __?__.

3. When $A = s^2$, A is called the __?__ of s.

4. Name the two square roots of 16.

5. Compute in your head. **a.** $\sqrt{100}$ **b.** $-\sqrt{81}$ **c.** $16\sqrt{81}$

6. a. Approximate $\sqrt{407}$ to the nearest hundredth using your calculator.
b. Check your answer.

7. Between what two integers is $\sqrt{15}$?

8. The area of a square is 289m². What is the length of a side of the square?

9. Every positive number has __?__ square root(s).

In 10–15, find all solutions.

10. $x^2 = 121$

11. $301 = b^2$

12. $\dfrac{9}{x} = \dfrac{x}{16}$

13. $\sqrt{n} = 20$

14. $\sqrt{t} = 9$

15. $-16 = m^2$

In 16 and 17, what will the computer program in the lesson cause the computer to print for the inputted value of A?

16. $A = 0$

17. $A = -13$

Applying the Mathematics

n	$\sqrt{n}$
1	*1*
2	*1.414*
3	
4	
5	
6	
7	
8	
9	
10	
11	
12	
13	
14	
15	
16	
17	
18	
19	
20	*4.472*

18. Find both solutions to $m^2 + 64 = 100$.

19. Use a calculator to complete this table of three-place approximations to positive square roots of the whole numbers from 1 to 20.

20. For clarity, the product of a number y and the square root of x is written $y\sqrt{x}$ instead of $\sqrt{x} \cdot y$. ($\sqrt{x} \cdot y$ can be mistaken for $\sqrt{xy}$.) When $x = 4$ and $y = 9$, evaluate
a. $y\sqrt{x}$ **b.** $\sqrt{xy}$ **c.** $x\sqrt{y}$

21. Solve $\sqrt{r} = 0.81$.

22. Simplify $2\sqrt{15} \cdot 8\sqrt{15}$.

23. Find the area of this rectangle.

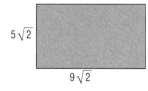
$5\sqrt{2}$
$9\sqrt{2}$

24. Evaluate: **a.** $\sqrt{3^2 + 4^2}$; **b.** $\sqrt{3^2 \cdot 4^2}$.

25. Solve: $|x - 3| = 4$. *(Lesson 7-3)*

26. Write an expression for the distance between the points (p, m) and (a, m). *(Lesson 7-3)*

27. Here are three points: $(0, -3)$, $(0, 0)$, $(0, 3)$.
    a. Write an equation of the line containing the three points.
    b. On which axis do they lie? *(Lesson 7-2)*

28. Graph all pairs of solutions to $y = -2x + 3$. *(Lesson 7-1)*

29. Solve for g. $\dfrac{g - 3}{5} = \dfrac{2g + 8}{3}$ *(Lessons 5-7, 6-6, 6-8)*

30. Solve $2p + 5k = 17k$ for p. *(Lesson 6-6)*

31. If $x + 5 = 12$ and $8 + y + 3y = 0$, then find the value of $y - 2x$. *(Lesson 6-3)*

32. What value cannot replace y in this expression? *(Lessons 5-2, 6-1)*

$$\frac{9}{8 - 4y}$$

33. A circle is inscribed in a square as shown at the right. If a point is selected at random from the square region, what is the probability it lies outside the circle? *(Lesson 5-5)*

34. Trace $\overline{AB}$. Find points C and D so that $ABCD$ is a square. *(Previous course)*

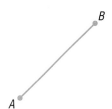

35. Use your calculator to try to evaluate $\sqrt{-4}$.
    a. Write the key sequence you are using.
    b. What does your calculator display?
    c. Why does the calculator display what it does?

7-5

Pythagorean Theorem

A very famous statement in geometry, the Pythagorean Theorem, gets its name from the Greek mathematician Pythagoras, who lived about 2500 years ago. This formula relates the lengths of the three sides of a right triangle.

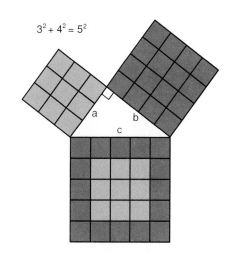

$3^2 + 4^2 = 5^2$

In a right triangle, one of the angles must be 90°. In the diagram the right angle is formed by the sides of length a and b. Those sides are called the **legs** of the right triangle. The longest side of the triangle is across from the right angle and is called the **hypotenuse.** The diagram shows that the square of the hypotenuse equals the sum of the squares of the squares of the two legs.

Pythagorean Theorem:

In a right triangle with legs a and b and hypotenuse c,
$$a^2 + b^2 = c^2.$$

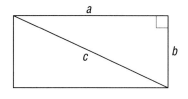

You can think of a right triangle as being formed by cutting a rectangle in half. Two sides of the rectangle become the triangle's legs and the diagonal becomes its hypotenuse.

■ ■ ■ ■ ■ ■ ■ ■

Example 1 What is the length of the hypotenuse of the right triangle drawn at right?

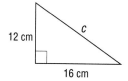

12 cm

16 cm

Solution Substitute the lengths of the legs of the triangle for a and b in the formula. The hypotenuse is c.

Substituting,
$$a^2 + b^2 = c^2$$
$$12^2 + 16^2 = c^2$$
$$144 + 256 = c^2$$
$$400 = c^2$$

The solution for c is either the positive or negative square root of 400.

$$c = 20 \text{ or } c = -20$$

However, c represents the length of the hypotenuse, and its length cannot be negative. So use only the positive solution. The hypotenuse is 20 cm.

Take the square root of each side of the Pythagorean Theorem. You get
$$\sqrt{a^2 + b^2} = \sqrt{c^2}.$$
That is,
$$\sqrt{a^2 + b^2} = c.$$

To evaluate $\sqrt{a^2 + b^2}$, you must use the correct order of operations. The radical sign $\sqrt{}$ acts as a grouping symbol. The powers must be done first, and then the addition, before the square root is evaluated.

Example 2   Show that $\sqrt{3^2 + 4^2} \neq 3 + 4$.

Solution   Follow the order of operations.
$$\sqrt{3^2 + 4^2} = \sqrt{9 + 16}$$
$$= \sqrt{25}$$
$$= 5$$

But $3 + 4 = 7$ and $5 \neq 7$. So $\sqrt{3^2 + 4^2} \neq 3 + 4$.

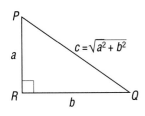

Notice that $\sqrt{3^2 + 4^2} < 3 + 4$. For *all* nonzero a and b, $\sqrt{a^2 + b^2} < a + b$. This fact can be visualized using the right triangle shown here. The hypotenuse is the direct path from P to Q and has length $\sqrt{a^2 + b^2}$. The sum $a + b$ is the length of the path from P to R to Q. The path with length $\sqrt{a^2 + b^2}$ is shorter than the path with length $a + b$.

The following program finds the length of the hypotenuse of a right triangle if the lengths of the legs are entered.

```
10 PRINT "FIND LENGTH OF HYPOTENUSE"
20 PRINT "GIVE LENGTH OF ONE LEG"
30 INPUT A
40 PRINT "GIVE LENGTH OF OTHER LEG"
50 INPUT B
60 LET C = SQR(A * A + B * B)
70 PRINT "LENGTH OF HYPOTENUSE IS"; C
80 END
```

Notice the parentheses for grouping in line 60. When 12 and 16 are entered for A and B, the computer calculates A * B + B * B to be 400. Then it takes just the positive square root of 400. So C = SQR(400)=20.

Sometimes you know the hypotenuse and one leg, and you want to find the length of the other leg of a right triangle.

Example 3   The bottom of a ladder is 3 feet from a wall. The ladder is 12 feet long. How far above the ground does the top of the ladder touch the wall?

Solution   Let x stand for the distance (in feet) from the ground to the top of the ladder. This is one of the legs of the right triangle shown above. The other leg is 3 ft and the hypotenuse is 12 ft (the length of the ladder). According to the Pythagorean Theorem,

$$x^2 + 3^2 = 12^2$$
$$x^2 + 9 = 144$$
$$x^2 = 135$$
$$x = \sqrt{135} \text{ or } -\sqrt{135}$$

Length is positive; ignore the negative solution.
So the ladder touches the wall $\sqrt{135} \approx 11.62$ feet above the ground.

Questions

Covering the Reading

1. State the Pythagorean Theorem.

2. Examine the right triangle.
   a. Which sides are the legs?
   b. Which side is the hypotenuse?

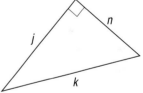

3. A right triangle can be formed by slicing what figure along its diagonal?

4. *True or false*   The longest side of a right triangle is always a leg.

5. Evaluate when $x = 20$ and $y = 21$.
   a. $\sqrt{x^2 + y^2}$ **b.** $x + y$

6. What will the computer program in this lesson print if 5 and 12 are entered for the values of A and B?

In 7 and 8, find the value of the variable. If the answer is not a whole number, find both its exact value and an approximation rounded to the nearest hundredth.

7.

8.

9. In Example 3, if the bottom of the ladder is moved out so that it is 5 feet away from the wall, how far up the building will the top of the ladder reach?

10. Explain why $\sqrt{a^2 + b^2} < a + b$ whenever a and b are lengths.

11. Find x.

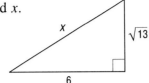

12. Some pedestrians want to get from point A to point B. The two roads shown at right meet at right angles. **a.** If they follow the roads, how far will they walk? **b.** Instead of walking along the road, they take the shortcut. Use the Pythagorean Theorem to find the length of the shortcut, rounded to the nearest tenth of a km. **c.** How much distance do they save?

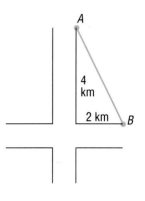

13. The area of a square is 81 square centimeters.
   a. Find the length of a side.
   b. Use the Pythagorean Theorem to find the length of the diagonal.

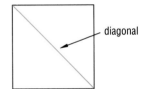

14. a. Solve $a^2 + b^2 = c^2$ for a, if a is positive. **b.** Write a computer program to find the length of one leg of a right triangle if the lengths of the other leg and the hypotenuse are entered.

15. If an object is dropped, it takes about $\sqrt{\dfrac{d}{16}}$ seconds to fall d feet. If an object is dropped from a 100-foot building, about how long does it take to hit the ground? *(Lesson 7-4)*

16. Simplify in your head. *(Lesson 7-4)*
   a. $3\sqrt{25}$ **b.** $10 + 4\sqrt{144}$ **c.** $2\sqrt{9} + -6a\sqrt{100}$

In 17 and 18, use your calculator. Fill in the blanks with $<$, $=$, or $>$. *(Lesson 7-4)*

17. $\sqrt{2} + \sqrt{8}$ _?_ $\sqrt{10}$ **18.** $\sqrt{2} \cdot \sqrt{8}$ _?_ $\sqrt{16}$

19. Sean guessed there were 823 beans in a jar. The actual number of beans was 1000. What is the error of Sean's guess? *(Lesson 7-3)*

20. Which of these are perfect squares? *(Lesson 7-4)*
   a. 441   **b.** 1030   **c.** 1296

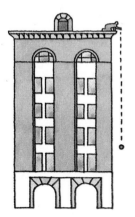

21. Solve. *(Lessons 7-3, 7-4)*
    a. $x^2 = 121$ **b.** $\sqrt{y} = 121$ **c.** $|z| = 121$

22. Find the distance between (80, 60) and (-200, 60). *(Lesson 7-3)*

23. a. Solve $7x + 2y = 14$ for y. *(Lesson 6-1)*
    b. Graph $7x + 2y = 14$. *(Lesson 7-1)*

24. Beth begins with $10 in her account and saves $5 a week. Seth begins with $30 in his account and saves $3 a week. After how many weeks will they have the same amount of money in their accounts? *(Lesson 6-6)*

25. Which is faster, reading w words in m minutes or reading $2w$ words in $3m$ minutes? *(Lesson 5-2)*

26. In the figure to the left, a square is drawn around a 3″ square and a 5″ square which overlap in a 1″ square. What is the probability that a randomly selected point falls in a colored area? *(Lesson 5-5)*

27. *Multiple choice* Which equation says that current sales C are 8% more than last year's sales y? *(Lesson 6-3)*
    (a) $1.08C = y$ (b) $1.08y = C$
    (c) $.92C = y$ (d) $.92y = C$

28. The rectangles below are similar. *(Lesson 5-7)*
    a. Use a proportion to find a.
    b. What is the perimeter of the smaller rectangle?

29. Find the area of a square with side $10\sqrt{3}$. *(Lesson 7-4)*

Exploration

30. Six of the segments in the figure below have length one unit.
    a. Find AB, AD, AE, AF, and AG.
    b. Copy the drawing and add to it to make a segment whose length is $\sqrt{8}$.

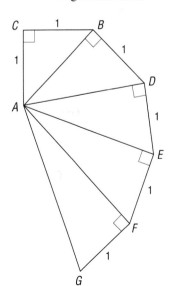

Square Roots of Products

Consider $\sqrt{ab}$, the positive square root of the product ab. Squaring gives

$$(\sqrt{ab})^2 = \sqrt{ab} \cdot \sqrt{ab} = ab.$$

Now consider $\sqrt{a} \cdot \sqrt{b}$, the product of two positive square roots. Squaring gives

$$(\sqrt{a} \cdot \sqrt{b})^2 = (\sqrt{a} \cdot \sqrt{b}) \cdot (\sqrt{a} \cdot \sqrt{b})$$
$$= \sqrt{a} \cdot \sqrt{b} \cdot \sqrt{a} \cdot \sqrt{b}$$
$$= \sqrt{a} \cdot \sqrt{a} \cdot \sqrt{b} \cdot \sqrt{b}$$
$$= ab.$$

Since $\sqrt{ab}$ and $\sqrt{a} \cdot \sqrt{b}$ are positive numbers and have the same square, $\sqrt{ab} = \sqrt{a} \cdot \sqrt{b}$. This is a useful property, so we give it a name.

Square Root of a Product Property:

If $a \geq 0$ and $b \geq 0$, then

$$\sqrt{ab} = \sqrt{a} \cdot \sqrt{b}.$$

This property can easily be checked. For instance, let $a = 9$ and $b = 100$. Then the Square Root of a Product Property indicates

$$\sqrt{9 \cdot 100} = \sqrt{9} \cdot \sqrt{100}$$
or $\sqrt{900} = 3 \cdot 10$, which is known to be true.

For another example, let $a = 4$ and $b = 3$. Substituting in the property

$$\sqrt{4 \cdot 3} = \sqrt{4} \cdot \sqrt{3}$$
thus, $\sqrt{12} = 2\sqrt{3}.$
Check with a calculator. $3.464\ldots = 2 \cdot 1.732\ldots$

This example shows that $\sqrt{12}$ is twice $\sqrt{3}$. In this case $\sqrt{12}$ has been rewritten with a smaller integer under the radical sign. This is called **simplifying the radical.** The key to simplifying is to find a perfect square factor of the number under the radical sign.

Example 1 Simplify $\sqrt{50}$.

Solution A perfect square factor of 50 is 25.

$$\begin{aligned} \sqrt{50} &= \sqrt{25 \cdot 2} \\ &= \sqrt{25} \cdot \sqrt{2} \qquad \text{Square Root of a Product Property} \\ &= 5\sqrt{2} \qquad\quad \text{since } \sqrt{25} = 5 \end{aligned}$$

Check Use your calculator. 50 $\boxed{\sqrt{\ }}$ gives 7.0710678.

$$5 \boxed{\times} 2 \boxed{\sqrt{\ }} \boxed{=} \text{ gives 7.0710678.}$$

Is $5\sqrt{2}$ really simpler than $\sqrt{50}$? It all depends. For most estimating and calculations, $\sqrt{50}$ is simpler. But for seeing patterns, $5\sqrt{2}$ may be easier. In the next example, the answer $8\sqrt{2}$ is related to the given information in a simple way. You would not see that without "simplifying" $\sqrt{128}$.

Example 2 Each leg of a right triangle is 8 inches long. Find the exact length of the hypotenuse.

Solution Use the Pythagorean Theorem to find the length of the hypotenuse.

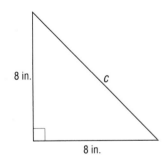

8 in.

8 in.

c

$$\begin{aligned} c^2 &= 8^2 + 8^2 \\ c^2 &= 128 \\ c &= \sqrt{128} \\ c &= \sqrt{64 \cdot 2} \qquad \text{64 is a perfect square factor of 128.} \\ &= \sqrt{64} \cdot \sqrt{2} \\ &= 8\sqrt{2} \end{aligned}$$

The exact length of the hypotenuse is $\sqrt{128}$ or $8\sqrt{2}$ inches. Can you see how the lengths of the legs and the hypotenuse are related? You are asked to generalize the pattern as an Exploration.

The Square Root of a Product Property applies to square roots containing variables.

Example 3 If x and y are both positive, simplify $\sqrt{9x^2y^2}$.

Solution $\begin{aligned}[t] \sqrt{9x^2y^2} &= \sqrt{9} \cdot \sqrt{x^2} \cdot \sqrt{y^2} \\ &= 3 \cdot x \cdot y \\ &= 3xy \end{aligned}$

Check Let $x = 2$, $y = 4$: $\sqrt{9(2)^2(4)^2} = \sqrt{9 \cdot 4 \cdot 16} = \sqrt{576} = 24$.
Does this equal $3 \cdot 2 \cdot 4$? Yes.

Solutions to some equations can be rewritten by using the Square Root of a Product Property.

Example 4 Solve the equation $2x^2 = 150$.
Solution

$$2x^2 = 150$$
$$x^2 = 75$$
$$x = \sqrt{75} \text{ or } -\sqrt{75}$$

Now rewrite $\sqrt{75}$. A perfect square factor is 25.

$$\sqrt{75} = \sqrt{25 \cdot 3}$$
$$= \sqrt{25} \cdot \sqrt{3}$$
$$= 5\sqrt{3}$$

So the two solutions to the equation are $5\sqrt{3}$ and $-5\sqrt{3}$.

Check First check $5\sqrt{3}$.

Does

$$2(5\sqrt{3})^2 = 150?$$
$$2 \cdot 5\sqrt{3} \cdot 5\sqrt{3} = 150?$$
$$2 \cdot 5 \cdot 5 \cdot \sqrt{3} \cdot \sqrt{3} = 150?$$
$$50 \cdot 3 = 150? \text{ Yes, it checks.}$$

Now check $-5\sqrt{3}$.

Does

$$2(-5\sqrt{3})^2 = 150?$$
$$2 \cdot -5\sqrt{3} \cdot -5\sqrt{3} = 150?$$
$$2 \cdot -5 \cdot -5 \cdot \sqrt{3} \cdot \sqrt{3} = 150? \text{ Yes, it checks.}$$

Questions

Covering the Reading

1. By the Square Root of a Product Property, $\sqrt{p} \cdot \sqrt{q} = \underline{\ ?\ }$.

2. To use the Square Root of a Product Property to rewrite $\sqrt{ab}$, both a and b must be $\underline{\ ?\ }$.

3. With a calculator, estimate to two decimal places:
   a. $\sqrt{7}$ **b.** $\sqrt{5}$ **c.** $\sqrt{7} \cdot \sqrt{5}$ **d.** $\sqrt{35}$

In 4–6, **a.** find the perfect square factor of the number under the radical sign; **b.** simplify.

4. $\sqrt{20}$ 5. $\sqrt{50}$ 6. $\sqrt{700}$

7. With your calculator, as a check to Example 2, verify that $\sqrt{128} = 8\sqrt{2}$.

8. If a and b are both positive, simplify $\sqrt{36a^2b^2}$.

9. Each leg of a right triangle is 10 cm. Find the exact length of the hypotenuse.

10. Solve $3x^2 = 2100$.

Applying the Mathematics

In 11 and 12, find the exact value for the variable. Rewrite your answer so the number under the radical has no perfect square factors.

11.

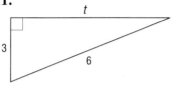

12.

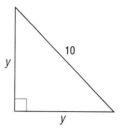

In 13 and 14, solve and check. Give your answer in simplified form.

13. $(2a)^2 = 24$

14. $\dfrac{9}{x} = \dfrac{x}{6}$

15. Simplify each radical expression.
   a. $\sqrt{75}$ **b.** $\sqrt{12}$ **c.** $\sqrt{75} + \sqrt{12}$

16. Find the area of the rectangle at the left.

17. The area of a square is $20w^2$. Write the exact length of one side.

In 18 and 19, rewrite the numerator and simplify the fraction.

18. $\dfrac{\sqrt{175}}{5}$

19. $\dfrac{\sqrt{300}}{5}$

In 20 and 21, simplify in your head.

20. $\sqrt{5^2 \cdot 11^2}$
   $25 \cdot 121)$

21. $\sqrt{100 \cdot 81 \cdot 36}$

Review

22. How high is the kite in the picture below? *(Lesson 7-5)*

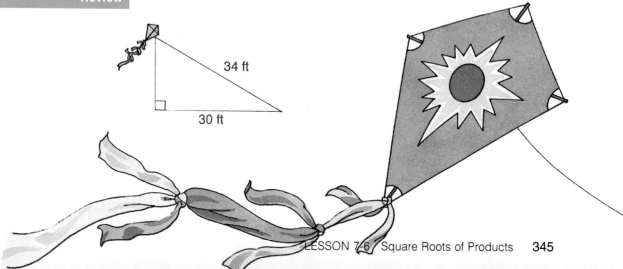

23. Graph $x < -13$ **a.** on a number line and **b.** on a coordinate plane. *(Lesson 7-2)*

24. Find both solutions of $|x + 2| = 14$. *(Lesson 7-3)*

25. If $A = \{1, 2, 3, 4, 5, 6, 7, 8\}$, $B = \{2, 4, 6, 8, 10, 12\}$, and $C = \{1, 3, 5, 7, 9, 11\}$, find each of the following:
a. $A \cup B$
b. $B \cap C$
c. $A \cup (B \cap C)$ *(Lessons 3-4, 3-5)*

26. Refer to the graph below.
a. ℓ is a _?_ line.
b. m is a _?_ line.
c. The coordinates of $\ell \cap m$ are (_?_, _?_). *(Lesson 7-2)*

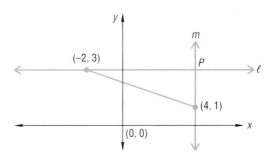

27. If $k = -3.5$ calculate **a.** k^2, **b.** $(-k)^2$, and **c.** $|k|^2$. *(Lessons 1-4, 7-3)*

28. From March 27, 1987 to June 29, 1987, the price of Texas intermediate crude oil went from $18.62 to $20.35 a barrel. Find the percent increase. *(Lesson 5-4)*

29. Betsy has q quarters and d dimes. She has at least $5.20. Write an inequality that describes this situation. *(Lesson 6-7)*

In 30 and 31, solve. *(Lesson 6-6)*

30. $9y - 2 = 11y + 54$

31. $50,000x - 10,000 = 30,000x + 50,000$

32. The diameter of a metal rod is to be 1.5 cm, with an allowable error of 0.05 cm. Is a rod with diameter 1.496 cm in the allowable range? *(Lesson 7-3)*

Exploration

33. Generalize Example 2 on page 343 and Question 9 on page 345.

Streets of many cities are laid out in a grid pattern like the coordinate plane. To get from point A to point C by car you have to travel along streets, so the shortest distance from A to C is not the hypotenuse of triangle ABC, but the sum of the legs $AB + BC$. But a bird or helicopter could go directly along segment $\overline{AC}$ **"as the crow flies."**

Example 1   **a.** How far is it from A to C traveling by car along $\overline{AB}$ and $\overline{BC}$? (Each unit is a city block.)

b. How far is it from A to C flying in a straight line?

Solution

a. Count the city blocks between points A and B. There are 4 blocks. Count the city blocks between B and C. There are 6 of them.

$$AB + BC = 4 + 6 = 10$$

Traveling by car, the distance between A and B is 10 blocks.

b. Use the Pythagorean Theorem.

$$\begin{aligned} AC^2 &= AB^2 + BC^2 \\ &= 4^2 + 6^2 \\ &= 16 + 36 \\ &= 52 \\ AC &= \sqrt{52} \\ &\approx 7.2 \end{aligned}$$

The flying distance from A to C is $\sqrt{52}$ blocks or approximately 7.2 blocks.

The map above looks like part of the coordinate plane. The idea used in Example 1 can be applied to find the distance between any points in the coordinate plane.

Example 2 The streets in the Joneses' neighborhood are laid out in a grid oriented North-South and East-West. Suppose that a museum is 2 miles east and 4 miles south of the Joneses' house. A post office is 5 miles west and 3 miles north of the Joneses' house.
a. What is the street distance from the post office to the museum?
b. How far is it from the post office to the museum as the crow flies?

Solution Since the distances are given from the Joneses' house, make a graph with the Joneses' house *J* at (0, 0), the museum *M* at (2, -4), and the post office *P* at (-5, 3).

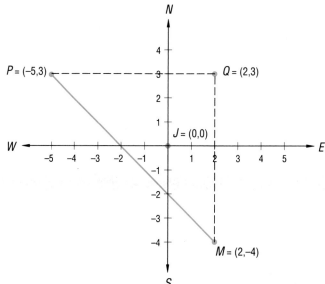

a. The key idea is to put point *Q* in the picture to form right triangle *PQM*.
Q = (2, 3). The street distance is *PQ* + *QM* miles.

$$PQ = |\text{-}5 - 2| = 7 \qquad\qquad QM = |3 - (\text{-}4)| = 7$$

So the street distance from the post office to the museum is 7 + 7 = 14 miles.

b. "As the crow flies" means the distance *PM* using the Pythagorean Theorem.

$$\begin{aligned}
PM^2 &= PQ^2 + QM^2 \\
&= 7^2 + 7^2 \\
&= 98 \\
PM &= \sqrt{98} \\
&\approx 9.9 \text{ miles}
\end{aligned}$$

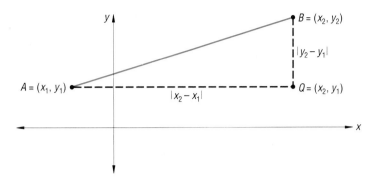

The calculations done in Example 2b can be generalized. To find the distance between $A = (x_1, y_1)$ and $B = (x_2, y_2)$, use $Q = (x_2, y_1)$ to form a right triangle ABQ.

Then $AQ = |x_2 - x_1|$
and $\quad QB = |y_2 - y_1|$.

By the Pythagorean Theorem,

$$(AB)^2 = (AQ)^2 + (QB)^2$$
$$= |x_2 - x_1|^2 + |y_2 - y_1|^2.$$

Now take the square root of each side. The result is a formula for the distance between any two points in the plane.

Pythagorean Distance Formula:

The distance AB between points $A = (x_1, y_1)$ and $B = (x_2, y_2)$ is $AB = \sqrt{|x_2 - x_1|^2 + |y_2 - y_1|^2}$.

Example 3   Let $D = (4, 3)$ and $F = (6, -2)$. Find DF.

Solution   Use the Pythagorean Distance Formula with $x_1 = 4$, $y_1 = 3$; $x_2 = 6$, $y_2 = -2$.

$$DF = \sqrt{(4 - 6)^2 + (3 - -2)^2}$$
$$= \sqrt{(-2)^2 + 5^2}$$
$$= \sqrt{29} \approx 5.4$$

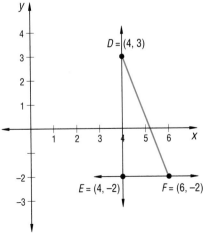

1. In the diagram below, each grid represents a city block.
   a. How many blocks does it take to go from *P* to *Q* by way of *R*?
   b. Use the Pythagorean Theorem to find the distance from *P* to *Q* as the crow flies.

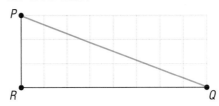

2. Find *MN* if $M = (40, 60)$ and $N = (40, 18)$.

3. Use the graph at the right to:
   a. Find *XY*.
   b. Find *YZ*.
   c. Use the Pythagorean Distance Formula to find *XZ*.

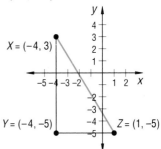

In 4 and 5, find *DF*.

   4. $D = (5, 1)$; $F = (11, -7)$ **5.** $D = (-3, 5)$; $F = (-1, -8)$

In 6–9, refer to Example 2. A school is at point *S*, which is 1 mile east and 2 miles north of the Joneses' house. Remember that $J = (0, 0)$.

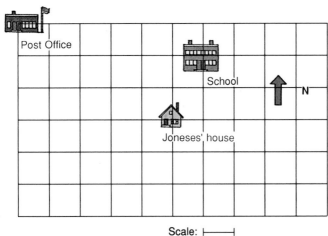

Scale: ⊢———⊣
1 mile

6. What are the coordinates of *S*?

7. Find the distance one would travel by car to get from the post office to the school.

8. Find the street distance from the Joneses' house to the school.

9. How far is it from the Joneses' house to the school as the crow flies?

10. What is the Pythagorean distance between (x_1, y_1) and (x_2, y_2)?

In 11 and 12, use the diagram below. It shows part of a town laid out in a rectangular grid. Each block is 200 m long in an east-west direction and 100 m long in a north-south direction.

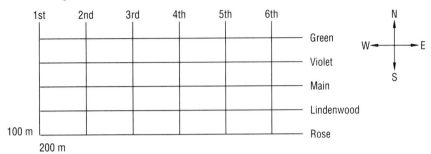

11. Locate the point at 5th and Green and the point at 2nd and Rose. What is the shortest street distance between these points?

12. As the crow flies, what is the distance from 5th and Green to 2nd and Rose?

In 13 and 14, use this computer program that asks for coordinates of points in the coordinate plane, (x_1, y_1) and (x_2, y_2), and then calculates the distance between them.

```
10 PRINT "DISTANCE IN THE PLANE"
20 PRINT "BETWEEN (X1, Y1) AND (X2, Y2)"
30 PRINT "ENTER X1, Y1"
40 INPUT X1, Y1
50 PRINT "ENTER X2, Y2"
60 INPUT X2, Y2
70 LET HOZ = ABS(X1 − X2)
80 LET VERT = ABS(1 − Y2)
90 LET DIST = SQR(HOZ * HOZ + VERT * VERT)
100 PRINT "THE DISTANCE IS "; DIST
110 END
```

13. What value of DIST will the computer print when $(x_1, y_1) = (19, -5)$ and $(x_2, y_2) = (-3, 15)$?

14. What value of DIST will the computer print when $(x_1, y_1) = (0, 0)$ and $(x_2, y_2) = (1, 1)$?

15. *Skill sequence* If $y + 4 = 10$, find the value of each expression. *(Lessons 2-6, 7-4)*
  a. $2y + 8$
  b. $(y + 4)^2$
  c. $2(y + 4)^2 - \dfrac{y + 4}{2}$
  d. $(\sqrt{y + 4})^2$

16. Simplify: **a.** $\sqrt{18}$; **b.** $\sqrt{50}$; **c.** $\sqrt{18} + \sqrt{50}$. *(Lesson 7-6)*

17. Find the exact value of the variable in the triangle at left. *(Lessons 7-5, 7-6)*

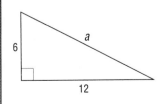

18. Simplify $(8\sqrt{10})^2$ without using a calculator. *(Lesson 7-4)*

19. The distance from C to D on the number line below is twice the distance from A to B. Write an equation and solve to find x. (Assume that A is the point with the smallest coordinate.) *(Lesson 7-3)*

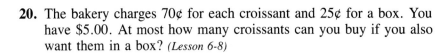

20. The bakery charges 70¢ for each croissant and 25¢ for a box. You have $5.00. At most how many croissants can you buy if you also want them in a box? *(Lesson 6-8)*

21. In the triangle at the left, find the value of x. *(Lessons 3-8, 6-2)*

In 22–25, solve without writing intermediate steps. *(Lessons 6-1, 6-6)*

22. $5y - 7 = 28$

23. $8a + 4a - 2a = 30$

24. $10n = 8n - 30$

25. $-3m = -2m$

26. Graph the image of quadrilateral $WXYZ$ under a size change of $-\frac{1}{2}$. *(Lesson 5-6)*

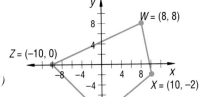

27. *Skill sequence*   *(Lessons 4-4, 5-7, 6-1)*

  a. Solve for m. $w = \dfrac{m}{a}$

  b. Solve for t. $z = \dfrac{s}{t}$

  c. Solve for k. $x + 1 = \dfrac{k}{b}$

Exploration

28. The grid below pictures streets 1 unit apart in a city. Graph all points at a street distance of 5 units from the intersection of Main and Central.

Central

Main

The following problem may look complex.

$$\text{Solve } (k + 6)^2 = 81.$$

However, it can be solved using *chunking*.

Chunking is the process of grouping some small bits of information into a single piece of information. For instance, when reading the word "store," you don't think "s, t, o, r, e." You chunk the five letters into one word. In algebra, chunking can be done by viewing an entire algebraic expression as one variable.

Example 1 If $3y = 8.5$, find $6y + 5$.

Solution This can be done without solving for y. Think of $3y$ as a chunk. Since you know $3y$, you can double it to get $6y$. Then add 5.

$$3y = 8.5$$
Double it. $\qquad 6y = 17$
Add 5. $\qquad 6y + 5 = 22$

Check If $6y + 5 = 22$, then $6y = 17$, so $3y = 8.5$.

The equation at the beginning of the lesson is $(k + 6)^2 = 81$. To solve an equation like this, think what value the expression inside the parentheses must have. Also, remember that an equation like $x^2 = 81$ has two solutions.

Example 2 Solve $(k + 6)^2 = 81$.

Solution What number squared is 81? There are two such numbers, 9 and -9. So $k + 6$ can be 9 or -9.

$$k + 6 = 9 \qquad \text{or} \qquad k + 6 = -9$$
$$k \quad = 3 \qquad \text{or} \qquad k \quad = -15$$

There are two solutions, 3 and -15.

Check Check 3. $\qquad$ Does $\quad (3 + 6)^2 = 81$?
$$9^2 \quad = 81? \text{ Yes, it checks.}$$
Check -15. $\quad$ Does $(-15 + 6)^2 = 81$?
$$(-9)^2 \quad = 81? \text{ Yes, it checks.}$$

Equations in which the variable (or an expression involving the variable) is squared often have two solutions. You need to find both!

Another operation that often leads to equations with two solutions is absolute value.

Example 3   Solve $|6x - 3| = 21$.

Solution   The absolute value of two numbers, 21 and -21, is 21. So $6x - 3$ can be 21 or it can be -21.

$$|6x - 3| = 21$$

$$6x - 3 = 21 \qquad \text{or} \qquad 6x - 3 = -21$$
$$6x \quad = 24 \qquad\qquad\qquad 6x \quad = -18$$
$$x \quad = 4 \qquad\qquad\qquad x \quad = -3$$

Check   Check 4.   Does $|6 \cdot 4 - 3| = 21$?
$$|24 - 3| = 21?$$
$$|21| = 21? \text{ Yes, it checks.}$$

Check -3.   Does $|6 \cdot -3 - 3| = 21$?
$$|-18 - 3| = 21?$$
$$|-21| = 21? \text{ Yes, it checks.}$$

In Lesson 7-3, you used the idea of distance to find and graph solutions to inequalities with absolute value. Chunking makes it possible to do this algebraically.

Example 4   Solve and graph $|x - 9| < 5$.

Solution   Consider the easier problem of solving $|x| < 5$.
Its solution is $-5 < x < 5$.
Now think of $x - 9$ as a chunk. Put this chunk in place of x.

The solution to $|x - 9| < 5$ is $-5 \quad < \quad x - 9 \quad < 5.$

Add 9 to each part of the inequality. $-5 + 9 < x - 9 + 9 < 5 + 9$

$$4 \quad < \quad x \quad < 14$$

The graph is:

Check   Try the endpoints 4 and 14 in the original sentence. They should make the sides equal.
Does $|4 - 9| = 5$?
$$|-5| \quad = 5$$

354

You can also see that 14 works. Try a number between 4 and 14. We try 10.

Does $|10 - 9| < 5$?

$\qquad |1| \quad < 5$? Yes. It checks.

Example 5 Solve $|y - 30| > .002$.

Solution Consider the easier problem $|y| > .002$.

Its solution is: $\qquad\qquad\qquad\qquad\qquad\qquad y < -.002$ or $y > .002$

Now think of $y - 30$ as a chunk.

$|y - 30| > .002$ has the solution $\quad y - 30 < -.002$ or $y - 30 > .002$.

Solve each inequality. $\qquad\qquad\qquad\qquad y < 29.998$ or $y > 30.002$

Check Graph the solutions.

29.998   30   30.002

The graph pictures all points whose distance from 30 is greater than .002. This is exactly what is meant by $|y - 30| > .002$.

You have already used chunking to simplify expressions. For instance, consider the addition $\dfrac{4}{3x + 1} + \dfrac{3}{3x + 1}$. If you think of $(3x + 1)$ as a single chunk C, then the addition looks like $\dfrac{4}{C} + \dfrac{3}{C}$ and its sum is $\dfrac{7}{C}$. So the original sum is $\dfrac{7}{3x + 1}$.

Questions

Covering the Reading

1. What is chunking?

2. If $3x = 8.5$, find $6x - 1$.

3. If $4y = 13$, find $12y + 7$.

4. **a.** If $(m - 11)^2 = 64$, what two values can $m - 11$ have?
   b. Find both solutions to $(m - 11)^2 = 64$.

5. Find all solutions to $(p + 3)^2 = 225$.

6. **a.** In $|.5a + 17| = 29$, what two values are possible for $.5a + 17$?
   b. Solve $|.5a + 17| = 29$.

7. Find both solutions to $|6y + 3| = 45$.

8. **a.** In adding $\dfrac{11}{2y - 6} + \dfrac{4}{2y - 6}$, what should you think of as a chunk?

   b. Perform the addition.

9. Simplify $\dfrac{x}{x + 8} - \dfrac{8}{x + 8}$.

In 10–20, you may find chunking helpful.

10. Simplify $\dfrac{t + 4}{t + 3} + \dfrac{t + 2}{t + 3}$.

11. If $18y - 12t = 25$, find $9y - 6t$.

12. If $5x + 4y = 27$ and $2y = 1$, find $5x$.

13. Solve $(3x^2)^2 = 144$.   14. Solve $|8n + 20| = 4$.

15. Approximate the solutions of $(d + 11)^2 = 57$ to two decimal places.

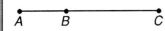

16. $AB = 7m + 1$ and $BC = 21m + 3$. Find the ratio of AB to BC.

17. Simplify: $3\sqrt{5} + 6\sqrt{5} - 2\sqrt{5}$.

18. If $\sqrt{3y + 1} = 3$, find **a.** $3y + 1$ **b.** y

19. If $\sqrt{2y + 8n} = 6$, find **a.** $2y + 8n$ **b.** $y + 4n$ **c.** $5y + 20n$

20. Simplify $3(x^2 - 2y) + 6(x^2 - 2y) - 4(x^2 - 2y)$.

21. Compute.
$$\frac{8(x + 7)}{5a} \cdot \frac{3a}{2(x + 7)}$$

22. A mast on a sailboat is strengthened by a wire (called a *stay*) as shown at the left. The mast is 35 ft tall. The wire is 37 ft long. How far from the base of the mast does the wire reach? (Round to the nearest inch.) *(Lesson 7-5)*

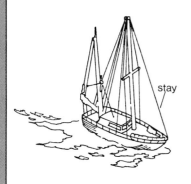

stay

23. These two right triangles are similar. Corresponding sides are parallel.
a. Find x.
b. Find y. *(Lesson 7-5)*

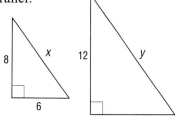

24. Find the distance between $(-3, 2)$ and $(11, 6)$. *(Lesson 7-7)*

25. Evaluate $\dfrac{6 + 2\sqrt{16 - 9}}{3}$ to the nearest tenth. *(Lesson 7-4)*

26. Simplify $(5\sqrt{3})^2$. *(Lesson 7-4)*

27. Graph $2y - 3x = 12$. *(Lesson 7-1)*

28. Graph $x = 1$ in the coordinate plane. *(Lesson 7-2)*

29. You buy s sweaters at d dollars a sweater. The total cost is T dollars. Write an equation that relates s, d, and T. *(Lesson 4-2)*

30. Suppose a laser printer prints 3 pages/min. How long will it take to print 25 documents with 8 pages per document? *(Lesson 4-2)*

31. The figure at the left below is operated on by a size change of magnitude 8.
a. Find the length of each side of the image. *(Lesson 5-6)*

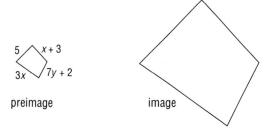

preimage   image

b. Find the perimeter of each figure. Put your answer in simplest form. *(Lesson 6-3)*

Exploration

32. Often people use chunking as a help in memorization. Here are the first twenty decimal places of π.

$$3.14159265358979323846$$

Try memorizing these 20 digits by memorizing the following four chunks of five digits each.

14159   26535   89793   23846

Summary

This chapter covers a variety of ideas related to lines and distance. Among them are absolute value, square root, and square. The absolute value of a number is its distance from the origin. If a square has area A, the positive square root of A, $\sqrt{A}$, is the length of a side of the square. The Pythagorean Theorem relates lengths of sides in a right triangle. If the legs have lengths a and b and the hypotenuse has length c, then

$$a^2 + b^2 = c^2.$$

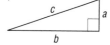

The distance between two points on a number line with coordinates x_1 and x_2 on a number line is $|x_2 - x_1|$. Using the Pythagorean Theorem, the distance between two points on the coordinate plane can be determined. If the points are (x_1, y_1) and (x_2, y_2), the Pythagorean distance is

$$\sqrt{(x_2 - x_1)^2 + (y_2 - y_1)^2}.$$

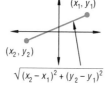

Square roots can be simplified using the Square Root of a Product Property:

$$\sqrt{ab} = \sqrt{a}\,\sqrt{b}\cdot$$

Here are the simplest equations involving these ideas, along with their solutions. Here $a > 0$.

Equation	Solution(s)
$\lvert x \rvert = a$	$a, \; -a$
$x^2 = a$	$\sqrt{a}, \; -\sqrt{a}$
$\sqrt{x} = a$	a^2

Here are the simplest sentences involving absolute value, along with their graphs. Again, $a > 0$.

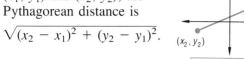

Sentence	Graph
$\lvert x \rvert = a$	
$\lvert x \rvert < a$	
$\lvert x \rvert > a$	

Chunking is a problem-solving technique by which an expression is considered as a single number. Chunking allows equations like these to be solved when more complicated expressions are substituted for x.

Solutions to a linear equation in two variables may be graphed by making a table of values and plotting the resulting points. The result is a line. If the equation has only one variable, the graph is a horizontal line $y = k$ or a vertical line $x = h$.

Vocabulary

Below are the most important terms and phrases for this chapter. You should be able to give a definition or general description and a specific example of each.

Lesson 7-1
linear equation

Lesson 7-2
horizontal line $y = k$
vertical line $\quad x = h$
boundary line
half-plane

Lesson 7-3
absolute value, $\lvert x \rvert$, ABS( )
Distance Formula on a number line

Lesson 7-4
square root
radical sign, $\sqrt{}$, SQR( )
perfect square

Lesson 7-5
leg of a right triangle
hypotenuse
Pythagorean Theorem

Lesson 7-6
Square Root of a Product Property
simplifying radicals

Lesson 7-7
"as the crow flies"
Pythagorean Distance Formula

Lesson 7-8
chunking

Progress Self-Test

Take this test as you would take a test in class. You will need graph paper and a calculator. Then check your work with the solutions in the Selected Answers section in the back of the book.

In 1 and 2, simplify.

1. $|-21| + |9|$

2. Evaluate $4 |n - 17|$ if $n = 35$.

In 3 and 5, find the distance between two points on a number line with coordinates:

3. -40 and -13.

4. $5x$ and $-8x$.

5. Approximate $\sqrt{210}$ to the nearest hundredth using a calculator.

6. The area of a square is 29 cm². What is the exact length of one side?

7. Evaluate $\sqrt{\dfrac{x + y}{2}}$ if $x = 120$ and $y = 42$.

8. If a and b are positive, simplify $\sqrt{64a^2b^2}$.

9. Write $\sqrt{45}$ with a smaller integer under the radical sign.

10. Simplify $\sqrt{44}$.

11. If $4y = 6.5$, find $12y - 5$.

In 12–19 find all solutions.

12. $n^2 = 576$

13. $-15 = k^2$

14. $\sqrt{y} = 4$

15. $3y^2 = 48$

16. $\dfrac{7}{x} = \dfrac{x}{14}$

17. $|4n - 18| = 22$

18. $|x - 3| > 7$

19. $|y - 4| < 2$

20. How long is the ladder below?

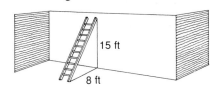
15 ft
8 ft

21. Find the value of y.

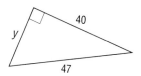
40
y
47

22. Write an equation for the line containing the points $(-1, 6)$, $(3, 6)$, and $(0.5, 6)$.

In 23 and 24, use the diagram below. It shows part of a town laid out in a rectangular grid. The distance between consecutive streets is 200 meters for E-W streets and 100 meters for N-S streets.

23. What is the shortest street distance from 4th and Pine to 2nd and Olive?

24. As the crow flies, what is the distance from 4th and Pine to 2nd and Olive?

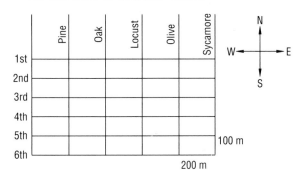

25. $A = (-3, 9)$ and $B = (6, 1)$. Find AB.

In 26 and 27, use the computer program below.

```
10 PRINT "TABLE OF (X,Y) VALUES"
20 PRINT "X VALUE", "Y VALUE"
30 FOR X = -2 TO 2
40 LET Y = 2 - 3 * X
50 PRINT X,Y
60 NEXT X
70 END
```

26. What ordered pairs will the computer print?

27. Graph all solutions to the equation in line 40.

28. Graph $y = -4x$ in the coordinate plane.

29. Graph the points in the coordinate plane satisfying $y \le 5$.

Chapter Review

Questions on SPUR Objectives

SPUR stands for **S**kills, **P**roperties, **U**ses, and **R**epresentations.
The Chapter Review questions are grouped according to the
SPUR Objectives for this chapter.

SKILLS deal with the procedures used to get answers.

Objective A: *Calculate absolute values. (Lesson 7-3)*

1. Give the absolute value of -43.

2. Simplify -|-1|.

In 3–8, simplify.

3. |13 − 19|

4. 5 · |4.2 − 3.8|

5. |3| − |-8|

6. ABS(3 − 8)

7. 5 |4 − 6|

8. |-20| − |15| + |-2|

In 9–11, find all solutions.

9. |d| = 16

10. |k| = 0

11. |300 − x| = 23

Objective B: *Evaluate and simplify expressions involving square roots. (Lessons 7-4, 7-6)*

12. **a.** Approximate $\sqrt{105}$ to the nearest hundredth using your calculator.
    b. Check your answer.

13. Between what two integers is $\sqrt{20}$?

14. $4 + \sqrt{\frac{50}{2}}$

15. -SQR(100)

16. $\sqrt{\frac{20}{5} + 5}$

17. Evaluate $45 - \sqrt{4 + 2n}$ when $n = 70$.

18. Evaluate $20 - 3(2 + 3\sqrt{x})$ when $x = 16$.

Objective C: *Solve equations involving squares and square roots. (Lessons 7-4, 7-6)*

In 19–26, solve. Give the exact answer, simplified.

19. $5x^2 = 200$

20. $q^2 = 121$

21. $\frac{16}{z} = \frac{z}{5}$

22. $\frac{2x}{5} = \frac{40}{x}$

23. $x^2 = -100$

24. $\sqrt{v} = 9$

25. $\sqrt{w} + 3 = 8$

26. $4\sqrt{t} = 1$

Objective D: *Use chunking to evaluate expressions and solve equations. (Lesson 7-8)*

27. If $7y = 21.2$, find $21y + 4$.

In 28–35, solve.

28. $(m + 2)^2 = 64$

29. $(z - 4)^2 = 144$

30. $\sqrt{n + 5} = 25$

31. $\sqrt{3q + 2} = 11$

32. $|2y - 3| = 9$

33. $-7 = |2 - x|$

34. $|6 - x| \geq 9$

35. $|x - 3| < 4$

■ **Objective E:** *Apply the Square Root of a Product Property.* *(Lesson 7-6)*

In 36–41, simplify.

36. $\sqrt{20^2 + 20^2}$

37. $\sqrt{500}$

38. $3\sqrt{72}$

39. $\dfrac{\sqrt{150}}{5}$

40. $\dfrac{\sqrt{99}}{12}$

41. $\sqrt{4} \cdot \sqrt{9} - \sqrt{3} \cdot \sqrt{48}$

42. If a and b are positive, simplify $\sqrt{49a^2b^2}$.

43. Simplify $\sqrt{5x^2}$ when $x > 0$.

USES deal with applications of mathematics in real situations.

■ **Objective F:** *Use squares and square roots in measurement problems.* *(Lessons 7-4, 7-5, 7-6)*

44. The jacket for a record album is a square with area 150 sq in. Find the length of the side of the jacket, rounded to the nearest hundredth.

45. The area of a square is $147y^2$. Write the exact length of one side.

46. A side of a square is $3\sqrt{11}$. Find the area of the square.

47. A rectangular field is 300 feet long and 100 feet wide. To the nearest foot, how far will you walk if you cut across the field diagonally?

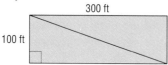

48. Ben uses a guy wire to support a young tree. He attaches it to a point 6 ft up the tree trunk. The wire is 10 ft long. How far away from the trunk will the wire reach?

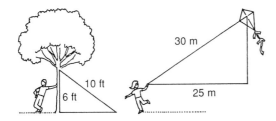

49. Megan is flying a kite. She has unrolled 30m of string and the kite is 25m from her along the ground. How high is the kite?

■ **Objective G:** *Find distance on a number line. (Lesson 7-3)*

In 50–52, find the distance between points with the given coordinates.

50. 72 and 18.

51. 13 and -11

52. 15 and s.

53. Find the distance between the two points with coordinates d and j.

54. Write a simplified expression for the distance between points with coordinates $5x - 3$ and $2x + 7$.

■ **Objective H:** *Graph equations for straight lines by making a table of values. (Lesson 7-1)*

55. a. Draw the graph of $y = 2x + 1$.
   b. For what value of y is $(-5, y)$ on this line?

56. Graph the set of points in the plane with $y = -2x$.

57. Suppose a person begins with $15 and adds $2 to an account each week.
   a. Write an equation that represents the amount of money t in the account after w weeks.
   b. Draw the graph of the equation in part a.

58. The level in a swimming pool is dropping at the rate of 0.5 feet per hour. Suppose the level in the pool is at the 6 foot mark. The level L after h hours is given by $L = 6 - .5h$. Graph this relationship.

■ **Objective I:** *Graph horizontal and vertical lines. (Lesson 7-2)*

In 59 and 60, answer true or false.

59. The graph of all points in the plane satisfying $x = -0.4$ is a vertical line.

60. The graph of all points in the plane satisfying $y = 73$ is a horizontal line.

In 61 and 62, graph the points satisfying each equation.

61. $x = 4$

62. $y \geq 1$

63. Give a sentence describing the colored region in the graph below.

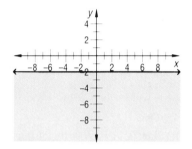

64. Write an equation for the line containing the points $(5, 11)$, $(5, 4)$, and $(5, -7)$.

■ **Objective J:** *Find the lengths of the sides of a right triangle using the Pythagorean Theorem. (Lesson 7-5)*

65. Find the value of n.

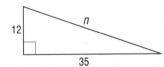

66. What is the length of the hypotenuse of the triangle on the left below?

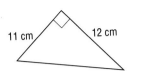

67. Find the value of k in the triangle at right above.

68. A diagonal of a rectangle has length 10 and one side has length 6. What are the lengths of the other sides of the rectangle?

▨ **Objective K:** *Calculate distances in the plane.*
(Lesson 7-7)

In 69 and 70, use the graph below.

69. a. Find CB.
  b. Find AB.
  c. Find AC.

70. Find the distance from A to the origin.

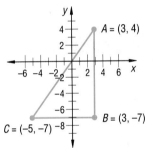

In 71–74, find AB.

71. $A = (50, 10)$ and $B = (50, 4)$

72. $A = (14, -20)$ and $B = (-2, -20)$

73. $A = (8, -3)$ and $B = (4, 12)$

74. $A = (-2, 9)$ and $B = (-5, -1)$

In 75 and 76, **a.** find the points on the line where $x = -2$ and $x = 8$. **b.** Find the distance between the points in part a.

75. $y = 2x - 3$

76. $\frac{1}{2}x - y = 5$

In 77 and 78, suppose the school the Conners' children attend is 3 miles east and 5 miles south of their home. The library is 4 miles west and 2 miles north of the Conners' house. The streets in their neighborhood are laid out in a rectangular grid.

77. What is the street distance from the school to the library?

78. How far is it from the school to the library as the crow flies?

▨ **Objective L:** *Graph solutions to sentences of the form* $|x - a| < b$, *or* $|x - a| > b$, *where a and b are real numbers. (Lesson 7-3)*

In 79–84 graph all solutions on a number line.

79. $|x - 2| = 4$

80. $|y - 7| = 1\frac{1}{2}$

81. $|z + 5| \leq 12$

82. $|a + 6| > 9$

83. $3 \geq |s - 2|$

84. $.03 < |r - 5|$

Slopes and Lines

Below is a graph of the population of Manhattan Island (part of New York) every ten years from 1790 to 1980. Coordinates of some of the points are shown.

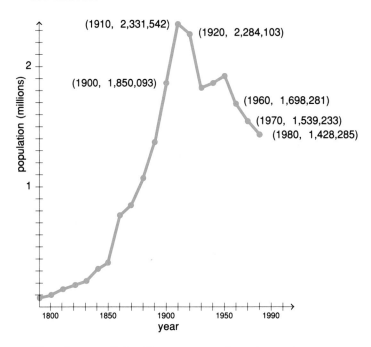

The slopes of the lines connecting the points tell how fast the population went up or down. In this chapter, you will study many examples of lines and slopes.

Rates of Change

At age 9 Karen was 4′3″ tall. At age 11 she was 4′9″ tall. How fast did she grow from age 9 to age 11? To answer this question, we calculate the *rate of change* of Karen's height per year.

Rate of change in height per year (from age 9 to age 11)

$$= \frac{\text{change in height}}{\text{change in age}} = \frac{4'9'' - 4'3''}{(11 - 9)\text{ years}} = \frac{6''}{2\text{ years}} = 3\frac{\text{inches}}{\text{year}}.$$

At age 14 Karen was 5′4″ tall. How fast did she grow from age 11 to age 14? Again we calculate. Rate of change in height per year (from age 11 to age 14)

$$= \frac{\text{change in height}}{\text{change in age}} = \frac{5'4'' - 4'9''}{(14 - 11)\text{ years}} = \frac{7''}{3\text{ years}} = 2.\overline{3}\frac{\text{inches}}{\text{year}}.$$

Karen grew at a faster rate from age 9 to age 11 than from age 11 to 14.

The data can be graphed and points connected. The rate of change then measures how fast the graph goes up as you read from left to right along the *x*-axis.

The segment connecting (9, 4′3″) to (11, 4′9″) is steeper than the one connecting (11, 4′9″) and (14, 5′4″) since the rate of change is greater.

■ ■ ■ ■ ■ ■ ■ ■

Example The graph on page 364 shows the population of Manhattan. Find the rate of change of the population (in people per year):
a. between 1900 and 1910;
b. between 1910 and 1920;
c. between 1960 and 1970.

Solution Every rate of change is found by dividing two changes.
a. Between 1900 and 1910:

$$\frac{2{,}331{,}542 - 1{,}850{,}093}{1910 - 1900} = \frac{481{,}449 \text{ people}}{10 \text{ years}}$$
$$= 48{,}144.9 \text{ people per year}$$

Notice that the population increased and the rate of change is positive. Between 1900 and 1910, from left to right, the graph slants up.

b. Between 1910 and 1920:

$$\frac{2,284,103 - 2,331,542}{1920 - 1910} = \frac{\text{-}47,439 \text{ people}}{10 \text{ years}}$$

$$= \text{-}4743.9 \text{ people per year}$$

c. Between 1960 and 1970:

$$\frac{1,539,233 - 1,698,281}{1970 - 1960} = \frac{\text{-}159,048 \text{ people}}{10 \text{ years}}$$

$$= \text{-}15,904.8 \text{ people per year}$$

In both parts b and c, the population decreased and therefore the rate of change is negative. Between those dates, from left to right, the graph slants down.

In Example 1, since a number of people is divided by a number of years, the unit of the rate of change is $\frac{\text{people}}{\text{year}}$. The unit of a rate of change is always a rate unit.

The following table summarizes the relationship between rates of change and their graphs.

Situation	Rate of Change	Graph (from left to right)
increase	positive	upward slant
no change	zero	horizontal
decrease	negative	downward slant

Another way of thinking of rate of change is in terms of coordinates. In the above example, the year is the x-coordinate and the population size is the y-coordinate. The **rate of change** between two points is calculated by dividing the difference in the y-coordinates by the difference in the x-coordinates.

The rate of change between points (x_1, y_1) and (x_2, y_2) is

$$\frac{y_2 - y_1}{x_2 - x_1}.$$

Questions

Covering the Reading

1. The rate of change in height per year is the change in __?__ divided by the change in __?__.

2. Which is steeper: the segment connecting (9, 4′3″) to (11, 4′9″) or the segment connecting (11, 4′9″) to (14, 5′4″)?

In 3 and 4, suppose Karen in the lesson is 5′7″ tall at age 18 and 5′7″ tall at age 19.

3. **a.** What is the rate of change of her height from age 14 to age 18?
   b. Was the rate of change of Karen's height greater from age 9 to age 11 or from age 14 to age 18?

4. **a.** What is the rate of change of her height from age 18 to age 19?
   b. What is the unit of the rate of change of her height from age 18 to age 19?

In 5–8, use the graph of the population of Manhattan on page 364.

5. If the rate of change in population is positive, does the population increase or decrease?

6. Between 1950 and 1960, did the population increase or decrease?

7. Find the rate of change in population between 1970 and 1980.

8. Find the rate of change in population from 1900 to 1980.

9. In terms of coordinates, the rate of change between two points is the __?__ of the y-coordinates divided by the difference of the __?__.

10. Describe the graph of the segment connecting two points when the rate of change between them is **a.** positive; **b.** negative; **c.** zero.

Applying the Mathematics

11. Below are heights (in inches) for a boy from age 9 to 15.

Age	9	10	11	12	13	14	15
Height	51″	53″	58″	61″	63″	64″	65″

   a. Accurately graph these data and connect the points as was done in the lesson.
   b. Using the graph, in which two-year period did the boy grow the fastest?
   c. Calculate his rate of growth in that two-year period.

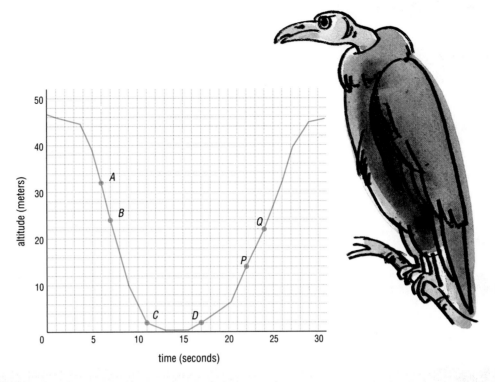

In 12–15, use the graph above. This graph shows the altitude of a vulture's flight over a period of time.

12. a. Give the coordinates of point *A*.
   b. Give the coordinates of point *B*.
   c. What is the rate of change of altitude (in meters per second) between points *A* and *B*?

13. What is the rate of change of altitude in meters per second between points *C* and *D*?

14. After about how many seconds does the altitude of the vulture begin to increase?

15. Is the rate of change between *P* and *Q* positive or negative?

16. Older people tend to lose height. Tim reached his full height at age 20, when he was 74″ tall. He stayed that height for 35 years, then started losing height. At age 65 his height is 73″. What was the rate of change of his height from age 55 to age 65?

17. Suppose Karen in the lesson was *h* inches tall at age 7. Write an expression for the rate of change in her height from age 7 to age 9.

18. In the graph at the right, find the rate of change in feet per minute.

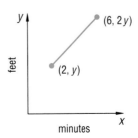

19. a. Graph $y = 3x + 2$.
  b. Graph $y = -3x + 2$ on the same axes.
  c. Where do they intersect? *(Lesson 7-1)*

20. What is the total cost of p pencils at 12¢ each and e erasers at 20¢ each? *(Previous course)*

In 21 and 22, evaluate. *(Lesson 1-4)*

21. $\dfrac{y_2 - y_1}{x_2 - x_1}$ when $y_2 = 5$, $y_1 = 6$, $x_2 = -2$, and $x_1 = -4$

22. $\dfrac{y_2 - y_1}{x_2 - x_1}$ when $y_2 = 8$, $y_1 = 4$, $x_2 = 2$, and $x_1 = \frac{1}{3}$

23. *Skill sequence* Solve for y. *(Lesson 6-6)*
  a. $33 - 4y = 12$
  b. $3x - 4y = 12$
  c. $ax - 4y = 12$

24. Simplify $\dfrac{-\frac{2}{5}}{4}$ *(Lesson 5-1)*

25. *Skill sequence*   *(Lessons 2-2, 5-7, 7-8)*
  a. Simplify $\dfrac{2}{a} - \dfrac{1}{a}$.   **b.** Simplify $\dfrac{2}{3 + 4x} - \dfrac{1}{3 + 4x}$.
  c. Solve $\dfrac{2}{3 + 4x} - \dfrac{1}{3 + 4x} = \dfrac{4}{x}$.

26. Suppose a stamp collection now contains 10,000 stamps. If it grows at 1000 stamps a year, how many stamps will there be in x years? *(Lesson 6-4)*

27. a. What was the population of the town or city where you live (or nearest where you live) in 1940, 1950, 1960, 1970, and 1980?
  b. In which 10-year period did the population grow the most?

Constant Rates of Change

Consider the following situation. An ant is 12 feet high on a flagpole. The ant walks down the flagpole at a rate of 8 inches (which is $\frac{2}{3}$ foot) per minute.

This is a **constant decrease** situation. Each minute the height of the ant decreases by $\frac{2}{3}$ foot. You can see the constant decrease by graphing the height of the ant after 0, 1, 2, 3, and 4 minutes of walking. Below are the ordered pairs (time, height) charting the ant's progress.

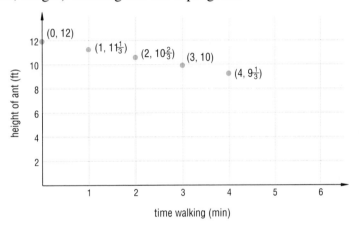

After 1 minute the ant is $11\frac{1}{3}$ feet high. After 4 minutes the ant is $9\frac{1}{3}$ feet high. Now calculate the rate of change of height between these points.

$$\frac{\text{change in height}}{\text{change in time}} = \frac{11\frac{1}{3} \text{ feet} - 9\frac{1}{3} \text{ feet}}{1 \text{ minute} - 4 \text{ minutes}} = -\frac{2 \text{ feet}}{3 \text{ minutes}}$$

The rate of change is $-\frac{2}{3}$ foot per minute, the same as the ant's rate. As long as the ant has a constant rate of walking, the rate of change in its height will always equal that constant rate.

Note that all the points on the graph of the ant's height lie on a line. In *any* situation when there is a constant rate of change between points, the points lie on a line. The constant rate of change is called the slope of the line.

Definition:

The slope of the line through (x_1, y_1) and (x_2, y_2) is

$$\frac{y_2 - y_1}{x_2 - x_1}.$$

Example 1 **a.** Show that (0, 3), (4, 1), and (-8, 7) lie on the same line.
b. Give the slope of that line.

Solution

a. Pick pairs of points and calculate the rate of change between them. The rate of change determined by (0, 3) and (4, 1) is

$$\frac{1 - 3}{4 - 0} = \frac{-2}{4} = -\frac{1}{2}.$$

The rate of change determined by (4, 1) and (-8, 7) is

$$\frac{7 - 1}{-8 - 4} = \frac{6}{-12} = -\frac{1}{2}.$$

The rate of change determined by (0, 3) and (-8, 7) is

$$\frac{7 - 3}{-8 - 0} = \frac{4}{-8} = -\frac{1}{2}.$$

Since the rate of change between any pair of the given points is $-\frac{1}{2}$, the points lie on the same line.

b. The slope of the line is the constant rate of change, $-\frac{1}{2}$.

Given an equation for a line, it is easy to find the slope of the line. Just find two points on it and calculate the rate of change between them.

Example 2 Find the slope of the line with equation $3x + 4y = 6$.

Solution Find two points on the line. If $x = 2$, then $y = 0$. If $x = 0$, then $y = \frac{3}{2}$. So find the rate of change between (2, 0) and (0, $\frac{3}{2}$).

$$\text{slope} = \frac{\frac{3}{2} - 0}{0 - 2} = \frac{\frac{3}{2}}{-2} = -\frac{3}{4}$$

Check Find another point on the line, say (10, -6) and calculate the rate of change between it and (2, 0). This gives $\frac{0 - -6}{2 - 10} = -\frac{3}{4}$. The slope is the same, so it checks. [You could have checked using (10, -6) and (0, $\frac{3}{2}$) as well.]

The program below determines the slope of a line given two points.

```
10 PRINT "DETERMINE SLOPE FROM TWO POINTS"
20 PRINT "GIVE COORDINATES OF FIRST POINT"
30 INPUT X1, Y1
40 PRINT "GIVE COORDINATES OF SECOND POINT"
50 INPUT X2, Y2
60 LET M = (Y2 − Y1)/(X2 − X1)
70 PRINT "THE SLOPE IS "; M
80 END
```

For instance, to compute the first slope in the solution to Example 1a, at line 30 you input 0, 3. At line 50 you input 4, 1.

Recall the population of Manhattan in Lesson 8-1. That population has been increasing and decreasing at different rates. Since the rate of change has not been constant, the points on the graph of the population are not on a straight line.

Questions

Covering the Reading

1. In a constant increase or decrease situation, all points lie on the same __?__.

2. What is the constant rate of change between any two points on a line called?

3. An ant starts 5 feet from the base of a flagpole and climbs $\frac{1}{3}$ foot up the pole each minute.
   a. Graph the ant's progress using ordered pairs (time, height).
   b. Find the rate of change between any two points on the graph.

4. a. Calculate the slope determined by (1, 2) and (6, 11).
   b. Calculate the slope determined by (6, 11) and (-10, -16).
   c. Do the points (1, 2), (6, 11), and (-10, -16) lie on the same line?

5. Using two different sets of points, show that the slope of the line at the left is 1.

6. Find the slope of the line with equation $5x - 2y = 10$.

7. An equation for the height y of the ant in this lesson after x minutes is $y = -\frac{2}{3}x + 12$.
   a. Find two points on this line not graphed in this lesson.
   b. Find the rate of change between those points.

8. Using the computer program from this lesson, what would you input at lines 30 and 50 to determine the slope of the line in Question 4a?

Applying the Mathematics

In 9–11, the points $(2, 5)$ and $(3, y)$ are on the same line. Find y when the slope of the line is:

9. 3   **10.** -2   **11.** $\frac{1}{5}$

In 12–14, use the figure at the right.

12. The slope determined by A and B seems to equal the slope determined by B and what other point?

13. The slope determined by __?__ and __?__ is negative.

14. The slope determined by __?__ and __?__ is zero.

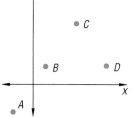

15. a. Find the slope of the horizontal line $y = 2$ below left.
   b. Find the slope of the horizontal line $y = -3$ below left.
   c. From parts a and b, what do you think can be said about the slope of all horizontal lines?

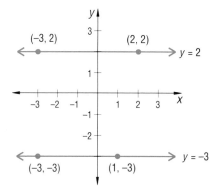 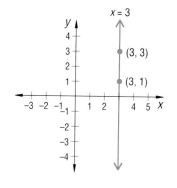

16. Consider the vertical line $x = 3$ graphed above right.
   a. What happens when you try to find its slope?
   b. From your answer to part a, do you think that vertical lines have slope?

17. Consider lines p, q, and r graphed at the right.
   a. Which line has slope $\frac{4}{3}$?
   b. Which line has slope $-\frac{4}{3}$?
   c. Which line has slope zero?

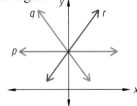

18. Consider the following set of points. High-rise apartment monthly rents: (13th floor, $325), (17th floor, $365), (19th floor, $385).
   a. Calculate the slope of the line through these points.
   b. What does the slope represent in this situation?

Review

19. Below is a graph showing the cost (in cents) of sending a one-ounce first-class letter. The graph is shown in 5-year intervals beginning in 1960. *(Lesson 8-1)*

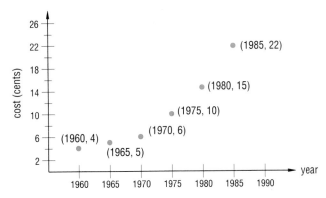

a. During which 5-year period did the cost of postage increase the fastest?
b. During which two 5-year periods was the increase in postage the same?
c. Calculate the average increase per year from 1965 to 1975.
d. Give a reason why the points on the graph are not connected.

In 20–22, graph the line. *(Lessons 7-1, 7-2)*

20. $y = 5x + 1$ **21.** $y = -2$ **22.** $x = -2$

In 23 and 24, write an expression for the amount of money the person has after x weeks. *(Lesson 6-4)*

23. Eddie is given $100 and spends $4 a week.

24. Gretchen owes $350 on a stereo and is paying it off at $5 a week.

25. *Skill sequence* Solve. *(Lessons 5-7, 6-6, 7-4)*
   a. $\dfrac{w}{3} = \dfrac{5}{9}$ **b.** $\dfrac{w+2}{3} = \dfrac{5}{9}$ **c.** $\dfrac{w}{3} = \dfrac{5}{w}$

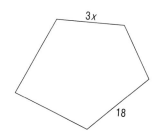

26. The two figures at the left are similar. Corresponding sides are parallel. Find the value of x. *(Lessons 5-7, 5-8)*

27. Simplify. *(Lesson 7-6)*
   a. $\sqrt{3600}$ **b.** $\sqrt{25b^2}$ **c.** $\sqrt{18}$

Exploration

28. Find a record of your height at some time over a year ago. Compare it with your height now. How fast has your height been changing from then until now?

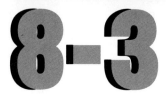
Properties of Slope

A General Motors test ramp for bulldozers goes down 0.6 foot for each foot it goes across, as illustrated at right.

The part of the ramp with the triangle is enlarged and diagramed in the graph below.

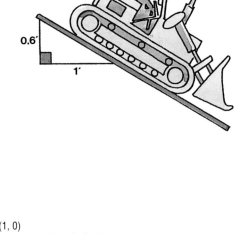

There are two points shown on the graph. The rate of change between these points is the slope of the line.

$$\text{slope} = \frac{\text{change in height}}{\text{change in length}} = \frac{0 - 0.6}{1 - 0} = \text{-}0.6$$

This verifies an important property of slopes of lines.

The slope of a line is the amount of change in the height of the line as you go 1 unit to the right.

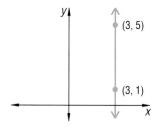

There is one kind of line for which this property of slopes does not apply. You cannot go to the right on a vertical line. Furthermore, if you try to calculate the slope, the denominator will be zero. At left, you would be calculating $\frac{5 - 1}{3 - 3}$. Thus the slope of a vertical line is not defined.

Example 1   Graph the line which passes through (3, -1) and has a slope of -2.

Solution 1   Plot point (3, -1). Since the line has slope -2, move from (3, -1) one unit to the right and two units down. Plot that point and draw the line.

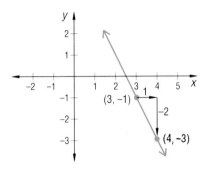

To graph lines having slopes that are given as fractions, it helps to make the unit intervals on the coordinate axes bigger.

Example 2   Graph the line through (-2, 1) with slope $\frac{1}{4}$.

Solution 1   Draw the axes with the unit interval split into fourths. Plot (-2, 1), then move right 1 unit and up $\frac{1}{4}$ unit. Plot the point (-1, $1\frac{1}{4}$) and draw the line.

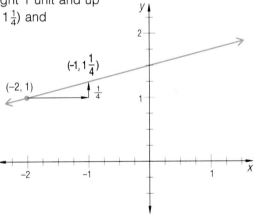

Solution 2   Plot (-2, 1) as in Solution 1. However, instead of going across 1 and up $\frac{1}{4}$, go across 4 · 1 and up 4 · $\frac{1}{4}$. That is, go across 4 and up 1 to the point (2, 2).

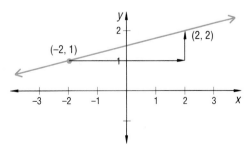

The idea in Examples 1 and 2 helps in drawing graphs of repeated addition situations.

Example 3 In 1988, postage rates for first-class mail were raised to $.25 for the first ounce and $.20 for each additional ounce. Graph the relation between weight and cost for whole number weights.

Solution The starting point is (1, .25). Because the rate goes up $.20 for each ounce, the slope is $.20/oz. The points are on a line because the situation is a constant increase situation.

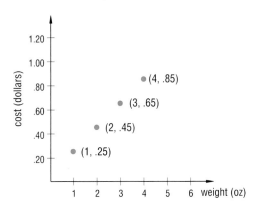

An equation for the cost C to mail a first-class letter weighing w whole ounces is $C = .25 + .20(w - 1)$. This is an equation for the line that contains the points graphed in Example 3. Notice that the given numbers .25 and .20 both appear in this equation. In the next lesson, you will learn how the slope can easily be seen from the equation.

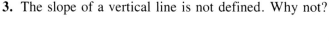

Questions

Covering the Reading

1. A test track for bulldozers goes down 0.5 unit for each unit across. What is its slope?

2. Slope is the amount of change in the __?__ of a graph as you go __?__ unit to the right.

3. The slope of a vertical line is not defined. Why not?

378

4. Graph the line which passes through (0, 2) and has slope 3.

5. Graph the line which passes through (-1, 2) and has slope $-\frac{2}{5}$.

6. In 1987, postal rates were 22¢ for the first ounce and 17¢ for each additional ounce. Graph the relation between weight and cost for whole number weights.

7. Using 1988 postal rates, what would it cost to mail a letter weighing 9 oz?

In 8 and 9, give the slope of the line.

8.
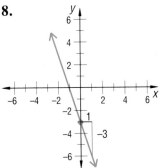

9.
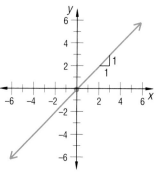

Applying the Mathematics

10. Graph the line through (-2, 1) with a slope of 0.

In 11–13, name one other point on the line.

11. through (-5, 2) with a slope of 7

12. through (6, -11) with a slope of -8

13. through (0, 0) with a slope of $\frac{5}{4}$

14. Mae spends $65 a week on food. After 3 weeks she has $415 left in her food budget. Draw a graph to represent Mae's food budget as weeks go by.

15. Kareem collects magazines. He receives 3 magazines a week. This year, after 10 weeks he has 66 magazines in his collection. Draw a graph to represent the growth of Kareem's collection over the weeks.

Review

16. A rental truck costs $39 for a day plus $.25 a mile. After x miles, the total cost will be y dollars.
   a. Write an equation relating x and y. *(Lesson 6-4)*
   b. Graph the line. *(Lesson 7-1)*
   c. Find the rate of change between any two points on the graph.
   (Lesson 8-2)

In 17 and 18, calculate the slope determined by the two points. *(Lesson 8-2)*

17. (5, 3) and (8, -2)   **18.** (-6, -9) and (-13, 5)

19. Find the slope of the line $3x - y = 15$. *(Lesson 8-2)*

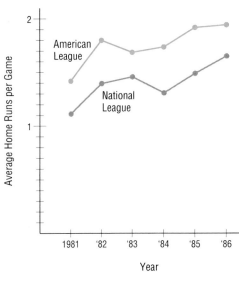

Average Home Runs per Game

Year	American League	National League
1981	1.42	1.12
1982	1.83	1.34
1983	1.68	1.44
1984	1.75	1.32
1985	1.92	1.47
1986	1.95	1.64

In 20–22, refer to the graph above. *(Lesson 8-1)*

20. Between what two consecutive years was there the biggest increase in average home runs per game for the
   a. American League;   **b.** National League?

21. Between what two consecutive years was there the biggest decrease in average home runs per game for the
   a. American League;   **b.** National League?

22. What is the rate of change from 1981 to 1986 in average home runs per game for the
   a. American League;   **b.** National League?

23. If $y = mx + b$ and $y = -4$, $m = 3$, and $x = 2$, find b. *(Lessons 1-5, 2-6)*

24. If the triangle at left has perimeter 38, what is z? *(Lesson 6-6)*

25. Solve and check. $(y + 1)^2 = 64$ *(Lesson 7-8)*

26. If $4a = 15$, find $12a - 5$. *(Lesson 7-8)*

27. *Skill sequence*   *(Lessons 2-2, 5-7, 6-9, 7-8)*
   a. Simplify $(T - 3) - (5 - T)$.   **b.** Simplify $\dfrac{T - 3}{T} - \dfrac{5 - T}{T}$.
   c. Solve $\dfrac{T - 3}{T} - \dfrac{5 - T}{T} = \dfrac{4}{3}$.

28. Calculate in your head. *(Lesson 5-4)*
   a. 15% of $8.00   **b.** 200% of x dollars

Exploration

29. Here is a famous puzzler. Beware of the trick. Enrique Escargot, a snail, is 30 feet deep in a well. Every day he climbs 3 feet up the walls of the well. At night the walls are damper and he slips down 2 feet. How many days will it take him to climb out of the well?

LESSON
8-4

Slope-Intercept Equations for Lines

Quincy has $225 saved for a used car and saves an additional $8 a week. After 3 weeks, Quincy will have 225 + 3 · 8 dollars, or $249. After x weeks, Quincy will have y dollars, where $y = 225 + 8x$.

The line $y = 225 + 8x$ is graphed at right. There are two key numbers in the equation for this line. The number 8 is the slope of the line. The number 225 indicates where the line crosses the y-axis. That number is called the **y-intercept** of the line.

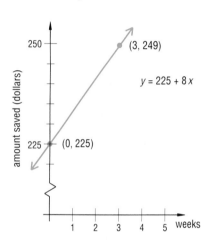

Definition:

When a graph intersects the y-axis at the point $(0, b)$, the number b is a y-intercept for the graph.

Quincy's situation, the graph of the line, and the equation $y = 225 + 8x$ are completely determined by the slope and y-intercept. When the equation is rewritten in the form $y = 8x + 225$, it is said to be in **slope-intercept form.**

Slope-Intercept Property:

The line with equation $y = mx + b$ has slope m and y-intercept b.

Example 1   Give the slope and y-intercept of $y = \frac{1}{2}x + 4$.

Solution   The equation is already in slope-intercept form. Compare $y = \frac{1}{2}x + 4$ to $y = mx + b$. $m = \frac{1}{2}$ and $b = 4$. So the slope is $\frac{1}{2}$ and the y-intercept is 4.

Check   A y-intercept of 4 means the line must contain (0, 4). Does (0, 4) satisfy $y = \frac{1}{2}x + 4$? Yes, because $4 = \frac{1}{2} \cdot 0 + 4$.

The advantage of slope-intercept form is that it tells you so much about the line. So it is often useful to convert other equations for lines into slope-intercept form.

Example 2   Give the slope and y-intercept of $y = -8 - 3x$.

Solution   Rewrite $y = -8 - 3x$ in the form $y = mx + b$ using the Commutative Property of Addition and the definition of subtraction.

$$y = -3x + -8$$

The slope is -3 and the y-intercept is -8.

Example 3   Write the equation $3x + 4y = 9$ in slope-intercept form. Give the slope and y-intercept.

Solution   Solve the given equation $3x + 4y = 9$ for y.

$$4y = -3x + 9$$
$$\frac{1}{4} \cdot 4y = \frac{1}{4}(-3x + 9)$$
$$y = -\frac{3}{4}x + \frac{9}{4}$$

The slope is $-\frac{3}{4}$. The y-intercept is $\frac{9}{4}$ or $2\frac{1}{4}$.

Equations for all nonvertical lines can be written in slope-intercept form.

Example 4 Write an equation for the line with slope 5 and y-intercept -1 and graph the line.

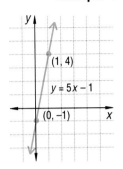

Solution Substitute 5 for m and -1 for b in $y = mx + b$.
So $y = 5x + -1$ or $y = 5x - 1$ is an equation for the line.
Since the intercept is -1, the graph contains (0, -1). Since the slope is 5, the graph goes up 5 units for each unit it goes to the right.

Check Find two points on $y = 5x - 1$. We find (0, -1) and (1, 4). The slope determined by the points is 5.

Every constant increase or constant decrease situation can be described by an equation whose graph is a line. The y-intercept of that line can be interpreted as the starting amount. The slope of that line is the amount of increase or decrease per unit.

Example 5 Pam received $100 for graduation and spends $4 of it a week.
a. Find an equation for the amount y she has after x weeks.
b. What is the slope and y-intercept of the graph?

Solution
a. The equation is found by methods you have learned in previous chapters.

$$y = 100 - 4x$$

b. Rewrite in slope-intercept form.

$$y = -4x + 100.$$

The slope is -4 and the y-intercept is 100.

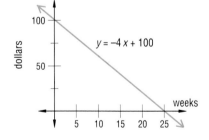

Check A graph of $y = -4x + 100$ is shown at the left. In 25 weeks, she will have $0, which agrees with the given information.

Recall that every vertical line has an equation of the form $x = h$ where h is a fixed number. Equations of this form clearly cannot be solved for y. Thus equations of vertical lines cannot be written in slope-intercept form. This confirms that the slope of vertical lines cannot be defined.

Questions

Covering the Reading

1. What form of an equation of a line is $y = mx + b$?

2. Finish the check of Example 1.

In 3 and 4, give **a.** the slope and **b.** the y-intercept.

3. $y = 4x + 2$

4. $y = -\frac{1}{3}x + 6$

In 5–7, consider these three graphs. **a.** Match the situation with its graph.
b. Give the slope of the line. **c.** Give the y-intercept.

(i)

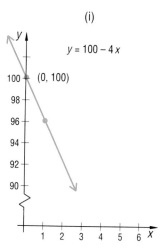

(ii)

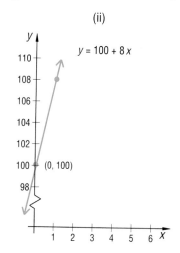

(iii)

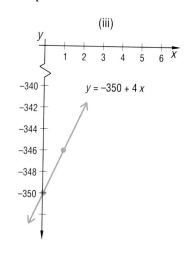

5. Quincy began with $100 and earns $8 a week.

6. Pam was given $100 for graduation and spends $4 a week.

7. The Carter family owes $350 on a refrigerator and is paying it off at $4 a week.

8. When a constant increase situation is graphed, the y-intercept can be interpreted as the ___?___.

9. Equations of ___?___ lines cannot be written in slope-intercept form.

In 10 and 11, write an equation of the line with the given characteristics.
10. slope -3, y-intercept 5   **11.** slope $\frac{2}{3}$, y-intercept -1

In 12 and 13, **a.** write in slope-intercept form; **b.** give the slope.
12. $y = 7.3 - 1.2x$ **13.** $x + 6y = 7$

In 14 and 15, give **a.** the slope and **b.** the y-intercept.
14. $3x + 2y = 10$ **15.** $y = x$

16. Write an equation of a horizontal line with y-intercept -4.

17. Graph $y = -\frac{1}{2}x + 5$ by using the slope and y-intercept.

18. Match each line n, p, q, and r with its equation below.
   a. $y = x$
   b. $y = x + 1$
   c. $y = x + 3$
   d. $y = x - 3$

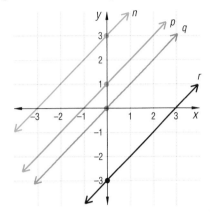

19. Match each line s, t, u, and v with its equation below.
   a. $y = 2x - 2$
   b. $y = \frac{2}{3}x - 2$
   c. $y = {}^-2x - 2$
   d. $y = -\frac{2}{3}x - 2$

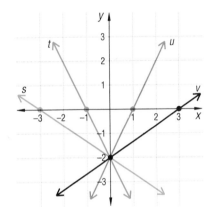

In 20 and 21, each situation naturally leads to an equation of the form $y = mx + b$. **a.** Give the equation. **b.** Graph the equation.

20. Begin with $8.00. Collect $.50 per day.

21. A sailboat is 9 miles from you. It travels towards you at a rate of 6 mph.

In 22 and 23, use this information: nonvertical lines are parallel if they have the same slope.

22. *True or false*   The lines $y = \frac{1}{4}x + 7$ and $x + 4y = 2$ are parallel.

23. Write an equation of a line with y-intercept -7 that is parallel to the line $y = 3x + 1$.

Review

24. Find the slope of the line through $(5, -3)$ and $(2, 7)$. *(Lesson 8-2)*

25. *Multiple choice*   A line through which of the following pairs of points has no slope? *(Lesson 8-2)*
(a) $(4, 6)$, $(6, 6)$ (b) $(3, -1)$, $(3, 4)$
(c) $(0, 0)$, $(5, 6)$ (d) $(0, 0)$, $(5, 0)$

26. Find the mean, median and range of each set. *(Lessons 1-2, 1-3)*
 a. 34, 73, 21, 95, 86
 b. 3.4, 7.3, 2.1, 9.5, 8.6

27. A brick wall is 20 feet high. How far away from the base of the wall will a 24-ft ladder be located when its top is at the top of the wall? *(Lesson 7-5)*

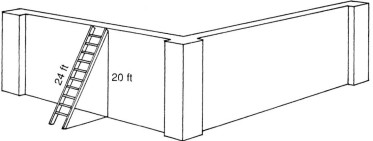

28. A rectangle is 5 cm longer than twice its width. Its perimeter is 58 cm.
 a. Find its width and length. *(Lesson 6-1)*
 b. What is its area? *(Lesson 4-1)*

In 29–31, solve. *(Lessons 5-7, 6-9)*

29. $\dfrac{3x + 5}{2 - 4x} = \dfrac{2}{3}$ **30.** $\dfrac{1}{2(1 + y)} = \dfrac{3}{y}$ **31.** $\dfrac{z + 2}{2} = 2z - 2$

Exploration

32. A graphing calculator or computer graphing program may be helpful in this question.
 a. Graph the lines with equations $y = 3x + 5$, $y = 4x + 5$, and $y = 5x + 5$.
 b. Describe the graph of $y = mx + 5$.
 c. Graph $y = -2x + 5$. Is your description in part b true when m is negative?

8-5

Equations for Lines with a Given Point and Slope

In Lesson 8-3, you graphed lines given the slope and any point on the line. In the last lesson you found an equation of a line given its slope and a particular point on the line, its *y*-intercept. You can, in fact, find an equation in slope-intercept form of a line given its slope and any point on the line, not necessarily the *y*-intercept.

Example 1 Find an equation in slope-intercept form for the line through (-3, 5) with a slope of 2.

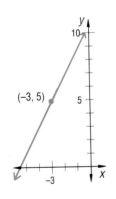

Solution You know that $y = mx + b$ is the slope-intercept equation of a line. In this case you are given $m = 2$. All that is needed is b.
1. Substitute $m = 2$ and the coordinates (-3, 5) into $y = mx + b$. This gives $5 = 2 \cdot -3 + b$.
2. Solve this equation for b. $5 = -6 + b$ so $b = 11$.
3. Substitute the values for m and b in $y = mx + b$. So the equation is $y = 2x + 11$.

Check Draw a rough graph of the line through (-3, 5) with slope 2. The graph shows that the *y*-intercept 11 seems reasonable.

If the line has a slope, the method of Example 1 will always work.

To find an equation of a nonvertical line given one point and the slope:
1. Substitute the slope m and the coordinates of the point (x, y) in the equation $y = mx + b$.
2. Solve the equation from step 1 for b.
3. Substitute the value for m and the value you found for b in $y = mx + b$.

Example 2 A line crosses the x-axis at (6, 0) and has slope -4. Find an equation for the line in slope-intercept form.

Solution Follow the steps above with $m = -4$, $x = 6$, and $y = 0$.
1. Substitute for m, x, and y. $0 = -4 \cdot 6 + b$
2. Solve for b. $24 = b$
3. Substitute for m and b. $y = -4x + 24$

This procedure is often useful for finding an equation to describe a real-life situation.

Example 3 The population of the province of Ontario in Canada was 8,543,000 in 1980. If the population is increasing at a rate of 80,000 people per year, find an equation relating the population y to the year x in Ontario.

Solution This is a constant increase situation so it can be described by a line with equation $y = mx + b$. The rate of 80,000 people per year is the slope, so $m = 80{,}000$. The population of 8,543,000 in 1980 is described by the point (1980, 8,543,000). You now have the slope and one point so you can follow the steps above.
1. Substitute for m, x, and y. $8{,}543{,}000 = 80{,}000 \cdot 1980 + b$
2. Solve for b. $8{,}543{,}000 = 158{,}400{,}000 + b$
   $-149{,}857{,}000 = b$
3. Substitute for m and b. $y = 80{,}000x - 149{,}857{,}000$

Check When $x = 1980$, does $y = 8{,}543{,}000$? Yes, so it checks.

Questions

Covering the Reading

1. Describe the steps for finding an equation for a line given the slope and one point on the line.

In 2–5, given the slope and one point, find an equation for the line.

2. point (2, 3), slope 4 3. point (-10, 3), slope -2

4. point (-6, 0), slope $\frac{1}{3}$ 5. point (4, $-\frac{1}{2}$), slope 0

6. The population of the province of Quebec in Canada was 6,398,000 in 1980. If the population is increasing at a rate of 40,000 people per year, find an equation relating the population of Quebec to the year.

7. Marty is spending money at the average rate of $3 a day. After 14 days he has $68 left. Write an equation relating the amount left y to the number of days x.

8. Diane knows a phone call to a friend costs 25¢ for the first 3 minutes and 10¢ for each additional minute. Write an equation to describe the cost y of a call of x-minutes duration. Check your answer.

9. Refer to Example 3. If the rate of population increase stays constant, what will be the population of Ontario in the year 2000?

10. The *x-intercept* of a line is the point where the line crosses the x-axis. The slope of a line is -4 and its x-intercept is 7. Find an equation for the line.

11. The slopes of two lines are reciprocals. The equation of one of the lines is $y = 2x + 1$.
    a. Find the slope of the second line.
    b. Find an equation for the second line if it passes through the point $(2, 3)$.

12. Stephanie, a 4-week-old baby, weighs 142 oz. She is gaining weight at the rate of 3 oz per week. Give her weight y at age 15 weeks.

13. Match each of lines m, n, p, and q with its equation. *(Lesson 8-4)*
    a. $y = -\frac{1}{4}x$
    b. $y = -\frac{1}{4}x + 3$
    c. $y = -1 - \frac{1}{4}x$
    d. $y = -4 - \frac{1}{4}x$

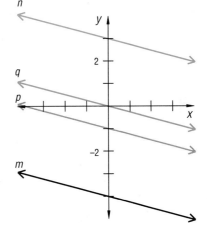

14. Match each of lines s, t, u, and v with its equation. *(Lesson 8-4)*
    a. $y = x + 2$
    b. $y = -x + 2$
    c. $y = 2 - 3x$
    d. $y = \frac{3}{2}x + 2$

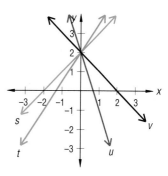

15. Do the points (1, 3), (-3, -5), and (3, 6) lie on the same line? *(Lesson 8-2)*

16. The following two points give information about an overseas telephone call: (5 minutes, $5.91), (10 minutes, $10.86). Calculate the slope and describe what it stands for. *(Lesson 8-2)*

17. Which section, or sections, of this graph shows: **a.** the fastest increase? **b.** the slowest decrease? **c.** no change? *(Lesson 8-1)*

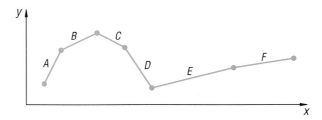

18. A rock climber starts at an elevation of 11,565 feet. He goes up $2 \frac{\text{ft}}{\text{min}}$ for 2 minutes, then down $1 \frac{\text{ft}}{\text{min}}$ for 3 minutes. He stands still for one minute, then descends 5 feet in one minute. Graph his elevation during this time. *(Lesson 8-1)*

19. Graph $y = -x$. *(Lesson 8-4)*

20. A survey showed that 7 out of 10 adults drink coffee in the morning. If 650 adults live in an apartment complex, how many of them would you expect to drink coffee in the morning? *(Lesson 5-7)*

21. Solve for n. *(Lessons 4-9, 7-4)*
   a. $n^2 = 24$ **b.** $\sqrt{n} = 24$ **c.** $n! = 24$

22. Solve. *(Lessons 4-4, 7-3, 7-4, 7-8)*
   a. $3x = 12$
   b. $3\sqrt{y} = 12$
   c. $3 \cdot |z - 5| = 12$

23. Simplify. *(Lesson 5-1)*
   a. $\frac{15}{2} \div \frac{15}{3}$ **b.** $\frac{a}{2} \div \frac{a}{3}$ **c.** $\frac{a}{b} \div \frac{a}{c}$

Exploration

24. Examine Example 3.
   a. Find the population at the last census of the state where you live.
   b. Estimate or find out how fast your state is growing per year.
   c. Using your estimate, find a linear equation relating the population y to the year x.
   d. Use this equation to estimate what the population of your state will be when you are 50 years old.

8-6

Equations for Lines Through Two Points

A phone book might tell you that a call to a friend some distance away costs 25¢ for the first 3 minutes and 10¢ for each additional minute. This is like being given the point (3, 25) and the slope 10. If you know an equation of the line through this point with this slope, then you have a formula which can help you know the cost of any phone call to that friend. In Lesson 8-5 you learned how to find such an equation.

Sometimes the given information does not include the slope, but includes two points. For instance, you might be told that a 5-minute call overseas costs \$5.91 and a 10-minute call costs \$10.86. You have two data points, (5, 5.91) and (10, 10.86). To obtain an equation for the line through these points, first calculate the slope. Then work as before.

Before finding this telephone cost equation, we work through an example with simpler numbers.

Example 1 Find an equation for the line through (5, -1) and (-3, 3).

Solution
1. First find the slope.

$$m = \frac{3 - {-1}}{-3 - 5} = \frac{4}{-8} = -\frac{1}{2}$$

Now you have the slope and two points. This is more information than you need. Pick one of these points. We pick (5, -1).

2. Substitute $-\frac{1}{2}$ and the coordinates of (5, -1) into $y = mx + b$.

$$-1 = -\frac{1}{2}(5) + b$$

3. Solve for b.　　$-1 = -\frac{5}{2} + b$

$$\frac{3}{2} = b$$

4. Substitute the values of m and b into the equation.

$$y = -\frac{1}{2}x + \frac{3}{2}$$

Check Substitute the coordinates of the point not used to see if they work.

Point (-3, 3): Does $3 = -\frac{1}{2}(-3) + \frac{3}{2}$?

Does $3 = \frac{3}{2} + \frac{3}{2}$? Yes.

The procedure involves just one more step than the procedure of Lesson 8-5.

To find an equation for a nonvertical line given two points on it:
1. Find the slope determined by the two points.
2. Substitute the slope m and the coordinates of one of the points (x, y) in the equation $y = mx + b$.
3. Solve for b.
4. Substitute the values you found for m and b in $y = mx + b$.

This method of finding the equation of a line was developed by René Descartes in the early 1600s. We now apply it to the telephone cost problem.

Example 2 If a 5-minute overseas call costs $5.91 and a 10-minute call costs $10.86, what is the cost y of a call of x-minutes duration? (Assume that this is a constant increase situation and x is a positive integer.)

Solution What you want is the line through (5, 5.91) and (10, 10.86).
1. Find the slope.

$$m = \frac{10.86 - 5.91}{10 - 5} = \frac{4.95}{5} = 0.99$$

2. Substitute 0.99 and one of the points into $y = mx + b$. We pick (5, 5.91).

$$5.91 = 0.99(5) + b$$

3. Solve for b.
$$5.91 = 4.95 + b$$
$$b = 0.96$$

4. Substitute the values for m and b into $y = mx + b$.

$$y = 0.99x + 0.96$$

Check Does this equation give the correct cost for a 10-minute call? Substitute 10 for x in the equation. Then y = 0.99(10) + 0.96 = 9.9 + 0.96 = 10.86, as needed.

The equation $y = 0.99x + 0.96$ of Example 2 says that a call costs 0.96 to make and then 0.99 for each minute. The letters x and y might be replaced by t (for time) and c (for cost), in which case the formula would become $c = 0.99t + 0.96$.

Linear relationships occur in many places. Some of them may surprise you.

Example 3 Biologists have found that the number of chirps some crickets make per minute is related to the temperature. The relationship is very close to being linear. When crickets chirp 124 times a minute, it is 68°F. When they chirp 172 times a minute, it is 80°F. Below is a graph of this information.
a. Find an equation for the line through the two points.
b. About how warm is it if you hear 100 chirps in a minute?

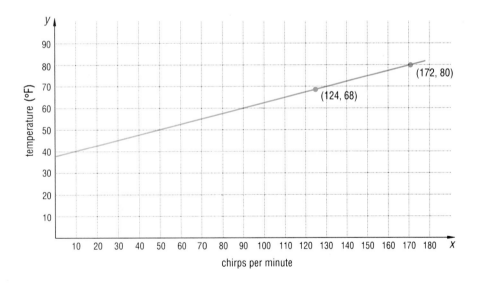

Solution

a. 1) First find the slope.

$$m = \frac{80 - 68}{172 - 124} = \frac{12}{48} = \frac{1}{4}$$

2) Substitute $\frac{1}{4}$ and the coordinates of (124, 68) into $y = mx + b$.

$$68 = \tfrac{1}{4}(124) + b$$

3) Solve for b.

$$68 = 31 + b$$
$$37 = b$$

4) Substitute for m and b.

An equation is $y = \tfrac{1}{4}x + 37$.

Check   Substitute the coordinates of the point (172, 80) that was not used in finding the equation.

$$\text{Does } 80 = \tfrac{1}{4}(172) + 37?$$
$$\text{Does } 80 = 43 + 37? \text{ Yes.}$$

Solution

b. Substitute 100 for x.

$$y = \tfrac{1}{4}(100) + 37$$
$$y = 25 + 37$$
$$y = 62$$

It is about 62° when you hear 100 chirps in a minute.

Check   Plot (100, 62). It should lie on the line.

The equation in Example 3 enables you to find the temperature for any number of chirps. By solving for x in terms of y, you could get a formula for the number of chirps to expect at a given temperature. Formulas like these seldom work far from the data points. Crickets tend not to chirp at all below 50°F, yet the formula $y = \tfrac{1}{4}x + 37$ predicts about 52 chirps a minute at 50°F.

Questions

Covering the Reading

In 1–3, find an equation for the line through the two given points. Check your answer.

1. (1, 9), (7, 3)

2. (6, -3), (-8, -10)

3. (0, 11), (13, 0)

4. Who developed the method used in this lesson for finding the equation of a line?

5. In Example 2, what would an 8-minute call cost?

6. If a 5-minute overseas call to Bonn costs $4.50 and a 10-minute call costs $8.50, find a formula relating time and cost.

In 7–9, refer to Example 3.

7. The number of times a cricket chirps in a minute and the temperature is very close to what kind of relationship?

8. When the number of chirps is 90, about what is the temperature? Use the graph to find the answer.

9. By substituting in the equation $y = \frac{1}{4}x + 37$, about how many chirps per minute would you expect if the temperature is 70°F?

Applying the Mathematics

10. **a.** Show that $A = (-4, 7)$, $B = (1, 5)$, and $C = (16, -1)$ lie on the same line.
    b. Find an equation for the line.

11. The sum of the measures of the angles in a triangle is 180°. In a quadrilateral, the sum is 360°. Find an equation relating n, the number of sides in a polygon, and the sum of its angles, S.
    (Hint: $S = mn + b$.)

12. It was reported in 1960 that the total length y and the tail length x of females of the snake species *Lampropeltis polyzona* have close to a linear relationship. When $x = 60$ mm, $y = 455$ mm. When $x = 140$ mm, $y = 1050$ mm.
    a. Find the slope of the relationship between x and y. (Approximate your answer to the nearest tenth.)
    b. Find an equation for the relationship between x and y using (60, 455) and your slope from part a.
    c. Check your equation using (140, 1050).

13. The graph below shows the linear relationship between Fahrenheit and Celsius temperatures. The freezing point of water is 32 °F and 0 °C. The boiling point of water is at 212 °F and 100 °C.
   a. Find an equation that relates Celsius and Fahrenheit temperatures. (Hint: $C = mF + b$.)
   b. When it is 150 °F, what is the temperature in degrees Celsius?
   c. When it is 150 °C, what is the temperature in degrees Fahrenheit?

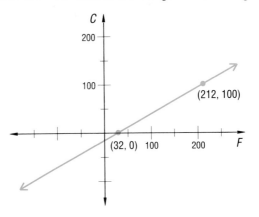

Review

14. A cab company charges a base rate plus $1.10 a mile. A 12-mile cab ride costs $14. Write an equation relating the number of miles driven to the cost of the cab ride. *(Lesson 8-5)*

15. Graph $y = 4x - 3$ by using the slope and y-intercept. *(Lesson 8-4)*

16. A road leads from the town of Salida. There are signs every 5 miles that tell the elevation. *(Lesson 8-1)*

miles from Salida:	elevation (feet)
0	1744
5	1749
10	1749
15	1759
20	1757

   a. Graph these data.
   b. Calculate the rate of change of elevation for the entire distance.

17. Triangles *ABC* and *DEF* below are similar. Corresponding sides are parallel. **a.** Solve for *x*.
   b. What is the ratio of similitude? *(Lesson 5-8)*

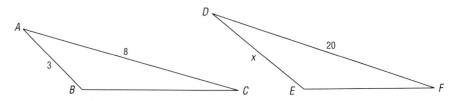

18. What question does the following computer program answer? *(Lesson 8-5)*

```
10 INPUT M, X, Y
20 LET B = Y − M * X
30 PRINT B
```

19. A student has quiz scores of 88, 94, 85, and 80.
   a. What quiz score does the student need in order to average 88 for the five quizzes?
   b. What quiz score is needed in order to have a median score of 88 for the five quizzes?
   c. What quiz score is needed in order to have a mode score of 88 for the five quizzes? *(Lesson 1-2)*

20. *Skill Sequence* Solve. *(Lessons 7-4, 7-8)*
   a. $x^2 = 64$
   b. $(y − 7)^2 = 64$
   c. $(3y − 7)^2 = 64$

21. Simplify. *(Lesson 1-8)*

   a. $\frac{12}{144}$ **b.** $\frac{3}{3^2}$ **c.** $\frac{d}{d^2}$

Exploration

22. A ride in a taxicab costs a fixed number of dollars plus a constant charge per mile.
   a. Find a rate for taxi rides in your community.
   b. Graph your findings on coordinate axes like the ones at right.

8-7

Fitting a Line to Data

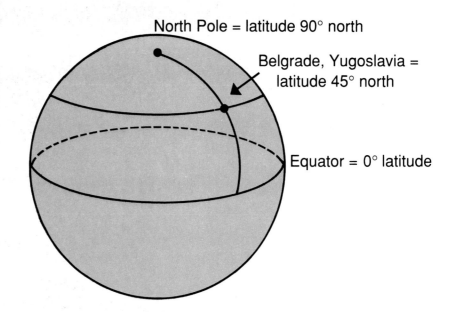

North Pole = latitude 90° north

Belgrade, Yugoslavia = latitude 45° north

Equator = 0° latitude

In many situations, there are more than two points. The points may not all lie on a line. Still an equation may be found that closely describes the coordinates. Finding that equation is the subject of this lesson.

The latitude of a place on the earth tells how far the place is from the equator. Latitudes in the Northern Hemisphere range from 0° at the equator to 90° at the North Pole.

In the table below are the latitude and mean high temperatures in April, for selected cities in the Northern Hemisphere. (The mean high temperature is the mean of all the high temperatures for the month.) Although in all of these cities temperature is measured in degrees Celsius, we have converted the temperatures to Fahrenheit for you.

Latitude and Temperature in Selected Cities

City	North Latitude	April Mean High Temperature (°F)
Lagos, Nigeria	6	89
San Juan, Puerto Rico	18	84
Calcutta, India	23	97
Cairo, Egypt	30	83
Tokyo, Japan	35	63
Rome, Italy	42	68
Belgrade, Yugoslavia	45	45
London, England	52	56
Copenhagen, Denmark	56	50
Moscow, USSR	56	47

To graph the data points in the table on page 398, let the latitude be the *x*-value. The *y*-value is the Fahrenheit temperature. Both units are degrees. For instance the data point for Tokyo is (35, 63). Here is the scattergram.

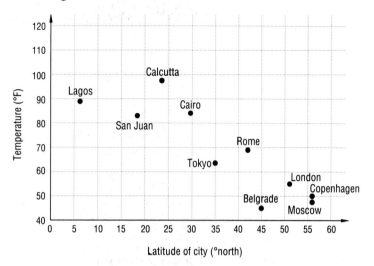

The points on the graph suggest that the higher the latitude, the lower the temperature. What does the data suggest for a city like Quito, Equador, at the equator (0° latitude)? What would you predict for the temperature at a city at 19° latitude, like Mexico City or Bombay?

To answer these questions, it helps to "fit a line" to the data. No line will pass through all the data points, but you can find a line that describes the trend of higher latitude, lower temperature. The simplest way is to take a ruler and draw a line that seems closest to all the points. This is called "fitting a line by eye," and one such line is graphed below.

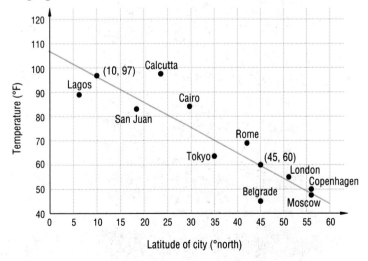

Once a line has been fitted to the data, the idea is to find an equation for the line and then use the equation to predict temperatures for different latitudes.

Example 1   Find a line to the latitude and temperature data points.

Solution   We use the line from the previous page. It contains the points (10, 97) and (45, 60). First find the slope of the line.

$$\text{slope} = \frac{60 - 97}{45 - 10} = \frac{-37}{35} \approx -1.06$$

Now substitute this slope and the coordinates of one of the points into $y = mx + b$ and solve. We use (10, 97).

$$97 = -1.06 \cdot 10 + b$$
$$97 = -10.6 + b$$
$$107.6 = b$$

So an equation of the line is $y = -1.06x + 107.6$.

Check   Substitute (45, 60) into the equation.
Does $60 = -1.06(45) + 107.6$?
$60 \approx 59.9$ so it checks.

The negative slope -1.06 means that as you move 1° north in latitude, the April mean high temperature is about 1°F lower. The *y*-intercept 107.6 means that the high temperature at the equator (0° latitude) should be about 108° Fahrenheit. Using the equation $y = -1.06x + 107.6$, you can estimate the temperature for cities not listed on the chart.

Example 2   Use the equation for the fitted line to predict the April mean high temperature for Madrid, Spain, which is at 40° north latitude.

Solution   Use 40° for *x* in the equation.

$$y = -1.06 \cdot x + 107.6$$
$$y = -1.06 \cdot 40 + 107.6$$
$$y = 65.2$$

You can predict that Madrid would have an April mean high temperature of about 65°F.

Calahora Castle in Andalucia, Spain

The mean high temperature in April for Madrid is actually 64°. The predicted temperature is remarkably close. This is not always the case. For Mexico City, at a latitude of 19° north, the line predicts a temperature of 87°. The actual April mean temperature for Mexico City is 78°. The prediction is high because Mexico City is at an altitude of about one mile, and temperatures at high altitudes are lower.

Questions

Covering the Reading

1. What does the latitude of a place on earth signify?

2. What is the latitude of the North Pole?

3. What is the latitude of the equator?

4. Which city is farther north, Calcutta or Cairo?

5. What does it mean to "fit a line by eye" to a scattergram?

6. Use the graph of the fitted line in this lesson to predict the temperature in a city at 25° north latitude.

7. Once a line is fitted, what is the first step toward getting an equation for the line?

8. **a.** Acapulco, Mexico, is at 17°N latitude. Use the equation in Example 2 to estimate its average April high temperature.
   b. The actual value for Acapulco is 87°. Give a reason why the answer in part a is closer to the actual value for Acapulco than the prediction was for Mexico City.

9. For which of the set of points below is fitting a line appropriate?

a.   b.

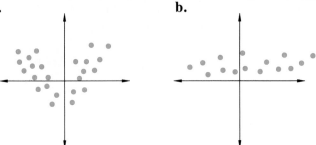

c.   d.

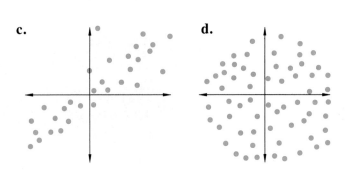

10. Use the following data. The temperatures are the mean low temperatures in January.

City	North Latitude	January Mean Low Temperature (°F)
Lagos, Nigeria	6	74
San Juan, Puerto Rico	18	67
Calcutta, India	23	55
Cairo, Egypt	30	47
Tokyo, Japan	35	29
Rome, Italy	42	39
Belgrade, Yugoslavia	45	27
London, England	52	35
Copenhagen, Denmark	56	29
Moscow, USSR	56	9

a. Draw a scattergram showing a point for each city.
b. Use a ruler to fit a line to the data.
c. Estimate the coordinates of two points on the line you drew.
d. Find an equation for the line through the points in part c.
e. Complete the following sentence: "As you go one degree north, the January low temperature tends to __?__."
f. What does the equation predict for a January mean low temperature at the equator?
g. Use your equation to predict the January mean low temperature for Acapulco. (Note: the actual mean low is 70°.)
h. Predict the January mean low temperature for the North Pole.

11. *Yes or no* By using negative values of x in the equation in Example 1, would you expect to predict temperatures for cities south of the equator?

12. Give the slope and y-intercept of the line $3x + 5y = 2$. *(Lesson 8-4)*

13. Find an equation for the line through $(3, 2)$ with a slope of $\frac{3}{5}$. *(Lesson 8-5)*

14. Find an equation for the line through $(0, 7)$ and $(4, 0)$. *(Lesson 8-6)*

15. Graph the points satisfying $x \leq 3$: **a.** on the number line; **b.** in the coordinate plane. *(Lessons 1-1, 7-2)*

16. An integer from 1 through 25 is chosen at random. Find the probability the integer is even or greater than 20. *(Lesson 3-6)*

17. Seattle scored 17 points in the first nine minutes of a game. At that rate about how many points would they score in a 48-minute game? *(Lesson 5-7)*

18. Solve for y. *(Lesson 6-6)*
a. $8y = 2 + y$
b. $By = C + y$

19. Refer to the drawing below. About how far, to the nearest mile, is it from Marshall to Union? *(Lesson 7-5)*

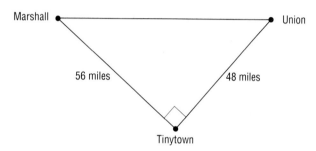

20. The scale on a map reads 1 inch = 2 miles. Two buildings are 3.6 miles apart. How far apart should they be on the map? *(Lesson 5-7)*

21. *Skill sequence*   When $x = -4.2$ and $y = -0.7$, evaluate: *(Lesson 1-4)*
a. $-\dfrac{-x}{y}$ 　　　　　 **b.** $\dfrac{-x}{-y}$ 　　　　　 **c.** $-\dfrac{-x}{-y}$

22. The volume of a cube is 1000 cm³. Calculate in your head. *(Lesson 4-1)*
a. the length of an edge
b. the area of a face

23. a. Find the latitude of your school to the nearest degree.
b. Predict the April mean high temperature in °F for your latitude using the line in Example 1.
c. Check your prediction in part b against some other source of the April mean high temperature. (Newspapers or TV weather records are possible sources.)

24. What is meant by *longitude*?

8-8

Equations for All Lines

In this chapter, lines have been used to describe situations involving constant increase or decrease. They have also been used to fit a line to data. The slope-intercept form $y = mx + b$ arises naturally from these applications. All lines except vertical lines have equations in slope-intercept form.

Some situations naturally lead to equations of lines in a different form.

> The Ramirez family bought 4 adult and 3 children's tickets to the play. They spent $24.00. If x is the cost of an adult ticket and y the cost of a children's ticket, then
>
> $$4x + 3y = 24.$$

The pairs of values of x and y that work in the equation above are possible costs of the ticket. For instance, since $3(4.50) + 4(2) = 24$, the adult tickets could have cost $4.50 and the children's tickets $2. This yields the point (4.50, 2).

This equation has the form

$$Ax + By = C.$$

All terms with variables are on one side of the equation. The constant term is on the other. This is the **standard form** for an equation of a line. Any equation for a line is either in standard form or can be rewritten in standard form.

Example 1 Rewrite $5y + 8 = 3x$ in standard form and find the values of A, B, and C.

Solution In standard form, the terms with variables are on the left side, so add $-3x$ to both sides.

$$-3x + 5y + 8 = -3x + 3x$$
$$-3x + 5y + 8 = 0$$

Now add -8 to both sides to get the constant on the right side.

$$-3x + 5y = -8.$$

This is standard form with $A = -3$, $B = 5$, and $C = -8$.

Some people prefer that the coefficient A be positive in standard form. To rewrite the equation in Example 1 with a positive A, you could multiply each side by -1.

$$-1(-3x + 5y) = -1(-8)$$
$$3x - 5y = 8$$

Both sides of this equation can be multiplied by any number. For instance, multiplying by 2 yields

$$6x - 10y = 16.$$

This is another equation for the same line. Every line has many equations in standard form. The equation $3x - 5y = 8$ is usually considered simplest because 3, 5, and 8 have no common factors.

Example 2 Rewrite $y = .25x$ in standard form with integer values of A, B, and C.

Solution Since $.25 = \frac{25}{100} = \frac{1}{4}$, multiply both sides by 4.

$$4y = x$$

Now, add $-x$ to get both variables on the left side.

$$-x + 4y = -x + x$$
$$-x + 4y = 0$$

Here $A = -1$, $B = 4$, and $C = 0$.

Graphing a line is sometimes easier when its equation is given in standard form. This is particularly the case when A and B are factors of C. The strategy is to find the intercepts. Recall that the *y*-intercept is the value of *y* when $x = 0$. The **x-intercept** is the value of *x* when $y = 0$. It is the *x*-coordinate of the point of intersection of the graph and the *x*-axis.

Example 3   Graph $4x + 3y = 24$, the equation of the ticket costs of the Ramirez family.

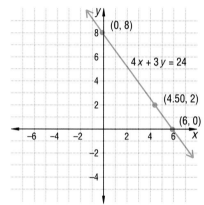

Solution   Since 4 and 3 are factors of 24, it is easy to find the intercepts. When $x = 0$, $3y = 24$, so $y = 8$. This gives the point (0, 8). When $y = 0$, $4x = 24$, so $x = 6$. This gives the point (6, 0). Plot the points and connect them.

Check   Find a third point satisfying the equation. Earlier we noted that (4.50, 2) satisfies the equation. Is it on the graph? Yes.

Lines in standard form often arise naturally from real situations.

Example 4   Roast beef sells for $6 a pound. Shrimp is $12 a pound. Andy has $96 to buy beef and shrimp for a party. Write an equation in standard form to describe the different possible combinations Andy could buy.

Solution
Let x = pounds of roast beef bought, so $6x$ = cost of roast beef.
Let y = pounds of shrimp bought, so $12y$ = cost of shrimp.

$$\begin{array}{ccccc}
\text{Cost of roast beef} & + & \text{cost of shrimp} & = & \text{total cost} \\
6x & + & 12y & = & 96
\end{array}$$

This equation on page 406 is in standard form. It can be further simplified by multiplying both sides by $\frac{1}{6}$.

$$\tfrac{1}{6}(6x + 12y) = \tfrac{1}{6} \cdot 96$$
$$x + 2y = 16$$

Check To check the equation, suppose Andy buys 6 pounds of roast beef. He will have spent $6 \cdot \$6 = \$36$. He will have $\$96 - \$36 = \$60$ left to spend on shrimp. At \$12 per pound he can buy $\frac{60}{12} = 5$ pounds of shrimp.

Does $(6, 5)$ work in $x + 2y = 16$? Yes, $6 + 2 \cdot 5 = 16$.

The lines in all these Examples are *oblique,* meaning they are not horizontal or vertical. Recall from Lesson 7-2, if a line is vertical, then it has an equation of the form $x = h$. A horizontal line has an equation $y = k$. These are already in standard form. For example,

$$x = 3 \qquad \text{is equivalent to} \qquad 1 \cdot x + 0 \cdot y = 3$$
$$A = 1, B = 0, C = 3$$
$$y = -\tfrac{1}{2} \qquad \text{is equivalent to} \qquad 0 \cdot x + 1 \cdot y = -\tfrac{1}{2}.$$
$$A = 0, B = 1, C = -\tfrac{1}{2}$$

Thus *every line* has an equation in the standard form $Ax + By = C$.

Questions

Covering the Reading

1. The equation $y = mx + b$ arises naturally from what two applications covered in this chapter?

2. **a.** If each adult ticket cost the Ramirez family (at the beginning of the lesson) \$3.75, how much would a child's ticket have cost?
   b. Give the coordinates of the point. Can you find the point on the graph of Example 3?

3. The form of the equation $Ax + By = C$ is called __?__.

4. __?__ lines cannot be written in slope-intercept form.

5. Every line can be written in the form __?__.

In 6 and 7, the line is in the form $Ax + By = C$. Give the values of A, B, and C.

6. $4x + 2y = 5$ 7. $x - 8y = 2$

In 8–10, write the equation in standard form. Give values of A, B, and C.

8. $2x = 3y - 12$ 9. $y = 4x$ 10. $-4y = 20$

In 11 and 12, graph using the method of Example 3.

11. $3x + 5y = 30$ **12.** $2x - 3y = 12$

13. Refer to Example 4. Find three different combinations of roast beef and shrimp Andy could buy.

14. *Multiple choice*   Which of the following is *not* an equation for a line?
(a) $x + 7y = 3$ (b) $y = x + 3$
(c) $5x = 3$ (d) $xy = 3$

15. A 100-point test has x questions worth 2 points apiece and y questions worth 4 points apiece. Write an equation that describes all possible numbers of questions.

16. Carl has $36 in five-dollar bills and singles. How many of each kind of bill does he have?
a. Write an equation to describe this situation.
b. Give three solutions.
c. Graph all possible solutions. (The graph will be discrete.)

17. Write an equation in standard form for the line through (0, 4) and (2, 0).

18. Find an equation for the line through the points (-1, 7) and (1, 1). *(Lesson 8-6)*

19. The length of the winning toss in the men's discus throw has shown a steady increase since the first Olympics in 1896. The scattergram below shows the winning tosses in meters. *(Lesson 8-7)*

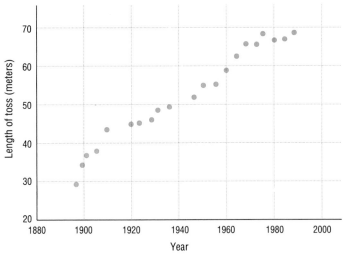

a. Trace the graph and fit a line to the data.
b. Find an equation for the line in part a.
c. Use your equation to predict the length of the winning toss in 1992.
d. Why might the prediction for 1992 be incorrect?

20. Refer to Question 19. What is the length of the Olympic Record discus throw in feet? (Recall that 1 in. = 2.54 cm) *(4-2)*

21. Graph in the plane: $y < 17$. *(Lesson 7-2)*

In 22–24, determine whether $(0, 0)$ is a solution to the sentence. *(Lesson 2-7)*

22. $x + y < \text{-}4$

23. $6x - y > \text{-}6$

24. $12x > y + 13$

25. A regular die is tossed once and a coin is flipped once. What is the probability of getting a 3 on the die and tails on the coin? *(Lesson 4-7)*

26. *Skill sequence*   Solve. *(Lessons 3-3, 4-5)*
   a. $x - 4 = \text{-}1$
   b. $\text{-}y - 4 = \text{-}1$
   c. $\text{-}4 - z = \text{-}1$

27.

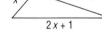 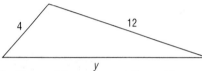

The triangles above are similar, with corresponding sides parallel. *(Lesson 5-8)*
   a. Find x.
   b. Find y.
   c. Knowing the actual lengths, redraw the triangles to make them more accurate.

Exploration

28. A graphing calculator or computer graphing program may be useful in this question.
   a. Graph the lines with equations:

$$3x + 2y = 6$$
$$3x + 2y = 12$$
$$3x + 2y = 18$$

   b. What happens to the graph of $3x + 2y = C$ as C gets larger?
   c. Try values of C that are negative. What can you say about the graphs of $3x + 2y = C$ then?

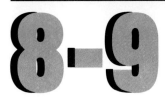

8-9

Graphing Linear Inequalities

You have graphed inequalities on the plane with horizontal and vertical lines in Lesson 7-2. This lesson extends linear inequalities to all lines in the plane.

Example 1 Draw the graph of $y \geq -3x + 2$.

Solution First graph the boundary line that separates the two regions. Here the boundary line is $y = -3x + 2$. Its y-intercept is 2. Then, since the slope is -3, plot the point 1 unit to the right and 3 units down. That point is (1, -1). Draw the line through (0, 2) and (1, -1).

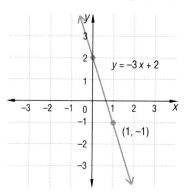

The $\geq$ sign in $y \geq -3x + 2$ indicates that points are desired whose y-coordinate is greater than or equal to the y values which satisfy $y = -3x + 2$. Since y values get larger as one goes higher, shade the entire region above the line.

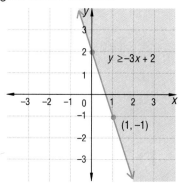

Check Try the point (0, 3), which is in the shaded region. Does it satisfy the inequality? Is $3 > -3 \cdot 0 + 2$? Yes, 3 is greater than 2. So the correct side of the line has been shaded.

The regions on either side of a line in a plane are called half-planes.

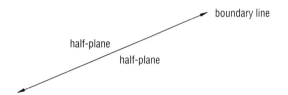

So the graph of $y \geq -3x + 2$ can be described as a line and one of its half-planes.

In general:
1. The graphs of $y > mx + b$ and $y < mx + b$ are the half-planes of the line $y = mx + b$.
2. The graphs of $Ax + By > C$ and $Ax + By < C$ are the half-planes of the line $Ax + By = C$.

When an inequality is in the standard form $Ax + By > C$, you cannot use the inequality sign to determine which side of the line to shade. A more direct method, the testing of a point, is usually used. The point (0, 0) is often chosen if it is not on the boundary line.

With the inequalities $\leq$ or $\geq$, the boundary line is drawn as a *solid* line as in Example 1. But with $<$ or $>$, the boundary is *not* part of the solution. It is not included in the graph. In this case, the boundary is drawn as a *dotted* line.

Example 2 Graph $3x - 4y > 12$.

Solution Since 3 and 4 are factors of 12, it is easy to find the intercepts of the boundary line $3x - 4y = 12$. They are (0, -3) and (4, 0). Plot these points and graph a dotted line through them. To determine which side of the line is to be shaded, substitute (0, 0) into the original equation. Is $3 \cdot 0 - 4 \cdot 0 > 12$? Is $0 > 12$? No. Since (0, 0) is in the upper half-plane and is *not* a solution, shade the lower half-plane. The finished graph is at the right.

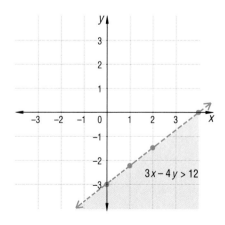

In summary, to graph any linear inequality:
1. Graph a dotted or solid line as the solution to the corresponding linear equation.
2. Shade the half-plane that satisfies the condition. [You may have to test a point. If possible, use (0, 0).]

Linear inequalities can help you to understand certain situations.

Example 3 Serena has less than $5.00 in nickels and dimes. How many nickels and dimes might she have?

Solution There are too many possibilities to list, so we graph them.
Let n = the number of nickels Serena has.
d = the number of dimes Serena has.
Since each nickel is worth .05 and each dime is worth .10,

$$.05n + .10d < 5.00.$$

Multiply both sides by 100 to make the sentence easier to work with.

$$5n + 10d < 500$$

This inequality is graphed below.

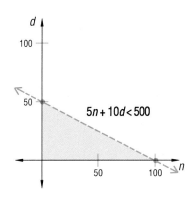

Since n and d cannot be negative, only the points in the first quadrant or on the axes are shaded. The graph is actually discrete, since n and d must be integers, but there are so many points that shading is easier. The graph thus shows that there are very many solutions.

Check (0, 0) is on the graph. If Serena has 0 nickels and 0 dimes, does she have less than $5.00? Of course.

Questions

Covering the Reading

1. The graphs of __?__ and __?__ are the half-planes of the line $y = mx + b$.

2. The graphs of __?__ and __?__ are the half-planes of the line $Ax + By = C$.

3. Name the two steps to graph any linear inequality.

4. In the graph of each inequality, tell whether the boundary line should be drawn as a solid or dotted line.
   a. $y > -4x - 7$ **b.** $y \leq -4x - 7$
   c. $4x + y < -7$ **d.** $4x + y \geq -7$

5. **a.** What point is usually chosen to test which half-plane is shaded?
   b. When would you not choose it?

In 6–9, graph the inequality.

6. $x + y > 4$

7. $y \geq -3x - 2$

8. $5x - y > 3$

9. $5x - y \leq 3$

10. Ted has less than $4.00 in quarters and dimes. How many quarters and dimes does he have?

Applying the Mathematics

11. "It will take at least 20 points to make the playoffs," the hockey team coach told the players. "We get 2 points for a win and 1 for a tie." Let W be the number of wins and T the number of ties.
   a. Write an inequality to describe the values of W and T that will enable the team to make the playoffs.
   b. Graph these values.

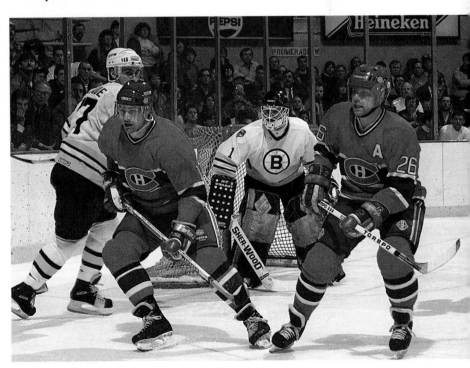

12. Suppose m and n are positive integers. How many points (m, n) satisfy $m + n < 5$?

13. a. Graph the points (x, y) for which $x + y > 5$ and $x + y < 8$.
   b. Describe the graph.

○ Review

14. Write $4x - 28 = 3y$ in standard form. Give the values of A, B, and C. *(Lesson 8-8)*

15. A travel agency bought 25 coach tickets and 5 first-class tickets for a flight. If the agency spent $55,000, write an equation that describes all possible costs of the tickets. *(Lesson 8-8)*

16. Find an equation for the line through (-6, 1) with slope 3. *(Lesson 8-5)*

17. A rectangular field is 100 yards wide by 300 yards long.

How much shorter is the distance from A to B if you walk diagonally across the field instead of around the outside edges? *(Lesson 7-5)*

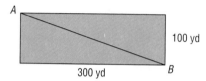

18. *Skill sequence*   Solve. *(Lessons 4-4, 7-8)*
   a. $3x = 1987$
   b. $14y - 11y = 1987$
   c. $14(z - 3) - 11(z - 3) = 1987$

19. Rewrite in decimal form. *(Appendix B)*
   a. 10^{-1} **b.** $3 \cdot 10^{-2}$ **c.** $2 \cdot 5 \cdot 10^{-6}$

20. Give the exact circumference. *(Lesson 1-5)*

a.

diameter = 8

b.

radius = 0.2

c.

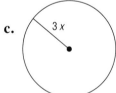

radius = 3x

In 21 and 22, use the graphs below.

21. Of the 10,886,000 males aged 25–29 in 1980 how many were single? *(Lesson 5-4)*

22. What was the average yearly increase in the percentage of females aged 20–24 staying single?
  a. from 1970 to 1980?
  b. from 1980 to 1987?
  c. Is the rate of increase constant from 1970 to 1987? *(Lesson 8-1)*

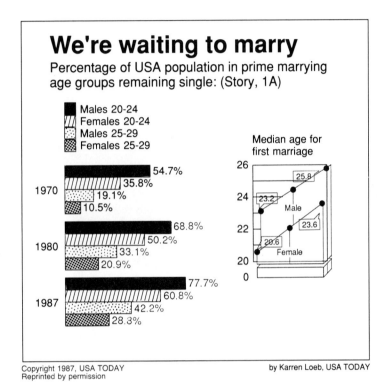

We're waiting to marry
Percentage of USA population in prime marrying age groups remaining single: (Story, 1A)

- ■ Males 20-24
- ▨ Females 20-24
- ▦ Males 25-29
- ▨ Females 25-29

1970
54.7%
35.8%
19.1%
10.5%

1980
68.8%
50.2%
33.1%
20.9%

1987
77.7%
60.8%
42.2%
28.3%

Median age for first marriage

26
25.8
24 — 23.2
Male
22
23.6
20.6
20
Female
0

Copyright 1987, USA TODAY
Reprinted by permission

by Karren Loeb, USA TODAY

23. Determine, any way you can, the number of solutions to Example 3.

Summary

The rate of change between two points (x_1, y_1) and (x_2, y_2) is $\dfrac{y_2 - y_1}{x_2 - x_1}$. When points all lie on the same line, the rate of change between them is constant and is called the slope of the line. The slope tells how much the line rises or falls as you move one unit to the right. When it is positive, the line goes up to the right. When the slope is negative, the line falls to the right. When the slope is 0, the line is horizontal. The slope of vertical lines is not defined.

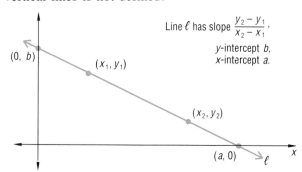

Line ℓ has slope $\dfrac{y_2 - y_1}{x_2 - x_1}$, y-intercept b, x-intercept a.

Constant increase or constant decrease situations lead naturally to linear equations of the form $y = mx + b$. The graph of the set of points (x, y) satisfying this equation is a line with slope m and y-intercept b. Other situations lead naturally to linear equations in the standard form $Ax + By = C$. When the $=$ sign in these equations is replaced by $<$ or $>$, the graph of the resulting linear inequality is a half-plane, the set of points on one side of the line.

A line is determined by any point on it and its slope, and its equation can be found from this information. Likewise, an equation can be found for the line containing two given points. If more than two points are given, then there may be more than one line determined. You can then find two points on a line that comes close to all the points and use these points to determine an equation for the line.

Vocabulary

Below are the most important terms and phrases for this chapter. You should be able to give a definition or general description and a specific example of each.

Lesson 8-1
rate of change

Lesson 8-2
constant decrease
slope

Lesson 8-4
y-intercept
slope-intercept form
Slope-Intercept Property

Lesson 8-7
fitting a line to data

Lesson 8-8
standard form for an equation of a line
x-intercept

CHAPTER 8

Progress Self-Test

Take this test as you would take a test in class. You will need graph paper, a ruler, and a calculator. Then check your work with the solutions in the Selected Answers section in the back of the book.

In 1 and 2, refer to the line graphed at the right.

1. Calculate its slope.
2. **a.** Give its y-intercept.
   b. Give its x-intercept.
3. Do the points (4, -5), (-2, 1) and (20, -20) all lie on the same line? Explain your answer.

In 4 and 5, find the slope and y-intercept of the line.

4. $y = 8 - 4x$.
5. $5x + 2y = 1$
6. Find an equation of the line with slope $\frac{3}{4}$ and y-intercept 13.
7. Find an equation of the line with slope -2 containing the point (-5, 6).
8. Rewrite the equation $y = 5x - 2$ in standard form $Ax + By = C$ and give the values of A, B, and C.
9. What is true about the slope of horizontal lines?
10. If a line has a slope of $\frac{3}{5}$, how much will the graph change as you go one unit to the right?

In 11 and 12, use the following data of total U.S. Army personnel during and after World War II.

Year	Total Personnel
1942	3,074,184
1943	6,993,102
1944	7,992,868
1945	8,266,373
1946	1,889,690

11. In which two-year period was there the greatest increase in U.S. Army personnel?
12. What is the rate of change for the entire period from 1942 to 1946?

13. A basketball team scored 67 points with x baskets worth 2 points each and y free throws worth 1 point each. Write an equation that describes all possible values of x and y.
14. A couple has already donated $100 to a charity. Now they have decided to donate $5 a week taken directly from their paycheck. After x weeks, they will have donated a total of y dollars. Give the slope and y-intercept of the line describing the relationship between x and y.
15. At age 12 Patrick weighed 43 kg; at 14 he weighed 50 kg. Find a linear equation relating Pat's weight y to his age x.
16. Graph $-3x + 2y = 12$.
17. Graph $y \le x + 1$.

In 18–21, the scattergram below illustrates the wingspan and length of 2-engine and 3-engine jet planes.

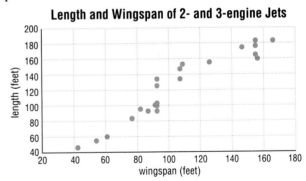

Length and Wingspan of 2- and 3-engine Jets

18. From the scattergram, about what length would you expect a jet to be if its wingspan is 100 feet?
19. **a.** Draw a line to fit the data and estimate the coordinates of two points on it.
    b. Find the slope of your line.
20. Write an equation for the line you have drawn in Question 19.
21. Use the equation from Question 20 to answer Question 18.

CHAPTER 8

Chapter Review

Questions on **SPUR** Objectives

SPUR stands for **S**kills, **P**roperties, **U**ses, and **R**epresentations.
The Chapter Review questions are grouped according to the
SPUR Objectives for this chapter.

SKILLS deal with the procedures used to get answers

Objective A: *Find the slope of the line through two given points. (Lesson 8-2)*

1. Calculate the slope of the line containing (2, 4) and (6, 2).

2. Calculate the slope of the line through (1, 5) and (-2, -2).

3. Find the slope of line A at the right.

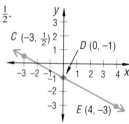

4. Using two different pairs of points, show that the slope of the line at the right is $-\frac{1}{2}$.

Objective B: *Find an equation for a line given its slope and any point on it. (Lessons 8-4, 8-5)*

5. Give an equation for the line with slope 4 and y-intercept 3.

6. What is an equation for the line with slope p and y-intercept q?

In 7–10, find an equation for the line:

7. through (-4, 1) with slope -2

8. through (6, 10) with slope 0

9. through $(3, \frac{1}{4})$ with slope 30

10. through (3, -1) with undefined slope

Objective C: *Find an equation for a line through two given points. (Lesson 8-6)*

In 11 and 12, find an equation for the line through the two given points.

11. (5, -2), (-7, -8)

12. (0.5, 6), (0, 4)

PROPERTIES deal with the principles behind the mathematics.

Objective D: *Recognize slope as a rate of change. (Lessons 8-2, 8-3)*

13. Slope is the amount of change in the __?__ of the graph as you go one unit to the __?__.

14. The slope determined by two points is the change in the __?__ coordinates divided by the __?__ in the x-coordinates.

15. As it ascends, a jet aircraft climbs 0.46 km for each km it travels away from its starting point.
    a. Draw a picture of this situation.
    b. What is the slope of the ascent?

16. The slope of the line through (a, b) and (c, d) is __?__.

■ **Objective E:** *Rewrite an equation for a line in standard form or slope-intercept form. (Lessons 8-4, 8-8)*

In 17 and 18, write in the form $Ax + By = C$. Then give the values of A, B, and C.

17. $x - 22 = 5y$

18. $y = 2x + 7$

In 19 and 20, rewrite the equations in slope-intercept form.

19. $2x + y = 4$

20. $x + 3y = 11$

■ **Objective F:** *Given an equation for a line, find its slope and y-intercept. (Lessons 8-2, 8-4)*

In 21–24, find the slope and y-intercept of the line.

21. $y = 7x - 3$

22. $4x + 5y = 1$

23. $y = \text{-}x$

24. $48x - 3y = 30$

USES deal with applications of mathematics in real situations.

■ **Objective G:** *Calculate rates of change from real data. (Lesson 8-1)*

In 25 and 26, use the average weights (in kg) for girls between birth and age 14, as given below.

Age (yr)	kg
birth	4.4
1	9.1
2	11.3
3	13.6
4	15.0
5	17.2
6	20.4
7	22.2
8	25.4
9	28.1
10	31.3
11	34.9
12	39.0
13	45.5
14	48.5

25. Find the average rate of change of weight from age 10 to 14.

26. a. According to these data, in which two-year period do girls gain weight fastest?
b. What is this rate of change?

In 27 and 28, use the temperatures (in °F) shown below recorded at O'Hare Airport in Chicago, starting at 10 P.M., November 21, 1985.

Temperatures Recorded at O'Hare Airport November 21, 1985

10 P.M.	20°	4 A.M.	14°
11 P.M.	18°	5 A.M.	13°
Mdnt.	17°	6 A.M.	13°
1 A.M.	16°	7 A.M.	14°
2 A.M.	15°	8 A.M.	17°
3 A.M.	15°	9 A.M.	20°

27. Find the average rate of change in temperature from 11 P.M. to 7 A.M.

28. a. During which three-hour period did the temperature decrease the most?
b. In that period, what was the average rate of change of the temperature?

Objective H: *Use equations for lines to describe real situations.* *(Lessons 8-4, 8-5, 8-6, 8-8)*

In 29 and 30, each situation can be represented by a straight line. Give the slope and y-intercept of the line describing this situation.

29. Julie rents a truck. She pays an initial fee of $15 and then $0.25 per mile. Let y be the cost of driving x miles.

30. Nick is given $50 to spend on a vacation. He decides to spend $5 a day. Let y be the amount Nick has left after x days.

In 31 and 32, each situation leads to an equation of the form $y = mx + b$. Find that equation.

31. A student takes a test and gets a score of 50. He gets a chance to take the test again. It is estimated that every hour of studying will increase his score by 3 points. Let x be the number of hours studied and y be his score.

32. A plane loses altitude at the rate of 5 meters per second. It begins at an altitude of 8500 meters. Let y be its altitude after x seconds.

33. Julio plans a diet to gain 0.2 kg a day. After 2 weeks he weighs 40 kg. Write an equation relating Julio's weight w to the number of days d on his diet.

34. Each month about 50 new people come to live in a town. After 3 months the town has 25,500 people. Write an equation relating the number of months m to the number of people n in the town.

35. The games of the 21st Modern Olympiad were in 1976. The games of the 20th Olympiad were 4 years earlier. Let y be the year of the nth summer Olympic games. Give a linear equation which relates n and y.

36. Robert babysat for $2.50 an hour and mowed lawns for $10 an hour. He earned a total of $25. Write an equation that describes the possible babysitting hours B and lawnmowing hours L he could have spent at these jobs.

REPRESENTATIONS deal with pictures, graphs, or objects that illustrate concepts.

Objective I: *Graph a straight line given its equation, or given a point on it and its slope.* *(Lessons 8-2, 8-3, 8-4)*

In 37–40, graph the line with the given equation.

37. $y = -2x + 4$

38. $y = \frac{1}{2}x - 3$

39. $8x + 5y = 400$

40. $x - 3y = 11$

41. Graph the line that passes through $(0, 4)$ and has a slope of 4.

42. Graph the line that passes through $(-2, 4)$ and has a slope of $-\frac{3}{4}$.

43. Graph the line with slope 0 and y-intercept 2.3.

44. Graph the line with slope 2 and y-intercept -4.

■ **Objective J:** *Given data which approximate a linear graph, find a linear equation to fit the graph.* *(Lesson 8-7)*

45. Olympic swimmers get faster and faster. The scattergram below shows the times for the winners of the Women's 400-meter freestyle for the years 1924 to 1988.

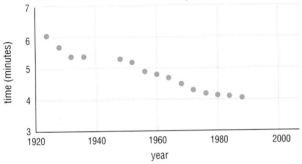

Women's 400-Meter Freestyle Olympic Winners

Year	Winner	Time	Converted Time
1924	Martha Norelius, U.S.	6:02.2	6.04
1928	Martha Norelius, U.S.	5:42.8	5.71
1932	Helene Madison, U.S.	5:28.5	5.48
1936	Hendrika Mastenbroek, Netherlands	5:26.4	5.44
1948	Anne Curtis, U.S.	5:17.8	5.30
1952	Valerie Gyenge, Hungary	5:12.1	5.20
1956	Lorraine Crapp, Australia	4:54.6	4.91
1960	Chris von Saltza, U.S.	4:40.6	4.84
1964	Virginia Duenkel, U.S.	4:43.3	4.72
1968	Debbie Meyer, U.S.	4:31.8	4.53
1972	Shane Gould, Australia	4:19.04	4.32
1976	Petra Thümer, E. Germany	4:09.89	4.17
1980	Ines Diers, E. Germany	4:08.76	4.15
1984	Tiffany Cohen, U.S.	4:07.10	4.12
1988	Janet Evans, U.S.	4:03.85	4.06

The times have been converted to decimal parts of a minute: 6:02.2 is graphed as 6.04 minutes. The result of converting all the times is listed in the table. Use the converted times to answer the questions.

a. Graph a line to fit the data.

b. Find the slope of the fitted line.

c. Find an equation for the line.

d. Predict the winning time in 1992.

■ **Objective K:** *Graph linear inequalities.* *(Lesson 8-9)*

46. The regions on either side of a line in a plane are called __?__.

47. If you have only x nickels and y quarters and a total of less than $2.00, graph all possible values of x and y.

In 48–51, graph.

48. $y \geq x + 1$

49. $y < -3x + 2$

50. $3x + 2y > 5$

51. $x - 8y \leq 0$

Janet Evans competing in the 1988 Olympics.

Exponents and Powers

A city of 100,000 people is planning for the future. The planners want to know how many schools the town will need during the next 50 years. They test three assumptions.

(1) The population stays the same.
(2) The population increases by 2000 people per year.
  (increases by a constant amount)
(3) The population grows by 2% a year.
  (increases at a constant growth rate)

Here is a graph of what would happen under the three assumptions. P is the population n years from now.

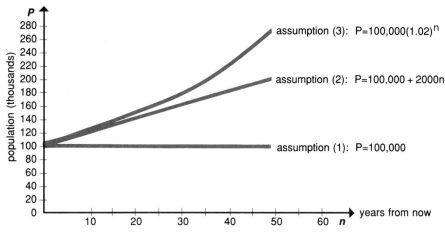

Assumption (3) is often considered the most reasonable of the three. Under this assumption, $P = 100{,}000(1.02)^n$. Because the variable n is an exponent, this equation is said to represent *exponential growth*. Exponential growth describes many other kinds of situations. To study exponential growth, you need to know how to compute and use powers and exponents.

Compound Interest

Recall that when n is a positive integer,

$$x^n = \underbrace{x \cdot x \cdot \ldots \cdot x.}_{n \text{ factors}}$$

The number x^n is the **nth power** of x and is read "x to the nth power" or just "x to the n." In x^n, x is the **base** and n is the **exponent**. In the expression $100{,}000(1.02)^n$ found on page 422, 1.02 is the base, n is the exponent, and 100,000 is the coefficient of 1.02^n. An important application of exponents and powers is *compound interest.*

When you save money, you have choices of where to put it. Of course, you can keep it at home. Banks, savings and loan associations, and credit unions will pay you to give them your money to save. The amount you give them is called the **principal.** The amount they pay you is called **interest.**

Interest is always a percent of the principal. The percent the money earns per year is called the **annual yield.**

▪ ▪ ▪ ▪ ▪ ▪ ▪ ■

Example 1　Suppose you deposit P dollars in a savings account upon which the bank pays an annual yield of 5.5%. If the account is left alone, how much money will be in it at the end of a year?

Solution 1　Total = principal + 5.5% of principal
$$= P + .055P$$
$$= (1 + .055)P$$
$$= 1.055P$$

Solution 2　Total = 100% of principal + 5.5% of principal
$$= 105.5\% \text{ of principal}$$
$$= 1.055P$$

So, if you deposited $1000 in a savings account for a year with an annual yield of 5.5%, a bank would pay you 5.5% of $1000, or $55 interest. Banks and other savings institutions pay you interest because they want money to loan to other people. They might charge someone 12% on a loan. Thus, if the bank could loan your $1000 (perhaps to someone buying a car or a house) at 12%, the bank would receive 12% of $1000, or $120 from that person. So the bank would earn $120 − $55 = $65 on your money. Part of that $65 goes for salaries to the people who work at the bank, part for other bank costs, and part as profit to the owners of the bank.

Savings accounts pay **compound interest,** which means that the interest earns interest.

Example 2   Suppose you deposit $100 in a savings account upon which the bank pays an annual yield of 5.5%. If the account is left alone, how much money will be in it at the end of *three* years?

Solution   Each year the amount in the bank is multiplied by $1 + .055 = 1.055$.

End of first year:
$$100(1.055) = 100(1.055)^1 = 105.50$$

End of second year:
$$100(1.055)(1.055) = 100(1.055)^2 \approx 111.3025 \approx 111.30$$

End of third year:
$$100(1.055)(1.055)(1.055) = 100(1.055)^3 \approx 117.4241 \approx 117.42$$

At the end of three years there will be $117.42 in your account.

Examine the pattern in the solution to Example 2. At the end of n years there will be
$$100(1.055)^n$$

dollars in the account. The general formula for compound interest uses this expression, but variables replace the principal and annual yield.

Compound Interest Formula:

If a principal P earns an annual yield of i, then after n years there will be a total T, where

$$T = P(1 + i)^n.$$

The compound interest formula is read "T equals P times the quantity 1 plus i, that quantity to the nth."

Example 3 $1500 is deposited in a savings account. What will be the total amount of money in the account after 10 years at an annual yield of 6%?

Solution   Here $P = \$1500$, $i = 6\%$, and $n = 10$.
Substitute the values into the compound interest formula. $6\% = .06$

$$T = P(1 + i)^n$$
$$= 1500(1 + .06)^{10}$$

Use the calculator key sequence

$$1500 \boxed{\times} 1.06 \boxed{y^x} 10 \boxed{=}.$$

Displayed will be 2686.2716, which is approximately $2686.27. In 10 years, at an annual yield of 6%, $1500 will increase to $2686.27.

Example 4 How much interest will be earned on a principal of $800 after 5 years if there is an annual yield of 5.8%?

Solution   The question asks for the *interest*, not the total, so first find the total amount. Then, to find the interest, subtract the original amount from the total.

Use the formula, with $P = \$800$, $i = 5.8\%$, and $n = 5$.

$$T = P(1 + i)^n$$
$$= 800(1 + 0.058)^5$$
$$= 800(1.058)^5$$
$$= 1060.52$$

The total amount in the savings account is $1060.52. The amount of interest earned is $1060.52 − $800, or $260.52.

Questions

Covering the Reading

1. Name three kinds of savings institutions.

In 2–4, match each term with its description.

2. money you deposit   A. annual yield

3. interest paid on interest   B. compound interest

4. yearly percentage paid on   C. principal
amount in a savings account

In 5 and 6, write an expression for the amount in the bank after one year if P dollars are in an account with an annual yield as given.

5. 6%

6. 7.2%

7. Write a calculator key sequence to evaluate $573(1.063)^{24}$.

8. In Question 7, identify the base, coefficient, and exponent of the number evaluated.

9. Consider the situation of Example 2. **a.** How much money will you have in your savings account at the end of 4 years? **b.** How much interest will you have earned?

10. a. Write the compound interest formula.
  b. What does T represent?
  c. What does P stand for?
  d. What is i?
  e. What does n represent?

11. How much interest will be earned in 7 years on a principal of $500 at an annual yield of 5.6%?

12. A bank advertises an annual yield of 6.8% on a 6-year CD (certificate of deposit). If the CD's original amount was $2000, how much will it be worth after 6 years?

Applying the Mathematics

13. Susan invests $100 at an annual yield of 5%. Jake invests $100 at an annual yield of 10%. They leave the money in the bank for 2 years.
  a. How much interest does each person earn?
  b. Does Jake earn exactly twice the interest that Susan does?

14. Which yields more money, **a.** an amount invested for 5 years at an annual yield of 6%, or **b.** the same amount invested for 3 years at an annual yield of 10%?

15. Use your calculator. If a principal of $1000 is saved at an annual yield of 8% and the interest is kept in the account, in how many years will it double in value?

In 16–18, evaluate. *(Lesson 1-4)*

Review

16. $-4x^5$ when $x = \frac{1}{2}$

17. $5m^2 - y^2$ when $m = 4$ and $y = -1$

18. $2(cy)^3$ when $c = 0.8$ and $y = 0.5$

19. Lonnie puts $7.00 in his piggy bank. Each week thereafter he puts in $2.00. (The piggy bank pays no interest.) **a.** Write an equation showing the total amount of dollars T after W weeks. **b.** Graph the equation. *(Lesson 8-2)*

20. *Skill sequence* Simplify. *(Lessons 4-1, 6-3)*
   a. $12(3n)$ **b.** $12(3n - 7)$
   c. $-12(3n - 7)$ **d.** $n(3n - 7)$

21. Solve: $(1 + x)^2 = 1.664$. *(Lesson 7-8)*

22. Find t: $2^t = 64$. *(Previous course)*

Exploration

23. Find out the yield for a savings account in a bank or other savings institution near where you live. (Often these yields are in newspaper ads.)

24. a. Enter 25 $\boxed{y^x}$ 0 $\boxed{=}$ on your calculator. What is displayed?

   b. Enter 0 $\boxed{y^x}$ 0 $\boxed{=}$ on your calculator. What is displayed?

   c. Enter -1 $\boxed{y^x}$ 0 $\boxed{=}$ on your calculator. What is displayed?

   d. Try other numbers and generalize what happens.

9-2

Exponential Growth

Another important application of powers and exponents is population growth. The next example concerns rabbit populations, which can grow quickly. Rabbits are not native to Australia, and in 1859, 22 rabbits were imported from Europe as a new source of food. Australia's conditions were ideal for rabbits, and they flourished. Soon, there were so many rabbits that they damaged grazing land. By 1887, the government was offering a reward for a population control technique.

Example 1 Twenty-five rabbits are introduced to an area. Assume that the rabbit population doubles every six months. How many rabbits are there after 5 years?

Solution Since the population doubles twice each year, in 5 years it will double 10 times. The number of rabbits will be

$$25 \cdot 2 \cdot 2 \cdot 2 \cdot 2 \cdot 2 \cdot 2 \cdot 2 \cdot 2 \cdot 2 \cdot 2.$$

10 factors

To evaluate this expression on a calculator rewrite it as

$$25 \cdot 2^{10}.$$

Use the y^x key. After 5 years there will be 25,600 rabbits.

The rabbit population in Example 1 is said to grow exponentially. In **exponential growth,** the original amount is repeatedly *multiplied* by a nonzero number called the **growth factor.**

Growth model for powering:

When an amount is multiplied by g, the growth factor, in each of x time periods, then after the x periods, the original amount will be multiplied by g^x.

In Lesson 6-4, you studied linear increase, in which a number is repeatedly *added* to the original amount. If the growth factor g is greater than 1, exponential growth always overtakes linear increase.

Example 2 Suppose you have $10. Your rich uncle agrees either to (1) increase each day by $50 what you had the previous day, or (2) multiply what you had the previous day by 1.5. Which is the better choice?

Solution Make a table to compare the two options.

	Linear increase: add 50	Exponential growth: multiply by 1.5
1st day	$10 + 50 \cdot \mathbf{1} = \$ 60$	$10 \cdot 1.5^1 = \$15$
2nd day	$10 + 50 \cdot \mathbf{2} = \$110$	$10 \cdot 1.5^2 = \$22.50$
3rd day	$10 + 50 \cdot \mathbf{3} = \$160$	$10 \cdot 1.5^3 = \$33.75$
4th day	$10 + 50 \cdot \mathbf{4} = \$210$	$10 \cdot 1.5^4 = \$50.63$
5th day	$10 + 50 \cdot \mathbf{5} = \$260$	$10 \cdot 1.5^5 = \$75.94$
6th day	$10 + 50 \cdot \mathbf{6} = \$310$	$10 \cdot 1.5^6 = \$113.91$
7th day	$10 + 50 \cdot \mathbf{7} = \$360$	$10 \cdot 1.5^7 = \$170.86$
nth day	$10 + 50n$	$10 \cdot 1.5^n$

As you can see, on the 7th day, you begin having your money grow faster with choice (2). By the end of 2 weeks (14 days) the options are as follows.

14th day $10 + 50 \cdot \mathbf{14} = \$710$ $10 \cdot 1.5^{14} = \$2919.29$

Here is how linear increase and exponential growth compare, in general.

Linear Increase	Exponential Growth
1. Begin with an amount b.	1. Begin with an amount b.
2. *Add m* (the slope) for x time periods.	2. *Multiply* by g (the growth factor) for x time periods.
3. After the x time periods, there will be $b + mx$.	3. After the x time periods, there will be $b \cdot g^x$.

The growth model applies even when $x = 0$. If there are no time periods, the original amount is unchanged, so $b \cdot g^0 = b$. That means $g^0 = 1$, regardless of the value of the growth factor g.

Zero Exponent Property:

If g is any nonzero real number, then $g^0 = 1$.

In words, the zero power of any nonzero number equals 1. For example, $4^0 = 1$, $(-2)^0 = 1$, and $\left(\frac{5}{7}\right)^0 = 1$.

Example 3 Evaluate $4 \cdot 10^2 + 5 \cdot 10^1 + 7 \cdot 10^0$.

Solution Apply the correct order of operations. Do powers first.

$$= 4 \cdot 100 + 5 \cdot 10 + 7 \cdot 1$$
$$= 400 + 50 + 7$$
$$= 457$$

Putting it altogether, $4 \cdot 10^2 + 5 \cdot 10^1 + 7 \cdot 10^0 = 457$. Notice that the coefficient of the second power of 10 becomes the hundreds digit, the coefficient of the first power of 10 becomes the tens digit, and the coefficient of the zero power of 10 becomes the units digit.

The Compound Interest Formula is an application of the growth model of powering. In

$$T = P(1 + i)^n$$

the growth factor is $1 + i$. If $100 is invested for 0 years at 5%, then $P = 100$, $i = 0.05$, and $n = 0$. But there is no time for the money to earn anything. So $T = 100$. Substituting all these values,

$$100 = 100(1.05)^0.$$

So $1.05^0 = 1$. The yield of 5% has no effect. Again the Zero Exponent Property is confirmed.

Questions

In 1–3, refer to Example 1.

Covering the Reading

1. After 7 years, **a.** how many times has the rabbit population doubled? **b.** How many rabbits are there?

2. How many rabbits will there be after 10 years?

3. If the rabbit population triples in 6 months, rather than doubles, how many rabbits will there be after 5 years?

4. Give the general formula for **a.** linear growth; **b.** exponential growth.

5. Refer to Example 2. After one month (30 days), how much would you have with choice (1)? with choice (2)?

6. State the growth model for powering.

7. a. If $2000 is invested at a rate of 5.5%, how much money will be in the account after 0 years?

   b. What is the significance of part **a**?

8. If b is any nonzero real number, then b to what power equals 1?

In 9 and 10, simplify.

9. $6 \cdot 10^2 + 3 \cdot 10^1 + 8 \cdot 10^0$ **10.** $9 \cdot 10^4 + 4 \cdot 10^2 + 10^0$

Applying the Mathematics

11. The following chart describes the exponential growth of a colony of bacteria. You can see that this strain of bacteria grows very fast. In only one hour it grows from 2000 to 54,000 bacteria.

Time Intervals from now	Time (min)	Number of Bacteria
0	0	2000
1	20	6000 $= 2000 \cdot 3^1$
2	40	18,000 $= 2000 \cdot 3^2$
3	60	54,000 $= 2000 \cdot 3^3$

   a. How long does it take the population to triple?

   b. How many times will the population triple in two hours?

   c. How many bacteria will be in the colony after two hours?

   d. How many bacteria will be in the colony after four hours?

In 12 and 13, town A has 20,000 people and is adding 800 people each year. Town B has 20,000 people and is growing exponentially by 4% each year.

12. After 1 year, how many more people will live in town B than town A?

13. After 10 years, how many more people will town B have than town A?

14. The gross income of a company is growing at the rate of 15% per year. The company's gross income this year is $2,000,000. If the growth rate remains constant, what will be the company's gross income at the end of 3 years?

15. In 1986 the U.S. national debt was about 1.7 trillion dollars and was growing at a rate of about 7% per year. At this rate, find the national debt in 1994.

16. Jamaica's population of 2,347,000 in 1986 was expected to grow exponentially by 1.2% each year for the next twenty years.

   a. With this growth, what will the population be in 1989?

   b. With this growth, what will the population be in 2006?

17. Calculators and computers are getting smaller and smaller. It has been estimated that the amount of information that can be stored on a silicon chip is doubling every 18 months. If today's chips have 5000 memory locations, how many might the same size chip have in 6 years?

In 18–21, simplify.

18. $(4y)^0$ when $y = \frac{1}{2}$

19. $7^0 \cdot 7^1 \cdot 7^2$

20. $(x + y)^0$ when $x = 3$ and $y = -8$

21. $\left(\frac{1}{2}\right)^0 + \left(\frac{2}{3}\right)^2$

Review

22. $2200 is deposited in a savings account.
  a. What will be the total amount of money in the account after 6 years at an annual yield of 6%?
  b. How much interest will have been earned in those 6 years? *(Lesson 9-1)*

23. Jeremy invests x dollars for 2 years at an annual yield of 7%. At the end of the 2 years he has $915.92 dollars in his account.
  a. Write an equation describing this situation.
  b. Find x. *(Lesson 9-1)*

24. A card is drawn randomly from a deck of 52 playing cards. Find
  a. P(a three); **b.** P(a king or an ace). *(Lessons 1-7, 3-5)*

25. Suppose a letter from the alphabet is chosen randomly. What is the probability that it is in the first half of the alphabet and is a vowel? *(Lesson 4-8)*

26. *Skill sequence* Solve. *(Lessons 7-4, 7-8)*
  a. $y^2 = 144$
  b. $(4y)^2 = 144$
  c. $(4y - 20)^2 = 144$
  d. $y^2 + 80 = 144$

In 27–30, simplify. *(Lessons 1-5, 6-9)*

27. $6(n + 8) + 4(2n - 1)$

28. $13 - (2 - x)$

29. $4s^9$ when $s = \frac{1}{2}$

30. $t^2 \cdot t^3$ when $t = 11$

31. *Skill sequence* Simplify. *(Lesson 5-1)*
  a. $\frac{2}{3} \div \frac{4}{3}$
  b. $\frac{x}{15} \div \frac{x}{5}$
  c. $\frac{a}{b} \div \frac{c}{b}$

Exploration

32. An old story is told about a man who did a favor for a king. The king wished to reward the man and asked how he could do so. The man asked for a chessboard with one kernel of wheat on the first square of the chessboard, two kernels on the second square, four on the third square, eight on the fourth square, and so on for the entire sixty-four squares of the board. Find how many grains of wheat would be on the whole chessboard. (The answer may amaze you. It is about 500 times the total present yearly wheat production of the world.)

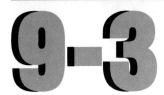

LESSON 9-3

Exponential Decay

bonjour	vous	avec
au revoir	étudie	mais
salut	je suis	mal
voici	français	oui
voilà	anglais	non
franc	parle	est-ce que
aujourd'hui	bien	pourquoi
mercredi	très	ils sont

A student crams for a Friday French test, learning 100 vocabulary words Thursday night. Each day the student expects to forget 10% of the words known. If the test is delayed from Friday to Monday, what will happen if the student does not review?

To answer this question, a table is convenient. Since 10% of the words are forgotten, 90% are remembered.

Day	Day Number	Words Known
Thursday	0	100
Friday	1	$100(.90) = 90$
Saturday	2	$100(.90)(.90) = 100(.90)^2 = 81$
Sunday	3	$100(.90)(.90)(.90) = 100(.90)^3 = 72.9 \approx 73$
Monday	4	$100(.90)(.90)(.90)(.90) = 100(.90)^4 = 65.61 \approx 66$

The pattern is just like that of the growth model or compound interest. After d days, this student will know about $100(.90)^d$ words. Because the growth factor .90 is less than 1, the number of known words decreases. The situation is of a type called **exponential decay.**

Exponential decay can occur when populations are decreasing.

Example   A town with population 67,000 is losing 5% of its population each year. At this rate, how many people will be left in the town after 10 years?

Solution   If 5% of the population is leaving, 95% is staying. Every year, the population is multiplied by 0.95.

$$\text{population} = 67{,}000 \cdot (0.95)^{10}$$
$$\approx 40{,}115$$

After ten years, the population will be about 40,115.

434

Exponential decay is similar to compound interest. If the town starts with P people and loses $r\%$ of its population each year, then after n years there will be $P(1 - r\%)^n$ people. The percent loss is like an annual yield loss.

Questions

Covering the Reading

1. In the French test situation of this lesson, if the test is delayed a week until the next Friday, about how many words will the student know?

2. Evaluate $100x^5$ when $x = .90$

In 3 and 4, refer to the Example.

3. What is the population of this town after 1 year?

4. What is the population of this town after n years?

5. If the population of a town declines by 6.3% each year, by what number would you multiply to find the population each year?

6. A school has 2500 students. The number of students is decreasing by 3% each year.
   a. By what number would you multiply to find the number of students after each year?
   b. If this rate continues, write an expression for the number of students after n years.
   c. If this rate continues, how many students will the school have after 10 years?

Applying the Mathematics

7. The original size of a diagram is 8 inches by 10 inches. It is put through a photocopy machine six times. Each time it is reduced to 75% of its previous dimensions. What is its final size?

8. Bertha Bigbucks had 12 children. In her will she left her first child $300,000. The second child got $\frac{1}{4}$ of what the first child did. The third child got $\frac{1}{4}$ of what the second child did, and so on. How much did the last child get?

9. The following program finds the result when an amount grows or decays exponentially. The BASIC statement for n^x is N ^ X or N ↑ X.

```
10 PRINT "EXPONENTIAL GROWTH"
20 PRINT "WHAT IS THE AMOUNT AT BEGINNING?"
30 INPUT B
40 PRINT "WHAT ARE YOU MULTIPLYING BY?"
50 INPUT G
60 PRINT "HOW MANY TIME PERIODS?"
70 INPUT T
80 LET AMT = B * G ^ T
90 PRINT "TOTAL AMOUNT IS "
100 PRINT AMT
110 END
```

  a. What will the total be if 8, 3, and 6 are entered?
  b. What will the total be if 8, 0.33, and 6 are entered?
  c. What should be entered to find the amount to which a city of 100,000 will grow in 10 years if the growth rate is 2% a year?

10. In the 1980s, the growth rate of the population of the Central American country Honduras was about 3.4% per year (a very high rate). The 1985 population of Honduras was estimated at 4,393,000. At this rate, what will the population be (to the nearest thousand) in the year 2000? *(Lesson 9-2)*

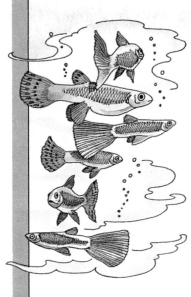

11. Robert buys six guppies. Every month the guppy population doubles. Assume the population continues to grow at this rate.
  a. How many guppies will there be after 4 months?
  b. How many will there be after a year? *(Lesson 9-2)*

12. Calculate the interest paid on $500 at a 6.1% annual yield for 5 years. *(Lesson 9-1)*

13. When $n = 3$ and $x = 4$, which is larger, x^n or n^x? *(Lesson 9-1)*

14. Given the equation $y = -\frac{1}{2}x + 20$, find the slope and the y-intercept. *(Lesson 8-4)*

15. Graph the line which passes through (-6, -1) with slope 5. *(Lesson 8-2)*

16. If $a + b = 75$, find $(a + b)^2$. *(Lesson 7-8)*

17. The sum of two numbers is 36. One of the numbers is k. Write an expression for the product of the numbers. *(Lesson 4-1)*

18. Calculate 230% of 60. *(Lesson 5-4)*

19. Calculate in your head. *(Lesson 6-3)*
   a. The total cost of 6 cans of beans at \$.98 per can.
   b. The total cost of 4 tickets at \$15.05 per ticket.
   c. A 15% tip for a \$40.00 dinner bill.

20. Write an expression for the volume of the box pictured at the right. *(Lesson 4-1)*

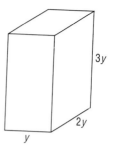

In 21 and 22, solve.

21. $2x = 512$ *(Lesson 4-4)*

22. $486 = 18 \cdot 3^t$ *(Lessons 1-4, 4-4)*

23. *Skill sequence Lesson 2-3)*
   a. Simplify $\frac{2}{9} + \frac{5}{9}$.
   b. Simplify $\frac{2}{3} + \frac{5}{9}$.
   c. Solve $\frac{2}{3}x + \frac{5}{9}x = 33$.

Exploration

24. Exponential decay gets its name from the decay of elements in nature, called radioactive decay. Radioactive decay is used to approximate the age of archaeological objects that were once alive. This can be done because all living things contain radioactive $carbon_{14}$, which has a half-life of 5600 years.
   a. What is meant by the half-life of an element?
   b. A fossil animal bone is found to have $\frac{1}{16}$ of the $carbon_{14}$ that it had as living animal bone. How old is the bone?

The age of this Tyrannosaurus rex was determined to be about 70 million years.

Graphing Exponential Growth and Decay

The situations of the preceding lessons can all be graphed. In Lesson 9-2, you saw linear increase and exponential growth contrasted. In linear increase a constant amount is added and in exponential growth a constant amount is multiplied. In exponential growth, the multiplier is greater than one, and so such growth will eventually overtake linear increase. The difference can be seen in the graph of each type.

Linear Increase
$y = mx + b, m > 0$

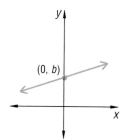

$(0, b)$

Exponential Growth
$y = b \cdot g^x, g > 1$

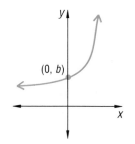

$(0, b)$

Example 1 You invest $100 at an annual yield of 6%. **a.** Graph your savings if you take the interest out of the bank and put it in a piggy bank each year. **b.** On the same set of axes graph your savings if you leave the interest in the bank. Use values at the end of 0, 5, 10, 15, 20, 25, and 30 years.

Solution 1

a. 6% of the original $100 (or $6) is earned in interest each year. If y is the amount after x years, $y = 100 + 6x$. Make a table.

number of years	0	5	10	15	20	25	30
value	$100	$130	$160	$190	$220	$250	$280

b. Make a table for the amount saved at 6% compound interest. Use the formula $y = 100(1.06)^x$.

number of years	0	5	10	15	20	25	30
value	$100	$134	$179	$240	$321	$429	$574

Plot the points and connect them for each graph. In the graph at right, the piggy bank graph (blue) is linear. The compound interest graph (green) is exponential.

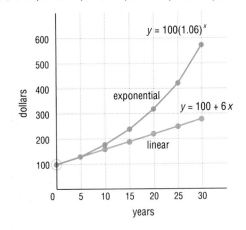

Solution 2   Use a graphing calculator or a computer with a function grapher. Set the domain from 0 to 30 and the range from 0 to 600. First graph $y = 100 + 6x$, then $y = 100(1.06)^x$. You should get graphs like those in Solution 1.

As you can see in the graph in Example 1, exponential growth graphs curve upward. They do not follow a straight line.

■ ■ ■ ■ ■ ■ ■ ■

Example 2   The number of bacteria per square millimeter in a certain culture doubles every 6 hours. There are 100 bacteria per square millimeter at first count. So 6 hours later there are 200, 12 hours later there are 400, and so on. After x 6-hour intervals, there will be B bacteria in the culture, where $B = 100 \cdot 2^x$. Draw 5 points on the graph of this equation and connect them with a smooth curve.

Solution   Let the 5 values of x be 0, 1, 2, 3, and 4. This leads to the points (0, 100), (1, 200), (2, 400), (3, 800), (4, 1600). These are graphed at the right.

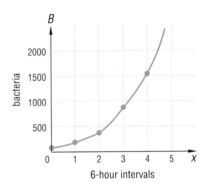

■ ■ ■ ■ ■ ■ ■ ■

Example 3   Use the graph in Example 2. **a.** Estimate the number of bacteria per square millimeter at the end of 15 hours. **b.** Estimate how long it takes for the number of bacteria to grow to 1000 per square millimeter.

Solution   **a.** The number of 6-hour intervals in 15 hours is $\frac{15}{6} = 2.5$. Read where $x = 2.5$ on the graph. At this point $N \approx 550$.
   b. On the graph find where $N = 1000$. At this point $x \approx 3.3$ intervals, so it should take about $3.3 \cdot 6 \approx 20$ hours.

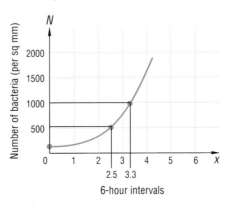

Check Use $N = 100 \cdot 2^x$ and your calculator.

a. $100 \boxed{\times} 2 \boxed{y^x} 2.5 \boxed{=} 566$ is close to 500 on the graph.

b. $100 \boxed{\times} 2 \boxed{y^x} 3.3 \boxed{=} 985$ is close to 1000.

The next example deals with an exponential decay situation found in Lesson 9-3.

Example 4 A student memorizes 100 words and then forgets 10% of those words each day. If w is the number of words known after d days, then $w = 100(.90)^d$. Graph this equation for values of d between 0 and 10.

Solution From the formula, obtain the coordinates of points.

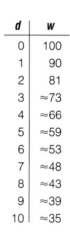

d	w
0	100
1	90
2	81
3	≈ 73
4	≈ 66
5	≈ 59
6	≈ 53
7	≈ 48
8	≈ 43
9	≈ 39
10	≈ 35

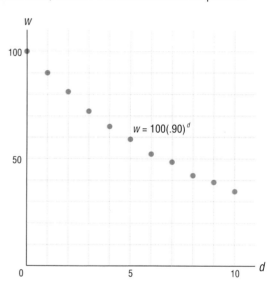

$w = 100(.90)^d$

The graph of Example 4 differs from those in the previous examples. It goes down as x increases. The others go up as x increases.

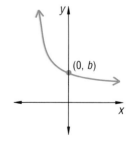

Linear Decrease
$y = mx + b, \ m < 0$

$(0, b)$

Exponential Decay
$y = b \cdot g^x, \ 0 < g < 1$

$(0, b)$

Questions

Covering the Reading

1. In linear increase, a constant amount is __?__ to the total, while in exponential growth, a constant amount is __?__ by the total.

2. As x increases, which increases more rapidly, $y = 100 + 6x$ or $y = 100(1.06)^x$?

3. What is the difference in shape of the graphs of linear decrease and exponential decay?

In 4 and 5, refer to Example 1.

4. How much more will the amount be in 30 years at compound interest than with the piggy bank?

5. Use the graph to estimate when the investment will be worth $500.

6. If you start with 100 bacteria per square millimeter and they double every 6 hours, what is the growth formula?

7. From the graph in Example 2, estimate the number of bacteria at the end of 21 hours.

8. From the graph in Example 2, estimate the number of hours it takes to produce 2000 bacteria.

9. Refer to Example 4.
   a. About how many words did the student forget the first 3 days?
   b. About how many more words did the student forget the second 3 days?

Applying the Mathematics

10. Tell whether each of the following graphs is linear or exponential.
    a. $y = 3x - 2$ **b.** $y = 3x$
    c. $y = 200(1.05)^x$ **d.** $2y = x + 1$

11. The graph of which of the following equations goes up to the right?
    a. $y = 5^x$ **b.** $y = .075^x$
    c. $y = 100 \cdot (2.3)^x$ **d.** $y = \frac{1}{2}(10)^x$

12. If the growth rate of the 1980s continues, the U.S. population will double every 75 years. Since the U.S. population was 226.5 million in 1980, the number of people y (in millions) in the U.S. after x 75-year periods from 1980 is given by the formula $y = 226.5 \cdot 2^x$.
    a. Graph six points on the graph, letting x equal the integers 0 to 5, and connect them with a smooth curve.
    b. Estimate the value of x when the U.S. population is one billion.
    c. In what year is the U.S. population expected to be one billion?

13. When one plate of glass allows 60% of the light through, the amount of light y passing through x panes of tinted glass is described by the formula $y = (.6)^x$. Draw 5 points on the graph of this equation.

14. Consider the equation $y = (\frac{1}{2})^x$.
   a. Find the value of y as both a fraction and a decimal when x takes on each value in this replacement set: $\{0, 3, 10, 20\}$.
   b. Draw the graph using the points from part a.
   c. Does the graph ever touch the x-axis?

Review

15. Forgottonia's population of 680,000 is decreasing at a rate of 25% per year. What will the town's population be in 20 years? *(Lesson 9-3)*

In 16 and 17, consider the equations $k = 30 + 1.05n$ and $k = 30(1.05)^n$. *(Lesson 9-2)*

16. In which equation does k increase more rapidly?

17. Which equation could represent the value of $30 invested at 5% compound interest for n years?

18. Evaluate $4x^n$
   a. when $x = 1.2$ and $n = 3$;
   b. when $x = 1.2$ and $n = 0$. *(Lessons 1-4, 9-2)*

19. Evaluate $(2x)^2(3y)^3$ when $x = 4$ and $y = -2$. *(Lesson 1-4)*

20. The graph at the right shows the average diameter of rocks in a stream at half-mile intervals from the stream's source.
   a. Fit a line to the data.
   b. Give two points on your line in part a.
   c. Find the equation of your line using the points in part b.
   d. Predict the average diameter of rocks 7.5 miles from the stream's source.
   (Lesson 8-7)

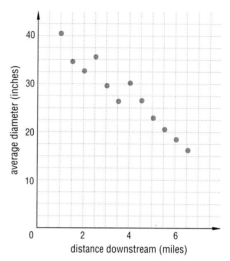

21. a. Copy and complete the computer program below to find the total T in the bank when $100 is invested at an annual yield of 6% for x years.

```
10 PRINT "$100 AT 6 PERCENT"
20 PRINT "YEARS", "AMOUNTS"
30 FOR YEAR = 0 TO 20
40    LET AMOUNT = _?_
50    PRINT YEAR, AMOUNT
60 NEXT YEAR
70 END
```

   b. When this program is run, write the last line that will be printed.
   c. How would you modify line 30 to print the table for $x = 1, 2, 3, \ldots, 100$?
   d. After the change in part c is made, what will be the last line printed when the program is run? *(Lesson 9-1)*

22. *Skill sequence* Add and simplify. *(Lesson 2-3)*
   a. $\dfrac{4}{x} + \dfrac{5}{x}$ **b.** $\dfrac{4}{y} + \dfrac{5}{2y}$ **c.** $4 + \dfrac{5}{2z}$

In 23 and 24, simplify. *(Lesson 6-3)*

23. $4a^2(a + 3) + 2a(a^2 - 1)$

24. $b(b + 3) - b(b + 1)$

25. Recall that if an item is discounted $x\%$, you pay $(100 - x)\%$ of the original price. Calculate in your head the amount you pay for a jacket originally costing $200 and discounted.
   a. 10%.   **b.** 20%.   **c.** 33%. *(Lesson 5-4)*

Exploration

26. Calculate values of $y = 100(1.06)^x$ when $x = -5$ and $x = -10$. Refer to Example 1. What do the answers mean?

9-5

Products of Powers

You already know two meanings for b^n.

(1) Repeated multiplication: $b^n = b \cdot b \cdot \ldots \cdot b$

$\underbrace{\hspace{6em}}_{n \text{ factors}}$

(2) Growth: b^n is the growth factor after n years if there is growth by a factor b in each one-year interval.

Each meaning shows that powers are closely related to multiplication.

Example 1 Multiply $x^7 \cdot x^5$.

Solution Use the repeated multiplication meaning of x^n.

$$x^7 \cdot x^5 = \underbrace{x \cdot x \cdot x \cdot x \cdot x \cdot x \cdot x}_{7 \text{ factors}} \cdot \underbrace{x \cdot x \cdot x \cdot x \cdot x}_{5 \text{ factors}}$$

$$= \underbrace{x \cdot x \cdot x \cdot x \cdot x \cdot x \cdot x \cdot x \cdot x \cdot x \cdot x \cdot x}_{12 \text{ factors}}$$

$$= x^{12}$$

So, $x^7 \cdot x^5 = x^{7+5} = x^{12}$.

Check Test a special case. Let $x = 3.2$.
Does $(3.2)^7 \cdot (3.2)^5 = (3.2)^{12}$? Try it with a calculator.
The key sequences

3.2 $\boxed{y^x}$ 7 $\boxed{=}$ $\boxed{\times}$ 3.2 $\boxed{y^x}$ 5 and 3.2 $\boxed{y^x}$ 12 $\boxed{=}$ give the same value.

Example 2 Suppose a colony of bacteria doubles every hour. Then, if there were 2000 bacteria in the colony at the start, after h hours there will be T bacteria, where

$$T = 2000 \cdot 2^h.$$

How many bacteria will there be after the 5th hour? How many bacteria will there be 3 hours after that?

Solution 1 There will be $2000 \cdot 2^5$ bacteria at the end of the 5th hour. Three hours later the bacteria will have doubled three more times. There will be $(2000 \cdot 2^5) \cdot 2^3$ bacteria. This equals $2000 \cdot (2^5 \cdot 2^3)$ bacteria.

Solution 2 Three hours after the 5th hour is the 8th hour. There will be $2000 \cdot 2^8$ bacteria.

These examples suggest the Product of Powers Property.

Product of Powers Property:

When b^m and b^n are defined,

$$b^m \cdot b^n = b^{m+n}.$$

The Product of Powers Property tells how to multiply two powers with the *same base*. An expression with different bases like $a^3 \cdot b^4$ usually cannot be simplified.

Example 3 Simplify $r^4 \cdot s^3 \cdot r^5 \cdot s^8$.

Solution Use the properties of multiplication to group factors with the same base.

$$r^4 \cdot s^3 \cdot r^5 \cdot s^8 = r^4 \cdot r^5 \cdot s^3 \cdot s^8$$

Apply the Product of Powers Property. $= r^9 \cdot s^{11}$.

$r^9 \cdot s^{11}$ cannot be simplified further because the bases are different.

Check Look at the special case when $r = 2$, $s = 3$.
Does $\quad 2^4 \cdot 3^3 \cdot 2^5 \cdot 3^8 = 2^9 \cdot 3^{11}$?
Does $16 \cdot 27 \cdot 32 \cdot 6561 = 512 \cdot 177{,}147$?
Yes, they each equal 90,699,264.

Chunking is useful for simplifying an expression in which a power is raised to a power.

Example 4 Simplify $(x^3)^4$.

Solution Think of x^3 as a single number, that is, chunk x^3.

$$(x^3)^4 = x^3 \cdot x^3 \cdot x^3 \cdot x^3$$
$$= x^{3 + 3 + 3 + 3}$$
$$= x^{12}$$

Check Use a special case. Let $x = 3$. $(x^3)^4 = (x^3)^4 = 27^4 = 531{,}441$. $3^{12} = 531{,}441$. Since the two expressions are equal, it checks.

Example 4 is an instance of the Power of a Power Property:

Power of a Power Property:

When $b > 0$, for all m and n,

$$(b^m)^n = b^{mn}.$$

For instance, $(b^0)^3 = b^{0 \cdot 3} = b^0 = 1$. This checks, because $1^3 = 1$. The Power of a Power Property implies that $(b^m)^n = (b^n)^m$. So the square of x^{10} is the same as the 10th power of x^2.

Questions

Covering the Reading

1. Explain the meaning of x^6:
   a. using repeated multiplication;
   b. using growth.

2. State the Product of Powers Property.

3. Why can't $x^8 \cdot y^2$ be simplified?

4. a. Simplify $a^2 \cdot a^3$.
   b. Check your answer by letting $a = -2$.

5. Refer to Example 2. Give answers in exponential form.
   a. How many bacteria will there be after 11 hours?
   b. How many bacteria will there be 6 hours after that?

6. Suppose a population P of bacteria triples each day.
   a. Write an expression for the number of bacteria after 5 days.
   b. How many days after the fifth day will the bacteria population be $P \cdot 3^{17}$?

7. State the Power of a Power Property.

In 8–19, simplify.

8. $(k^{10})^3$

9. $(n^2)^6$

10. $(x^2)^3 - (x^3)^2$

11. $x^5 \cdot x^{50}$

12. $13^{12} \cdot 17^{10}$

13. $(1.5)^8 \cdot (1.5)^2$

Applying the Mathematics

14. $a^3 \cdot a^5 \cdot b^0 \cdot a^2 \cdot b^9$

15. $2(x^3 \cdot x^4)$

16. $(x^3 \cdot x^4)^2$

17. $3m^4 \cdot 5m^2$

18. $b^3(a^3b^5)$

19. $12 \cdot 12^{100}$

20. a. Simplify $2^3 \cdot 2^x$.
    b. Check your answer by letting $x = 4$.

In 21 and 22, use the Distributive Property, then simplify. Recall that unlike terms cannot be combined.

21. $a^2(a^3 + 4a^4)$ **22.** $y(y^7 - y^2)$ (Remember, $y = y^1$.)

23. a. Calculate $(-1)^n$ for $n = 1, 2, 3, 4, 5, 6, 7$, and 8.
   b. What is $(-1)^{100}$?

24. You are descended from two natural parents each of whom had two natural parents, each of whom had two natural parents, and so on.
   a. If you traced your family tree back through ten generations of natural parents, at most how many ancestors would you have?
   b. Suppose you marry, and you and your spouse have a child. At most how many ancestors could your child have back through eleven generations?

Review

25. Graph $y = 2^x$ for values of x from 0 to 5. *(Lesson 9-4)*

26. Write an expression for the population of a city y years from now whose current population is 2,500,000 when
   a. the population is growing at 4.5% per year,
   b. the population is decreasing 3% per year,
   c. the population is decreasing by 2500 people each year. *(Lessons 6-4, 9-2, 9-3)*

27. Rewrite $7y - 2x - 7 = 19 - 9x$
   a. in standard form. *(Lesson 8-8)*
   b. in slope-intercept form. *(Lesson 8-4)*

In 28–31, write in scientific notation. *(Appendix B)*

28. 4,000,000,000   **29.** 2,439,000

30. 0.00036   **31.** 0.897

32. *Skill sequence*   *(Lessons 2-2, 6-9, 7-8)*
   a. Simplify $x - (4x + 6)$
   b. Simplify $\dfrac{x}{2 - x} - \dfrac{4x + 6}{2 - x}$
   c. Solve $\dfrac{x}{2 - x} - \dfrac{4x + 6}{2 - x} = 1$

Exploration

33. Metric measures involving powers of 10^3 have special names. For instance, 10^3 meters is one kilometer. Find the name of each of the following expressions.
   a. 10^6 grams   **b.** 10^{-6} meters
   c. 10^{-3} seconds   **d.** 10^9 watts

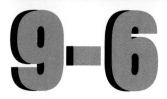

9-6

Negative Exponents

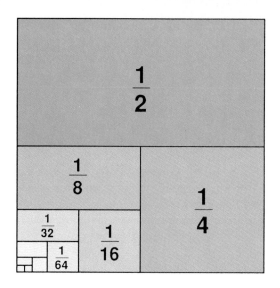

You have used the base 10 with a negative exponent to represent small numbers in scientific notation.

$$10^{-1} = 0.1, \ 10^{-2} = 0.01, \ 10^{-3} = 0.001, \text{ and so on}$$

Now consider other bases with negative exponents:

What is the value b^{-n}?

The following pattern of the powers of 2 helps to answer this question.

$$2^5 = 32$$
$$2^4 = 16$$
$$2^3 = 8$$
$$2^2 = 4$$

Each exponent is one less than the one above it and each number on the right is half the number above. This pattern continues.

$$2^1 = 2$$
$$2^0 = 1$$
$$2^{-1} = \frac{1}{2}$$
$$2^{-2} = \frac{1}{4} = \frac{1}{2^2}$$
$$2^{-3} = \frac{1}{8} = \frac{1}{2^3}$$
$$2^{-4} = \frac{1}{16} = \frac{1}{2^4}$$
$$2^{-5} = \frac{1}{32} = \frac{1}{2^5}$$

The pattern is simple: $2^{-n} = \frac{1}{2^n}$. That is, 2^{-n} is the reciprocal of 2^n.

We call the general property the **Negative Exponent Property.**

Negative Exponent Property:

For any nonzero b,

$b^{-n} = \dfrac{1}{b^n}$, the reciprocal of b^n.

The Negative Exponent Property can be deduced by multiplying b^n and b^{-n}.

$$
\begin{aligned}
b^n \cdot b^{-n} &= b^{n + -n} & \text{Product Property of Powers} \\
&= b^0 \\
&= 1 & \text{Zero Exponent Property}
\end{aligned}
$$

Since the product of b^n and b^{-n} is 1, b^{-n} must be the reciprocal of b^n.

Example 1   Write 3^{-4} as a simple fraction.

Solution   Using the Negative Exponent Property, $3^{-4} = \dfrac{1}{3^4} = \dfrac{1}{81}$.

Notice that 3^{-4} is a positive number. This surprises many people, but it fits the pattern of all applications. For instance, examine this graph of $y = 2^x$, extended to include negative values of x.

x	$y = 2^x$
-1	$\dfrac{1}{2} = 0.5$
-2	$\dfrac{1}{2^2} = 0.25$
-3	$\dfrac{1}{2^3} = 0.125$

Even when x is negative, the number 2^x is still positive. *All* powers of positive numbers are positive.

The Negative Exponent Property also agrees with the negative powers of 10 you already know.

$$
10^{-6} = .000001 = \frac{1}{1,000,000} = \frac{1}{10^6}
$$

Also, as a special case $b^{-1} = \dfrac{1}{b}$. That is, the -1 power of a number equals its reciprocal. You can use the Negative Exponent Property to rewrite expressions so they have no negative exponents.

Example 2 Rewrite $q^5 \cdot t^{-3}$ without negative exponents.

Solution Substitute $\dfrac{1}{t^3}$ for t^{-3}.

$$q^5 \cdot t^{-3} = q^5 \cdot \dfrac{1}{t^3}$$
$$= \dfrac{q^5}{t^3}$$

Negative exponents work in the growth model for powering. They stand for time years ago.

Example 3 Recall the compound interest formula

$$T = P(1 + i)^n.$$

Three years ago, Mr. Cabot put money in a CD at an annual yield of 7%. If the CD is worth $3675 now, what was it worth then?

Solution Here $P = 3675$, $i = .07$, and $n = -3$ (for three years ago). Use a calculator.

3675 $\boxed{\times}$ 1.07 $\boxed{y^x}$ 3 $\boxed{\pm}$ $\boxed{=}$

gives 2999.8947.
So Mr. Cabot probably started with $3000.

Questions

Covering the Reading

1. Write the next three equations in this pattern:
$$4^3 = 64$$
$$4^2 = 16$$
$$4^1 = 4$$
$$4^0 = 1$$

2. *Multiple choice* $x^{-n} =$

(a) $-x^n$ 　　　(b) $(-x)^n$ 　　　(c) $\dfrac{1}{x^{-n}}$ 　　　(d) $\dfrac{1}{x^n}$

In 3–5, write as a simple fraction.

3. 5^{-2} **4.** 3^{-6} **5.** 7^{-1}

6. b^{-1} is the __?__ of b.

In 7–9, write without negative exponents.

7. $x^5 y^{-2}$ **8.** $3a^{-2}b^{-4}$ **9.** 2^{-n}

10. Refer to the graph of $y = 2^x$.
   a. Can the value of x be negative? **b.** Can the value of y be negative?

11. Write 10^{-9} **a.** as a decimal; **b.** as a simple fraction.

12. *True or false* If x is positive, x^{-4} is positive.

13. Refer to Example 3. Find what Mr. Cabot's CD was worth
   a. one year ago; **b.** two years ago.

Applying the Mathematics

14. Theresa has $1236.47 in a savings account that has had an annual yield of 5.25% since she opened the account. Assuming no withdrawals or deposits were made, how much was in the account 8 years ago?

15. If the reciprocal of $(1.06)^{11}$ is $(1.06)^n$, what is n?

In 16 and 17, simplify.

16. $c^j \cdot c^{-j}$ **17.** $t^{-2} \cdot t^{-4}$

18. a. Evaluate $3 \cdot 10^4 + 5 \cdot 10^2 + 6 \cdot 10^1 + 2 \cdot 10^0 + 4 \cdot 10^{-1} + 7 \cdot 10^{-3}$

   b. Evaluate $9 \cdot 10^3 + 8 \cdot 10^2 + 7 \cdot 10^1 + 6 \cdot 10^0 + 5 \cdot 10^{-1} + 4 \cdot 10^{-2} + 9 \cdot 10^{-4}$

19. Consider the equation $y = 3^x$.
   a. Complete the table. **b.** Graph $y = 3^x$.

x	y
2	
1	
0	
-1	
-2	
-3	

In 20–22, use this information. The human population P of the earth, x years from 1985, can be estimated by the formula

$$y = 5 \text{ billion } (1.017)^x.$$

Use this formula to estimate the earth's population in

20. 1990; **21.** 1980; **22.** 1970.

In 23–25, simplify. *(Lesson 9-5)*

23. $a^2 \cdot a^5$ **24.** $(b^3)^{10}$ **25.** $3c^3 \cdot 4c^4$

26. A certain kind of virus doubles its population every 3 hours.
    a. In two days, how many times does the population double?
    b. If a biologist begins an experiment with 25 virus organisms, how many does he or she have after two days? *(Lesson 9-2)*

27. *Multiple choice* The graph at the right is the graph of points on
    (a) $y = 0.3^x$.
    (b) $y = -0.3^x + 1$
    (c) $y = 0.8^x$.
    (d) $y = -0.8^x + 1$
    (Lesson 9-4)

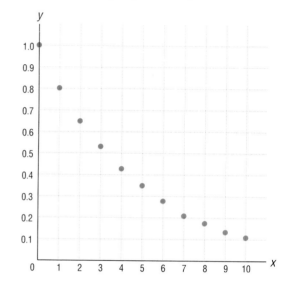

28. Consider the line $y = 2x - 3$.
    a. Give its slope.
    b. Give its y-intercept.
    c. Graph the line. *(Lesson 8-4)*

29. Write an equation for the line with slope -5 that passes through the point $(2, -1)$. *(Lesson 8-5)*

30. *Skill sequence (Lessons 6-1, 6-9)*
    a. Simplify $4(7a - 2)$.
    b. Simplify $4(7a - 2) - 3(5a + 1)$.
    c. Solve $4(7a - 2) - 3(5a + 1) = 15$.

31. Simplify $\dfrac{3x + 6x}{12x}$. *(Lesson 6-3)*

In 32 and 33, a number n is chosen at random from $\{-10, -9, -8, \ldots, -1, 0, 1, 2, \ldots, 8, 9, 10\}$. Find the probability that

32. $n > -4$; *(Lesson 1-7)*

33. $n \leq -9$ or $n \geq 9$. *(Lesson 3-6)*

34. $(\frac{1}{2})^2 = \frac{1}{2} \cdot \frac{1}{2} = \frac{1}{4}$. Find other positive and negative integer powers of the number $\frac{1}{2}$.
    a. How do the powers of $\frac{1}{2}$ compare with the powers of 2?
    b. Generalize to other pairs of reciprocal bases.

9-7

Quotients of Powers

Here is part of a list of the positive integer powers of 2.

2, 4, 8, 16, 32, 64, 128, 256, 512, 1024, 2048, 4096, ...

Multiply any two of these numbers and you will find that the product is on the list. For instance,

$$32 \cdot 128 = 4096.$$

This can be explained using the Product of Powers Property.

$$2^5 \cdot 2^7 = 2^{12}$$

It may surprise you that if you *divide* any number on the list, the quotient is on the list. For instance, consider 512 and 16.

Dividing larger by smaller	*Dividing smaller by larger*
$\dfrac{512}{16} = 32$	$\dfrac{16}{512} = \dfrac{1}{32}$

As powers, $\quad \dfrac{2^9}{2^4} = 2^5,$ $\qquad\qquad \dfrac{2^4}{2^9} = \dfrac{1}{2^5} = 2^{-5}$

These examples illustrate the Quotient of Powers Property.

Quotient of Powers Property:

When b^m and b^n are defined and $b \neq 0$, then

$$\frac{b^m}{b^n} = b^{m-n}$$

When simplifying $\dfrac{b^m}{b^n}$, if the larger power is in the numerator, the result is a positive power of b. If the larger power is in the denominator, then the result is a negative power of b.

Example 1 Simplify $\dfrac{y^{12}}{y^5}$

Solution Use the Quotient of Powers Property.

$$\frac{y^{12}}{y^5} = y^{12-5}$$
$$= y^7$$

Check Use repeated multiplication,

$$\frac{y^{12}}{y^5} = \frac{\cancel{y} \cdot \cancel{y} \cdot \cancel{y} \cdot \cancel{y} \cdot \cancel{y} \cdot y \cdot y \cdot y \cdot y \cdot y \cdot y \cdot y}{\cancel{y} \cdot \cancel{y} \cdot \cancel{y} \cdot \cancel{y} \cdot \cancel{y}}$$
$$= y^7$$

■ ■ ■ ■ ■ ■ ■ ■ ■

Example 2   Simplify $\dfrac{3^{16}}{3^{27}}$.

Solution $\dfrac{3^{16}}{3^{27}} = 3^{16-27}$

$\qquad\qquad = 3^{-11}$

$\qquad\qquad = \dfrac{1}{3^{11}}$

Either 3^{-11} or $\dfrac{1}{3^{11}}$ can be considered simpler than the given fraction.

Check   Use a calculator. Does $\dfrac{3^{16}}{3^{27}} = \dfrac{1}{3^{11}}$?

Yes, 3 ⓨˣ 16 ÷ 3 ⓨˣ 27 ⓐ= gives 0.0000056.

$\qquad$ 3 ⓨˣ 11 ⓐ± ⓐ= also gives 0.0000056.

To use the Quotient of Powers Property, both bases must be the same. For instance, $\dfrac{a^9}{b^4}$ cannot be simplified.

■ ■ ■ ■ ■ ■ ■ ■

Example 3   In 1985, there were approximately 3.6 billion one-dollar bills in circulation and about 227 million people in the U.S. How many dollar bills was this per person?

Solution   Since dollars per person is a rate unit, the answer is found by division.

$$\dfrac{\text{number of dollar bills}}{\text{number of persons}} = \dfrac{3.6 \text{ billion}}{227 \text{ million}}$$

Change the words to their power of 10 equivalents.
$$= \dfrac{3.6 \cdot 10^9}{227 \cdot 10^6}$$

This is a product of fractions.
$$= \dfrac{3.6}{227} \cdot \dfrac{10^9}{10^6}$$

By the Quotient of Powers Property
$$\approx 0.0158 \cdot 10^3$$

$$\approx 16 \dfrac{\text{dollar bills}}{\text{person}}$$

Check   Change the numbers to decimals and simplify the fraction.

$$\dfrac{3,600,000,000}{227,000,000} = \dfrac{3,600}{227} \approx 16$$

In Example 3, it was easier to rewrite the fraction $\dfrac{3.6 \cdot 10^9}{227 \cdot 10^6}$ as the product of two fractions before using the Quotient of Powers Property. This technique can be especially helpful when an expression with several bases is involved.

Example 4 Simplify $\dfrac{7a^3b^2c^6}{28a^2b^5c}$.

Solution $\dfrac{7a^3b^2c^6}{28a^2b^5c} = \dfrac{7}{28} \cdot \dfrac{a^3}{a^2} \cdot \dfrac{b^2}{b^5} \cdot \dfrac{c^6}{c}$

$= \dfrac{1}{4} \cdot a^{3-2} \cdot b^{2-5} \cdot c^{6-1}$ (Recall that $c = c^1$.)

$= \dfrac{1}{4} \cdot a \cdot b^{-3} \cdot c^5$

$= \dfrac{1}{4} \cdot a \cdot \dfrac{1}{b^3} \cdot c^5$

$= \dfrac{ac^5}{4b^3}$

Experts do all of this work in one step.

Questions

Covering the Reading

1. Rewrite the multiplication problem $64 \cdot 256 = 16{,}384$ using powers of 2.

2. State the Quotient of Powers Property.

3. In 1985, there were approximately 929 million five-dollar bills in circulation and 227 million people in the U.S. Convert these words to powers of 10 and find the number of five-dollar bills per person.

In 4–8, use the Quotient of Powers Property to simplify the fraction.

4. $\dfrac{y^5}{y^5}$

5. $\dfrac{3^2}{3^8}$

6. $\dfrac{9.5 \cdot 10^{12}}{1.9 \cdot 10^4}$

7. $\dfrac{w^2z^6}{42w^2z^3}$

8. $\dfrac{4abc^{10}}{28a^2b^5c}$

9. $\dfrac{r^{10}}{s^7}$ cannot be simplified because ___?___.

10. If $m = n$, $\dfrac{b^m}{b^n} = $ ___?___.

© 1987 King Features Syndicate, Inc.

11. $4 \cdot 10^6$ pounds of plastic are used each year in the U.S. to produce dry cleaning bags, plastic cups, plates and the like. About how many pounds of plastic are used per person?

12. About $1.13 \cdot 10^6$ pounds of spinach are consumed per year in the United States. How much is this per person?

In 13–18, simplify.

13. $\dfrac{x^{12n}}{x^{3n}}$

14. $\dfrac{(7m)^2}{(7m)^3}$

15. $\dfrac{(x + 3)^6}{(x + 3)^6}$

16. $\dfrac{3x^2}{y^3} \cdot \dfrac{y^5}{x^9}$

17. $\dfrac{3p^5 + 2p^5}{p^4}$

18. $\dfrac{x^{-2}}{x^{-3}}$

19. Alaska has a population of $4.79 \cdot 10^5$ and an area of $1.48 \cdot 10^6$ square km. Find Alaska's population per square km.

20. Other than the sun, the star nearest to us, Alpha Centauri, is about $6.2 \cdot 10^{12}$ miles away. Earth's moon is about $2.4 \cdot 10^5$ miles from us. If it took astronauts about 3 days to get to the moon in 1969, at that speed how long would it take to get to Alpha Centauri?

21. Write 2^{-6} as a decimal. *(Lesson 9-6)*

In 22–24, simplify. *(Lessons 9-5, 9-6)*

22. $4^x \cdot 4^y$

23. $(y^4)^3$

24. $(y^4)^3 \cdot y^4$

25. Simplify $x^5 \cdot x^5 \cdot x^5$. Check by letting $x = 3$. *(9-5)*

26. In 1976, Milo invested $6000 for 10 years at 5% compound interest. In 1981, Sylvia invested $6000 for 5 years at 10%. By the end of 1986, who had more? *(Lesson 9-1)*

27. In your head, find the ratio of the areas of two squares, the first with a side of length 2 and the second with a side of length 3. *(Lesson 5-3)*

28. A box with dimensions 4 by 6 by 8 will hold how many times as much as one with dimensions 2 by 3 by 4? *(Lesson 5-6)*

29. a. Find the perimeter of the rectangle at the right. *(Lesson 2-1)*
b. Write a simplified expression for the area of this rectangle. *(Lesson 6-3)*
c. If the area is 312, find the value of x. *(Lesson 6-1)*

12

$2x + 4$

30. An elevator is y floors high in a tall building. After x seconds, $y = 54 - 3x$.
  a. Give the slope and y-intercept of $y = 54 - 3x$ and describe what they mean in this situation.
  b. Graph the line. *(Lesson 8-4)*

In 31–33, solve. *(Lessons 4-6, 6-7)*

31. $-4x \leq 25$

32. $-4x > 6x + 25$

33. $5x - 6 \geq 9(x + 2)$

Exploration

34. The average 8th grader has a volume of about 3 cubic feet.
  a. If you took all the students in your school, would their volume be more or less than the volume of one classroom that is 10 feet high, 30 feet long, and 30 feet wide?
  b. Assume the population of the world to be 5 billion people. Is the volume of all the people more or less than the volume of a cubic mile? How much more or less? Assume the average volume of a person equals $\frac{4}{3}$ the average volume of an eighth grader. (There are 5280^3 cubic feet in a cubic mile.)

9-8

Powers of Products and Quotients

$(3x)^4$ is an example of a **Power of a Product.** It can be rewritten without parentheses using repeated multiplication.

$$(3x)^4 = (3x) \cdot (3x) \cdot (3x) \cdot (3x)$$
$$= 3 \cdot x \cdot 3 \cdot x \cdot 3 \cdot x \cdot 3 \cdot x \qquad \text{Associative Property}$$
$$= 3 \cdot 3 \cdot 3 \cdot 3 \cdot x \cdot x \cdot x \cdot x \qquad \text{Commutative Property}$$
$$= 3^4 \cdot x^4 \qquad\qquad\qquad\qquad \text{meaning of } x^n$$
$$= 81x^4$$

You can check this. Consider the special case when $x = 2$. Then

$$(3x)^4 = 6^4 = 1296 \text{ and } 81x^4 = 81 \cdot 2^4 = 81 \cdot 16 = 1296.$$

When you want a positive integer power of a product, you can always use repeated multiplication.

$$(ab)^n = \underbrace{(ab) \cdot (ab) \cdot \ldots \cdot (ab)}_{n \text{ factors}}$$

$$= \underbrace{a \cdot a \cdot \ldots \cdot a}_{n \text{ factors}} \cdot \underbrace{b \cdot b \cdot \ldots \cdot b}_{n \text{ factors}}$$

$$= a^n \cdot b^n$$

This result holds when n is any integer, positive, negative or zero.

Power of a Product Property:

> For all nonzero a and b, and any integer n,
> $$(ab)^n = a^n \cdot b^n$$

Example 1 Suppose one cube has edges twice the size of another. The volume of the second cube is how many times the volume of the first?

Solution The volume of a cube is the cube of its edge.

$$\text{Volume of larger cube} = (2e)^3$$
$$= 2^3 e^3$$
$$= 8e^3$$
$$= 8 \cdot \text{volume of the smaller cube}$$

So the larger cube has eight times the volume of the smaller.

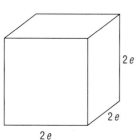

The Power of a Product Property is quite convenient when taking powers of numbers written in scientific notation.

Example 2 The radius of the earth is approximately $6.36 \cdot 10^3$ km. The volume of a sphere of radius r is given by the formula

$$V = \tfrac{4}{3}\pi r^3.$$

Calculate the approximate volume of the earth.

Solution Substitute $6.36 \cdot 10^3$ for r.

$$V = \tfrac{4}{3}\pi(6.36 \cdot 10^3)^3$$

Apply the Power of a Product Property.

$$V = \tfrac{4}{3}\pi(6.36)^3 \cdot (10^3)^3$$

$$\approx \tfrac{4}{3}\pi(257.26) \cdot 10^9$$

$$\approx 1077.60 \cdot 10^9 \quad \text{(Note: } (10^3)^3 = 10^3 \cdot 10^3 \cdot 10^3 = 10^9\text{)}$$

$$\approx 1.08 \cdot 10^{12} \text{ km}^3 \text{ in scientific notation}$$

$$\approx 1{,}080{,}000{,}000{,}000 \text{ km}^3 \text{ written as a decimal}$$

The Power of a Quotient Property is very similar to the Power of a Product Property. It enables powers of fractions to be easily found.

Power of a Quotient Property:

For all a and nonzero b, and any integer n,

$$\left(\frac{a}{b}\right)^n = \frac{a^n}{b^n}.$$

Example 3 Write $\left(\tfrac{2}{3}\right)^5$ as a simple fraction.

Solution Use the Power of a Quotient Property. $\left(\dfrac{2}{3}\right)^5 = \dfrac{2^5}{3^5}$

$$= \frac{32}{243}$$

Check Change the fractions to decimals. $\left(\dfrac{2}{3}\right)^5 \approx (.\overline{6})^5 \approx 0.1316872\ldots$
$\frac{32}{243} \approx 0.1316872\ldots$ also.

Example 4 Rewrite $3 \cdot \left(\dfrac{x}{2y}\right)^4$ as a single fraction.

Solution First rewrite the power using the Power of a Quotient Property.

$$3 \cdot \left(\frac{x}{2y}\right)^4 = 3 \cdot \frac{x^4}{(2y)^4}$$

Now use the Power of a Product Property.

$$= 3 \cdot \frac{x^4}{2^4 y^4}$$

$$= \frac{3x^4}{16y^4}$$

Check By repeated multiplication,

$$3 \cdot \left(\frac{x}{2y}\right)^4 = 3 \cdot \frac{x}{2y} \cdot \frac{x}{2y} \cdot \frac{x}{2y} \cdot \frac{x}{2y} = \frac{3 \cdot x^4}{16y^4}.$$

The properties in this lesson help to calculate powers of negative numbers. You know that $(-1)^2 = 1$, $(-1)^3 = -1$, $(-1)^4 = 1$, and so on. Think of $-x$ as $-1 \cdot x$. Here are some examples.

$(-3)^4$	$\left(-\frac{4}{5}\right)^7$	$(-xy)^6$
$= (-1 \cdot 3)^4$	$= \left(-1 \cdot \frac{4}{5}\right)^7$	$= (-1 \cdot x \cdot y)^6$
$= (-1)^4 \cdot 3^4$	$= (-1)^7 \cdot \left(\frac{4}{5}\right)^7$	$= (-1)^6 \cdot x^6 \cdot y^6$
$= 1 \cdot 3^4$	$= -1 \cdot \dfrac{4^7}{5^7}$	$= 1 \cdot x^6 \cdot y^6$
$= 81$	$= -\dfrac{4^7}{5^7}$	$= x^6 y^6$

Experts remember that odd powers of negatives are negative. Even powers of negatives are positive. They do all of the intermediate steps at once.

Caution: Powers take precedence over opposites. Although $(-b)^2 = b^2$, it happens that when $b \ne 0$, $-b^2 \ne b^2$. The number $-b^2$ is negative, whereas b^2 is positive.

Questions

Covering the Reading

1. **a.** Rewrite $(5x)^3$ without parentheses.
   b. Check your answer by letting $x = 2$.

2. Calculate $(1.3 \cdot 10^4)^5$.

3. In Example 1, suppose the length of each side of the smaller cube is 12.5 feet.
a. Find the volume of the cube.
b. Find the volume of the larger cube.

4. What happens to the volume of a cube when each edge is doubled?

5. The radius of Earth's moon is approximately $(1.773 \cdot 10^3)$ km. Calculate its approximate volume. Give your answer **a.** as a decimal, and **b.** in scientific notation.

In 6–8, write as a simple fraction.

6. $(\frac{1}{2})^4$

7. $(\frac{7}{10})^3$

8. $(\frac{2}{3})^6$

In 9–11, rewrite each expression without parentheses.

9. $(8y)^3$

10. $\left(\dfrac{m}{n}\right)^2$

11. $4L \cdot \left(\dfrac{k}{2L}\right)^2$

Applying the Mathematics

In 12 and 13, answer *true or false*.

12. $-5^2 = (-5)^2$

13. $(-7)^2 = 7^2$

In 14–19, first choose the property from this list needed to simplify the expression, then simplify.
a. Product of Powers   **b.** Quotient of Powers   **c.** Power of a Power
d. Power of a Product   **e.** Power of a Quotient

14. $(ab)^3$

15. $x^5 x^8$

16. $\left(\dfrac{I}{S}\right)^3$

17. $\dfrac{k^{12}}{k^9}$

18. $y \cdot y^3$

19. $(v^{-2})^3$

In 20–25, rewrite without parentheses and simplify.

20. $(\frac{2}{7}z)^4$

21. $\left(\dfrac{a^{-1}}{b^5}\right)^3$

22. $\frac{1}{2}(6x)^2$

23. $(pqr)^0$

24. $\left(\dfrac{u}{3}\right)^t$

25. $2(kq)^5(3kq^4)^2$

26. If $x = 4$, what is the value of $\dfrac{(4x)^8}{(4x)^5}$?

27. The edge of one cube is k in. The edge of a second cube is 5 times as long.
a. Write an expression for the volume of the first cube.
b. Write a simplified expression for the volume of the second cube.

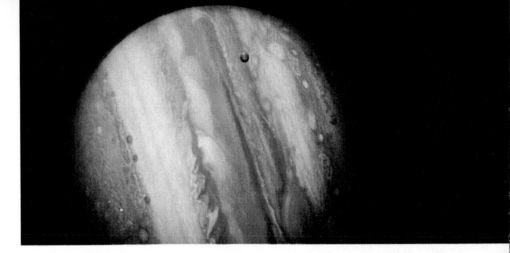

28. The radius of Jupiter is 11 times that of Earth. The volume of Jupiter is how many times the volume of Earth?

Review

29. Simplify $\dfrac{5n^2 - 3n^2}{10n^2}$. *(Lesson 9-7)*

30. Which is larger, $(5^4)^3$ or $5^4 \cdot 5^3$? *(Lesson 9-5)*

31. *Skill sequence*   Simplify. *(Lesson 9-5)*
   a. $2(x \cdot x^4)$
   b. $x \cdot (x^4)^2$
   c. $(x \cdot x^4)^2$

32. A tree casts a shadow 24 ft long. A yardstick casts a shadow 2.25 ft long.
   a. Draw a diagram illustrating this situation.
   b. Write a proportion that describes this situation.
   c. How tall is the tree? *(Lesson 5-8)*

33. A store advertises a sweater for $45 dollars for the first 2 months it is in stock, then discounts the price 8% in each 2-month period thereafter. What is the selling price of the sweater after 10 months? *(Lesson 9-3)*

34. List all the perfect squares under 150. *(Lesson 7-4, Previous course)*

35. Find the area of the colored region at the right. (Recall that for a triangle, Area $= \frac{1}{2} \cdot$ base $\cdot$ height.) *(Lessons 3-2, 4-1)*

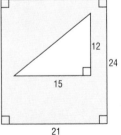

Exploration

36. Each of these numbers can be written in more than one way in the form a^n, where a and n are positive integers from 2 to 20. For each, find two pairs of values of a and n.
   a. 81   **b.** 256   **c.** 32,768   **d.** 43,046,721

LESSON

9-9

Remembering the Properties

Seven properties of powers have been studied in this chapter. They are, for all integers n and nonzero a and b:

Zero Exponent	$b^0 = 1$
Negative Exponent	$b^{-n} = \dfrac{1}{b^n}$
Product of Powers	$b^m \cdot b^n = b^{m+n}$
Quotient of Powers	$\dfrac{b^m}{b^n} = b^{m-n}$
Power of a Power	$(b^m)^n = b^{mn}$
Power of a Product	$(ab)^n = a^n b^n$
Power of a Quotient	$\left(\dfrac{a}{b}\right)^n = \dfrac{a^n}{b^n}$

It is easy to confuse these properties. Fortunately, mathematics is *consistent*. As long as you apply properties correctly, the results you get using some properties will not disagree with the results you get using other properties. We begin with what may look like a new problem: a negative power of a fraction.

Example 1 Write $\left(\dfrac{2}{3}\right)^{-4}$ as a simple fraction.

Solution 1: Think: The problem asks for the power of a quotient. So use that property.

$$\left(\frac{2}{3}\right)^{-4} = \frac{2^{-4}}{3^{-4}}$$

Now evaluate the numerator and denominator.

$$= \frac{\dfrac{1}{2^4}}{\dfrac{1}{3^4}} = \frac{\dfrac{1}{16}}{\dfrac{1}{81}} = \frac{1}{16} \cdot 81 = \frac{81}{16}$$

Solution 2: Think: The problem asks for a negative exponent. So use the Negative Exponent Property.

$$\left(\frac{2}{3}\right)^{-4} = \frac{1}{\left(\dfrac{2}{3}\right)^4}$$

Now use the Power of a Quotient Property.

$$= \frac{1}{\dfrac{2^4}{3^4}} = \frac{3^4}{2^4} = \frac{81}{16}$$

Solution 3: Think: $-1 \cdot 4 = -4$. Use the Power of a Power Property.

$$\left(\frac{2}{3}\right)^{-4} = \left(\left(\frac{2}{3}\right)^{-1}\right)^4$$

Now use the Negative Exponent Property.

$$= \left(\frac{3}{2}\right)^4 = \frac{3^4}{2^4} = \frac{81}{16}$$

Check Use a calculator. $2 \boxed{\div} 3 \boxed{=} \boxed{y^x} 4 \boxed{\pm} \boxed{=}$ gives 5.0625, which is $\frac{81}{16}$.

Before the days of hand-held calculators (before the early 1970s), problems with large exponents could not be approached in a first-year algebra course. So it would be difficult to check some answers. With a calculator, a strategy called **testing a special case** is often useful.

Example 2 Norm was asked to simplify $x^8 \cdot x^6$. He forgot whether the answer should be x^{14}, x^{48}, or $2x^{14}$. How can he be helped by a special case?

Solution Let $x = 3$. That is a special case. Now calculate $3^8 \cdot 3^6$ (with a calculator) and see if it equals 3^{14} or 3^{48} or $2 \cdot 3^{14}$. A calculator shows

$$3^8 \cdot 3^6 = 4{,}782{,}969$$
$$3^{14} = 4{,}782{,}969$$
$$3^{48} = 7.9766 \cdot 10^{22}$$
$$2 \cdot 3^{14} = 9{,}565{,}938.$$

The answer is 3^{14}. So $x^8 \cdot x^6 = x^{14}$.

In the test of a special case, the number tested must not be too special. A pattern may work for a few numbers, but not for all. A **counterexample** is a special case for which a pattern is false. To show that a pattern is not always true, it is enough to find *one* counterexample.

Example 3 Ali noticed that $2^3 = 2^2 + 2^2$ since $8 = 4 + 4$. She guessed that, in general, there is a property

$$x^3 = x^2 + x^2.$$

She tested a second case by letting $x = 0$. She found that $0^3 = 0^2 + 0^2$. She concluded that her *property* was always true. Is Ali right?

Solution No. Try a different number. Let $x = 5$.

$$\text{Does } 5^3 = 5^2 + 5^2?$$
$$\text{Is } 125 = 25 + 25? \text{ No.}$$

$x = 5$ is a counterexample which shows that Ali's *property* does not *always* hold.

Two is a very special number. It has properties that other numbers do not. For instance, squaring it gives the same result as doubling it. So beware of using 2 as a special case. Also avoid 0 and 1. Some patterns are true for whole numbers but not fractions. Some patterns are true for positive numbers, but not negative numbers.

You can test a special case even in applications. In the next example, a special case is used to solve a problem involving compound interest.

Example 4   In 1985, Monica wanted to invest some money for 10 years. She had two choices. Either (1) invest it all at 7% annual yield, or (2) invest half at 6% and half at 8%. Would she earn the same amount each way?

Solution   Let the amount Monica wants to invest be P. Then under plan (1) she will have $P(1.07)^{10}$. Under plan (2) she will have $\frac{1}{2}P(1.06)^{10} + \frac{1}{2}P(1.08)^{10}$. These are hard to compare, so we select a value of P, a special case. We pick $4000.
Plan (1): $4000 at 7%

$$T = 4000(1.07)^{10}$$
$$\approx 7868.61$$

Plan (2): $2000 at 6% and $2000 at 8%

$$T = 2000(1.06)^{10} + 2000(1.08)^{10}$$
$$\approx 3581.70 + 4317.85$$
$$= 7899.55$$

Monica would earn $7868.61 under the first plan and $7899.55 under the second. The yield on the plans is not the same. Splitting her money between the investments gives in plan (2) a greater return.

In the questions, if you have trouble remembering a property or are not certain that you have simplified an expression correctly, try this technique of testing a special case.

Questions

Covering the Reading

1. a. Write $\left(\frac{5}{2}\right)^{-3}$ as a fraction.
   b. Check your answer using a calculator.

In 2–4, *choose* the correct choice and check by testing a special case.

2. $\dfrac{x^6}{x^3} =$

   (a) x^2 (b) x^3 (c) 2   (d) 1

3. $\left(\dfrac{m}{n}\right)^2 =$

   (a) $\dfrac{2m}{n}$ (b) $\dfrac{m}{n}$ (c) $\dfrac{2m}{2n}$ (d) $\dfrac{m^2}{n^2}$

4. $\left(\dfrac{3x}{y}\right)^{-2} =$

   (a) $\dfrac{y^2}{3x^2}$ (b) $-6x^2y^2$ (c) $\dfrac{y^2}{9x^2}$ (d) $\dfrac{-6x^2}{y^2}$

5. Consider the equation $x^4 = 4x^2$. Tell if the equation is true for the special case indicated.
   a. $x = 0$ **b.** $x = 2$ **c.** $x = -2$ **d.** $x = 3$

6. *True or false* If more than two special cases of a pattern are true, then the pattern is true.

7. Define counterexample.

8. *True or false* If one special case of a pattern is not true, then the general pattern is not true.

9. Suppose Monica can invest money at (1) 10% annual yield for 5 years or at (2) 5% annual yield for 10 years. Let her principal be $4000. Will the two investments give the same total amount?

Applying the Mathematics

10. Consider the pattern $\dfrac{1}{z} + \dfrac{1}{y} = \dfrac{y + z}{yz}$.
   a. Is the pattern true when $y = 3$ and $z = 4$?
   b. Test a special case when $y = z$.
   c. Test another special case. Let $y = 5$ and $z = 2$. Convert the fractions to decimals to check.
   d. Do you think this pattern is true?

11. Find a counterexample for the pattern $(2x)^3 = 2x^3$.

12. Consider the pattern $a^2 + b^2 = (a + b)^2$. Test special cases to decide whether this pattern is true.

13. *Multiple choice* Use a special case to answer. Suppose a, b, and c are all positive numbers.

If $\dfrac{a}{b} = c$ and $b = c$, then b equals which of the following:

(a) $\dfrac{a}{2}$ (b) $\sqrt{a}$ (c) a (d) $2a$ (e) a^2

14. Use special cases to answer. If a price is discounted 30% and then the sale price is discounted 10%, what percent of the original price is the sale price?

Review

In 15–17, simplify. *(Lessons 9-7, 9-8)*

15. $\left(\dfrac{3}{5x}\right)^4 \cdot \left(\dfrac{2}{3}\right)^2$ **16.** $(7ny)^3$ **17.** $x^2 \cdot \left(\dfrac{3}{x}\right)^2$

18. If $\dfrac{6n^5}{x} = 3n$, what is x? *(Lesson 9-7)*

19. Which is largest: 2^{1492}, $(2^{14})^{92}$, or $2^{((14)^{92})}$? *(Lesson 9-5)*

20. John bought a used car for $1700. It lost 10% of its value each year for the first 5 years. Then its value stayed the same for 5 years. At this point it became a collector's item and increased in value 20% each year. Find the value of the car 20 years after John bought it. *(Lessons 9-2, 9-3)*

21. Use the Pythagorean Theorem to find the length of the longest segment that can be drawn on a 3-inch by 5-inch card. Check by measuring an index card. *(Lesson 7-5)*

22. The rectangle below is enlarged to a size twice as long and 3 times as wide.

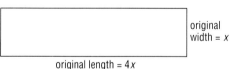

original width = x

original length = $4x$

     a. What is the new length?
     b. What is the new width?
     c. What is the new area?
     d. The new area is how many times as big as the old area? *(Lessons 4-1, 5-3)*

23. *Skill sequence* Solve for y. *(Lesson 6-8)*
     a. $2x + 3y = 4$ **b.** $4x + 6y = 8$
     c. $6x + 9y = 12$ **d.** $2ax + 3ay = 4a,\ (a \neq 0)$.

24. In a studio audience, there are *m* men and *w* women. Suppose one contestant is chosen at random from the audience. Find P(the contestant is a woman). *(Lesson 1-7)*

25. There is a $6 profit on an item selling for $20. What percent of the selling price is the profit? *(Lesson 5-4)*

26. A motorist used 8 gallons of gas in traveling 275.5 miles. At this rate, how far could the motorist travel on 10 gallons of gas? *(Lesson 5-7)*

27. If a point from the square at right is picked at random, what is the probability that it is not in the colored rectangle? *(Lesson 5-5)*

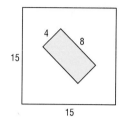

28. *Skill sequence* Evaluate $\frac{4}{3}\pi r^3$ for the given *r*. Leave your answer in terms of π. *(Lessons 9-5, 9-8)*
   a. $r = 3$
   b. $r = 3k$
   c. $r = 3k^2$

Exploration

29. If you do not have a calculator, testing a special case can be difficult. So it helps to know the small positive integer powers of 2, 3, 4, 5, and 6. Copy and fill in this table of values of x^n.

$n =$	2	3	4	5	6	7	8	9	10
$x = 2$									
3									
4									
5									
6									

How can you use the properties of powers to make it easier to remember these powers?

Summary

The nth power of x is written x^n. The number n is called the exponent. Thus, whenever there is an exponent, there is a power. The expression x^n may mean $x \cdot x \cdot \ldots \cdot x$, where there are n factors. Because powers are related to multiplication, the basic properties of powers involve multiplication and division. For all integers m and n and all non-zero x and y,

$$x^m \cdot x^n = x^{m+n} \qquad \frac{x^m}{x^n} = x^{m-n}$$

$$(xy)^n = x^n y^n \qquad \left(\frac{x}{y}\right)^n = \frac{x^n}{y^n}$$

x^n can also be the growth factor in a time interval of length n, when the growth factor in the unit interval is x. Important applications of exponential growth and decay are population growth and compound interest. In compound interest, the growth factor is the quantity $1 + i$, where i is the annual yield. So at an annual yield of i, after n years an amount P grows to $P(1 + i)^n$. When the growth factor is between 0 and 1, the amount gets smaller, and exponential decay occurs.

The growth model allows x^n to be interpreted when n is not a positive integer. The number x^0 is a growth factor in 0 time, so x^0 is the identity under multiplication. Thus $x^0 = 1$. The number x^{-n} is a growth factor going back in time, and

$$x^{-n} = \frac{1}{x^n}.$$

Vocabulary

Below are the most important terms and phrases for this chapter. You should be able to give a general description and a specific example of each.

Lesson 9-1
nth power, base, exponent
principal
interest
annual yield
compound interest
Compound Interest Formula

Lesson 9-2
exponential growth
growth factor
growth model for powering
Zero Exponent Property

Lesson 9-3
exponential decay

Lesson 9-5
Product of Powers Property
Power of a Power Property

Lesson 9-6
Negative Exponent Property

Lesson 9-7
Quotient of Powers Property

Lesson 9-8
Power of a Product Property
Power of a Quotient Property

Lesson 9-9
testing a special case
counterexample

Progress Self-Test

Take this test as you would take a test in class. You will need graph paper and a calculator. Then check your work with the solutions in the Selected Answers section in the back of the book.

In 1–6, simplify.

1. $b^7 \cdot b^{11}$

2. $\dfrac{5x^2y}{15x^{10}}$

3. $7 \cdot 10^1 + 3 \cdot 10^0 + 6 \cdot 10^{-1}$

4. $(-8)^{-5}$

5. $\dfrac{3z^6}{12z^4}$

6. -5^4

7. Rewrite $\left(\dfrac{3}{x}\right)^2 \cdot \left(\dfrac{x}{3}\right)^4$ as a simple fraction.

8. Write $(3y^2)^4$ without parentheses.

9. If $q = 11$, then $6q^0 = \underline{\ ?\ }$.

10. Find a counterexample to the pattern $x^{-1} = -x$.

11. Consider the pattern $x^2 - y^2 = (x - y)(x + y)$.
   a. Test three special cases with positive and negative numbers.
   b. Does it appear the pattern is true?

12. Name the general property that justifies the simplification $2^{10} \cdot 2^3 = 2^{13}$.

13. Felipe invests $6500 in an account with an annual yield of 5.71%. Without any withdrawals or more deposits, how much will be in the account after 5 years?

14. Darlene invests $1900 for three years at an annual yield of 5.8%. At the end of the three years, how much interest has she earned?

In 15 and 16, the present population P of a city has been growing exponentially at 3% a year.

15. What will the population be in fifteen years?

16. If $P = 100,000$, what was the population 2 years ago?

17. Which of the following equations are exponential?
   a. $y = (\frac{1}{3})^x$ **b.** $y = 27 + 14x$
   c. $y = \frac{1}{3}x$ **d.** $y = 27 \cdot 14^x$

18. Graph $y = 3^x$ for $0 \le x \le 3$.

19. A duplicating machine enlarges a picture 30%. If that enlarger is used 3 times, how many times as large as the original picture will the final picture be?

20. Recall that the volume V of a sphere with radius r is $V = \frac{4}{3}\pi r^3$. The radius of the sun (roughly a sphere of gas) is about $4.33 \cdot 10^5$ miles. Estimate the volume of the sun.

Chapter Review

Questions on **SPUR** Objectives

SPUR stands for **S**kills, **P**roperties, **U**ses, and **R**epresentations.
The Chapter Review questions are grouped according to the
SPUR Objectives for this chapter.

SKILLS deal with the procedures used to get answers.

■ **Objective A:** *Evaluate integer powers of real numbers. (Lessons 9-2, 9-5, 9-6, 9-7, 9-8, 9-9)*

1. Evaluate. **a.** 3^4 **b.** -3^4 **c.** $(-3)^4$

2. Simplify $-2^5 \cdot (-2)^5$.

3. If $y = 7$, then $4y^0 =$ __?__.

4. If $x = 2$, then $3x^3 - x^2 =$ __?__.

In 5 and 6, simplify.

5. $7 \cdot 10^2 + 3 \cdot 10^0 + 4 \cdot 10^{-2}$

6. $3 \cdot 10^1 + 8 \cdot 10^0 + 9 \cdot 10^{-1}$

In 7 and 8, rewrite without an exponent.

7. -5^{-3} **8.** 2^{-5}

In 9–12, write as a simple fraction.

9. $(\frac{2}{7})^3$ **10.** $(-\frac{4}{3})^4$

11. $(\frac{1}{3})^{-4}$ **12.** $10 \cdot (\frac{2}{5})^{-3}$

■ **Objective B:** *Simplify products, quotients, and powers of powers. (Lessons 9-5, 9-7)*

In 13–18, simplify.

13. $x^4 \cdot x^7$

14. $r^3 \cdot t^5 \cdot r^8 \cdot t^2$

15. $y^2(x^3 y^{10})$

16. $\dfrac{3a^4 c}{3a^5}$

17. $\left(\dfrac{2^{19}}{2^{19}}\right)^2$

18. $(x^5)^3 + (x^3)^5$

19. Rewrite $\dfrac{4m^6}{20m^2}$ without fractions.

20. Simplify $\dfrac{(x + 8)^5}{(x + 8)^2}$.

21. Rewrite xy^{-2} without a negative exponent.

22. Rewrite $2m^{-1}n^4 p^2$ without a negative exponent.

■ **Objective C:** *Rewrite powers of products and quotients. (Lesson 9-8)*

In 23–28, rewrite without parentheses.

23. $\left(\dfrac{y}{x}\right)^3$

24. $(4x)^m$

25. $\left(\dfrac{a}{b}\right)^5$

26. $(3m^2 n)^3$

27. $4 \cdot \left(\dfrac{k}{3}\right)^3$

28. $2(4x)^2$

PROPERTIES deal with the principles behind the mathematics.

■ **Objective D:** *Test a special case to determine whether a pattern is true. (Lesson 9-9)*

29. For which of the following special cases is the pattern $x = x^2$ true?
  a. $x = 0$ **b.** $x = 1$
  c. $x = 2$ **d.** $x = -1$

30. Consider the pattern $(x^2)^y = x^{2y}$.
  a. Is the pattern true when $x = 3$ and $y = 4$?
  b. Is the pattern true when $x = 5$ and $y = 1$?
  c. Based on your answers to parts a and b is the pattern true?

In 31 and 32, find a counterexample to the pattern.

31. $(a + b)^3 = a^3 + b^3$

32. $(x^3)^2 = x^{(3^2)}$

■ **Objective E:** *Identify properties of exponents.* (*Lessons 9-2, 9-5, 9-6, 9-7, 9-8*)

Here is a list of the power properties in this chapter. For all integers m and n and nonzero a and b:

Zero Exponent Property:	$b^0 = 1$
Product of Powers Property:	$b^m \cdot b^n = b^{m+n}$
Power of a Product Property:	$(ab)^n = a^n \cdot b^n$
Negative Exponent Property:	$b^{-n} = \dfrac{1}{b^n}$
Quotient Property of Powers:	$\dfrac{b^m}{b^n} = b^{m-n}$
Power of a Power Property:	$(b^m)^n = b^{mn}$
Power of a Quotient Property:	$\left(\dfrac{a}{b}\right)^n = \dfrac{a^n}{b^n}$

In 33–40, name the general property or properties that justify the simplification.

33. $a^7 \cdot b^7 = (ab)^7$

34. $a^7 \div a^2 = a^5$

35. $(4.36)^0 = 1$

36. $4^6 \cdot 4^9 = 4^{15}$

37. $\left(\dfrac{7}{g}\right)^y = \dfrac{7^y}{g^y}$

38. $6^3 \cdot 2^0 = 6^3$

39. $14^{-2} = \dfrac{1}{14^2}$

40. $\left(\dfrac{x}{y}\right)^{-2} = \dfrac{y^2}{x^2}$

USES deal with applications of mathematics in real situations.

■ **Objective F:** *Calculate compound interest.* (*Lesson 9-1*)

In 41 and 42, use the advertisement below.

```
┌─────────────────────────┐
│  GUARANTEED             │
│  7.7% YIELD             │
│  $2,500 MINIMUM         │
└─────────────────────────┘
```

41. Using the annual yield, calculate how much money there will be in your account if you deposit $2500 for 3 years.

42. Using the annual yield, calculate how much interest $3000 will earn if deposited for 4 years.

43. Susan invested $1200 in a bank at an annual yield of 6%. Without any withdrawals, how much money would she have in the account after 2 years?

44. *Multiple choice* Which yields more money?
   (a) an amount invested for 2 years at an annual yield of 10%
   (b) the same amount invested for 10 years at an annual yield of 2%

■ **Objective G:** *Solve problems involving exponential growth and decay.* (*Lessons 9-2, 9-3*)

45. Jennifer earns $7.25 an hour. She gets a 12% raise each year. How much will she earn per hour after 4 years on the job?

46. In 1986, the United States had an inflation rate of about 4% per year. That means an article costing $100 in 1986 would cost $104 in 1987. Consider a hardcover book that sells for $16.95 in 1986. At the rate of inflation above, how much would the same book cost in 1990?

In 47 and 48, after a few hours a colony of bacteria that doubles every hour has 8000 bacteria. After n more hours there will be T bacteria where

$$T = 8000 \cdot 2^n.$$

47. a. Find the value of T when $n = 4$.
   b. In words describe the meaning of your answer to part a.

48. a. Find T when $n = -3$.
   b. Describe the meaning of your answer to part a.

In 49 and 50, the population in a city of 1,500,000 is decreasing exponentially at a rate of 3% per year. The population P in n years can be described by

$$P = 1,500,000 \cdot (0.97)^n.$$

49. What will the population be in 10 years' time?

50. What is the population when $n = 0$? What does your answer mean?

■ **Objective H:** *Use and simplify expressions with powers in everyday situations.* *(Lessons 9-6, 9-7, 9-8)*

51. A certain photographic enlarger can make any picture $\frac{3}{2}$ times its original size. By how many times will a picture be enlarged if the enlarger is used
   a. twice   **b.** 5 times?

52. Water blocks out light. (At a depth of 10 meters it is not as bright as on the surface.) Suppose 1 meter of water lets in $\frac{9}{10}$ of the light. How much light will get through x meters of water?

53. In 1990, there were about $5 \cdot 10^9$ people on earth. The land area of the earth is about $1.46 \cdot 10^8$ sq km. How many people are there per sq km?

54. The moon is nearly a sphere with radius of $1.05 \cdot 10^3$ miles. The volume of a sphere is $\frac{4}{3}\pi r^3$. To the nearest billion cubic miles, what is the volume of the moon?

REPRESENTATIONS deal with pictures, graphs, or objects that illustrate concepts.

■ **Objective I:** *Graph exponential growth and decay relationships.* *(Lesson 9-4)*

55. Graph $y = 2^x$ for $-4 \le x \le 4$.

56. Graph $y = 4^x$ for $-3 \le x \le 3$.

57. When x is large, the graph of which equation increases faster,
   $y = 56 + .04x$ or $y = 56 \cdot (1.04)^x$?

58. Tell whether the graph of each equation is linear or exponential.
   a. $y = 4x$ **b.** $y = 4^x$
   c. $y = 100 \cdot (3.4)^x$ **d.** $y = \frac{3}{4}x + 100$

59. Suppose $200 is invested in a bank at a 5% annual yield. Make a table for the amount in the bank after 0, 5, 10, and 15 years if
   a. the interest is removed each year;
   b. the interest is kept in the bank and compounded.
   c. Graph smooth curves through the amounts in parts a and b on the same set of axes.

60. Match each graph with its description.
   a. constant increase
   b. constant decrease
   c. exponential growth
   d. exponential decay

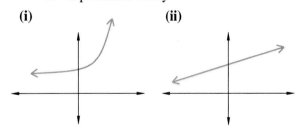

(i)   (ii)

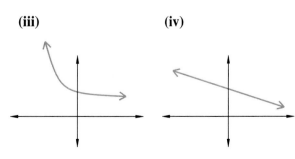

(iii)   (iv)

Polynomials

A *polynomial* is formed by adding, subtracting, or multiplying (but not dividing) numbers and variables. The expressions

$$3x^2$$
$$3a + 4a^2 - 5c$$
$$\tfrac{1}{2}\, y^2 z t^5 - 17$$

are polynomials. Many of the algebraic expressions you have studied in this book are polynomials.

Polynomials arise naturally in certain situations. In this chapter you will learn that polynomial

$$Ax^3 + By^2 + Cz$$

tells how much you would have if you invested A dollars 3 years ago at a yield $x - 1$, added B dollars to it 2 years ago at a yield $y - 1$, and added C dollars 1 year ago at a yield $z - 1$.

Polynomials also can approximate other expressions. Here is a graph of the polynomial equation

$$y = x - \tfrac{1}{5}x^3 + \tfrac{1}{120}x^5.$$

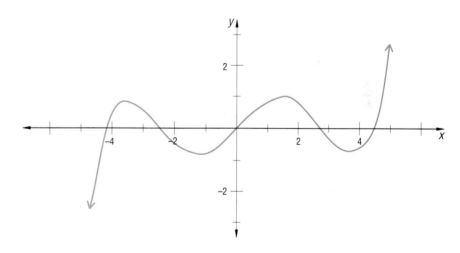

For values of x from -4 to 4, it approximates the motion of a wave.

In this chapter you will learn about situations that give rise to polynomials, and how to add, subtract, and multiply them.

LESSON
10-1

How Polynomials Arise

The largest money matters most adults deal with are

> salary or wages,
> savings,
> payments on loans for cars or trips or other items,
> insurance,
> social security payments and benefits, and
> home mortgages or rent.

Each of these items involves paying or receiving money each month or every few months or every year. But what is the total amount paid or received? The answer is not easy to calculate because interest is involved. Here is an example of this kind of situation.

Example 1 Each birthday, from age 12 on, Mary has received $50 from her grandparents. She saves the money and can get an annual yield of 7%. How much will she have by the time she is 16?

Solution It helps to write down how much Mary has on each birthday. On her 12th birthday she has $50. She then receives interest on that $50 and an additional $50 on her 13th birthday.

$$50(1.07) + 50 = \$103.50$$

Each year interest is paid on all the money previously accumulated and each year another $50 gift is added. The totals for her 12th through 16th birthdays are summarized below.

Birthday		Total
12th	50	= $50
13th	$50(1.07) + 50$	= $103.50
14th	$50(1.07)^2 + 50(1.07) + 50$	= $160.75
15th	$50(1.07)^3 + 50(1.07)^2 + 50(1.07) + 50$	= $222.00
16th	$50(1.07)^4 + 50(1.07)^3 + 50(1.07)^2 + 50(1.07) + 50$	= $287.54

from 12th birthday   from 13th birthday   from 14th birthday   from 15th birthday   from 16th birthday

The total of $287.54 she has by her 16th birthday is $37.54 more than the total $250 she received as gifts because of the interest earned.

476

Letting $x = 1.07$, you can write the amount of money Mary has with the expression

$$50x^4 + 50x^3 + 50x^2 + 50x + 50 \text{ dollars.}$$

This expression is useful because if the interest rate changes, you only have to substitute a different value for x. We call x a **scale factor**.

■ ■ ■ ■ ■ ■ ■ ■

Example 2   Suppose Erin's parents gave her $50 on her 12th birthday, $60 on her 13th, $70 on her 14th, and $80 on her 15th. If she invests all the money in an account with a yearly scale factor x, how much will she have on her 15th birthday?

Solution   The money from her 12th birthday will earn three years worth of interest, from her 13th, two years of interest, and from her 14th, one year. So the total is

$$50x^3 + 60x^2 + 70x + 80.$$

The expression $50x^3 + 60x^2 + 70x + 80$ is a **polynomial in the variable x.** Recall that a term is a number or a number multiplied by a variable. So this polynomial has four terms. The 80 is a **constant term.** All other terms consist of a number, called the *coefficient*, multiplied by a variable. Sometimes the variable has an exponent. Almost always the coefficient of a term is written first.

The largest exponent in a polynomial with one variable is the **degree of the polynomial**. For instance, the polynomial $17y^3 + 8y^5 - y + 6$ is a polynomial in y with degree 5. It is common practice to arrange terms of a polynomial in descending order of exponents ending with the constant term. If the terms are arranged in descending order of exponents, the polynomial is

$$8y^5 + 17y^3 - y + 6.$$

The coefficient of y^5 is 8. Since there is no y^4 term, we say that the coefficient of y^4 is 0. The coefficient of y^3 is 17. The coefficient of y^2 is 0. The coefficient of y is -1. The constant term is 6.

Polynomials with 1, 2, or 3 terms have special names.

Name	Number of terms	Examples
monomial	1 term	$3x^2, \frac{1}{2}xy^3, -500m^{14}$
binomial	2 terms	$c - 7d, 4x^2 - 1, w + 2$
trinomial	3 terms	$b^2 + 2b + 1$
		$9y^2 + 30yz + 25z^2$
		$-4x^4 - 7.2x^3 + .019$

1. Mary's grandfather will receive $200 on each of his birthdays from age 61 on. If he saves the money and can get an annual yield of 7%, how much will he have by the time he retires at age 65?

In 2–5, Huey, Dewey and Louie are triplets. They received the following cash presents on their birthdays.

	H	D	L
In 1982	$10	$15	$5
In 1983	$20	$15	$40
In 1984	$15	$15	nothing

Each year they put all their money into a bank account which paid a 6% annual yield.

2. How much money did Huey have on his 1983 birthday?

3. How much did Dewey have on his 1984 birthday?

4. How much did Louie have on his 1984 birthday?

5. In 1985, Huey received $30 on his birthday. If he had invested all the money received from years 1982–85 into an account with scale factor x, how much would he have had by his birthday in each of the following years?
   a. 1982   b. 1983   c. 1984   d. 1985

6. For the polynomial $2m^2 - 5m^4 + m - 80$, give the coefficient of:
   a. m^4 b. m^3 c. m^2 d. m

7. a. Write the polynomial $3 - 90x^3 + 4x^2 + x^5$ in descending order of exponents.
   b. Give its degree.

8. How many terms does each of the following have?
   a. trinomial   b. monomial   c. binomial

9. *True or false* $7xy$ is a trinomial.

10. Give the value of $x^3 + 2x^2 - 9x + 2$ when x is:
    a. 1   b. $\frac{1}{2}$ c. -1   d. 0

11. A wood harvester plants trees each spring in a forest. The first spring 16,000 trees were planted. The second spring 22,000 trees were planted, the third spring 18,000, and the fourth spring 25,000. Suppose that each tree contains $\frac{1}{100}$ of a cord of wood when it is planted. Then suppose each year the amount of wood grows by a scale factor y. How many cords of wood are in the forest the fourth spring?

12. The formula below gives the sum of the consecutive integers from 1 to n.

$$1 + 2 + 3 + \ldots + n = \tfrac{1}{2}n^2 + \tfrac{1}{2}n$$

For instance, when $n = 6$,

$$1 + 2 + 3 + 4 + 5 + 6 = \tfrac{1}{2} \cdot 6^2 + \tfrac{1}{2} \cdot 6$$
$$= \tfrac{36}{2} + \tfrac{6}{2}$$
$$= 21.$$

Use the formula to find the sum of the integers from 1 to 100.

13. The computer program below can be used to evaluate the polynomial $x^2 + 2x - 3$ for integer values of x from -3 to 3.

```
10   PRINT "VALUES OF POLYNOMIAL X ^ 2 + 2X − 3"
20   PRINT "X", "VALUE"
30   FOR X = -3 TO 3
40   LET V = X * X + 2 * X − 3
50   PRINT X, V
60   NEXT X
70   END
```

a. Write the table that this program will produce.
b. Rewrite the program to evaluate $x^2 - 8x + 16$ for integer values of x from -6 to 6.

14. Suppose that in 1985 Barry received $100 on his birthday. From 1986–1989 he received $150 on his birthday. He put the money in a shoebox. The money is still there.
a. How much money did Barry have after his 1989 birthday?
b. How much more would he have had if he had invested his money at an annual yield of 8% each year?

15. a. The price of a $100 object is marked up 10%. What is the new price? *(Lesson 6-3)*
  b. The price from part a. is discounted 10%. What is the new price? *(Lesson 6-3)*

16. *Multiple choice* Which is a pair of like terms? *(Lesson 2-2)*
  (a) $50x^3$ and $50x^3$
  (b) $50x^2$ and $50y^2$
  (c) $50x^4$ and $50x$
  (d) $50x$ and 50

17. In 1981, of 3,020,000 high school graduates, 1,537,000 were women.
  a. What was the relative frequency of female high school graduates in 1981?
  b. What is the probability that a randomly chosen high school graduate will be male? *(Lesson 1-8)*

18. Simplify $-5n + 2n + k + k + 10n$. *(Lesson 2-2)*

19. Simplify $(9x + 8) - (3x - 5)$. *(Lesson 6-9)*

20. Solve $-(x - 6.5) = 13.4$. *(Lesson 6-9)*

21. If the area of the colored part of the larger rectangle below is 20, what is x? *(Lessons 4-1, 6-3)*

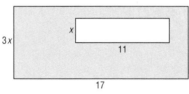

22. *Skill sequence* Simplify. *(Lesson 6-3)*
  a. $8(3y)$
  b. $8(3y + 7)$
  c. $8(3y + 7 - 4x)$

In 23–25, solve in your head. *(Lessons 3-3, 4-4)*

23. $x - \frac{1}{2} = \frac{5}{2}$
24. $\frac{1}{2}y = \frac{5}{2}$
25. $2z = \frac{2}{5}$

26. *Multiple choice* One of the graphs has the equation
$$y = x^4 - 4x^3 + 3x^2 + 4x - 4.$$

  a. Which one is it?

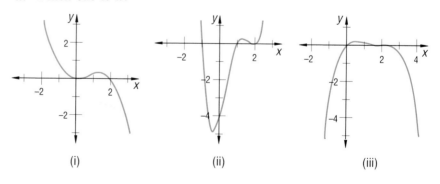

(i)   (ii)   (iii)

  b. Explain how you determined your answer to part a.

Adding and Subtracting Polynomials

In the previous lesson, Erin's parents gave her $50, $60, $70 and $80 for her 12th, 13th, 14th, and 15th birthdays. She put the money into a savings account with a scale factor x. The polynomial

$$50x^3 + 60x^2 + 70x + 80$$

expresses the amount she had on her 15th birthday. Suppose Erin's aunt gave her $20 on each of these 4 birthdays. If she put this money into the same account, the amount from the aunt's gifts would be

$$20x^3 + 20x^2 + 20x + 20.$$

The total amount she would have from all these gifts is found by adding these two polynomials.

$$(50x^3 + 60x^2 + 70x + 80) + (20x^3 + 20x^2 + 20x + 20)$$

This new expression contains some like terms. Recall that like terms are terms that have identical powers on identical variables. $50x^3$ and $20x^3$ are like terms, $60x^2$ and $20x^2$ are like terms, and so on. Polynomials can be simplified by using the Distributive Property to add like terms.

Example 1   Simplify $(50x^3 + 60x^2 + 70x + 80) + (20x^3 + 20x^2 + 20x + 20)$.

Solution 1   Use the Associative and Commutative Properties of Addition to rearrange the polynomials so that like terms are together.

$$(50x^3 + 20x^3) + (60x^2 + 20x^2) + (70x + 20x) + (80 + 20)$$

Use the Distributive Property to add like terms.

$$= (50 + 20)x^3 + (60 + 20)x^2 + (70 + 20)x + (80 + 20)$$
$$= 70x^3 + 80x^2 + 90x + 100$$

Solution 2 Experts add the pairs of like terms without rearranging them on paper. Here is an expert solution after two steps. The like terms are underlined after they have been combined so no term will be overlooked.

$$(\underline{50x^3} + \underline{60x^2} + 70x + 80) + (\underline{20x^3} + \underline{20x^2} + 20x + 20)$$
$$70x^3 + 80x^2 +$$

After two more combinations, all the original terms are accounted for.

$$= \underline{50x^3} + \underline{60x^2} + \underline{70x} + \underline{80} + \underline{20x^3} + \underline{20x^2} + \underline{20x} + \underline{20}$$
$$= 70x^3 + 80x^2 + 90x + 100$$

Check Think of the answer in relation to Erin's birthday presents. The first year she got $70 ($50 from her parents, $20 from her aunt). The $70 has 3 years to earn interest. The $80 from her next birthday earns interest for 2 years. And so on.

Polynomials can be subtracted as well as added. Differences are also simplified by collecting like terms, but special care must be taken with signs.

Example 2 Find the colored area in the figure below.

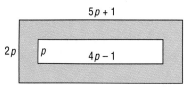

Solution 1 The colored area is the difference of the areas of the rectangles.

$$2p(5p + 1) - p(4p - 1)$$

Here we put in many steps.

$$= (10p^2 + 2p) - (4p^2 - p)$$
$$= (10p^2 + 2p) + (-4p^2 + p) \quad \text{recalling that } -(a - b) = -a + b$$
$$= 10p^2 + -4p^2 + 2p + p$$
$$= 6p^2 + 3p$$

Solution 2 An expert may write the following.

$$2p(5p + 1) - p(4p - 1)$$
$$= 10p^2 + 2p - 4p^2 + p \quad \text{remembering to distribute } -p$$
$$= 6p^2 + 3p$$

Check Test a special case. Let $p = 10$. The dimensions of the figure are now as follows.

The colored area $= 51 \cdot 20 - 10 \cdot 39 = 1020 - 390 = 630$.
When $p = 10$, $6p^2 + 3p = 6(10)^2 + 3(10) = 6 \cdot 100 + 30 = 630$.
So the answer checks.

■ ■ ■ ■ ■ ■ ■ ■

Example 3 Simplify $(12y^2 + 6y^3 - 5) - (2 + 6y^3)$.

Solution To combine the expressions, add the opposite of the terms in the second polynomial.

$$12y^2 + 6y^3 - 5 + {}^-2 + {}^-6y^3$$

Arrange in descending order.

$$= 6y^3 + {}^-6y^3 + 12y^2 - 5 + {}^-2$$

Combine the like terms.

$$= 12y^2 - 7$$

Check Pick a value for y. We let $y = 3$.
Does $(12(3)^2 + 6(3)^3 - 5) - (2 + 6(3)^3) = 12(3)^2 - 7$?
$(108 + 162 - 5) - (2 + 162) = 108 - 7$?
$265 - 164 = 101$?
Yes, it checks.

Questions

In 1 and 2, refer to Example 1.

Covering the Reading

1. Consider the polynomial $70x^3 + 80x^2 + 90x + 100$. What coefficient indicates the amount Erin received on her 14th birthday?

2. Suppose Erin gets $15, $25, $35, and $45 from cousin Lilly on her four birthdays. She also puts this money into her account.
   a. By her 15th birthday, how much money will Erin have from just her cousin?
   b. What is the total Erin will have received and saved from all of her birthday presents?

In 3–6, simplify the expressions.

3. $(12y^2 + 3y - 7) + (4y^2 - 2y - 10)$

4. $(6w^2 - w + 14) - (4w^2 + 3)$

5. $(3 + 5k^2 - 2k) + (2k^2 - 3k - 10)$

6. $(x^3 - 4x + 1) - (5x^3 + 4x - 8)$

7. A rectangle with dimensions $3p$ and $p + 1$ is contained in a rectangle with dimensions $8p$ and $4p + 2$, as in the figure below.
  a. Write an expression for the area of the big rectangle.
  b. Write an expression for the area of the little rectangle.
  c. Write a simplified expression for the area of the colored region.

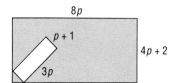

In 8 and 9, simplify.

8. $(12y^4 - 3y^3 + y) + (5y^3 - 7y^2 - 2y + 1) + (2y - 3y^2 + 6)$

9. $(x^2 - 4x + 1) - (3x^2 - 2x) - 2(7x + 4)$

In 10–12, Ashok, Barbara and Cam have the following amounts of money:

Ashok: $90y^3 + 20y^2 + 60y + 100$
Barbara: $60y^4 + 25y^3 + 80y^2 + 75y + 12$
Cam: $50y^2 + 100y + 150$

10. How much do Ashok and Barbara have together?

11. If Betsy has the same amount of money as Barbara, how much do the two girls have together?

12. If $y = 1.08$, who has more, Ashok or Cam?

13. The total surface area T of a prism with a square base is given by the formula $T = 2s^2 + 4sh$, where s is the length of a side of the base and h is the height. Prism Q has a square base with side of length n and height 5. Prism R has a square base with edge $3n$ and height 10.
  a. What is the surface area of Prism Q?
  b. What is the surface area of Prism R?
  c. Find the difference of the surface areas of Prism Q and Prism R.

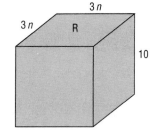

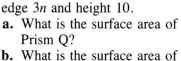

Applying the Mathematics

484

14. After five years of birthdays, Wanda has received and saved

$$80x^4 + 60x^3 + 70x^2 + 45x + 50$$

dollars, having put the money in a savings account at a scale factor x.
a. How much did Wanda get on her most recent birthday?
b. How much did Wanda get on the first birthday?
c. Give an example of a reasonable value for x in this problem.

d. If $x = 1$, how much has Wanda saved?
e. What does a value of 1 for x mean? *(Lesson 10-1)*

15. *Skill sequence* Simplify. *(Lessons 2-2, 9-5, 9-7)*
a. $x^m + x^n$ **b.** $x^m \cdot x^n$ **c.** $x^m \div x^n$

16. Four consecutive integers (like 7, 8, 9, 10) can be represented by the expressions n, $n + 1$, $n + 2$, and $n + 3$. The sum of four consecutive integers is 250. What are the integers? *(Lesson 6-1)*

17. There are 6 girls, 8 boys, 4 women, and 3 men on a community youth board. How many leadership teams consisting of one adult and one youngster could be formed from these people? *(Lesson 4-7)*

18. *Skill sequence* Solve for y. *(Lesson 6-1)*
a. $x - y = 5$ **b.** $x - 2y = 5$ **c.** $x - ay = 5$

19. In the equation $8m + 5m = 4(6 + tm)$, what is the value of t when m is equal to 3? *(Lesson 6-1)*

20. Solve and graph on a number line: $4.6 - 3.1w \geq 20.1$. *(Lesson 6-7)*

21. The length and width of a rectangle are integers. The area is 18. What are the possible perimeters? *(Lesson 4-1)*

22. *Skill sequence* *(Lesson 7-5)*
a. The area of a square is 49 in.^{2}. Find the length of a side.
b. The area of a square is 49 in.^{2}. Find the length of a diagonal.
c. The area of a square is 50 in.^{2}. Find the length of a diagonal.

In 23–25, simplify. Assume $a \geq 0$, $b \geq 0$, and $c \geq 0$. *(Lesson 7-4)*
23. $\sqrt{400}$ **24.** $\sqrt{9 \cdot 25}$ **25.** $\sqrt{a^2 b^4 c^2}$

26. At the right is a 5-by-8 rectangle with a hole in the center in the shape of a 1-by-4 rectangle. With a single cut (not necessarily straight), divide the colored shape into two pieces that, when rearranged, can form a 6-by-6 square.

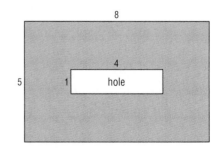

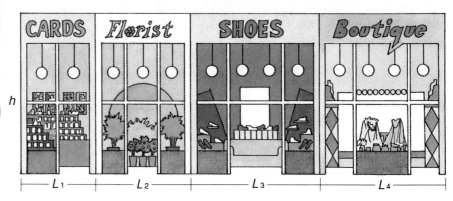

LESSON

10-3

h

Multiplying a Polynomial by a Monomial

The pictures show a view of a section of storefronts at a shopping mall. The displays in the windows are used to attract shoppers, so store owners and mall managers are interested in the areas of store-fronts. The total area of the four windows can be computed in two ways.

The first way is to consider the union of the windows, one big rectangle with length $(L_1 + L_2 + L_3 + L_4)$ and height h. Thus,

$$\text{Area} = h \cdot (L_1 + L_2 + L_3 + L_4).$$

The second way is to compute the area of each storefront and add the results. Thus,

$$\text{Area} = hL_1 + hL_2 + hL_3 + hL_4.$$

These areas are equal, so

$$h \cdot (L_1 + L_2 + L_3 + L_4) = hL_1 + hL_2 + hL_3 + hL_4.$$

Note that h is a monomial while $(L_1 + L_2 + L_3 + L_4)$ is a polynomial. Multiplying a monomial by a polynomial is simply an extension of the Distributive Property.

■ ■ ■ ■ ■ ■ ■ ■ ■

Example 1   Multiply $7x(x^2 + 2x - 1)$.

Solution   Multiply each term in the polynomial by the monomial.

$$= 7x \cdot x^2 + 7x \cdot 2x - 7x \cdot 1$$

Simplify using the Product of Powers Property.

$$= 7x^3 + 14x^2 - 7x$$

Check   Test a special case. We let $x = 3$.

Does $7 \cdot 3(3^2 + 6 - 1) = 7 \cdot 3^3 + 14 \cdot 3^2 - 7 \cdot 3$?

$$21 \cdot 14 = 189 + 126 - 21?$$

Yes, $294 = 294.$

486

When multiplying with polynomial expressions, care must be taken with the signs.

Example 2 Multiply $-4x^2(2x^3 + y - 5)$.

Solution: Distribute the $-4x^2$ over each term of the polynomial.

$$= \mathbf{-4x^2} \cdot 2x^3 + \mathbf{-4x^2} \cdot y - \mathbf{-4x^2} \cdot 5$$
$$= -8x^5 - 4x^2y - -20x^2$$

Simplify to avoid opposites when possible.

$$= -8x^5 - 4x^2y + 20x^2$$

In the last lesson you saw how the birthday gift polynomials were added and subtracted. Example 3 shows how the birthday situation can illustrate multiplying a monomial by a polynomial.

Example 3 Anwar received $50 on his 16th birthday, $70 on his 17th birthday, and $60 on his 18th birthday, which he was able to invest at a scale factor x. He received no more gifts, but continued to invest this money at the same scale factor for 3 more years. How much money did he have then?

Solution 1 The birthday gifts with interest have the value $50x^2 + 70x + 60$. This polynomial (as a chunk) is invested at scale factor x for three more years. Therefore the total is $(50x^2 + 70x + 60)x^3$.

Solution 2 Look at how long each gift individually earned interest: $50 for five years; $70 for four years; and $60 for three years. This leads to the polynomial $50x^5 + 70x^4 + 60x^3$.

Check The two solutions must be equal. $(50x^2 + 70x + 60)x^3 = 50x^5 + 70x^4 + 60x^3$. This is exactly what you would expect using the Distributive Property.

Questions

1. Suppose the height of the stores at the beginning of this lesson is $2h$. Write the area of the complete section of storefronts in two different ways.

2. Using the Distributive Property, $a(b + c + d) = \underline{\ ?\ }$.

In 3–6, multiply.

3. $z(q + r - 13)$

4. $5x^2(x^2 - 9x + 2)$

5. $-3wy(4y - 2w - 1)$

6. $-rt(r^2 - t^2)$

In 7 and 8, refer to Example 3.

7. Suppose Anwar kept his money invested at a scale factor x for 6 years after his 18th birthday. How much money would he have at the end of this period?

Applying the Mathematics

8. Suppose that after his 18th birthday, Anwar took the total amount and invested it at a new scale factor y for 4 years. How much money would he have after this?

In 9–12, simplify.

9. $2(x^2 + 3x) - 3x^2$

10. $a(y^2 - 2y) + y(a^2 + 2a)$

11. $m^3(m^2 - 3m + 2) - m^2(m^3 - 5m^2 - 6)$

12. $v(v + 1) - v(v + 2) + v(v + 3) - v(v + 4)$

13. At the right is a circle in a square.
  a. What is the area of the square?
  b. What is the area of the circle?
  c. What is the area of the colored region?

  d. If a person had 5 copies of the colored region, how much area would be colored?

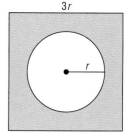

3r

14. Maryalice is now 21. She received $120 on her 19th birthday, $125 on her 20th birthday, and $200 on her 21st birthday.
  a. If she invested this amount at a scale factor x, how much does she have now?
  b. If she keeps the amount in part a invested at the same scale factor for two more years, how much will she have at age 23?
  c. If she invests the amount in part b at a different scale factor y, how much will she have at age 28?

Review

In 15–18, simplify. *(Lesson 10-2)*

15. $(4n^4 + 6n^3 - 9n^2 + 12) + (9n^3 + 7n^2 - 9n + 13)$

16. $(21y^2 - 3y + 1) - (4y^2 + 5y - 14) - 3y^2$

17. $\dfrac{20x^2y}{2xy}$ *(Lesson 2-3)*

18. $\dfrac{3}{2v} + \dfrac{4v}{2v}$ *(Lesson 9-7)*

19. On each of three successive birthdays, Mary received $25 from her grandparents. Mary also received $17 from other relatives on the first of these birthdays, $24 on the second and $31 on the third. Mary invested all the money at a scale factor of x each year. On the last of these birthdays,

    a. how much money did Mary have from her grandparents' gifts? *(Lesson 10-1)*

    b. How much money did she have from the gifts of her other relatives? *(Lesson 10-1)*

    c. How much money did she have in all? *(Lesson 10-2)*

20. a. Three years ago you invested A dollars at a rate of $x - 1$. How much would you have now?

    b. Two years ago you invested B dollars in a different investment at a rate of $y - 1$. Including the investment in part a, how much would you have now? *(Lesson 10-1)*

In 21–23 write an equation for the line with the following characteristics:

21. slope -2, y-intercept 5 *(Lesson 8-4)*

22. slope -4, contains the point $(-1, 6)$ *(Lesson 8-5)*

23. contains the points $(-5, 8)$, and $(-1, 16)$ *(Lesson 8-6)*

24. Calculate the speed of the object and identify a possible object. *(Lesson 5-2)*

    a. It flew 200 miles in $\frac{1}{2}$ hour.

    b. It slithered 14 meters in 7 seconds.

    c. It took 3 days to creep 15 inches.

25. *Skill sequence*   Simplify. *(Lesson 2-3)*

    a. $\dfrac{c}{3} + \dfrac{5c}{13}$ **b.** $\dfrac{c}{13q} + \dfrac{5c}{13}$ **c.** $\dfrac{c}{13q} + \dfrac{5c}{12}$

Exploration

26. In Example 3, Anwar has $50x^2 + 70x + 60$ dollars on his 18th birthday. Suppose this amount equals $200. By trial and error, determine the value of x to the nearest hundredth.

27. With a function grapher or graphing calculator, graph the equation $y = 50x^2 + 70x + 60$ and estimate where it crosses the horizontal line $y = 200$, and thus answer Question 26.

10-4

Common Monomial Factoring

The process that reverses multiplication is called factoring.

Multiplication: Begin with $37 \cdot 3$, end up with 111.
Factoring: Begin with 111, end up with $37 \cdot 3$.

The same idea is used with polynomials.

Multiplication: Begin with $50(x^4 + x^3 + x^2 + x + 1)$,
end up with $50x^4 + 50x^3 + 50x^2 + 50x + 50$.
Factoring: Begin with $50x^4 + 50x^3 + 50x^2 + 50x + 50$,
end up with $50(x^4 + x^3 + x^2 + x + 1)$.

Factoring a number or polynomial provides flexibility. The factored form may be easier to evaluate or use.

For instance, the expression $50x^4 + 50x^3 + 50x^2 + 50x + 50$ arose in Lesson 10-1 from a compound interest situation. To evaluate this expression on a calculator for a given value of x, you need to enter 11 numbers and perform 11 operations: 3 powerings, 4 multiplications, and 4 additions. A calculator key sequence shows this; the operations are shown below in the first 11 squares and the total number of steps is 23.

$50 \boxed{\times} x \boxed{y^x} 4 \boxed{+} 50 \boxed{\times} x \boxed{y^x} 3 \boxed{+} 50 \boxed{\times} x \boxed{x^2} \boxed{+} 50 \boxed{\times} x \boxed{+} 50 \boxed{=}$

In the above sequence, 50 has been entered five times. That is unnecessary. The factored form $50(x^4 + x^3 + x^2 + x + 1)$ contains 50 only once and only has 1 multiplication. This reduces the work to 8 operations and the total number of steps is 19.

$50 \boxed{\times} \boxed{(} x \boxed{y^x} 4 \boxed{+} x \boxed{y^x} 3 \boxed{+} x \boxed{x^2} \boxed{+} x \boxed{+} 1 \boxed{)} \boxed{=}$

The number 50 is a monomial factor of each term in the original polynomial. So the rewriting of $50x^4 + 50x^3 + 50x^2 + 50x + 50$ as $50(x^4 + x^3 + x^2 + x + 1)$ is called **common monomial factoring**. The idea in common monomial factoring is to isolate the largest common factor from each individual term.

Example 1 Find the largest common factor of $-30x^3y$, $20x^4$, and $50x^2y^5$.

Solution First find the largest common factor of the coefficients -30, 20, and 50. That factor is 10. Since each term has the variable x in it, the largest common factor will include an x. The lowest exponent of x in any of the terms is x^2, so x^2 is part of the largest common factor. The variable y does not appear in all terms so y does not appear in the largest common factor. Thus, the largest common factor is $10x^2$.

Example 2 Factor out the largest common monomial factor in $-30x^3y + 20x^4 + 50x^2y^5$.

Solution In Example 1, the largest common monomial factor of the three terms of this polynomial was found to be $10x^2$. Thus

$$-30x^3y + 20x^4 + 50x^2y^5 = 10x^2(\underline{~?~} + \underline{~?~} + \underline{~?~}).$$

Now divide to get the terms in the parentheses.

$$\frac{-30x^3y}{10x^2} = -3xy \qquad \frac{20x^4}{10x^2} = 2x^2 \qquad \frac{50x^2y^5}{10x^2} = 5y^5$$

So $-30x^3y + 20x^4 + 50x^2y^5 = 10x^2(-3xy + 2x^2 + 5y^5)$.

To generalize from Examples 1 and 2, the largest common factor of an expression will include the largest common factor of the coefficients of the terms. It will also include any common variable raised to the *lowest* exponent of that variable found in any of the terms.

Example 3 Factor $24x^2y + 6x$.

Solution The largest common factor of $24x^2y$ and $6x$ is $6x$ itself.

$$24x^2y + 6x = 6x(\underline{~?~} + \underline{~?~})$$

Now divide each term by $6x$ to fill in the factors.

$$= 6x(4xy + 1).$$

Check Test a special case. Let $x = 3$, $y = 4$. Follow the order of operations.

$$24x^2y + 6x = 24 \cdot 3^2 \cdot 4 + 6 \cdot 3 = 864 + 18 = 882.$$
$$6x(4xy + 1) = 6 \cdot 3(4 \cdot 3 \cdot 4 + 1) = 18(49) = 882. \text{ It checks.}$$

Factoring provides an alternate way of simplifying some fractions.

Example 4 Simplify $\dfrac{5n^2 + 3n}{n}$.

Solution 1 Factor the numerator and simplify the fraction.

$$\frac{5n^2 + 3n}{n} = \frac{n(5n + 3)}{n}$$
$$= 5n + 3$$

Solution 2 Separate the given expression into the sum of two fractions.

$$\frac{(5n^2 + 3n)}{n} = \frac{5n^2}{n} + \frac{3n}{n}$$

Divide.
$$= 5n + 3$$

Check The solutions give the same answer, so they check each other.

Questions

Covering the Reading

1. **a.** How many operations are needed to calculate
   $40x^4 + 40x^3 + 40x^2 + 40x + 40$?
   b. How many operations are needed to calculate
   $40(x^4 + x^3 + x^2 + x + 1)$?
   c. Which is more efficient, part a or part b?
   d. The process that converts the expression in part a to the expression in part b is called __?__.

2. What factoring question and answer are suggested by
   $3t(t^2 + 4) = 3t^3 + 12t$?

In 3–5, copy and complete.

3. $8x^3 + 40x = 8x(\underline{\ ?\ } + \underline{\ ?\ })$.

4. $4p^2 - 3p = p(\underline{\ ?\ } - \underline{\ ?\ })$.

5. $7x^2y - 3x - x^3 = (\underline{\ ?\ } - \underline{\ ?\ } - \underline{\ ?\ }) \cdot x$.

In 6–9, **a.** find the largest common factor of the terms of the polynomials;
b. factor the polynomial.

6. $27b^3 - 27c^3 + 27bc$ 7. $23a^3 + a^2$

8. $15x^2 - 21xy$ 9. $30xy^4 - 12y^3 + 18x^3y^5$

In 10 and 11, simplify.

10. $\dfrac{4x^2 - 2x}{x}$ 11. $\dfrac{12n^2 + 15n}{3n}$

Applying the Mathematics

12. **a.** Factor $3y^2 + 5y$.
    b. If a rectangle has area $3y^2 + 5y$, and its length is y, what is its width?

$5y$ | Area = $100y^3 - 55y^2 + 30y$

L

13. The area of the rectangle above is $100y^3 - 55y^2 + 30y$. Write an expression for the length L of the rectangle.

14. *Multiple choice* One factor of $6n^2 + 12n$ is $6n$. The other factor is a
(a) monomial   (b) binomial   (c) trinomial.

15. *True or false* If n is an integer, $n^2 + n$ can be written as the product of two consecutive integers.

16. Computers often have to do millions of calculations, so anything that reduces the number of calculations saves money. The following program asks the computer to evaluate

$$50x^4 + 50x^3 + 50x^2 + 50x + 50$$

for the 20,000 values of x from 1.00001 to 1.20000, increasing by steps of 0.00001.

```
10  FOR X = 1.00001 TO 1.2 STEP 0.00001
20    LET P = 50 * X ^ 4 + 50 * X ^ 3 + 50 * X ^ 2 + 50 * X + 50
30  NEXT X
40  END
```

Comment: We ignore other statements that would analyze or print values, but which are not necessary for this question.
a. If, on the average, each operation takes the computer a millionth of a second, how long will it take this program to run?

b. If the expression in line 20 is rewritten in factored form, how long will it take the program to run?
c. If each second of running time costs \$.25, how much will factoring save?

In 17 and 18, simplify.

17. $\dfrac{-3x^2y + 6xy - 9xy^2}{3xy}$ **18.** $\dfrac{28n^2 + 42n + 7}{14}$

19. On each birthday from age 16, Kate receives \$75 from her parents. She saves the money and gets 8% interest a year. How much will she have when she turns 21? *(Lesson 10-1)*

20. Add $3x^2 + 2x - 5$ to $8x^2 + 4 + 3x$. *(Lesson 10-2)*

21. Subtract $3x^2 + 2x - 5$ from $8x^2 + 4 + 3x$. *(Lesson 10-2)*

In 22–24, multiply and simplify. *(Lesson 10-3)*

22. $(3x + 5) \cdot x + (3x + 5) \cdot 2$

23. $8v^2(v + 1) - 4v(v^2 - 1)$

24. $4(x^2 - 3x + 1) - 2(x^2 + 5x + 3)$

In 25 and 26, find x if $AB = 12$. *(Lesson 2-6)*

25.

26.

27. On the third day of a diet, a person weighs 93.4 kg. On the 10th day, the person weighs 91.2 kg. What has been the average rate of change in weight per day? *(Lesson 8-1)*

In 28–30, solve for b.

28. $y = mx + b$ *(Lesson 2-6)*   **29.** $bx + c = d$ *(Lesson 6-1)*

30. $ax + by = c$ *(Lesson 6-1)*

31. Insert the following lines in the program of Question 16.

```
23   IF ABS(X−1.1)<0.00001 THEN PRINT X, P
26   IF ABS(X−1.2)<0.00001 THEN PRINT X, P
```

a. Run the program, timing how long it takes to run to the nearest tenth of a second.

b. Change line 20 to factored form and time the modified program.

c. Powering takes longer for computers to do than multiplication. Rewrite line 20 as follows:

```
20   LET P = 50 * (X * X * X * X + X * X * X + X * X + X + 1)
```

Run this modified program and time it.

d. Which is fastest and by how much?

10-5

Multiplying Polynomials

The Area Model for Multiplication provides a way of picturing multiplication. In this lesson, the area model is used to illustrate how to multiply two polynomials with many terms. For instance, to multiply $a + b + c + d$ by $x + y + z$, draw a rectangle with length $a + b + c + d$ and width $x + y + z$.

	a	b	c	d
x	ax	bx	cx	dx
y	ay	by	cy	dy
z	az	bz	cz	dz

The area of the biggest rectangle equals the sum of the areas of the twelve little rectangles.

Total area =
$$ax + ay + az + bx + by + bz + cx + cy + cz + dx + dy + dz$$

But the area of the biggest rectangle also equals the product of its length and width.

$$= (a + b + c + d) \cdot (x + y + z)$$

The Distributive Property can be used to justify this same result. Distribute the chunk $(x + y + z)$ over $(a + b + c + d)$ to get:

$$(a + b + c + d) \cdot (x + y + z) =$$
$$a(x + y + z) + b(x + y + z) + c(x + y + z) + d(x + y + z)$$

Four more applications of the Distributive Property lead to the same expansion found above.

$$= ax + ay + az + bx + by + bz + cx + cy + cz + dx + dy + dz$$

Because of the multiple use of the Distributive Property we call this an instance of the **Extended Distributive Property.**

Extended Distributive Property:

To multiply two sums, multiply each term in the first sum by each term in the second sum.

If one polynomial has m terms and the second n terms, there will be mn terms in their product. When possible, you should simplify the product by combining like terms.

Example 1 Multiply $(5x^2 + 4x + 3)(x + 7)$.

> **Solution** Multiply each term in the first polynomial by each in the second. There will be six terms in the product.
>
> $$= \mathbf{5x^2} \cdot x + \mathbf{5x^2} \cdot 7 + \mathbf{4x} \cdot x + \mathbf{4x} \cdot 7 + \mathbf{3} \cdot x + \mathbf{3} \cdot 7$$
> $$= 5x^3 + 35x^2 + 4x^2 + 28x + 3x + 21$$
>
> Now simplify by adding or subtracting like terms.
>
> $$= 5x^3 + 39x^2 + 31x + 21$$
>
> **Check** Let $x = 2$. The check is left to you as Question 7.

In these long problems it helps if you are neat and precise. Be extra careful with problems involving negatives or several variables.

Example 2 Multiply $(3x + y - 1)(x - 5y + 8)$.

> **Solution** Each term of $x - 5y + 8$ must be multiplied by $3x$, y, and -1. There will be nine terms.
>
> $$= \mathbf{3x} \cdot x + \mathbf{3x} \cdot -5y + \mathbf{3x} \cdot 8 + \mathbf{y} \cdot x + \mathbf{y} \cdot -5y + \mathbf{y} \cdot 8 + \mathbf{-1} \cdot x + \mathbf{-1} \cdot -5y + \mathbf{-1} \cdot 8$$
> Watch the signs!
>
> $$= 3x^2 - 15xy + 24x + xy - 5y^2 + 8y - x + 5y - 8$$
> $$= 3x^2 - 14xy + 23x - 5y^2 + 13y - 8$$
>
> **Check** A quick check of the coefficients can be found by letting all variables equal 1.
> Does $(3 + 1 - 1)(1 - 5 + 8) = 3 - 14 + 23 - 5 + 13 - 8$?
> Does $3 \cdot 4 = 12$? Yes.
> A better check requires using different values for both x and y, but if you have made an error, this quick check may find it.

After simplifying, the product of two polynomials can be a polynomial with fewer terms than one or both factors.

Example 3 Multiply $x^2 - 2x + 2$ by $x^2 + 2x + 2$.

Solution Each term of $x^2 + 2x + 2$ must be multiplied by x^2, $-2x$, and 2. Again there will be nine terms.

$$(x^2 - 2x + 2)(x^2 + 2x + 2)$$
$$= x^2(x^2 + 2x + 2) - 2x(x^2 + 2x + 2) + 2(x^2 + 2x + 2)$$
$$= x^4 + 2x^3 + 2x^2 - 2x^3 - 4x^2 - 4x + 2x^2 + 4x + 4$$
$$= x^4 + 4$$

Check Let $x = 10$. Then $x^2 - 2x + 2 = 82$ and $x^2 + 2x + 2 = 122$.
Now $82 \cdot 122 = 10{,}004$, which is $x^4 + 4$.
Ten is a nice value to use in checks, because powers of 10 are so easily calculated.

Questions

Covering the Reading

1. a. What multiplication is pictured below?
   b. Do the multiplication.

2. State the Extended Distributive Property.

In 3–6, multiply and simplify.

3. $(y^2 + 7y + 2)(y + 6)$

4. $(5c - 4d + 1)(c - 7d)$

5. $(m^2 + 10m + 3)(3m^2 - 4m - 2)$

6. $(x + 1)(2x + 3)$

7. Finish the check of Example 1.

8. Check Example 2 by letting $x = 10$ and $y = 2$.

9. Multiply $x^2 + 4x + 8$ by $x^2 - 4x + 8$.

10. Multiply $(n - 3)(n + 4)(2n + 5)$ by first multiplying $n - 3$ by $n + 4$, then multiplying their product by $2n + 5$.

Applying the Mathematics

In 11–13, multiply and simplify.

11. $(7 - 3x)(7 + 3x)$.

12. $(a + b + c)(a + b - c)$

13. $(x + 4)(x + 4) - (8x - 16)$

14. In Example 1, a 2nd degree polynomial is multiplied by a 1st degree polynomial. The product is a 3rd degree polynomial. Explain why this will always happen.

15. a. Show by a check that this multiplication is wrong.
$$(5x + 3)(5x - 2) = 25x^2 - 6$$

   b. Is there any value of x for which the multiplication in part a is correct?

Review

16. Suppose x is either 7, 6, 5, 3, -2, or -8. Find x given the clues.
Clue 1: $x > $ -3.
Clue 2: x is not the degree of $a^3 + 4a^2 + 7$.
Clue 3: x is not the coefficient of b^2 in the polynomial $4b^4 + 6b^2 - 3b + 9$.
Clue 4: x is not the number of terms in $2a^2 + a - 3b + c - 5f$.
(Lesson 10-1)

In 17–20, factor. *(Lesson 10-4)*

17. $4v^2 - 2v$

18. $11x^3 + 33x^2 + 22x$

19. $3x^2y^3 + 3xy^3$

20. $8m^2 - 4$

21. Subtract $3x^2 + 5$ from $2x^2 - 3x + 40$. *(Lesson 10-2)*

22. Graph $A \cup B$ on a number line where $A = $ set of solutions to $-4w > 100$, and $B = $ set of solutions to $3w + 8 > $ -40. *(Lessons 3-5, 6-7)*

23. If the area of a square is 64 square kilometers, find the length of a diagonal. *(Lesson 7-5)*

24. *Skill sequence* Solve. *(Lesson 6-7)*
   a. $4z - 7 = 13$
   b. $4(z + 8) - 7 = 13$
   c. $5(z + 8) - 7 = (z + 8) + 13$

In 25 and 26, simplify. *(Lesson 9-7)*

25. $\dfrac{14a^3b}{6ab^2}$

26. $\dfrac{-150m^5n^8}{100m^6n^3}$

Exploration

27. Multiply each of the following polynomials by $x + 1$.
   a. $x - 1$
   b. $x^2 - x + 1$
   c. $x^3 - x^2 + x - 1$
   d. $x^4 - x^3 + x^2 - x + 1$
   e. Look for a pattern and use it to multiply
   $(x + 1)(x^8 - x^7 + x^6 - x^5 + x^4 - x^3 + x^2 - x + 1)$.

28. Generalize the idea of Question 14.

Multiplying Binomials

In the last lesson, the Area Model for Multiplication was used to represent multiplying two polynomials. Recall that a polynomial with two terms is a binomial. You will multiply binomials often enough that we examine this as a special case.

The rectangle below has length $(a + b)$ and width $(c + d)$.

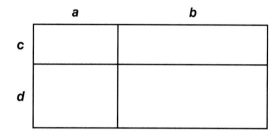

The area of the rectangle = length · width = $(a + b) \cdot (c + d)$. But the area of the rectangle also must equal the sum of the areas of the four small rectangles inside it.

	a	b
c	ac	bc
d	ad	bd

The sum of the areas of the four small rectangles = $ac + ad + bc + bd$. So,

$$(a + b) \cdot (c + d) = ac + ad + bc + bd.$$

Another way to show the pattern above is true is to chunk $(c + d)$ and distribute it over $(a + b)$ as follows:

$$(a + b) \cdot (c + d) = a(c + d) + b(c + d).$$

Now apply the Distributive Property twice more.

$$= ac + ad + bc + bd$$

This pattern shows how to multiply two binomials. The result has four terms. To get them, multiply each term in the first binomial by each term in the second binomial. The binomials $(a + b)$ and $(c + d)$ are each factors of the polynomial $ac + ad + bc + bd$.

Here is a way to remember which multiplications you have done. The name of the algorithm and the face might help.

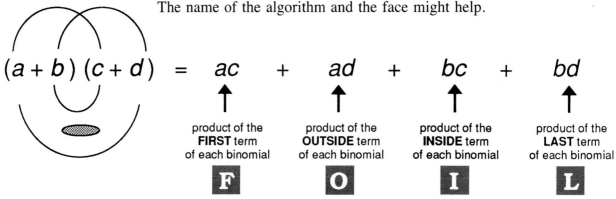

$$(a + b)(c + d) = ac + ad + bc + bd$$

product of the **FIRST** term of each binomial — **F**

product of the **OUTSIDE** term of each binomial — **O**

product of the **INSIDE** term of each binomial — **I**

product of the **LAST** term of each binomial — **L**

Example 1   Multiply $(m + 4)(m + 3)$.

Solution

$$\begin{array}{ccccc} & F & O & I & L \\ (m + 4)(m + 3) = & m \cdot m & + m \cdot 3 & + 4 \cdot m & + 4 \cdot 3 \end{array}$$
$$= m^2 + 3m + 4m + 12$$
$$= m^2 + 7m + 12$$

Check   Let $m = 10$. Then $m + 4 = 14$, $m + 3 = 13$, and $m^2 + 7m + 12 = 182$.
It checks, because $14 \cdot 13 = 182$.

After multiplying two binomials, you should simplify the product whenever possible by adding like terms. This was the last step in Example 1. If the binomials themselves involve expressions with like terms, then you will always be able to simplify the product. This happens even when subtraction is involved. You must, however, be careful with signs.

Example 2   Multiply $(2x - 5)(6x + 1)$.

Solution   Recall that $2x - 5 = 2x + \text{-}5$.

$$(2x - 5)(6x + 1) \qquad\qquad \text{FOIL algorithm}$$
$$= 2x \cdot 6x + 2x \cdot 1 - 5 \cdot 6x - 5 \cdot 1$$
$$= 12x^2 + 2x - 30x - 5$$
$$= 12x^2 - 28x - 5.$$

Check   We let $x = 3$ this time.
Does $(2 \cdot 3 - 5)(6 \cdot 3 + 1) = 12 \cdot 3^2 - 28 \cdot 3 - 5$?
Yes, each side equals 19.

Examples 1 and 2 illustrate an important property. The product of two first degree polynomials in a single variable is a polynomial of the second degree. This is true even when there is more than one variable in the polynomial, as Example 3 shows.

Example 3   Multiply $(7x + 5y)(x - 4y)$.

Solution
$$(7x + 5y)(x - 4y) = \overset{F}{7x \cdot x} + \overset{O}{7x \cdot -4y} + \overset{I}{5y \cdot x} + \overset{L}{5y \cdot -4y}$$
$$= 7x^2 - 28xy + 5yx - 20y^2$$

Notice that $-28xy$ and $5yx$ are like terms.

$$= 7x^2 - 23xy - 20y^2$$

Check   Let $x = 2$ and $y = 3$.
Does $(7 \cdot 2 + 5 \cdot 3)(2 - 4 \cdot 3) = 7 \cdot 2^2 - 23 \cdot 2 \cdot 3 - 20 \cdot 3^2$?
Does $\qquad 29 \cdot -10 \qquad = 28 - 138 - 180$?
Yes, each side equals -290.

You have seen that the product of two binomials generally has four terms, as in FOIL. After simplifying, the products in Examples 1–3 have three terms. Sometimes, after simplifying, there are only two terms.

Example 4   Multiply $4x - 3$ by $4x + 3$.

Solution
$$(4x - 3)(4x + 3) = \overset{F}{4x \cdot 4x} + \overset{O}{4x \cdot 3} - \overset{I}{3 \cdot 4x} - \overset{L}{3 \cdot 3}$$
$$= 16x^2 + 12x - 12x - 9$$
$$= 16x^2 - 9$$

Questions

Covering the Reading

1. a. What multiplication is pictured at the right?

b. Do the multiplication.

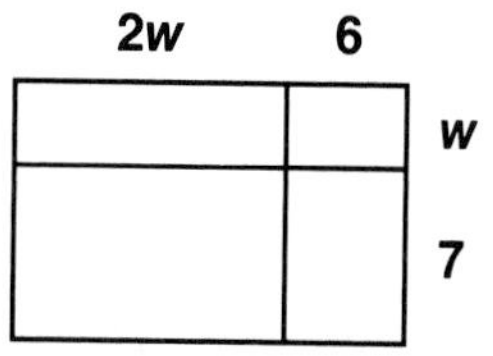

2. In the FOIL algorithm, what do the letters F, O, I, and L stand for?

In 3–10, multiply. Then simplify, if possible.

3. $(a + b)(c + d)$

4. $(2x - 3)(4y - 4)$

5. $(y + 5)(y + 6)$

6. $(3x + 2)(5x - 1)$

7. $(k - 3)(9k + 8)$

8. $(3m + 2n)(7m - 6n)$

9. $(a + b)(a - b)$

10. $(2x - 3y)(2x + 3y)$

Applying the Mathematics

11. a. One student is selected from the freshman or sophomore class as co-chairperson for the school dance committee. A second student is selected from the junior or senior class as co-chairperson. How many such twosomes are possible in a school with

550 freshmen,
500 sophomores,
450 juniors,
and 400 seniors?

b. Repeat part a, if there are f freshmen, p sophomores, j juniors, and s seniors.

12. a. Complete the following multiplication by using the FOIL algorithm. Simplify your answer.

$$4\tfrac{1}{2} \cdot 9\tfrac{3}{4} = (4 + \tfrac{1}{2})(9 + \tfrac{3}{4})$$

b. Check by converting to decimals and using a calculator.

13. a. Multiply $(x^2 + 2)(x^3 - 1)$.
b. Check your answer by letting $x = 4$.

In 14–16, multiply. Then simplify, if possible.

14. $(5c + 2)^2$ [Hint: $(5c + 2)^2 = (5c + 2)(5c + 2)$]

15. $n(n + 1)(n + 2)$ [Hint: Multiply $n + 1$ by $n + 2$ first]

16. $(2x - 3)(2x - 4)(2x - 5)$

Review

In 17–20, multiply and then simplify. *(Lessons 6-3, 10-5)*

17. $75(4q - 3r)$

18. $y(4y^2 - 3y + 2)$

19. $wz^3(4w^2z - 3wz^2 + 2)$

20. $(x^2 - 2)(x^2 + 3x + 2)$

21. Simplify $5(x^2 - 7) - 3(x^2 - 2x - 1) + 4(3x^2 + x - 7)$. *(Lesson 10-3)*

22. a. Simplify $(4x^2 - 7x - 8) + (6x^2 + 2x - 9)$.
b. Simplify $(4x^2 - 7x - 8) - (6x^2 + 2x - 9)$. *(Lesson 10-2)*

In 23 and 24, use this formula for the height h reached by a rocket after t seconds, when fired straight up from a launching pad 6 feet off the ground,

$$h = 6 + 96t - 16t^2.$$

23. The rocket reaches its maximum height at 3 seconds. What is its maximum height? *(Lesson 10-1)*

24. a. Find the height of the rocket after 7 seconds.
   b. What does this answer mean? *(Lesson 10-1)*

25. Factor $3x^3y + 2xy^3$. *(Lesson 10-4)*

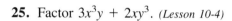

26. A point is selected at random from the big square at the left. Find the probability that the selected point is in the colored area. *(Lesson 5-5)*

27. Write an equation for the line which passes through the points $(-3, 7)$ and $(-9, 4)$. *(Lesson 8-6)*

28. Solve in your head. *(Lesson 7-8)*
   a. $3(k + 5) = 0$
   b. $6(m - \frac{1}{2}) = 0$
   c. $42(512 - x) = 0$

Exploration

29. The largest rectangular solid at right has length $(a + b)$, width $(c + d)$, and height $(e + f)$. Give at least two ways of expressing the volume of the box.

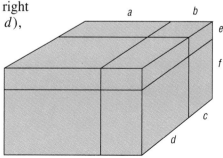

Squares of Binomials

The expression $(a + b)^2$ (read "a plus b, the quantity squared") is the square of the binomial $(a + b)$. So it can be written as the product $(a + b)(a + b)$. This product can be **expanded** using the Extended Distributive Property or its special case, the FOIL algorithm.

$$(a + b)(a + b) = a^2 + ab + ba + b^2$$
$$= a^2 + 2ab + b^2 \ (ab \text{ and } ba \text{ are like terms.})$$

Example 1 The area of a square with side $5n + 4$ is $(5n + 4)^2$. Expand this binomial.

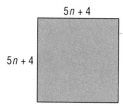

5n + 4

5n + 4

Solution 1 Change the square to multiplication.

$$(5n + 4)^2 = (5n + 4)(5n + 4)$$
$$= 25n^2 + 20n + 20n + 16$$
$$= 25n^2 + 40n + 16$$

Solution 2 Use the pattern for the square of $a + b$ with $a = 5n$ and $b = 4$.

$$(a + b)^2 = a^2 \quad + \quad 2ab \quad + b^2$$
$$(5n + 4)^2 = (5n)^2 + 2 \cdot 5n \cdot 4 + 4^2$$
$$= 25n^2 + \quad 40n \quad + 16$$

The square of a difference, $(a - b)^2$, can be expanded in the same way.

$$(a - b)^2 = (a - b)(a - b)$$
$$= a^2 - ab - ba + b^2$$
$$= a^2 - 2ab + b^2$$

504

Notice that after simplifying, the square of a binomial has three terms. It is a trinomial. Trinomials of the form $a^2 + 2ab + b^2$ and $a^2 - 2ab + b^2$ are called **perfect square trinomials** because each is the result of squaring a binomial.

Perfect Square Trinomial Patterns:

$$(a + b)^2 = a^2 + 2ab + b^2$$
$$(a - b)^2 = a^2 - 2ab + b^2$$

You need to know these patterns, and you should realize that they can be derived from the multiplication of binomials.

The algebraic description is short but many people remember these patterns in words.

The square of a binomial is the sum of
    (1) the square of its first term
    (2) twice the product of its terms
and (3) the square of its last term.

Example 2 Expand $(w - 3)^2$.

Solution 1 Follow the Perfect Square Trinomial Pattern. Here $a = w$ and $b = 3$.

$$(w - 3)^2 = w^2 - 2 \cdot w \cdot 3 + 3^2$$
$$= w^2 - 6w + 9$$

Solution 2 Change the square to multiplication of $w - 3$ by itself.

$$(w - 3)^2 = (w - 3)(w - 3)$$
$$= w^2 - 3w - 3w + 9$$
$$= w^2 - 6w + 9$$

Check Test a special case. Let $w = 5$. Then

$$(w - 3)^2 = (5 - 3)^2 = 2^2 = 4.$$
$$w^2 - 6w + 9 = 5^2 - 6 \cdot 5 + 9 = 25 - 30 + 9 = 4$$

It checks.

The Perfect Square Trinomial Patterns can be applied to find squares of certain numbers without a calculator. With practice, you can do these computations in your head.

Example 3 Compute: **a.** 43^2; **b.** 79^2; **c.** $(1.08)^2$.

Solution **a.** Think of 43 as 40 + 3.

$$43^2 = (40 + 3)^2$$
$$= 40^2 + 2 \cdot 40 \cdot 3 + 3^2$$
$$= 1600 + 240 + 9$$
$$= 1849$$

b. Think of 79 as 80 − 1.

$$79^2 = (80 - 1)^2$$
$$= 80^2 - 2 \cdot 80 \cdot 1 + 1^2$$
$$= 6400 - 160 + 1$$
$$= 6241$$

c. Think of 1.08 as 1 + .08.

$$(1.08)^2 = (1 + .08)^2$$
$$= 1^2 + 2 \cdot 1 \cdot .08 + (.08)^2$$
$$= 1 + .16 + .0064$$
$$= 1.1664$$

Questions

Covering the Reading

1. What is a perfect square trinomial?

2. Expand $(x + y)^2$.

3. Write an expression for the area of a square with side $3x + 7$,
   a. as the square of a binomial.
   b. as a perfect square trinomial.

4. **a.** Expand $(m - 6)^2$.
   b. Check your answer by letting $m = 10$.

5. Write $(x - y)^2$ as a perfect square trinomial.

In 6 and 7, give the middle term after expanding the binomials.
   6. $(w + 5)^2$ **7.** $(4p - 3)^2$

In 8–11, compute as in Example 3.

8. $(41)^2$ **9.** $(1.02)^2$

10. $(29)^2$ **11.** $(37)^2$

Applying the Mathematics

In 12–14, expand and simplify.

12. $(1 - x)^2$ **13.** $4(9 + y)^2$ **14.** $(a + b)^2 - (a - b)^2$

15. a. Evaluate $(x + y)^2$ and $x^2 + y^2$ for $x = 2$ and $y = 7$.
    b. By how much do the values in part a differ?
    c. Evaluate $2xy$ for $x = 2$, $y = 7$. Compare the answer with part b.

16. Show that $(x - 3)^2$ and $(3 - x)^2$ have the same expansion.

17. An object is falling at a rate of $32t + 20$ feet per second. If it falls for $t + 3$ seconds, write an expression for how far it drops.

18. Expand $(x + 2)^3$ by multiplying $(x + 2)(x + 2)(x + 2)$.

Review

In 19 and 20, multiply and simplify. *(Lesson 10-6)*

19. $(z - 11)(z + 8)$ **20.** $(2c - 7d)(c + 5d)$

21. *Skill sequence* Multiply and simplify. *(Lessons 6-3, 10-2, 10-3)*
    a. $4x^3(13x - 2)$
    b. $(4x^3 + 6x + 1)(13x - 2)$
    c. $4x^2(13x - 2) - (4x^2 - 7x + 1)$

22. Simplify $(3x^2 + 7x - 2) - 3(x^2 - 3x + 1)$ and give the degree of your answer. *(Lessons 10-1, 10-3)*

23. Factor $60x^2y + 60xy^2 + 60xy$. *(Lesson 10-4)*

24. *Multiple choice* It was reported in *USA Today* on July 12, 1985 that the Federal National Mortgage Association's first quarter earnings increased 363% to 11.1 million in the second quarter. Which equation can be used to find the first quarter earnings E in millions? *(Lesson 5-4)*
    (a) $3.63(11.1) = E$ (b) $4.63(11.1) = E$
    (c) $3.63E = 11.1$ (d) $4.63E = 11.1$

25. Solve $2x + 1 = 0$ in your head. *(Lesson 6-1)*

Exploration

26. The square of 41 is 81 more than the square of 40.
    The square of 51 is 101 more than the square of 50.
    The square of 61 is 121 more than the square of 60.
    Describe the general pattern using variables by filling in the blanks.
    The square of __?__ is __?__ more than the square of $10x$.

Recognizing Perfect Squares

Here is a table of the first twenty perfect square whole numbers.

n	1	2	3	4	5	6	7	8	9	10
n^2	1	4	9	16	25	36	49	64	81	100

n	11	12	13	14	15	16	17	18	19	20
n^2	121	144	169	196	225	256	289	324	361	400

You are probably able to recognize many of the numbers in the second row as perfect squares the moment you see them, but what about numbers you don't recognize? Are 528 and 3325 perfect squares? A calculator can help you answer these questions, but it can be handy to recognize perfect squares in your head—or at least to recognize numbers which cannot be perfect squares.

Look at the table above. All the perfect square numbers end with one of these digits:

$$1, 4, 5, 6, 9, 0.$$

It can be shown that *all* perfect square whole numbers end with one of these digits. So any whole number ending with 2, 3, 7, or 8 *cannot* be a perfect square.

There is another pattern hinted at in the table above. Note that $5^2 = 25$ and $15^2 = 225$. Here are some other squares of numbers ending in 5.

25^2	35^2	125^2
625	1225	15,625

All of these perfect squares end with 25. Also, the part of the number before the last two digits is always even and is the product of two consecutive whole numbers.

In 625, $6 = 2 \cdot 3.$
In 1225, $12 = 3 \cdot 4.$
In 15,625 $156 = 12 \cdot 13.$

Example 1 Determine if each number is a perfect square.
  a. 528   **b.** 3325   **c.** 4225

Solution
  a. Since 528 ends in 8, it cannot be a perfect square.
  b. 3325 ends in 25, but 33 is odd, so 3325 is not a perfect square.
  c. 4225 ends in 25, and 42 = 6 · 7, so 4225 is a perfect square. It is the square of 65.

Recognizing perfect square trinomials can also be a useful skill. If you switch the sides of each equation in the perfect square trinomial pattern, you get $a^2 + 2ab + b^2 = (a + b)^2$ and $a^2 - 2ab + b^2 = (a - b)^2$. The binomials $a + b$ and $a - b$ in the pattern are square roots of the original trinomials.

Recall, whenever a binomial is squared, the result is a trinomial in which:

  (1) The first and last terms are the squares
      of the terms of the binomial.
  (2) The middle term is twice the product of
      the terms of the binomial.

Example 2 Write the perfect square trinomial $9m^2 - 12mn + 4n^2$ as the square of a binomial.

Solution The first and last terms are the squares of 3m and 2n. Since the middle term of the trinomial is subtracted, each factor will be a difference. So $9m^2 - 12mn + 4n^2 = (3m - 2n)^2$.

Check Expand $(3m - 2n)^2$.

$$(3m - 2n)(3m - 2n) = 9m^2 - 6mn - 6mn + 4n^2$$
$$= 9m^2 - 12mn + 4n^2$$

It checks.

You must be sure to check your answers to be certain that all the signs are correct. The Perfect Square Trinomial Pattern will also tell you when a trinomial is *not* a perfect square.

Example 3 Is the trinomial $m^2 + 4m + 36$ a perfect square?

Solution The first and third terms are the perfect squares of m and 6. If the trinomial could be factored using the method in Example 2, it would equal $(m + 6)^2$. But checking,

$$(m + 6)^2 = m^2 + 12m + 36.$$

This does not equal $m^2 + 4m + 36$, so $m^2 + 4m + 36$ is not a perfect square trinomial.

Recognizing perfect squares can make solving some equations a simple task.

Example 4 Solve $x^2 + 6x + 9 = 2025$.

Solution Recognize that $x^2 + 6x + 9$ is a perfect square trinomial, the square of $x + 3$. Also note that 2025 is a perfect square, the square of 45. So the equation is

$$(x + 3)^2 = 2025.$$

Take a square root of each side. This gives two possibilities:

$$x + 3 = 45 \quad \text{or} \quad x + 3 = -45$$
$$x = 42 \quad \text{or} \quad x = -48$$

There are two solutions, 42 and -48.

Check Does $\quad 42^2 + 6 \cdot 42 + 9 = 2025$?
Yes. $\quad 1764 + \quad 252 \quad + 9 = 2025$

Does $(-48)^2 + 6 \cdot -48 + 9 = 2025$?
Yes. $\quad 2304 - \quad 288 \quad + 9 = 2025$

Questions

Covering the Reading

1. All perfect square whole numbers end with one of the digits __?__, __?__, __?__, __?__, __?__, or __?__.

In 2–5, determine if the number is a perfect square.

2. 324   **3.** 2125

4. 7225   **5.** 3063

6. Write $4m^2 - 4mn + n^2$ as the square of a binomial.

7. a. $x^2 + 10x + 25$ is a perfect square trinomial. What binomial squared equals $x^2 + 10x + 25$?
   b. Solve $x^2 + 10x + 25 = 9$.

8. Solve $(t - 5)^2 = 361$.

9. Refer to the table at the beginning of the lesson. The number in the tens place of the perfect squares ending with 6 is *odd*. It can be shown that no perfect square ending in 6 can have an even number in the tens place.
   a. Use this fact to decide which of the following might be perfect squares.
      (i) 576   (ii) 686   (iii) 776
   b. Test your answers to part a in order to decide if *all* whole numbers ending in 6 with an odd digit in the tens place are perfect squares.

In 10 and 11, determine whether or not the trinomial is a perfect square.

10. $4x^2 + 24x + 9$ **11.** $9w^2 + 18wz + 9z^2$

12. Solve $9t^2 - 12t + 4 = 100$ by rewriting the left side as the square of a binomial and then chunking.

13. This computer program uses a special case to test if $(4x - 2y)^2 = 16x^2 - 8xy + 4y^2$.

```
10 PRINT "ENTER VALUES FOR X AND Y"
20 INPUT X,Y
30 LET S = (4*X − 2*Y)*(4*X − 2*Y)
40 LET T = 16*X*X − 8*X*Y + 4*Y*Y
50 PRINT "S =" S, "T =" T
60 IF S = T THEN PRINT "THEY ARE EQUAL."
70 IF S<>T THEN PRINT "THEY ARE NOT EQUAL."
80 END
```

(Comment: S<>T means S<T or S>T.)

   a. What binomial does the program test?
   b. What will the computer print if you run the program for $x = 1$ and $y = 5$?

In 14–16, expand and simplify. *(Lesson 10-7)*

14. $(2p - q)^2$ **15.** $\frac{1}{2}(m + 3n)^2$ **16.** $(x^2 - y^2)^2 + 2x^2y^2$

In 17 and 18, multiply. *(Lesson 10-6)*

17. $(a - b)(a + b)$ **18.** $(2c - 3d)(2c + 3d)$

19. Write a simplified expression for the total area of the rectangles at the right. *(Lesson 10-3)*

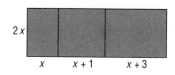

$2x$ $\quad$ x $\quad$ $x+1$ $\quad$ $x+3$

20. The formula $d = 0.042s^2 + 1.1s$ gives the approximate number of feet d needed to stop a car on a dry concrete road (including both reaction distance and braking distance) if the speed of the car is s miles per hour. How much farther will a car travel before stopping, if it is going 65 miles per hour instead of 55 miles per hour? *(Lesson 10-2)*

21. Jules bought 2 raffle tickets, one for a boat and one for a stereo. 5000 raffle tickets for the boat and 2250 tickets for the stereo were sold. What is the probability that Jules will win both the boat and the stereo? *(Lesson 4-8)*

22. If Aurora invested $1200 for three years at an annual yield of 5.7%, how much would she have at the end of three years? *(Lesson 9-1)*

23. Match each equation with its graph. *(Lesson 8-6)*

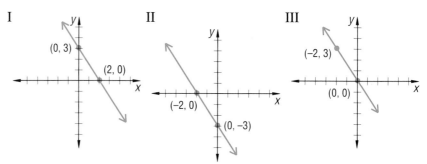

I $\qquad$ II $\qquad$ III

(0, 3) $\quad$ (2, 0) $\qquad$ (−2, 0) $\quad$ (0, −3) $\qquad$ (−2, 3) $\quad$ (0, 0)

a. $3x + 2y + 6 = 0$
b. $3x + 2y = 6$
c. $3x + 2y = 0$

24. Which holds more: a cube with edges of length 4 or a box with dimensions 3 by 4 by 5? *(Lesson 4-1)*

Exploration

25. a. How many perfect square whole numbers are there under 100? (Include zero.)
b. How many perfect square whole numbers are there under 500?
c. How many perfect square whole numbers are there under any given number n?

Difference of Squares

Consider two numbers n and 6. Multiply their sum $n + 6$ by their difference $n - 6$.

$$(n + 6)(n - 6) = n^2 + 6n - 6n - 36$$

The two middle terms are opposites, so add to 0.

$$= n^2 - 6^2$$

You have seen this pattern before. In words, the product of the sum and difference of two numbers is the **difference of squares** of the two numbers.

Difference of Two Squares Pattern:

$$(a + b)(a - b) = a^2 - b^2.$$

Example 1 Multiply $(3y + 7)(3y - 7)$.

Solution The binomial factors are the sum and difference of the same terms. Use the difference of squares pattern with $a = 3y$ and $b = 7$.

$$(3y + 7)(3y - 7) = (3y)^2 - 7^2$$
$$= 9y^2 - 49$$

Check Let $y = 4$.
$(3 \cdot 4 + 7)(3 \cdot 4 - 7) = 19 \cdot 5 = 95$.
Does $9 \cdot 4^2 - 49 = 95$? Yes

The Difference of Squares pattern can help you find the product of some pairs of numbers in your head.

Example 2 Compute $53 \cdot 47$.

Solution Find the mean of 53 and 47, which is 50.
Write 53 and 47 as a sum and difference involving 50.

$$53 \cdot 47 = (50 + 3)(50 - 3)$$

Notice that 3 is the other number involved in both cases. Use the Difference of Squares Pattern to do the binomial multiplication.

$$(50 + 3)(50 - 3) = 50^2 - 3^2$$
$$= 2500 - 9$$
$$= 2491$$

The shortcut in Example 2 works best when you can find the mean with little work and when the mean is easy to square. For example, to multiply 41 by 39:

(1) The mean is 40.
(2) 41 and 39 are both 1 away from 40.
(3) The product is $40^2 - 1^2 = 1600 - 1 = 1599$.

The shortcut uses the idea that $(40 + 1)(40 - 1) = 40^2 - 1^2$.

By reversing the pattern for the difference of squares you can factor a binomial in which one perfect square term is *subtracted* from another.

Example 3 Write the difference of squares $r^2 - 9s^2$ as the product of two binomials.

Solution Use the Difference of Two Squares Pattern with $a = r$ and $b = 3s$.

$$r^2 - 9s^2 = r^2 - (3s)^2$$
$$= (r + 3s)(r - 3s)$$

Check 1 Multiply. $(r + 3s)(r - 3s) = r^2 - 3rs + 3rs - 9s^2$
$$= r^2 - 9s^2$$

Check 2 Let $r = 5$, $s = 4$. Then

$(r + 3s)(r - 3s) = (5 + 3 \cdot 4)(5 - 3 \cdot 4) = 17 \cdot \text{-}7 = \text{-}119.$
$r^2 - 9s^2 = 5^2 - 9 \cdot 4^2 = 25 - 9 \cdot 16 = 25 - 144 = \text{-}119.$

It checks.

Example 4 Factor the difference of squares $25p^2q^2 - 1$.

Solution $25p^2q^2$ is a perfect square: $25p^2q^2 = (5pq)^2$.
1 is a perfect square: $\qquad\qquad\qquad 1 = 1^2$
So, $\qquad\qquad\qquad\qquad\qquad 25p^2q^2 - 1 = (5pq)^2 - 1^2$
$$= (5pq + 1)(5pq - 1).$$

Check Multiply. $(5pq + 1)(5pq - 1) = 25p^2q^2 - 5pq + 5pq - 1$
$$= 25p^2q^2 - 1$$

Questions

Covering the Reading

1. Fill in the blank: $a^2 - b^2 = (a - b) \cdot \underline{\ ?\ }$.

In 2–4, multiply and simplify.

2. $(x + 13)(x - 13)$ **3.** $(5y - 7)(5y + 7)$

4. $(w + kq)(w - kq)$

5. *Multiple choice*   Which is *not* the difference of two squares?
(a) $9 - w^2$ (b) $25k^2 + 36$
(c) $x^2y^2 - 1$ (d) $121m^2 - n^2$

In 6 and 7, compute using the method in Example 2.

6. $69 \cdot 71$ **7.** $85 \cdot 95$

8. *Multiple choice*   Which cannot be written as a product of two binomial factors?
(a) $x^2 + y^2$ (b) $x^2 - y^2$
(c) $x^2 + 2xy + y^2$ (d) $x^2 - 2xy + y^2$

In 9–12, write each difference of squares as the product of two binomials.

9. $x^2 - 1$ **10.** $a^2 - 64b^2$

11. $36m^2 - 25n^2$ **12.** $16p^2q^2 - 100$

Applying the Mathematics

13. Compute $8\frac{1}{2} \cdot 9\frac{1}{2}$ using the idea of Example 2.

14. Multiply $(x + \sqrt{11})(x - \sqrt{11})$.

15. Multiply and simplify $(\frac{1}{2} + 3m^2)(\frac{1}{2} - 3m^2)$.

16. A rectangle has 3 units more length and 3 units less width than the side of the colored square below.
a. Which has larger area, the square or the rectangle?
b. How much larger is that area?

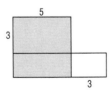

17. a. Write $x^4 - 81$ as the product of two binomials.
b. Write $x^4 - 81$ as the product of three binomials.

18. Write $(x + y)^2 - 16$ as the product of two trinomials.

19. *True or false* $49a^2 - 112a + 16$ is a perfect square trinomial. *(Lesson 10-8)*

20. The area of a square is $4b^2 + 4b + 1$. Find the length of its side. *(Lesson 10-8)*

21. Calculate in your head. *(Lesson 10-7)*
   a. 21^2 **b.** $(1.1)^2$

22. Multiply and simplify $(2x - y)(4x + 5y)$. *(Lesson 10-6)*

23. Factor $-9x^2 + 24x^3 + 18x^4$. *(Lesson 10-4)*

24. Expand $(a + b + c)^2$. *(10-5)*

25. **a.** Simplify $3(4 - 5x^2 + 2x) - 8(2x^2 + 1) + 3(x^2 - 4)$. *(Lesson 10-3)*
   b. Give the degree of the polynomial from part a. *(Lesson 10-1)*

26. Multiply $3b^2(4b^3 - 2by + 7)$. *(Lesson 10-3)*

27. Graph the line $2x + 5y = 30$. *(Lesson 8-8)*

28. Westinghouse Electric Corporation's second quarter earnings rose 12.3% to $143.9 million. What were the company's first quarter earnings? *(Lessons 5-4, 6-3)*

29. Solve **a.** $x = 2x$; **b.** $x > 2x$. *(Lessons 6-6, 6-7)*

30. **a.** Write down three consecutive integers.
   b. Square the second number.
   c. Find the product of the first and the third.
   d. Do this with three more sets of consecutive integers.
   e. What do you notice?
   f. Explain why this will always happen.

LESSON
10-10

The Zero Product Property

Suppose a baton is thrown upward from ground level at a speed of 35 feet per second. If there were no pull of gravity, the distance above the ground after t seconds would be $35t$, and the baton would never come down. Gravity pulls the baton down $16t^2$ feet in t seconds. So the distance d above the ground after t seconds is given by

$$d = 35t - 16t^2.$$

In this situation, how long before the baton hits the ground? At ground level the distance d is zero, so to answer this question you need to solve the equation $35t - 16t^2 = 0$. This equation can be solved using common monomial factoring and the **Zero Product Property.**

Zero Product Property:

For any real numbers a and b, if the product of a and b is 0, then $a = 0$ or $b = 0$.

Numbers may be represented by expressions. So in words, the Zero Product Property is: If two *expressions* multiply to zero, one or the other (or both) must be zero. This property is often combined with chunking to solve equations involving products of binomials or products of a monomial with binomials. Here are three examples.

Example 1   Solve for x: $5(x - 3) = 0$.

Solution 1   Use the Zero Product Property. Either $5 = 0$ or $x - 3 = 0$. We know $5 \neq 0$. So $x - 3 = 0$, which means $x = 3$.

Solution 2   Use the Distributive Property. $5x - 15 = 0$
$$5x = 15$$
$$x = 3$$

Example 2   For what values of y does $(y + 3)(y - 2) = 0$?

Solution   Using the Zero Product Property, either the chunk $(y + 3)$ equals zero, or the chunk $(y - 2)$ equals zero. Solve each equation for y.

$$y + 3 = 0 \quad \text{or} \quad y - 2 = 0$$
$$y = \text{-}3 \quad \text{or} \quad y = 2$$

So y is either 2 or -3.

Check   Let $y = 2$: $(2 + 3)(2 - 2) = 5 \cdot 0 = 0$.
Let $y = \text{-}3$: $(\text{-}3 + 3)(\text{-}3 - 2) = 0 \cdot \text{-}5 = 0$.

Example 3   Solve for m: $3m(4m - 1) = 0$.

Solution   Either the monomial $3m = 0$ or the binomial $(4m - 1) = 0$. Solving each equation for m,

$$3m = 0 \quad \text{or} \quad 4m - 1 = 0$$
$$m = 0 \quad \text{or} \quad 4m = 1$$
$$m = \tfrac{1}{4}.$$

So $m = 0$ or $\tfrac{1}{4}$.

Check   Let $m = 0$. $3 \cdot 0(4 \cdot 0 - 1) = 0 \cdot \text{-}1 = 0$
Let $m = \tfrac{1}{4}$. $3 \cdot \tfrac{1}{4}(4 \cdot \tfrac{1}{4} - 1) = \tfrac{3}{4} \cdot 0 = 0$

Common monomial factoring turns a polynomial into a product of factors. If the product of factors is zero, then the Zero Product Property can be used.

Example 4   Refer to the baton toss at the beginning of this lesson. How long was the baton in the air?

Solution   We need to solve $35t - 16t^2 = 0$.
Factoring, $t(35 - 16t) = 0$.
By the Zero Product Property,

$$\text{either } t = 0 \quad \text{or} \quad 35 - 16t = 0.$$
$$t = 0 \quad \text{or} \quad \text{-}16t = \text{-}35$$
$$t = 0 \quad \text{or} \quad t = \tfrac{\text{-}35}{\text{-}16} \approx 2.19.$$

The solution $t = 0$ means the baton was at ground level at the start. The solution $t \approx 2.19$ means that it hit the ground again after about 2.19 seconds. So it was in the air for about 2.19 seconds.

A word of caution: to use the Zero Product Property, the product must be *zero*. This property does *not* work for an equation such as $(x + 3)(x - 2) = 1$.

Questions

Covering the Reading

1. State the Zero Product Property.

2. For what values of k does $(k + 4)(k - 1) = 0$?

In 3–6, solve.

3. $2x(3x - 5) = 0$

4. $-18x(12 - 5x) = 0$

5. $(y - 15)(9y - 8) = 0$

6. $(p - 3)(2p + 4) = 0$

7. A ball is thrown upward from ground level at 45 feet per second. The distance d above the ground after t seconds is $d = 45t - 16t^2$. After how many seconds will the ball hit the ground?

8. Why can't the Zero Product Property be used on the equation $(w + 1)(w + 2) = 3$?

Applying the Mathematics

In 9 and 10, solve.

9. $4x(x - 11)(2x + 7) = 0$

10. $0 = (.4y - 2.1)(y - 18.62)(5.2 - .3y)$

11. A human cannonball in the circus is fired upward from the cannon at 80 feet per second. In t seconds, the cannonball will be $80t - 16t^2$ feet above where it was fired. In how many seconds will the cannonball reach a net that is the same height as the cannon?

12. *Multiple choice* In the equation $ax^2 + bx = 0$, one solution for x is always
(a) 1
(b) a
(c) b
(d) 0.

13. *Multiple choice* If $xy = 6$, then
(a) $x = 6$ or $y = 6$.
(b) $x = 3$ and $y = 2$.
(c) there is not enough information to tell.

14. Consider the equation $2v^2 = 3v$.
a. subtract $3v$ from both sides;
b. solve using the Zero Product Property.

15. Given the equation $t^2 - 6t + 9 = 9$,
a. add -9 to both sides and solve using the Zero Product Property;
b. recognize the left side as a perfect square and solve by taking square roots.

16. If $a \neq 0$, solve for x: $ax^2 - ax = 0$.

17. The product of four consecutive integers is zero. Find all possible values for the smallest of these integers.

Review

In 18 and 19, factor. *(Lessons 10-4, 10-9)*

18. $14x^2 - 7xy + 28x^3$.   **19.** $49p^2 - q^2$.

In 20–22, multiply and simplify. *(Lessons 6-3, 10-5, 10-9)*

20. $4y(3y - 7)$

21. $(3x + 7)(3x - 7)$

22. $(3z + 7)(5z + q - 8)$

23. Find the area of a square which is $4e + 1$ units on a side. *(Lesson 10-7)*

24. Simplify $\dfrac{13x^2 - 14x}{x}$. *(Lesson 10-4)*

25. A certain glass allows $\frac{9}{10}$ of the light hitting it to pass through. The fraction of light y passing through x thickness of glass is then

$$y = \left(\frac{9}{10}\right)^x.$$

Draw a graph of this equation for $0 \leq x \leq 6$. *(Lesson 9-4)*

26. An inchworm starts at a height of 11 ft on the trunk of an apple tree and crawls down at a constant rate of 4 inches ($\frac{1}{3}$ ft) per hour. After x hours, the inchworm will be y feet above the ground, where $y = -\frac{1}{3}x + 11$.
  a. Give the slope and y-intercept of this equation. *(Lesson 8-2)*
  b. How long will it take the inchworm to reach the ground? *(Lesson 6-1)*

In 27–29, solve. *(Lessons 5-7, 7-4)*

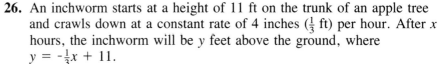

27. $\dfrac{3}{5} = \dfrac{x}{15}$ **28.** $1 = \dfrac{0.6}{m}$ **29.** $\dfrac{d}{9} = \dfrac{4}{d}$

30. In 1986, World Wide Wrench Company increased all their prices by 12% across the board. Their basic wrench now sells for $9.25. What did it sell for in 1985? *(Lessons 5-4, 6-3)*

Exploration

31. Given the equation $xy = 0$,
  a. give five different ordered pairs (x, y) that work in this equation.
  b. Graph this equation on a coordinate plane. Describe this graph in words.
  c. Compare the graph in part b with the graph of $xy = \frac{1}{100}$.

Recognizing Factors of Polynomials

Is $x^2 - 7x + 12$ equal to $(x - 3)(x - 4)$?

It is hard to answer this question just by looking because the trinomial $x^2 - 7x + 12$ has a different appearance than the product of the two binomials $(x - 3)$ and $(x - 4)$.

There are several ways of determining whether expressions are equal. In this lesson, you will study two of these methods, both of which are familiar to you. The first is simply to use the properties of algebra on one expression to make it look like the other. The second is to test a special case.

Example 1   Is $x^2 - 7x + 12$ equal to $(x - 3)(x - 4)$?

Solution 1   Multiply the binomials.

$$(x - 3)(x - 4) = x^2 - 4x - 3x + 12$$
$$= x^2 - 7x + 12$$

The expressions are equal, thus $(x - 3)$ and $(x - 4)$ are factors of $x^2 - 7x + 12$.

Solution 2   Test special cases. Substitute 5 for x in each expression.

$$(5)^2 - 7(5) + 12 = 25 - 35 + 12 = 2$$
$$(5 - 3)(5 - 4) = 2 \cdot 1 = 2$$

Let $x = -2$.

$$(-2)^2 - 7(-2) + 12 = 4 + 14 + 12 = 30$$
$$(-2 - 3)(-2 - 4) = -5 \cdot -6 = 30$$

The special cases work so it appears that the expressions are equal.

Testing a special case can easily show when expressions are *not* equal. Recall that if two expressions are not equal for only one special case, then you have found a *counterexample*. This means the expressions are not equal. But if you can't find a counterexample, it is likely (but not certain) that the expressions are equal.

When faced with a multiple choice question, it is sometimes possible to arrive at the right answer by eliminating wrong ones. This is called **ruling out possibilities**. When you have ruled out all answers but one, then that one is the correct choice. Many people like multiple-choice tests because even if you don't know the correct answer to a question you still have a chance of getting it right.

Example 2 *Multiple choice*   Which expression equals $x^2 + y^2$?
(a) $(x + y)^2$ (b) $(x + y)(x - y)$ (c) $(x - y)^2$ (d) none of (a)–(c)

Solution 1   Rule out answers by evaluating each of the choices using the FOIL algorithm.
(a) $(x + y)^2 = x^2 + 2xy + y^2$ (by the Perfect Square Trinomial Pattern). This does not equal $x^2 + y^2$.
(b) $(x + y)(x - y) = x^2 - y^2$ (by the Difference of Squares Pattern). This does not equal $x^2 + y^2$.
(c) $(x - y)^2 = x^2 - 2xy + y^2$ (again by the Perfect Square Trinomial Pattern). This does not equal $x^2 + y^2$.

None of the expressions equals $x^2 + y^2$, therefore (d) is the correct choice.

Solution 2   Test a special case. Let $x = 2$ and $y = 3$. Then $x^2 + y^2 = 4 + 9 = 13$ and the choices equal
(a) $(2 + 3)^2 = 25$ (b) $(2 + 3)(2 - 3) = $-5
(c) $(2 - 3)^2 = 1$ (d) none of (a)–(c).

Since none of (a)–(c) yields 13, (d) is the correct choice.

Questions

Covering the Reading

1. Name two ways of determining whether expressions are equal.

2. What are the factors of $x^2 - 7x + 12$?

3. Determine if $x^2 + 9x - 10$ is equal to $(x + 10)(x - 1)$ by
a. multiplying the binomials; **b.** testing two special cases.

4. When is testing a special case very useful?

5. In Example 2, why is only one special case needed?

Applying the Mathematics

Questions 6–13 are *multiple choice*.

6. $ax + bx + ay + by =$
(a) $(a + b)(x + y)$ (b) $(a + x)(b + y)$
(c) $a(x + b + y)$ (d) none of (a)–(c)

7. $3x^2 - 5x - 2 =$
(a) $(3x - 1)(x + 2)$ (b) $(3x - 5)(x - 2)$ (c) $(3x + 1)(x - 2)$
(d) $(3x - 2)(x + 1)$ (e) none of (a)–(d)

8. $2x^4 + x^3y + 2xy^2 + y^3 =$
(a) $(2x^2 + y)(x^2 + y^2)$ (b) $(x^3 + y^2)(2x + y)$
(c) $(x + y^2)(2x + y)$ (d) none of (a)–(c)

9. $x^4 - 10x^2 + 18 =$
(a) $(x - 3)(x - 6)$ (b) $(x - 2)(x - 9)$ (c) $(x^2 - 3)(x^2 - 6)$
(d) $(x^2 - 2)(x^2 - 9)$ (e) none of (a)–(d)

10. $4y^2 + 12y + 9 =$
(a) $(4y + 1)(y + 9)$ (b) $(4y + 9)(y + 1)$ (c) $(2y + 1)(2y + 9)$
(d) $(2y + 3)^2$ (e) none of (a)–(d)

11. $25x^2 - 81 =$
(a) $(5x - 9)^2$ (b) $(25x - 1)(x + 81)$ (c) $(5x - 9)(5x + 9)$
(d) $(25x - 81)(x + 1)$ (e) none of (a)–(d)

12. Which expression equals $m + n$?
(a) $\sqrt{m^2 + n^2}$ (b) $\left(\sqrt{m} + \sqrt{n}\right)^2$
(c) $\dfrac{m^2 + n^2}{m + n}$ (d) none of (a)–(c)

13. Sylvia was working on a physics project and had to calculate several estimates using the formula $y^2 - 6y - 7$. She knew that the calculations would go faster if this expression was factored. Which of the following is equal to $y^2 - 6y - 7$?
(a) $(y - 1)(y - 7)$ (b) $(y + 1)(y - 7)$ (c) $(y + 1)(y + 7)$
(d) $(y - 1)(y + 7)$ (e) none of (a)–(d)

Review

14. Foresters in the Allegheny National Forest were asked to estimate the volume of timber in the black cherry trees in the forest. They had to do it without cutting all the black cherry trees down. They did cut down 29 trees of varying sizes. They measured the diameter of each tree and the volume (in cubic feet) of wood that the tree produced.
  a. Fit a line to the data by eye.
  b. Is it a good fit?
  c. Find an equation for your line.
  d. Estimate the volume of a black cherry tree with diameter 15 inches. *(Lesson 8-7)*

Diameter	Volume
8.3	10.3
8.6	10.3
8.8	10.2
10.5	16.4
10.7	18.8
11.0	15.6
11.0	18.2
11.1	22.6
11.2	19.9
11.3	24.2
11.4	21.0
11.4	21.4
11.7	21.3
12.0	19.1
12.9	22.2
12.9	33.8
13.3	27.4
13.7	25.7
13.8	24.7
14.0	34.5
14.2	31.7
14.5	36.3
16.3	42.6
17.3	55.4
17.5	55.7
17.9	58.3
18.0	51.0
18.0	51.5
20.6	77.0

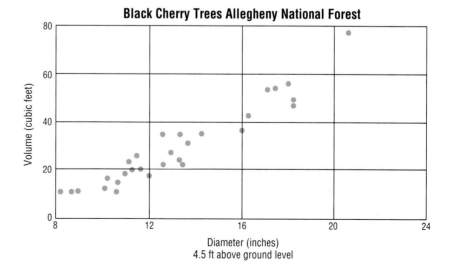

Black Cherry Trees Allegheny National Forest

Volume (cubic feet) vs. Diameter (inches) 4.5 ft above ground level

15. *Skill sequence* Solve. *(Lesson 10-10)*
  a. $3(y - 2) = 0$ **b.** $3y(y - 2) = 0$
  c. $(3y + 4)(y - 2) = 0$ **d.** $(y - 5)(3y + 4)(y - 2) = 0$

16. Solve $15t^2 - 55t = 0$. *(Lesson 10-10)*

17. Factor $100x - 100y + 100w - 100z$. *(Lesson 10-4)*

Exploration

18. The following diagram uses area to show the factoring of $x^2 + 4x + 3$.

$$x^2 + 4x + 3 = (x + 1)(x + 3)$$

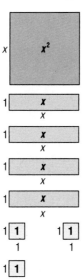

rearranged as

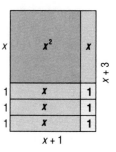

Make a drawing to show the factoring of
a. $x^2 + 7x + 10 = (x + 2)(x + 5)$;
b. $2x^2 + 9x + 4 = (2x + 1)(x + 4)$.

Summary

A term is a number, a product of numbers and variables, or a product of powers of numbers and variables. Polynomials are sums of terms and are classified by the number of terms. Polynomials arise from many situations. This chapter emphasizes situations where money is invested at the same rate but for different lengths of times. It also covers the use of polynomials in some formulas for areas, acceleration, and counting.

Like all algebraic expressions, polynomials can be added, subtracted, multiplied, and divided. All of these operations except division are studied in this chapter. Adding or subtracting polynomials is done by collecting like terms. Multiplying polynomials is done by multiplying each term of one polynomial by each term of the other, and then adding the products. Multiplication of polynomials can be represented by the area model.

Multiplication of binomials is a special case of multiplication of all polynomials and follows a pattern called FOIL.

$$(a + b)(c + d) = ac + ad + bc + bd$$

Certain special cases of this pattern are particularly important.

$$(a + b)(a + b) = a^2 + 2ab + b^2$$
$$(a - b)(a - b) = a^2 - 2ab + b^2$$
$$(a + b)(a - b) = a^2 - b^2$$

Reversing this process is called factoring. When the product of two or more factors is zero, then one of the factors must be zero. This Zero Product Property helps in solving certain equations.

Vocabulary

Below are the important new terms and phrases for this chapter. You should be able to give a general description and a specific example of each.

Lesson 10-1
polynomial
scale factor
polynomial in the variable x
constant term
degree of a polynomial
monomial, binomial, trinomial

Lesson 10-4
common monomial factoring

Lesson 10-5
Extended Distributive Property

Lesson 10-6
FOIL algorithm

Lesson 10-7
expanding a power of a polynomial
perfect square trinomial
Perfect Square Trinomial Patterns

Lesson 10-9
difference of squares
Difference of Two Squares Pattern

Lesson 10-10
Zero Product Property

Lesson 10-11
ruling out possibilities

Progress Self-Test

Take this test as you would take a test in class. You will need a calculator. Then check your work with the solutions in the Selected Answers section in the back of the book.

1. a. Simplify $4x^2 - 7x + 9x^2 - 12 - 11$.
   b. Give the degree of this polynomial.

In 2–11, perform the indicated operations and simplify.

2. Add $4k^3 - k^2 + 8$ and $7k^3 - 4k^2 + 6k - 2$.

3. Subtract $2v^2 + 5$ from $9v^2 + 2$.

4. Multiply $3v^2 - 9 + 2v$ by 4.

5. $6(3x^2 - 2x + 1) - 4(3x - 8)$

6. $-5z(z^2 - 7z + 8)$

7. $(x - 7)(x + 9)$

8. $(4y - 2)(3y - 16)$

9. $(q - 7)^2$

10. $(3x - 8)(3x + 8)$

11. $(2a - 3)(a^2 + 5ab + 7b^2)$

12. Is $4x^2 - 10x + 25$ a perfect square trinomial? If so, of what binomial is it the square?

In 13–16, factor.

13. $12m^3 - 2m^5$

14. $500x^2y + 100xy + 50y$

15. $z^2 - 81$

16. $y^2 + 12y + 36$

17. Simplify $\dfrac{8c^2 + 4c}{c}$.

18. Show how you can compute $29 \cdot 31$ mentally.

19. *Multiple choice* $3y^2 - 17y - 6 =$
   (a) $(3y + 1)(y - 6)$ (b) $(3y - 17)(y - 6)$
   (c) $3y(y - 17)$ (d) $(3y - 6)(y + 1)$
   (e) none of (a)–(d).

In 20 and 21, solve.

20. $7z(5z - 2) = 0$

21. $(x - 13)(2x + 15) = 0$

22. On his 18th birthday, Hank received $80. He received $60 on his 19th birthday and $90 on his 20th birthday. If he had invested all this money at a scale factor x, how much total money would he have on his 20th birthday?

23. A tennis ball bounces up from ground level at 8 meters per second. An equation that estimates the distance d above ground (in meters) after t seconds is $d = 8t - 5t^2$. After how many seconds will the ball return to the ground?

24. Represent the product $(a + b)(c + d + e)$ using areas of rectangles.

25. Write a simplified polynomial expression for the volume of the figure below.

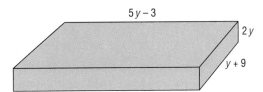

Chapter Review

Questions on **SPUR** Objectives

SPUR stands for **S**kills, **P**roperties, **U**ses, and **R**epresentations.
The Chapter Review questions are grouped according to the
SPUR Objectives for this chapter.

SKILLS deal with the procedures used to get answers.

Objective A. *Add and subtract polynomials.* *(Lessons 10-1, 10-2)*

In 1–2, **a.** simplify the expression; **b.** give its degree.

1. $5x^2 - 3x + 2x^2 + 7x + 1$
2. $(8m^4 - 2m^3) + (12m^3 - 6m^2 - 3m)$
3. Add $1.3x^2 + 14$, $4.7x - 1$, and $2.6x^2 - 3x + 6$.
4. Subtract $3y^5 - 2y^3 + 8y$ from $4y^5 - 6y^3 + 4y + 2$.

In 5–8, simplify.

5. $(k - 4) - (k^2 + 1)$
6. $(5p^2 - 1) - (6p^2 - p)$
7. $2(4x^2 - x - 4) + 4(3x - 7)$
8. $6(3x^2 + 4x - 7) - 5(2x^2 - x + 11)$

Objective B. *Multiply polynomials.* *(Lessons 10-3, 10-5, 10-6, 10-7, 10-9)*

In 9–22, multiply and simplify, if possible.

9. $4(x - x^3)$
10. $3k(k^2 + 4k - 1)$
11. $-2mn(3m - 2n + 4)$
12. $5xy(x + 3y^2)$
13. $(x - 3)(x + 7)$
14. $(y + 1)(y - 13)$
15. $(a - b)(c - d)$
16. $(4z + 1)(-z - 1)$
17. $(d - 1)^2$
18. $(2t + 3)^2$
19. $(a + 15)(a - 15)$

20. $(12b + m)(12b - m)$
21. $(a - 1)(a^2 + a + 1)$
22. $(x^2 + 3x - 2)(2x^2 + 5x + 4)$

Objective C. *Recognize perfect squares and perfect square trinomials.* *(Lesson 10-8)*

In 23 and 24, use a calculator to determine if the number is a perfect square.

23. 5625
24. 138,627

In 25–28, state whether the trinomial is a perfect square trinomial.

25. $16x^2 + 36x + 25$
26. $16y^2 - 8y + 1$
27. $4z^2 + 49$
28. $x^2 + 2x - 1$

Objective D. *Find common monomial factors of polynomials.* *(Lesson 10-4)*

29. Copy and complete: $7x^4 + 49x = 7x(\underline{\ ?\ } + \underline{\ ?\ })$.
30. Find the largest common factor of $20ay^3$, $-15y^4$, $35a^2y^6$.

In 31 and 32, factor.

31. $14m^4 + m^2$
32. $18b^3 - 21ab + 3b$

In 33 and 34, simplify.

33. $\dfrac{6z^3 - z}{z}$
34. $\dfrac{14x^2 + 12x}{2x}$

Objective E. *Factor perfect square trinomials and the difference of squares into a product of two binomials.* *(Lessons 10-8, 10-9)*

In 35 and 36, write each perfect square trinomial as the square of a binomial.

35. $m^2 + 16m + 64$

36. $9a^2 - 24ab + 16b^2$

In 37–40, write each difference of squares as the product of two binomials.

37. $a^2 - 4$

38. $b^2 - 81m^2$

39. $4x^2 - 1$

40. $25t^2 - 25$

PROPERTIES deal with the principles behind the mathematics.

Objective F. *Apply $(a + b)^2 = a^2 + 2ab + b^2$, $(a - b) = a^2 - 2ab + b^2$, and $(a + b)(a - b) = a^2 - b^2$ to multiply numbers in your head.* *(Lessons 10-7, 10-9)*

In 41–44, calculate in your head.

41. 41^2

42. $(.95)^2$

43. $63 \cdot 57$

44. $88 \cdot 92$

Objective G. *Recognize and use the Zero Product Property to solve equations.* *(Lesson 10-10)*

45. What is the Zero Product Property?

In 46 and 47, why can't the Zero Product Property be used on the given equation?

46. $(x + 3)(x + 4) = 5$

47. $(x + 3) + (x - 4) = 0$

In 48–51, solve.

48. $5q(2q - 7) = 0$

49. $(m - 3)(m - 1) = 0$

50. $(2w - 3)(3w + 5) = 0$

51. $(y - 3)(2y - 1)(2y + 1) = 0$

Objective H. *Recognize factors of polynomials using the properties of algebra, testing a special case and ruling out possibilities.* *(Lesson 10-11)*

52. Name two ways of determining whether expressions are equal.

53. Determine if $x^2 + 5x - 6$ is equal to $(x + 6)(x - 1)$ by
   a. multiplying the binomials;
   b. testing two special cases.

Questions 54 and 55 are *multiple choice.*

54. $11a^2 + 26a - 21 =$
   (a) $(11a - 7)(a - 3)$ (b) $(11a + 7)(a - 3)$
   (c) $(11a - 7)(a + 3)$ (d) $(11a + 7)(a + 3)$
   (e) none of (a)–(d)

55. $x^2 - 2xy + 13x - 26y =$
   (a) $(x - 13)(x + 2y)$ (b) $(x + 13)(x - 2y)$
   (c) $(x + 13y)(x - 2)$ (d) $(x - 13y)(x + 2)$
   (e) none of (a)–(d)

USES deal with applications of mathematics in real situations.

Objective I. *Translate real situations into polynomials.* *(Lessons 10-1, 10-2, 10-3, 10-6, 10-10)*

In 56 and 57, use the following information: Each birthday from age 11 on Katherine has received $250. She puts the money in a savings account with a scale factor of x.

56. Write an expression which shows how much Katherine will have after her 15th birthday.

57. If the bank pays 8% interest a year, calculate how much Katherine will have after her 13th birthday.

58. Recall that $d = .042s^2 + 1.1s$ gives the number of feet d needed to stop a car traveling at s mph on a concrete road. How much farther will a car travel before stopping, if it is going 50 mph instead of 30 mph?

59. José received $25 on his 12th birthday, $50 on his 13th birthday and $75 on his 14th birthday, which he invested at a scale factor x. He kept this money in the same account at the same scale factor for 4 more years. How much money did he have in this account?

60. At a family reunion there are b boys, m men, g girls, and w women. If each dance couple includes one male and female, how many different couples are possible?

61. An orange is thrown upward from ground level at 50 feet per second. Gravity pulls the orange down $16t^2$ feet in t seconds. An equation that gives the distance d above the ground after t seconds is $d = 50t - 16t^2$. After how many seconds will the orange hit the ground?

REPRESENTATIONS deal with pictures, graphs, or objects that illustrate concepts.

■ **Objective J.** *Represent areas of figures in terms of polynomials. (Lessons 10-2, 10-3, 10-5)*

62. Represent $(a + b + c)(d + e + f)$ using areas of rectangles

63. **a.** Write the area of rectangle *ABCD* below as the sum of 4 terms.
b. Write the area of *ABCD* as the product of 2 binomials.
c. Are the answers to parts a and b equal?

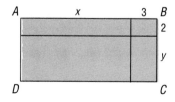

In 64 and 65, write a simplified expression for the area of the region in color.

64.

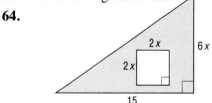

65.

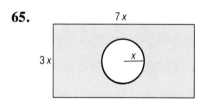

66. Write a polynomial for the volume of the figure below.

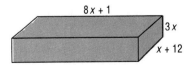

67. A box has dimensions x, $x + 1$, and $x - 1$. Write its volume as a polynomial.

Systems

Here are the men's and women's winning times in the Olympic 100-meter freestyle swimming race for all Olympic years from 1912 to 1988.

100-Meter Freestyle
Olympic Winning Time (seconds)

Year	Men's	Women's	Year	Men's	Women's
1912	63.2	72.2	1960	55.2	61.2
1920	61.4	73.6	1964	53.4	59.5
1924	59.0	72.4	1968	52.2	60.0
1928	58.6	71.0	1972	51.22	58.59
1932	58.2	66.8	1976	49.99	55.65
1936	57.6	65.9	1980	50.40	54.79
1948	57.3	66.3	1984	49.80	55.92
1952	57.4	66.8	1988	48.63	54.93
1956	55.4	62.0			

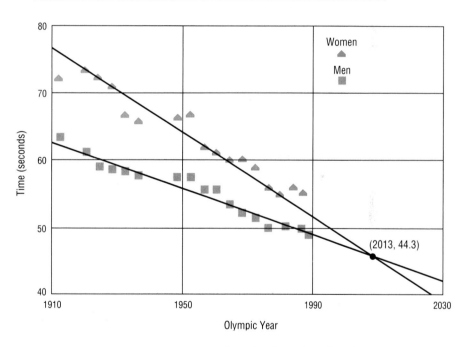

Matt Biondi, 1988 Olympic winner in men's 100-meter freestyle

The graph shows that the women's winning time has been decreasing faster than the men's. Lines have been fitted to the data. The lines intersect near (2013, 44.3). This means that if the times continue to decrease at a constant rate, the women's time would be faster than the men's in the Olympic year 2016. The winning times would each be about 44 seconds.

The finding of points of intersection of lines and other curves on the coordinate plane is called *solving a system*. In this chapter you will learn various ways of solving systems.

An Introduction to Systems

On page 531, if the time is y seconds in year x, equations of the lines are as follows:

$$\text{Men:} \quad y = -0.185x + 416.70$$
$$\text{Women:} \quad y = -0.313x + 674.41$$

The graphs of these lines give you a way of estimating the point of intersection. The point of intersection indicates the year when the men's and women's Olympic winning times in the 100-meter free-style will be the same. It also indicates the winning time. We now seek a way of finding the coordinates of this point.

The numbers in the equations above are quite complicated, so consider a simpler situation.

> The sum of two numbers is 22 and their difference is 8. What are the numbers?

To answer this question, first note that there are two numbers to be found. It is customary to name them x and y. The given information contains two *conditions* separated by the word *and*. Each condition can be translated into an equation.

Condition 1: The sum of the two numbers is 22.

$$x + y = 22$$

Condition 2: Their difference is 8.

$$x - y = 8$$

A **system** is a set of conditions each separated from the others by the word "and." In algebra systems, the conditions are usually equations and the word "and" is often signified with a single brace {. The system immediately above can be written

$$\begin{cases} x + y = 22 \\ x - y = 8. \end{cases}$$

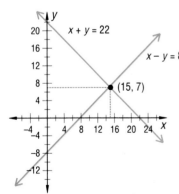

The *solution* to this system is the particular ordered pair (x, y) which satisfies both $x + y = 22$ and $x - y = 8$. The solution can be found by graphing the two lines. It is $(15, 7)$, the point of intersection shown at the left.

To check that $(15, 7)$ is a solution, the values for x and y must be checked in *both* conditions. Does $15 + 7 = 22$? Yes. Does $15 - 7 = 8$? Yes.

In general, the **solution set to a system** is the intersection of the solution sets for each of the conditions in the system.

Here are four ways to indicate the solution to the system graphed on the previous page.

as an ordered pair: (15, 7)
as an ordered pair identifying the variables: $(x, y) = (15, 7)$
by naming the variables individually: $x = 15$ and $y = 7$
as a set of ordered pairs: $\{(15, 7)\}$

When the sentences in a system have no solutions in common, there is no solution to the system. The solution set is the set with no elements, Ø.

Example   Find all solutions to the system $\begin{cases} y = 2x + 1 \\ y = 2x - 3. \end{cases}$

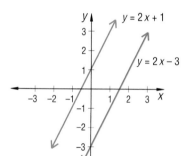

Solution   Draw the graph of each equation. Look for all intersection points.

Since these lines have the same slope and different y-intercepts, they are parallel and do not intersect. There is no solution to the system since there are no pairs of numbers that make both equations true.

Check   Twice a number plus one ($2x + 1$) could never give the same value as twice the same number minus three ($2x - 3$).

The systems discussed in this chapter all involve sentences whose graphs are lines. Accordingly, they are called **linear systems**. Graphing linear systems can help you find exact solutions, as in the solution to $\begin{cases} x + y = 22 \\ x - y = \ \ 8 \end{cases}$. If solutions do not have integer coordinates, however, it is likely that reading a graph will only give you an estimate. In the next lessons you will use algebra to find exact solutions to linear systems.

Questions

Covering the Reading

1. Define: system.

2. In algebra systems, the conditions are usually __?__ and the word __?__ is denoted by a brace.

3. Define: the solution set to a system.

4. When a system has two variables, each solution is an __?__.

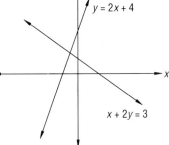

5. Refer to the graph at the right.
   a. What system is being graphed?
   b. What is the solution to the system?

6. Is (4, 8) a solution to the following system?
$$\begin{cases} 10x - y = 32 \\ y - x = 4 \end{cases}$$

7. The sum of two numbers is 18 and their difference is 8.
   a. If the numbers are x and y, translate the two conditions of the sentence above into two equations.
   b. Graph both of these equations on the same coordinate system.
   c. What are the numbers?

8. Find all solutions to the system $\begin{cases} y = 4x - 2 \\ y = 4x + 5 \end{cases}$.

9. The solution to the system $\begin{cases} y = 9x \\ y = 2x - 7 \end{cases}$ is (-1, -9). Write this solution in two other ways.

10. Why are the systems of Questions 8 and 9 called *linear systems*?

Applying the Mathematics

In 11 and 12, solve each system by graphing.

11. $\begin{cases} y = 2x + 1 \\ y = -3x + 6 \end{cases}$
12. $\begin{cases} y = x \\ 2x + 3y = -15 \end{cases}$

13. Use the equations, on page 532, of the system of the 100-meter freestyle times graphed on page 531. Verify that (2013, 44.3) is the solution to the system.

Review

14. Solve $2x + y = 7$ for y. *(Lesson 8-4)*

15. Solve $3x + 5(2x - 2) = 16$. *(Lesson 6-8)*

16. If a country has population P now and is increasing by I people per year, what will its population be in Y years? *(Lesson 6-4)*

17. Multiply $4x - y$ by 3. *(Lesson 6-3)*

18. Eight more than three times a number is two more than six times the number. What is the number? *(Lesson 6-6)*

19. a. How many quarts are there in 5 gallons?
b. How many quarts are there in n gallons? *(Previous course)*

20. Simplify $7\pi \div \left(\dfrac{2\pi}{3}\right)$. *(Lesson 5-1)*

21. In 1985, the cost for a large computer to process a typical data processing job was 4¢. The job took 0.4 seconds. What is the cost per second? *(Lesson 5-2)*

22. a. Evaluate $\dfrac{y_2 - y_1}{x_2 - x_1}$ where $y_2 = 7$, $y_1 = -1$, $x_2 = 8$, and $x_1 = 10$.
(Lesson 1-5)
b. What have you calculated in part a? *(Lesson 8-2)*

23. On a map, 1 inch represents 325 miles. The map distance from Los Angeles to Houston is $4\frac{3}{4}$ inches. Suppose you want the distance in miles from Los Angeles to Houston.
a. Write a proportion that will help solve the problem.
b. Find the distance in miles. *(Lesson 5-7)*

24. In the similar figures below, corresponding sides are parallel.

a. What is the ratio of similitude?
b. Find the value of x. *(Lesson 5-6)*

25. *Skill sequence* Solve. *(Lessons 6-1, 6-6, 6-8)*
a. $3x + 8 = -12$
b. $3x + 8 = x - 12$
c. $3(x + 8) = -4(x - 12)$

26. Find the mean of the three numbers. Calculate in your head.
(Lesson 1-2)
a. 3, 4, 5   **b.** 7, 10, 13   **c.** $20n$, $25n$, $30n$

Exploration

27. Some experts believe that even though the women's swim times are decreasing faster than the men's, it is the ratio of the times that is the key to predictions.
a. Compute the ratio of the men's time to the women's time for the 100-meter freestyle for each Olympic year.
b. Graph your results. (Plot Olympic year on the horizontal axis and the ratio of times on the vertical axis.)
c. What do you think the ratio will be in 2016? Does this agree with the prediction on page 531?

11-2

Solving Systems Using Substitution

Graphing is a useful technique, but it does not always show exact solutions to a system. However, exact solutions for some systems can be found algebraically. **Substitution** is a common method.

Example 1

From a car wash a service club made $109 that was to be divided between the Boy Scouts and the Girl Scouts. There were twice as many girls as boys so a decision was made to give the girls twice as much money. How much did each group receive?

Solution   Translate each condition into an equation. Suppose the Boy Scouts receive B dollars and the Girl Scouts receive G dollars. We number the equations in the system for reference.

The sum of amounts is $109.
Girls get twice as much as boys.

$$\begin{cases} B + G = 109 & (1) \\ \quad G = 2B & (2) \end{cases}$$

Since $G = 2B$ in equation (2), you can substitute $2B$ for G in equation (1).

$$B + 2B = 109$$
$$3B = 109$$
$$B = 36\tfrac{1}{3}$$

To find G, substitute $36\tfrac{1}{3}$ for B in either equation. We use equation (2).

$$G = 2B$$
$$= 2 \cdot 36\tfrac{1}{3}$$
$$= 72\tfrac{2}{3}$$

So the solution is $(B, G) = (36\tfrac{1}{3}, 72\tfrac{2}{3})$. The Boy Scouts will receive $36\tfrac{1}{3} \approx \$36.33$, and the Girl Scouts will get $72\tfrac{2}{3} \approx \$72.67$.

Check   Are both conditions satisfied? Will the groups receive a total of $109? Yes, $36.33 + $72.67 = $109. Will the girls get twice as much as the boys? Yes, it is as close as possible.

In the equation $G = 2B$, G is in terms of B. Substitution is a good method to use when one variable is given in terms of others. In the next example, two different variables are given in terms of a third.

Example 2 An orange punch is made by mixing two parts orange juice with three parts ginger ale. Six gallons of punch are needed. How many quarts of orange juice and how many quarts of ginger ale will it take?

Solution First, identify the unknowns.

Let J = the number of quarts of orange juice in the punch.
Let G = the number of quarts of ginger ale in the punch.
Let P = the number of quarts in one "part" of juice.

Remember that one gallon equals four quarts, so six gallons is twenty-four quarts. This problem has three given conditions.

Two parts are orange juice. $\quad\begin{cases} J = 2P & (1) \\ G = 3P & (2) \\ J + G = 24 & (3) \end{cases}$
Three parts are ginger ale.
The total is 24 (quarts).

To solve the system, substitute $2P$ for J and $3P$ for G in equation (3).

$$2P + 3P = 24$$
$$5P = 24$$
$$P = 4.8$$

Each part contains 4.8 quarts of liquid.
To find J, substitute 4.8 for P in equation (1).

$$J = 2P = 2 \cdot 4.8 = 9.6$$

To find G, substitute 4.8 for P in equation (2).

$$G = 3P = 3 \cdot 4.8 = 14.4$$

So it takes 9.6 quarts of orange juice and 14.4 quarts of ginger ale to make six gallons of this punch.

You might have been able to answer the questions of Examples 1 and 2 without algebra. Doing the algebra is then a check. But in the next example, algebra is necessary.

Example 3 Two lines have equations $3x + 2y = 10$ and $y = 4x + 1$. Where do they intersect?

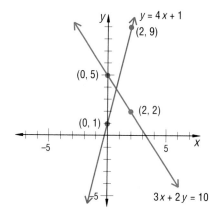

Solution The coordinates of the point of intersection are not integers, so reading the solution from the graph is difficult. Notice that the line $y = 4x + 1$ is in slope-intercept form. This makes substitution a natural method for solving the system. Substitute $4x + 1$ for y in the first equation and solve.

$$3x + 2(4x + 1) = 10$$
$$3x + 8x + 2 = 10$$
$$11x + 2 = 10$$
$$11x = 8$$
$$x = \frac{8}{11}$$

You have found the x-coordinate of the point of intersection. To find y, substitute for x in the second equation.

$$y = 4 \cdot \frac{8}{11} + 1$$
$$= \frac{32}{11} + 1$$
$$= \frac{43}{11}$$

The lines intersect at $\left(\frac{8}{11}, \frac{43}{11}\right)$.

Check This seems correct on the graph. You should substitute $\frac{8}{11}$ for x and $\frac{43}{11}$ for y in both equations to produce an exact check.

Does $3\left(\frac{8}{11}\right) + 2\left(\frac{43}{11}\right) = 10$?

Does $4\left(\frac{8}{11}\right) + 1 = \frac{43}{11}$?

You should verify that they do.

Questions

Covering the Reading

1. In Example 1 suppose the service club made $180. Also suppose the boys were to get three times as much money as the girls. How much would each group receive?

2. a. Solve the system.

$$\begin{cases} B = 7t \\ B - t = 30 \end{cases}$$

b. Check your answer.

3. Suppose the orange punch of Example 2 was made by mixing 3 parts orange juice to 1 part ginger ale. If 5 gallons of punch are made, how many quarts of orange juice and how many quarts of ginger ale will it take?

4. Solve for A, B, and K.

$$\begin{cases} A = 40K \\ B = 30K \\ A + B = 1400 \end{cases}$$

5. Complete the check of Example 3.

In 6–10, **a.** find the point of intersection of the lines algebraically, and **b.** check your answer.

6. $\begin{cases} y = x - 2 \\ -4x + 7y = 10 \end{cases}$

7. $\begin{cases} 12x - 5y = 30 \\ y = 2x - 6 \end{cases}$

8. $\begin{cases} x + y = 14 \\ x = 6y \end{cases}$

9. $\begin{cases} 2m - 3n = -15 \\ m = 4n \end{cases}$

10. $\begin{cases} 3a + 4b = -15 \\ b = 2a - 3 \end{cases}$

Applying the Mathematics

11. A will states that John is to get 3 times as much money as Mary. The total amount they will receive is $11,000.
a. Write a system of equations describing this situation.
b. Solve to find the amounts of money John and Mary will get.

12. A homemade sealer to use after furniture is stained can be made by mixing one part shellac with five parts denatured alcohol. To make a pint (16 fluid ounces) of sealer, how many fluid ounces of shellac and how many of denatured alcohol are needed?

13. Profits of a company were up $200,000 this year over last year. This was a 25% increase. If T and L the profits in dollars for this year and last year, then:

$$\begin{cases} T = L + 200{,}000 \\ T = 1.25L. \end{cases}$$

Find the profits for this year and last year.

Review

In 14 and 15, **a.** graph to find the solution to the system. **b.** Check your answer. *(Lesson 11-1)*

14. $\begin{cases} y = x + 1 \\ y = -2x + 13 \end{cases}$ **15.** $\begin{cases} y = 3x - 2 \\ y = 3x + 3 \end{cases}$

16. Solve $5x + 9y = 7$ for y. *(Lesson 8-4)*

17. Expand and simplify $x^2 + y^2$ when $y = x - 1$. *(Lessons 10-2, 10-5)*

18. Subtract $a + b + 7$ from $3a + b - 2$. *(Lesson 6-9)*

19. Graph $y \geq 2x + 2$ on a coordinate plane. *(Lesson 8-9)*

In 20–22, solve. *(Lesson 6-6)*

20. $3x + 11 = -2x - 1$

21. $-4y + 18 = 81 - 7y$

22. $15z - 3(4 + 8z) = 21$

Exploration

23. Sister cities Lovely and Elylov are building a joint conference center that will cost $10 million. Lovely's population in the last census was 35,729. Elylov's population was 74,212. How would you suggest they split the costs?

11-3

More with Substitution

Substitution is a good way to solve a system of two equations if both are solved for the same variable. This is the case if you have equations of two lines in slope-intercept form and want the exact intersection.

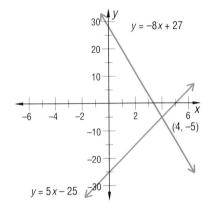

Example 1

In the graph above, two lines have equations $y = 5x - 25$ and $y = -8x + 27$. Where do they intersect? (Note that the units shown on the x-axis and y-axis are different.)

Solution   Solve the system.

$$\begin{cases} y = 5x - 25 & (1) \\ y = -8x + 27 & (2) \end{cases}$$

Substitute $-8x + 27$ for y in equation (1).

$$-8x + 27 = 5x - 25$$

Now solve for x as usual.

$$27 = 13x - 25$$
$$52 = 13x$$
$$4 = x$$

To find y, substitute 4 for x in either equation. We use equation (1).

$$y = 5 \cdot 4 - 25 = -5$$

The point of intersection is (4, -5).

Check   The point is on both lines, as substitution shows.

$$-5 = 5 \cdot 4 - 25 \text{ and } -5 = -8 \cdot 4 + 27$$

Suppose two quantities are increasing or decreasing at different constant rates. You may want to know when they are equal. The answer can be found by solving a system.

Example 2   A taxi ride in Burford costs 90¢ plus 10¢ each $\frac{1}{10}$ mile. In Spotswood, taxi rides cost 50¢ plus 15¢ each $\frac{1}{10}$ mile. For what distance do the rides cost the same?

Solution   Let d = the distance of a cab ride, in *tenths* of a mile. Let C = the cost of a cab ride of distance d.

In Burford: $C = .90 + .10d$ (1)
In Spotswood: $C = .50 + .15d$ (2)

The rides cost the same when the values of C and d in Burford equal the values in Spotswood. So there is a system to solve. Since both equations are expressed in terms of C, substitute $.90 + .10d$ for C in equation (2).

$$.90 + .10d = .50 + .15d$$

Now solve as usual.

$$.05d = .40$$
$$d = 8$$

So a ride of 8 tenths of a mile will cost the same in either city.

Check   Check to see if the cost will be the same for a ride of 8 tenths of a mile.

The cost in Burford is $.90 + .10 \cdot 8 = .90 + .80 = 1.70$.
The cost in Spotswood is $.50 + .15 \cdot 8 = .50 + 1.20 = 1.70$.
The cost is $1.70 in each city, so it checks.

Questions

Covering the Reading

1. *True or false*   Solving a system by substitution only approximates the answer.

In 2 and 3, find the point of intersection of the two lines.

2. $\begin{cases} y = 3x - 19 \\ y = x + 1 \end{cases}$ **3.** $\begin{cases} y = 12x + 50 \\ y = 10x - 60 \end{cases}$

In 4–8, refer to Example 2.

4. What would it cost for a 3-mile taxi ride in Burford? (Hint: convert to tenths of a mile)

5. What would it cost for a 2.5-mile taxi ride in Spotswood?

6. At what distances are taxi rides more expensive in Burford than in Spotswood?

7. At what distances are taxi rides more expensive in Spotswood than in Burford?

8. Suppose that in Manassas, a taxi ride costs $.70 plus 10¢ each $\frac{1}{10}$ mile.
   a. What does it cost to ride d tenths of a mile?
   b. At what distance does a ride in Manassas cost the same as a ride in Spotswood?

Applying the Mathematics

9. Check the answer to Example 2 by graphing the lines $C = .90 + .10d$ and $C = .50 + .15d$ on the same axes. (Let d be the first coordinate and C be the second coordinate.)

In 10–12, solve each system. Check your answers.

10. $\begin{cases} y = 21 - x \\ y = 3 + x \end{cases}$
11. $\begin{cases} a = 2b + 3 \\ a = 3b + 20 \end{cases}$

12. $\begin{cases} y = \frac{1}{2}x - 5 \\ y = -\frac{3}{4}x + 10 \end{cases}$

13. Jana has $290 and saves $5 a week. Dana has $200 and saves $8 a week.
   a. After how many weeks will they each have the same amount of money?
   b. How much money will each one have?

14. One plumbing company charges $45 for the first half-hour of work and $23 for each additional half-hour. Another company charges $35 for the first half-hour and then $28 for each additional half-hour. For how many hours will the cost of each company be the same?

15. In 1980, the population of Pittsburgh was about 424,000 and was decreasing by about 10,000 people a year. The 1980 population of Phoenix, Arizona, was about 790,000 and was increasing by 20,000 people a year.
    a. If these trends had been going on for some time, how many years before 1980 did each city have the same population?
    b. What was this population?

16. Frank weighs 160 lb and is on a diet to gain 2 lb a week so that he can make the football team. John weighs 208 lb and is on a diet to lose 3 lb a week so that he can be on the wrestling team at a lower weight. If they can meet these goals with their diets, when will Frank and John weigh the same, and how much will they weigh?

Review

17. Solve this system by substitution. *(Lesson 11-2)*

$$\begin{cases} a = 4b + 3 \\ 2b - 3a = 10 \end{cases}$$

18. Suppose 8 pounds of peanuts cost $15.90. At that rate, what is the cost of 10 pounds of peanuts? *(Lesson 4-2)*

19. Calculate the *y*-intercept in your head. *(Lesson 8-4)*
    a. $y = 7x - 2$ **b.** $2x + y = 7$ **c.** $7 + y = 2x$

20. Factor $3a^2b - 12ab^3 + 27ab$. *(Lesson 10-10)*

21. Graph all points for which $y > -x + 5$. *(Lesson 8-9)*

22. Simplify $\dfrac{7m^4n^5}{343m^3n^6}$. *(Lesson 9-8)*

23. *Skill sequence*   Solve for *x*. *(Lessons 6-1, 7-6, 10-10)*
    a. $2x - 18 = 0$
    b. $2x^2 - 18 = 0$
    c. $2x^2 - 18x = 0$

Exploration

24. Find the taxi rates where you live or in a nearby community. How do these rates compare to the cities in Example 2?

11-4

Solving Systems by Addition

The numbers $\frac{3}{4}$ and 75% are equal even though they do not look equal. So are $\frac{1}{5}$ and 20%. If you add these numbers (as seen below), the sums are equal.

$$\frac{3}{4} = 75\%$$

$$\frac{1}{5} = 20\%$$

So $\qquad \frac{3}{4} + \frac{1}{5} = 75\% + 20\%.$

Simplifying each side, $\qquad \frac{19}{20} = 95\%.$

This is one instance of the following generalization of the Addition Property of Equality.

Generalized Addition Property of Equality:

For all numbers or expressions a, b, c, and d:

If $\qquad\qquad\qquad\qquad a = b$
and $\qquad\qquad\qquad\qquad c = d,$
then $\qquad\qquad\qquad a + c = b + d.$

The Generalized Addition Property of Equality is quite useful in solving some systems. Consider the system

$$\begin{cases} 3x + 2y = 1 & (1) \\ x - 2y = 107 & (2). \end{cases}$$

If x and y satisfy both equations, they will satisfy the equation that results from adding the left and right sides. In this case $2y$ and $-2y$ add to 0, so the sum has only one variable.

$$\begin{array}{rll} 3x + 2y = & 1 & (1) \\ x - 2y = & 107 & (2) \\ \hline 4x \quad\quad = & 108 & (1) + (2) \end{array}$$

The line below equation (2) indicates the addition. Adding equation (1) and equation (2) gives equation (1) + (2). Solve $4x = 108$ as usual.

$$x = 27$$

To find y, substitute 27 for x in equation (1).

$$\begin{aligned} 3(27) + 2y &= 1 \\ 81 + 2y &= 1 \\ 2y &= -80 \\ y &= -40 \end{aligned}$$

Since (27, -40) checks in both equations, it is the solution. This **addition method** is an easy way to solve systems when coefficients of the same variable are opposites.

Example 1 Mr. Robinson flew his small plane the 80 miles from Tampa to Orlando in 40 minutes ($\frac{2}{3}$ hour) against the wind. He returned to Tampa in 32 minutes ($\frac{32}{60}$ hour) with the wind at his back. How fast was he flying? What was the speed of the wind?

Solution Let A be the average speed of the airplane without wind and W be the speed of the wind, both in miles per hour. His total speed against the wind is then $A - W$, and his speed with the wind is $A + W$. There are two conditions given on these total speeds.

From Tampa to Orlando his rate was $\dfrac{80\text{ miles}}{\frac{2}{3}\text{ hour}}$ or $120\,\dfrac{\text{miles}}{\text{hour}}$.

This was against the wind, so $A - W = 120$.

From Orlando to Tampa his rate was $\dfrac{80\text{ miles}}{\frac{32}{60}\text{ hour}}$ or $150\,\dfrac{\text{miles}}{\text{hour}}$.

This was with the wind, so $A + W = 150$.
Now solve the following system. Since the coefficients of W are opposites (1 and -1), add the equations.

$$
\begin{array}{ll}
A - W = 120 & (1) \\
A + W = 150 & (2) \\
\hline
2A \quad\;\; = 270 & (1) + (2) \\
A = 135 &
\end{array}
$$

Adding,

Substitute 135 for A in either equation. We choose (2).

$$
\begin{aligned}
135 + W &= 150 \\
W &= 15
\end{aligned}
$$

So the average speed of the airplane was 135 mph and the speed of the wind was 15 mph.

Check Refer to the original question. Against the wind, he flew at $135 - 15$ or 120 mph. In 40 minutes ($\frac{2}{3}$ of an hour) he would travel 80 miles, as desired. With the wind he was flying $135 + 15$ or 150 mph. At that rate he flew 80 miles in 32 minutes, which checks with the given conditions.

Sometimes the coefficients of the same variable are equal. In this case, use the Multiplication Property of Equality to multiply both sides of one of the equations by -1. This changes all the numbers in that equation to their opposites. Then you can use the addition method to find solutions to the system. Alternatively, you may subtract one equation from the other.

Example 2   Solve this system.

$$\begin{cases} 4x + 13y = 40 & (1) \\ 4x + 3y = -40 & (2) \end{cases}$$

Solution 1   Notice that the coefficients of x in (1) and (2) are equal. Multiply (2) by -1. Call this equation (3).

$$-4x - 3y = 40 \quad (3)$$

Now use the addition method with (1) and (3).

$$\begin{array}{rl} 4x + 13y = 40 & (1) \\ -4x - 3y = 40 & (3) \\ \hline 10y = 80 & (1) + (3) \\ y = 8 \end{array}$$

Substitute 8 for y in (1) to find the value for x.

$$\begin{array}{rl} 4x + 13(8) = 40 \\ 4x + 104 = 40 \\ 4x = -64 \\ x = -16 \end{array}$$

So $(x, y) = (-16, 8)$.

Solution 2   Subtract the second equation from the first. This again gives $10y = 80$ and you can continue as in Solution 1.

Check   Substitute in both equations.

(1) Does $4 \cdot -16 + 13 \cdot 8 = 40$? Yes.
(2) Does $4 \cdot -16 + 3 \cdot 8 = -40$? Yes.

■ ■ ■ ■ ■ ■ ■

Example 3   A resort hotel offers two weekend specials.

> Plan (1): 3 nights with 6 meals $132
> Plan (2): 3 nights with 2 meals $109

At these rates, what is the cost of one nights lodging and what is the average cost per meal? (Assume there is no discount for 6 meals.)

Solution   Let N = price of one nights lodging.
Let M = average price of one meal.

From Plan (1): $3N + 6M = 132$
From Plan (2): $3N + 2M = 109$

The coefficients of N are the same so subtract.

Subtract (2) from (1). $4M = 23$
$$M = 5.75$$

Substitute 5.75 for M in either equation. We select equation (1).

$$3N + 6(5.75) = 132$$
$$3N + 34.50 = 132$$
$$3N = 97.50$$
$$N = 32.50$$

Thus, $(N, M) = (32.50, 5.75)$.
One nights' lodging costs $32.50 and a meal averages $5.75.

Check   In Question 8, you are asked to check that at these rates the totals for Plans (1) and (2) are correct.

Questions

Covering the Reading

1. What is the goal in adding equations to solve systems?

2. Which property allows you to add corresponding sides of two equations to get a new equation?

3. Solve. $\begin{cases} 3x + 8y = 2 \\ -3x - 4y = 6 \end{cases}$ **4.** Solve. $\begin{cases} a + b = 11 \\ a - b = 4 \end{cases}$

5. In Example 1, suppose it took 50 minutes ($\frac{50}{60}$ hour) to fly to Orlando against the wind and 40 minutes ($\frac{2}{3}$ hour) for the return flight with the wind. Find the speed of the plane and the speed of the wind in miles per hour under these conditions.

6. When is it appropriate to multiply an equation by -1 or subtract the two equations as a first step in solving a system?

7. Solve. $\begin{cases} 2x - 3y = 5 \\ 5x - 3y = 11 \end{cases}$

8. Check Example 3.

9. A hotel offers the following specials: Plan (1) is two nights and one meal for $106. Plan (2) is 2 nights and 4 meals for $130. What price is the hotel charging per night and per meal?

Applying the Mathematics

10. As you know, $\frac{3}{4} = 75\%$ and $\frac{1}{5} = 20\%$. Is it true that $\frac{3}{4} - \frac{1}{5} = 55\%$?

In 11 and 12, solve.

11. $\begin{cases} 2x - 3y = 17 \\ 3y + x = 1 \end{cases}$ **12.** $\begin{cases} 4z - 5w = 15 \\ 2w + 4z = -6 \end{cases}$

13. Two eggs with bacon cost $1.35. One egg with bacon costs $.90. At these rates, what should bacon alone cost?

14. Five gallons of unleaded gas plus eight gallons of regular cost $13.87. Five gallons of unleaded plus two gallons of regular cost $7.93. Find the cost per gallon of each kind of gasoline.

15. Find two numbers whose sum is -1 and whose difference is 5.

16. Find two numbers whose sum is 4386 and whose difference is 2822.

Review

In 17 and 18, solve. *(Lessons 11-2, 11-3)*

17. $\begin{cases} y = 2x - 1 \\ y = 9x + 6 \end{cases}$ **18.** $\begin{cases} Q = 4z \\ R = -5z \\ 4R + Q = 40 \end{cases}$

19. Ida is playing with toothpicks. It takes 5 toothpicks to make a penta-gon and 6 toothpicks to make a hexagon. She has 100 toothpicks. She wants to make P pentagons and H hexagons. Give three different pos-sible values of P and H that use up all the toothpicks. *(Lesson 8-9)*

In 20 and 21, simplify by removing parentheses.

20. $3(2a^2 - 5)$ *(Lesson 6-3)*   **21.** $b^2(2b^2)$ *(Lesson 9-5)*

22. Suppose a 30-gram necklace is 14K gold. This means $\frac{14}{24}$ of the necklace is gold and the rest is other materials. How many grams of gold are in the necklace? *(Lesson 5-3)*

23. Graph the half-plane $x + 2y > 0$. *(Lesson 8-9)*

24. *Skill Sequence* Simplify. *(Lessons 2-2, 6-9, 7-8)*
    a. $100 - (80 - 4p) + 2(p + 5)$
    b. $\dfrac{100}{p + 2} - \dfrac{80 - 4p}{p + 2} + \dfrac{2(p + 5)}{p + 2}$

In 25–27, rewrite in the form $Ax + By = C$. *(Lesson 8-8)*

25. $-15 - 8x = y$

26. $2x + y = 17 - 7y$

27. $9y - 15 = 4x + 2y - 7$

Exploration

28. Multiplying equations by 100 can be used to find simple fractions for repeating decimals. For instance, to find a fraction for $.\overline{39}$, first let $d = .\overline{39}$. Then

$$
\begin{aligned}
100d &= 39.\overline{39} &\quad &(1)\\
d &= .\overline{39} &\quad &(2)\\
99d &= 39 &\quad &(1) - (2)\\
d &= \tfrac{39}{99} &\quad &\\
d &= \tfrac{13}{33} &\quad &
\end{aligned}
$$

Subtract (2) from (1).
Solve for d.
Simplify the fraction.

A calculator shows that $\frac{13}{33}$ has a decimal equivalent $0.393939...$
a. Use this process to find a simple fraction equal to $.\overline{81}$.
b. Modify the process to find a simple fraction equal to $.\overline{003}$.
c. Find a simple fraction equal to $3.89\overline{5}$.

11-5

Multiplying to Solve Systems

Recall that there are two common forms for equations of lines.

standard form: $Ax + By = C$

slope-intercept form: $y = mx + b$

The substitution method described in Lesson 11-3 is convenient for solving systems in which one or both equations are in slope-intercept form. The **multiplication method** outlined in the following example provides a way to solve any system where both equations are in standard form. The idea is to multiply one of the equations by a number chosen so that the addition method can be used on the result.

Example 1 Solve the following system.

$$\begin{cases} 5x + 8y = 21 & (1) \\ 10x - 3y = -15 & (2) \end{cases}$$

Solution 1 Multiply equation (1) by -2. This makes the coefficients of x opposites of one another. The code $(3) = -2 \cdot (1)$ in the solution below shows what has been done; equation (3) is found by multiplying both sides of equation (1) by -2. The solutions are unchanged because of the Multiplication Property of Equality.

$$\begin{cases} -10x - 16y = -42 & (3) = -2 \cdot (1) \\ 10x - 3y = -15 & (2) \end{cases}$$

Add the equations. $\quad -19y = -57 \quad (3) + (2)$

Solve for y. $\quad y = 3$

Substitute $y = 3$ in (1) to find x.

$$5x + 8 \cdot 3 = 21$$
$$5x + 24 = 21$$
$$5x = -3$$
$$x = -\tfrac{3}{5}$$

So the solution is $(x, y) = (-\tfrac{3}{5}, 3)$.

Solution 2 Multiply equation (2) by $-\tfrac{1}{2}$. This also makes the coefficients of x opposites.

$$\begin{cases} 5x + 8y = 21 & (1) \\ -5x + 1.5y = 7.5 & (3) = -\tfrac{1}{2} \cdot (2) \end{cases}$$

Add. $\quad 9.5y = 28.5 \quad (1) + (3)$

$$y = 3$$

The solution would proceed as in Solution 1 to find x.
Again $(x, y) = (-\frac{3}{5}, 3)$.

Check   Does $5 \cdot -\frac{3}{5} + 8 \cdot 3 = 21$? Yes, $-3 + 24 = 21$.

Does $10 \cdot -\frac{3}{5} - 2 \cdot 3 = -12$? Yes, $-6 - 6 = -12$.

Example 1 shows that the solution is the same no matter which equation is multiplied by a number. The goal is to obtain coefficients of one variable which are opposites. Then add the resulting equations to eliminate one of the variables.

Sometimes you must multiply *each* equation by a number before adding.

■ ▪ ▪ ▪ ▪ ▪ ▪ ▪

Example 2   Solve the system.

$$\begin{cases} 3a + 5b = 8 & (1) \\ 2a + 3b = 4.6 & (2) \end{cases}$$

Solution   To make the coefficients of a opposites, multiply equation (1) by 2 and equation (2) by -3.

$$\begin{cases} 6a + 10b = 16 & (3) = 2 \cdot (1) \\ -6a - 9b = -13.8 & (4) = -3 \cdot (2) \end{cases}$$

Now add. $\qquad\qquad\qquad b = 2.2 \qquad (3) + (4)$

Substitute in equation (1) to find the value of a.

$$3a + 5(2.2) = 8$$
$$3a + 11 = 8$$
$$3a = -3$$
$$a = -1$$

Therefore, $(a, b) = (-1, 2.2)$.

Check   Does $3(-1) + 5(2.2) = 8$? Yes, $-3 + 11 = 8$.
Does $2(-1) + 3(2.2) = 4.6$? Yes, $-2 + 6.6 = 4.6$.

Many situations lead naturally to linear equations in standard form. This results in a linear system that can be solved using the multiplication method.

Example 3 A marching band has 52 members and there are 24 in the pompon squad. They wish to form hexagons and squares like those diagrammed below. Can it be done with no people left over?

HEXAGON
*pompon person
in center*

SQUARE
*band member
in center*

Ⓑ   Ⓑ

Ⓑ   P   Ⓑ

Ⓑ   Ⓑ

P   P

Ⓑ

P   P

B = band member
P = pompon person

Solution Consider the entire formation to include h hexagons and s squares. There are two conditions in the system: one for band members and one for pompon people.

There are 6 $\frac{\text{band members}}{\text{hexagon}}$ and 1 $\frac{\text{band member}}{\text{square}}$, so $6h + s = 52$. (1)

There is 1 $\frac{\text{pompon person}}{\text{hexagon}}$ and 4 $\frac{\text{pompon people}}{\text{square}}$, so $h + 4s = 24$. (2)

To solve the system, multiply equation (1) by -4 and add to equation (2).

$$
\begin{array}{r}
-24h - 4s = -208 \\
\underline{h + 4s = 24} \\
-23h = -184 \\
h = 8
\end{array}
$$

Substitute 8 for h in (1).

$$6 \cdot 8 + s = 52$$
$$48 + s = 52$$
$$s = 4$$

Since h and s are both positive integers, the formations can be done with no people left over.

Check 8 hexagons would use 48 band members and 8 pompon people. 4 squares would use 4 band members and 16 pompon people. This setup uses exactly 52 band members and 24 pompon people.

A system can involve equations that are not in standard or slope-intercept form. Then it is wisest to rewrite the equations in one of these forms before proceeding. For example, to solve the system

$$\begin{cases} 5b = 8 - 3a & (1) \\ 2a + 3b = 4.6 & (2) \end{cases}$$

you could add $3a$ to both sides of equation (1). The result is the system that was solved in Example 2.

Questions

Covering the Reading

1. Which property allows you to multiply both sides of an equation in a system by a number without changing the solutions?

2. What is the goal of multiplying one or both equations in a system?

3. Consider this system. $\begin{cases} 6u - 5v = 2 & (1) \\ 12u - 8v = 5 & (2) \end{cases}$
   a. If equation (1) is multiplied by __?__, then adding the equations will eliminate u.
   b. Solve the system.

4. In Example 2, equation (1) could have been multiplied by 3 and equation (2) could have been multiplied by -5. Use this procedure to verify the solution.

5. Consider this system. $\begin{cases} 3a - 2b = 20 & (1) \\ 9a + 4b = 40 & (2) \end{cases}$
   a. If equation (1) is multiplied by -3, and the equations are added, what is the resulting equation?
   b. If equation (1) is multiplied by 2, and the equations are added, what is the resulting equation?
   c. Use one of these methods to solve the system.

6. Use the hexagon and square formations of Example 3. Will there be an exact fit if the marching band consists of 100 band members and 42 pompon people?

In 7–10, solve the system.

7. $\begin{cases} 6m - 7n = 6 \\ 7m - 8n = 15 \end{cases}$ 8. $\begin{cases} 3x = 4y + 2 \\ 9x - 5y = 7 \end{cases}$

9. $\begin{cases} 5x + y = 30 \\ 3x - 4y = 41 \end{cases}$ 10. $\begin{cases} 3a = 2b + 5 \\ a - 4b = 6 \end{cases}$

11. A marching band has 67 band members and 47 pompon people. They wish to form pentagons and rectangles like those diagrammed below. Will every person have a spot? If so, how?

	B	B			B	P	B
B	P	B		P	P	P
	B				B	P	B

12. The sum of two numbers is 45. Three times the first number plus seven times the second is 115. Find the two numbers.

13. A test has m multiple-choice questions and t true-false questions. If multiple-choice questions are worth 7 points each and the true-false questions 2 points each, the test will be worth a total of 185 points. If multiple-choice and true-false questions are each worth 4 points each, the test will be worth a total of 200 points. Find m and t.

In 14 and 15, solve the system.

14. $\begin{cases} 2m - 5n = 0 \\ 6m + n = 0 \end{cases}$

15. $\begin{cases} 4x - 3y = 2x + 5 \\ 8y = 5x - 13 \end{cases}$

In 16 and 17, solve the system.

16. $\begin{cases} 10x + 5y = 32 \\ 8x + 5y = 10 \end{cases}$ *(Lesson 11-4)*

17. $\begin{cases} y = \frac{1}{2}x + 3 \\ y = \frac{1}{3}x - 2 \end{cases}$ *(Lesson 11-3)*

18. Molly has \$400 and saves \$25 a week. Vince has \$1400 and spends \$25 a week.
  a. How many weeks from now will they each have the same amount of money?
  b. What will this amount be? *(Lesson 11-3)*

In 19 and 20, solve each equation or inequality.

19. $x^2 + 5 = 13$ *(Lesson 7-4)*

20. $3y - 2(4 + 2y) = 0$ *(Lesson 6-5)*

21. In December 1986, Jeana Yeager and Dick Rutan flew the Voyager airplane nonstop around the earth without refueling. The average rate for the 25,012-mile trip was 115 mph. How many days was this flight? *(Lesson 5-2)*

22. *Pythagorean triples* are three whole numbers A, B, and C, such that $A^2 + B^2 = C^2$. Here is a program that will generate a Pythagorean triple from two positive integers M and N where $M > N$. *(Lesson 7-5)*

```
10 PRINT "PYTHAGOREAN TRIPLES"
20 PRINT "ENTER M"
30 INPUT M
40 PRINT "ENTER N"
50 INPUT N
60 LET A = M * M − N * N
70 LET B = 2 * M * N
80 LET C = M * M + N * N
90 PRINT "A = ";A
100 PRINT "B = ";B
110 PRINT "C = ";C
120 END
```

a. What will the program print when M = 5 and N = 3?
b. What will the program print when M = 7 and N = 1?
c. Verify that the triples in part a and part b satisfy the Pythagorean Theorem.

23. Two dice are tossed once. What is the probability of getting a sum of 8? *(Lesson 1-7)*

24. *True or false (Lesson 1-8)*
a. Probabilities are numbers between 0 and 1, inclusive.
b. A probability of 1 means that an event must occur.
c. A relative frequency of -1 cannot occur.

Exploration

25. If your school has a band and pompon squad, determine if all members could fit exactly into the formations of Example 3.

26. Create formations that a band of 80 and pompon squad of 30 would fit exactly.

11-6

Weighted Averages and Mixtures

Mrs. Counts, a mathematics teacher at Eastwestern H.S., told her class that the final exam was worth 2 tests. Before the final exam she had given 6 tests.

Dennis averaged 91 on the 6 tests and got 84 on the final exam. He computed his average for the course by using a **weighted average.**

$$\tfrac{6}{8} \cdot 91 + \tfrac{2}{8} \cdot 84 = 89.25$$

Notice that 91 and 84 are scores per test. They are rates. The quantity $\tfrac{6}{8} \cdot 91$ comes from the fact that 6 tests out of a total of 8 averaged 91. The quantity $\tfrac{2}{8} \cdot 84$ comes from the fact that 2 tests (the final) out of a total of 8 scored 84. The ratios $\tfrac{6}{8}$ and $\tfrac{2}{8}$ are called weights. A **weight** is the part or percent of the total accounted for by each score.

Suppose two rates x_1 and x_2 have weights W_1 and W_2, with $W_1 + W_2 = 1$. Then the weighted average of the rates is

$$W_1 x_1 + W_2 x_2.$$

For Dennis's weighted average, his average scores $x_1 = 91$ and $x_2 = 84$ had weights $W_1 = \tfrac{6}{8}$ and $W_2 = \tfrac{2}{8}$. The weighted average 89.25 is closer to x_1 than to x_2. This is because x_1's weight is greater than x_2's weight.

Weighted averages are found when two or more quantities are mixed or put together to produce a new quantity.

Example 1

A store owner mixes 12 pounds of peanuts worth $1.59 per pound with 8 pounds of cashews worth $4.99 per pound. What is the resulting mixture worth per pound?

Solution Here, x_1 and x_2 are the rates $1.59/pound and $4.99/pound Now compute the weights in the mixture. There are altogether 12 + 8 = 20 pounds of nuts, so the weight for peanuts is $\frac{12}{20}$ and the weight for cashews is $\frac{8}{20}$. Compute the total value:

$$\text{value of peanuts} + \text{value of cashews} = \text{worth of mix}$$
$$\tfrac{12}{20} \cdot 1.59 \quad + \quad \tfrac{8}{20} \cdot 4.99 \quad = 2.95$$

The mixture is worth $2.95 per pound.

If you know speeds going and returning on a trip, the average speed for the entire trip is a *weighted* average of the speeds used on the trip. The weights are the amount of time traveled, not distance.

Example 2

Suppose towns A and B are 120 miles apart. If a car averages 60 mph from A to B and 40 mph on the way back, what is the average speed for the round trip?

Solution The rates are easy: 60 mph and 40 mph. To find the weights, determine the *time* traveled at each speed.

From A to B, $\quad t = \dfrac{d}{r} = \dfrac{120}{60} = 2$ hours.

From B to A, $\quad t = \dfrac{d}{r} = \dfrac{120}{40} = 3$ hours.

The entire round trip took 5 hours. Since 2 out of the 5 hours were at 60 mph and 3 out of the 5 hours were at 40 mph, the weights are $\frac{2}{5}$ and $\frac{3}{5}$. The average speed is

$$\tfrac{2}{5}(60) + \tfrac{3}{5}(40) = 48.$$

The average speed is 48 mph.

Check The round trip takes 5 hours. The distance traveled is 240 miles. This gives an average speed of 48 mph.

Pharmacists and other chemists often face situations involving weighted averages.

Example 3   Five ounces of a 30% alcohol solution are mixed with ten ounces of a 50% alcohol solution. What is the percent of alcohol in the result?

Solution   Here the rates are 30% and 50% (alcohol/total). The result is a solution with 15 ounces. The percent of alcohol in it is a weighted average of 30% alcohol and 50% alcohol.

$$\tfrac{5}{15} \cdot 30\% + \tfrac{10}{15} \cdot 50\%$$
$$= 10\% + 33\tfrac{1}{3}\%$$
$$= 43\tfrac{1}{3}\%$$

The resulting mixture is $43\tfrac{1}{3}\%$ alcohol.

Check   The 30% solution has 30% · 5 ounces, or 1.5 ounces of alcohol. The 50% solution has 50% · 10 ounces, or 5 ounces of alcohol. The resulting $43\tfrac{1}{3}\%$ solution has $43\tfrac{1}{3}\%$ · 15 ounces of alcohol. This should compute to 6.5 ounces of alcohol, which it does.

The weighted average of more than two quantities is found just like other weighted averages.

Example 4   In Mr. Vollmer's class, 3 students have no siblings (brothers or sisters), 8 students have 1 sibling per student, 10 have 2 siblings, 4 have 3 siblings, and 1 has 4 siblings. What is the average number of siblings per student?

Solution   The rates are the siblings per students. There are 26 students. Use a weighted average of 5 quantities.

$$\tfrac{3}{26} \cdot 0 + \tfrac{8}{26} \cdot 1 + \tfrac{10}{26} \cdot 2 + \tfrac{4}{26} \cdot 3 + \tfrac{1}{26} \cdot 4$$
$$= \tfrac{44}{26}$$
$$\approx 1.7$$

The students average 1.7 siblings.

Example 4 illustrates the following generalization of the definition of weighted average.

> Suppose n quantities $x_1, x_2, \ldots, x_n$ have weights $W_1, W_2, \ldots, W_n$ with $W_1 + W_2 + \ldots + W_n = 1$. Then the weighted average of the quantities is $W_1x_1 + W_2x_2 + \ldots + W_nx_n$.

Questions

Covering the Reading

1. Denise is a student in Mrs. Counts's class. She received a 75 average on the 6 tests and an 88 on the final.
   a. Fill in the missing weights in Denise's weighted average.

$$\underline{\ ?\ } \cdot 75 + \underline{\ ?\ } \cdot 88$$

   b. Compute her average for the course.

2. What is the weighted average of quantities x_1 and x_2 with weights W_1 and W_2?

3. What is the weighted average of quantities $x_1, x_2, \ldots, x_n$ with weights $W_1, W_2, \ldots, W_n$?

4. Refer to Example 1. Suppose the store owner mixes 15 pounds of peanuts with 10 pounds of cashews. What is the resulting mixture worth per pound?

5. Another store owner mixes 6 pounds of almonds worth $3.99 per pound with 9 pounds of peanuts worth $1.79 per pound. What is the resulting mixture worth per pound?

6. When average speeds are computed, what is the unit of the weights used?

7. Suppose towns A and B are 275 miles apart. If a car averages 55 mph from A to B and 25 mph on the way back, what is the average speed for the whole trip?

8. Six liters of a 10% alcohol solution are mixed with three liters of a 100% solution (that is, pure alcohol). What is the percent of alcohol in the resulting solution?

9. Refer to Example 4. A new student enrolled in Mr. Vollmer's class. This student has a brother and no sisters. Compute the average number of siblings per student in Mr. Vollmer's class including this new student. (Answer to the nearest tenth.)

Applying the Mathematics

10. A company puts out a party mix using 11 pounds of cereal at $.79 per pound, 6 pounds of raisins at $2.29 per pound and 13 pounds of sunflower seeds at $4.19 per pound. What is the resulting mixture worth per pound?

11. A plane makes a trip of 600 miles each way. Going, it travels at 400 mph, while returning it travels at 500 mph. What is the average speed for the round trip?

12. At left is a dot frequency diagram that shows the number of students in Mr. Roosevelt's homeroom who take various numbers of courses. What is the average number of courses taken by a student?

13. a. Five nights a week I get 7 hours of sleep. The other two nights I get 9 hours of sleep. What is my average number of hours of sleep a night in a week?

b. Five nights a week I get x hours of sleep. The other two nights I get y hours of sleep. What is my average number of hours of sleep a night in a week?

In 14 and 15, use this information. School grade point averages often use the following scale.

$A = 4.0$ $B = 3.0$ $C = 2.0$ $D = 1.0$ $F = 0.0$

14. Carla has the following transcript: 3A's, 6B's, 4C's, and 1D. What is her grade point average? (Answer to the nearest hundredth.)

15. Maurice needs a 2.0 grade point average to remain eligible for football. He has 2A's, 1B, 4C's, 4D's and 1F. Does he qualify?

Review

16. *Skill sequence*   Solve. *(Lessons 5-7, 6-1)*
 a. $\frac{3}{4}x + \frac{1}{4} = 15$
 b. $\frac{3}{4y} + \frac{1}{4y} = 15$
 c. $\frac{3}{z} \cdot 4 + \frac{1}{z} \cdot 4 = 15$

In 17 and 18, solve the system. *(Lessons 11-4, 11-5)*

17. $\begin{cases} 4x + 9y = -11 \\ -4x + y = -19 \end{cases}$ **18.** $\begin{cases} 3x - 6y = 15 \\ 2x + 5y = 0 \end{cases}$

19. Twice one number plus 3 times a second number is 22. The first number minus the second number is 31.
 a. Write a system describing the problem.
 b. Find the values of the two numbers. *(Lesson 11-5)*

20. Three events have probabilities of 0.75, 0.192, and 0.4. Which of these events is most likely to happen? *(Lesson 1-7)*

21. Solve $10x^2 = 320$. *(Lessons 7-4, 7-6)*

Exploration

22. a. What grading scale is used in your school?
 b. Compute your grade point average for last semester on this scale. (If your school has no grading scale, use that given in Questions 14–15.)

11-7

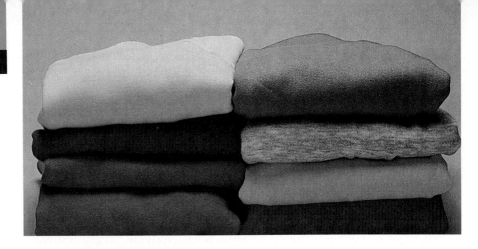

More with Mixtures

In the previous lesson you computed a weighted average when you knew weights and quantities. Often, you may be presented with a reverse situation. You know the average and want to find the amount of one of the quantities or weights.

Example 1 A retailer sold 400 sweatshirts at $10 retail and has 250 sweatshirts left. For what sale price should each of the remaining sweatshirts be sold to have an overall average price of $9 a sweatshirt?

Solution Let the sale price be S dollars per shirt. There are a total of 650 sweatshirts to start with. Think of the $10, S, and $9 as weights. Because the average is to be $9, S must be less than 9. Calculate the sales from each price.

$$\left(\begin{array}{c}\text{sales from}\\ \$10 \text{ shirts}\end{array}\right) + \left(\begin{array}{c}\text{sales from}\\ \$S \text{ shirts}\end{array}\right) = \left(\begin{array}{c}\text{sales from}\\ \$9 \text{ shirts}\end{array}\right)$$

$$400 \cdot 10 \quad + \quad 250 \cdot S \quad = \quad 650 \cdot 9$$

$$4000 + 250S = 5850$$
$$250S = 1850$$
$$S = \frac{1850}{250} = 7.4$$

The sale price should be $7.40.

Check 400 sweatshirts at $10 is $4000, and 250 sweatshirts at $7.40 is $1850; so the total sales are $5850. Does this equal 650 sweatshirts sold at $9? Yes, since $650 \cdot 9 = 5850$.

In Example 1, you were asked to find the amount of a weight in a weighted average situation. Example 2 is a similar kind of problem in which a system is used to find a missing quantity.

Example 2 A truck averages 35 mph for the first four hours of a trip. How long must the truck travel at 55 mph to have an overall average speed of 50 mph?

Solution You can use a system.
Let x = number of hours at 55 mph.
Let y = total number of hours for the trip.
Since the total number of hours is 4 more than the number of hours at 55 mph,

$$y = 4 + x. \qquad (1)$$

The other equation in the system comes from the total distance.

Because 4 out of a total y hours are traveled at 35 mph and x hours are traveled at 55 mph, the total distance is

$$4 \cdot 35 + x \cdot 55$$

The weighted average speed is given as 50 mph. So, in all, the driver traveled $50y$ miles.

$$140 + 55x = 50y \qquad (2)$$

Now the system is set up for solution.

$$\begin{cases} y = 4 + x & (1) \\ 140 + 55x = 50y & (2) \end{cases}$$

We use substitution. Substitute $4 + x$ for y in equation (2).

$$140 + 55x = 50(4 + x)$$
$$140 + 55x = 200 + 50x$$
$$5x = 60$$
$$x = 12$$

The truck must travel 12 hours at 55 mph to average 50 mph for the trip. (Notice that you did not need to find y for this solution.)

Check 4 hours at 35 mph is 140 miles, and 12 hours at 55 mph is 660 miles; so the total trip is 800 miles. Does this equal 16 hours at 50 mph? Yes, since $16 \cdot 50 = 800$.

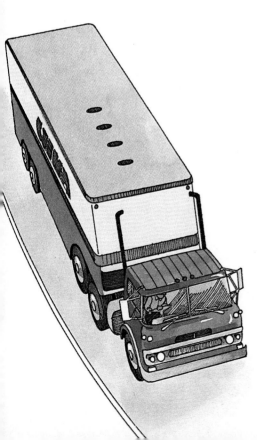

Questions

Covering the Reading

1. In Example 1, how do you know the sale price must be less than $9?

2. If the retailer of Example 1 sold 400 sweatshirts at $10 retail and had 150 sweatshirts left, at what sale price should the remaining sweatshirts be sold to have an overall average price of $9 a sweatshirt?

3. A retailer sold 130 pairs of sunglasses at $15 retail and has 70 pairs left. At what sale price should the remaining sunglasses be sold to have an overall average price of $13 a pair?

In 4 and 5, solve.

4. $\begin{cases} y = x + 8 \\ 200 + 12x = 20y \end{cases}$

5. $\begin{cases} q = z + 9 \\ 27z + 423 = 36q \end{cases}$

6. A truck averages 40 mph for the first 4 hours of a trip. How long must the truck travel at 55 mph to have an overall average speed of 50 mph?

Applying the Mathematics

7. A car averages 55 mph for the first 2 hours of a trip. How long can the car stop (travel at 0 mph) and still have an average speed of 40 mph when it returns to the road?

8. After 18 courses, Mahesh has a grade point average of 3.1. What must he average over the next 14 courses to have an overall grade point average of 3.25?

9. After 8 tests, Cyril has an 83 average. The final is worth 3 tests. What must he score on the final to have an overall average of 81?

10. After 6 tests, Janice has a 92 average. What can she average on the 4 remaining tests and still maintain a 90 overall average?

11. How many pounds of peanuts at $1.79 per lb must be mixed with 10 pounds of cashews at $3.99 per lb to produce a mixture worth $2.59 per lb?

12. How many pounds of peanuts at $1.79 per lb must be mixed with 10 pounds of cashews at $3.99 per lb and 2 pounds of pecans at $3.59 per lb to produce a mixture worth $2.59 per lb?

Review

13. Suppose 12 liters of a 40% alcohol solution are mixed with 3 liters of an 80% alcohol solution. What is the percent of alcohol in the new solution? *(Lesson 11-6)*

14. At the left is a diagram of numbers of barrels of various weights in a storehouse. What is the average weight of a barrel? *(Lesson 11-6)*

weight (kg) of barrels

15. A band has 94 members (B) with an additional 28 flag bearers (F). They plan to form the following formations:

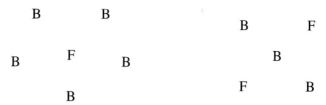

Will they exactly fit these formations? *(Lesson 11-5)*

16. On the same set of coordinate axes, **a.** graph $y = \frac{1}{2}x - 2$ and $2y - x = 10$ and **b.** give the coordinates of their point of intersection. *(Lessons 7-1, 11-1, 11-4)*

17. A point is selected at random from $\overline{AE}$ below. What is the probability that the point is on $\overline{AB}$? *(Lesson 5-5)*

```
    A      B   C      D     E
 ◄──┼──────┼───┼──────┼─────┼──────►  x
    2     10  15     23    32
```

18. A die is tossed once. What is the probability of tossing a number that is
a. even and less than 3?
b. odd and greater than 6? *(Lessons 1-7, 3-4)*

19. Write an inequality describing all numbers x such that $x > 0$ and $x < 12$. *(Lesson 3-4)*

20. Calculate in your head. *(Lesson 5-4)*
a. 5% of 200 **b.** 10% of 32 **c.** 200% of .5

Exploration

21. Weighted averages may be used to compare opinions about things. Cities are sometimes rated on the following factors:
(1) climate
(2) attractions (sports, museums, etc.)
(3) cost of living
(4) availability of good jobs.
a. Assign positive weights adding to 1 to these factors by how important each factor is to you. For instance, you could assign 30% to climate, 40% to attractions, 10% to cost of living, and 20% to jobs.
b. Ask an adult in your family what weights he or she would assign.
c. Rate your home town from 1 to 10 in each category.
d. Compute the weighted average.
e. We gave New York the following category ratings:
(1) 4   (2) 9   (3) 2   (4) 6.
Does your hometown do better on your ratings?

Parallel Lines

The idea behind parallel lines is that they "go in the same direction." All vertical lines are parallel to each other. So are all horizontal lines. But not all oblique lines are parallel. For oblique lines to be parallel, they must have the same slope.

If two lines have the same slope, then they are parallel.

You have learned that when two lines intersect in exactly one point, the coordinates of the point of intersection can be found by solving a system. But what happens when the lines are parallel? Consider this linear system.

$$\begin{cases} 2x + 3y = -6 & (1) \\ 4x + 6y = 24 & (2) \end{cases}$$

As usual, you can solve the system by multiplying the top equation by -2.

$$\begin{array}{ll} -4x - 6y = 12 & (3) = -2 \cdot (1) \\ \underline{4x + 6y = 24} & (2) \end{array}$$

If you add you get $\quad 0 = 36 \quad (3) + (2)$

which is impossible. This signals that the system has no solution, or no point of intersection. To check, graph the lines. The graph shows that the lines are parallel with no points in common.

As another check, put the equations for the lines in slope-intercept form.

line (1) $2x + 3y = -6$
$$3y = -2x - 6$$
$$y = -\tfrac{2}{3}x - 2$$

line (2) $4x + 6y = 24$
$$6y = -4x + 24$$
$$y = -\tfrac{2}{3}x + 4$$

Both lines (1) and (2) have the same slope of $-\tfrac{2}{3}$, but different y-intercepts; thus they are parallel.

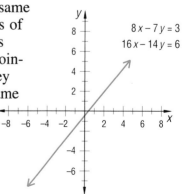

When an equation with no solution (such as $0 = 36$) results from correct work with equations, then the original conditions must also be impossible. There are no pairs of numbers that work in *both* equations (1) and (2).

In the example below, just the opposite occurs. A situation results that is always true.

■ ■ ■ ■ ■ ■ ■ ■

Example Solve.
$$\begin{cases} 8x - 7y = 3 & (1) \\ 16x - 14y = 6 & (2) \end{cases}$$

Solution Multiply the top equation by -2.

$$-16x + 14y = -6 \quad (3) = -2 \cdot (1)$$
$$\underline{16x - 14y = 6 \quad (2)}$$

Now add.

$$0 = 0 \quad (3) + (2)$$

This is always true, so any ordered pair that is a solution to equation (1) is also a solution to equation (2). There are infinitely many points that are on both lines.

You can check that (1) and (2) have the same slope and same y-intercept. So the graphs of equations (1) and (2) are the same line as shown at the right. Because the graphs coincide, the lines are called **coincident**. They too are parallel because they go in the same direction and have the same slope.

You have now studied all the possible number of points in common that two lines in the plane can have.

Number of solutions to system	Line	Slopes
one (the point of intersection)	intersecting	different
zero (no points of intersection)	parallel and nonintersecting	equal
infinitely many (same line)	parallel and coincident	equal

Questions

Covering the Reading

1. Give an example of a system of two nonintersecting lines.

2. Give an example of a system of two coincident lines.

3. Parallel lines have the same __?__.

4. Which two lines are parallel?
   a. $y = 3x + 5$ **b.** $y = 2x + 5$
   c. $y = 3x + 6$ **d.** $x = 2y + 5$

In 5 and 6, **a.** determine whether the system includes nonintersecting or coincident lines, and **b.** check by graphing.

5. $\begin{cases} 2x - 3y = 12 \\ 8x - 12y = 12 \end{cases}$ **6.** $\begin{cases} x - y = 5 \\ y - x = -5 \end{cases}$

In 7–9, match the orientation with the solution.

7. lines intersect   **a.** no solution

8. lines nonintersecting   **b.** infinitely many solutions

9. lines coincident   **c.** one solution

Applying the Mathematics

In 10–13, determine whether the system includes nonintersecting, intersecting, or coincident lines.

10. $\begin{cases} 3u + 2t = 7 \\ 14 - 2t = 6u \end{cases}$ **11.** $\begin{cases} 6a + 2b = 9 \\ 9a + 3b = 12 \end{cases}$

12. $\begin{cases} 2x - 5y = -3 \\ -4x + 10y = 6 \end{cases}$ **13.** $\begin{cases} \frac{1}{2}x - \frac{3}{2} = y \\ 2x + y = 3 \end{cases}$

In 14 and 15, could the given situation have happened?

14. A pizza parlor sold 39 pizzas and 21 gallons of soda for $396. The next day, at the same prices, they sold 52 pizzas and 28 gallons of soda for $518.

568

15. In an all-school survey of 900 students altogether 10% of the boys and 15% of the girls voted yes. In a different survey of 1080 students in the same school, 12% of the boys and 18% of the girls voted yes.

Review

16. Suppose 30 liters of a solution with an unknown percentage of alcohol is mixed with 5 liters of a 90% alcohol solution. If the resulting mixture is a 62% alcohol solution, what is the percentage of alcohol in the first solution? *(Lesson 11-7)*

17. A car travels at an average speed of 25 mph for an hour through a construction zone. How long must the car travel at 55 mph to achieve an overall average speed of 45 mph? *(Lesson 11-7)*

18. If 12 pounds of lettuce at $.79 per pound are mixed with 4 pounds of celery at $.59 per pound and 2 pounds of croutons at $1.69 per pound, how much is the mixed salad worth per pound? *(Lesson 11-6)*

19. Solve by any method. $\begin{cases} 4x + 3y = 7 \\ x - 1.5y = 7 \end{cases}$ *(Lessons 11-1, 11-3, 11-5)*

20. Draw the graph of $y \le 2x - 3$. *(Lesson 8-9)*

21. Write as the square of a binomial: $x^2 - 14x + 49$. *(Lesson 10-8)*

22. Write as the product of two binomials: $4t^2 - 81u^2$. *(Lesson 10-9)*

23. a. Give an equation of the horizontal line through (8, -12).
   b. Give an equation of the vertical line through (8, -12). *(Lesson 7-2)*

24. A sweater costs a store owner $15.50. If the profit is to be 40% of the cost, what is the selling price? *(Lessons 5-4, 6-3)*

Exploration

25. Here are more examples of parallel lines.

Nonintersecting	Coincident
$\begin{cases} 3x + 6y = -2 \\ 4x + 8y = 4 \end{cases}$	$\begin{cases} -2x + 8y = 30 \\ 5x - 20y = -75 \end{cases}$
$\begin{cases} 9x + 3y = 5 \\ 6x + 2y = 1 \end{cases}$	$\begin{cases} 12x - 4y = 80 \\ 15x - 5y = 100 \end{cases}$

Compare the ratios of the coefficients of x, the coefficients of y, and the constant terms. Use the pattern to find a way to recognize parallel lines directly from their equations.

11-9

Situations Which Always or Never Happen

Which job would you take?

Job 1	Job 2
Starting wage $5.60/hour; every 3 months the wage increases $.10/hour.	Starting wage $5.50/hour; every 3 months the wage increases $.10/hour.

Of course, the answer is obvious. Job 1 will always pay better than Job 2. But what happens when this is solved mathematically? Let n = number of 3-month periods worked.

Wage in Job 1
$5.60 + .10n$

Wage in Job 2
$5.50 + .10n$

When is the pay in Job 1 better than the pay in Job 2? The corresponding mathematical problem is to solve this inequality:

$$5.60 + .10n > 5.50 + .10n$$

Add $-.10n$ to each side.

$$5.60 + .10n - .10n > 5.50 + .10n - .10n$$
$$5.60 > 5.50$$

As was the case in solving systems of parallel lines, the variable has disappeared. Since $5.60 > 5.50$ is always true, n can be any real number. Job 1 will always pay a better wage than Job 2, as expected. So for any equations or inequalities we can now make the following generalization.

If, in solving a sentence, you get a sentence which is *always* true, then the original sentence is always true.

When does Job 1 pay *less* than Job 2? To answer this, you could solve:

$$5.60 + .10n < 5.50 + .10n$$
$$5.60 + .10n - .10n < 5.50 + .10n - .10n$$
$$5.60 < 5.50$$

It is never true that 5.60 is less than 5.50. So Job 1 never pays less than Job 2, something which was obvious from the pay rates.

If, in solving a sentence, you get a sentence which is *never* true, then the original sentence is never true.

Example 1 Solve: $5 + 3x = 3(x - 2)$

Solution
$$5 + 3x = 3x - 6$$
$$5 + 3x + \text{-}3x = \text{-}3x + 3x + \text{-}6$$
$$5 = \text{-}6$$

This is never true, so the original sentence has no solution. Write: *no solution*.

Example 2 Solve: $8(2y + 5) < 16y + 60$

Solution
$$16y + 40 < 16y + 60$$
$$\text{-}16y + 16y + 40 < \text{-}16y + 16y + 60$$
$$40 < 60$$

This is always true, so the original sentence is true for every possible value of y. Write: *y may be any real number*.

CAUTION: Here is a problem that looks like the one at the beginning of the lesson but is different.

Example 3 When will the population of the towns be the same?

Town 1
Present population 23,000
Growing 1000 a year

Town 2
Present population 23,000
Growing 1200 a year

Solution In t years:

Population of Town 1
$23,000 + 1000t$

Population of Town 2
$23,000 + 1200t$

The populations will be the same when
$$23,000 + 1000t = 23,000 + 1200t$$
$$23,000 = 23,000 + 200t$$
$$0 = 200t$$
$$0 = t$$

The solution $t = 0$ means their populations are the same *now*. *Having zero for a solution is different from having no solution at all.*

Questions

Covering the Reading

1. Job A pays a starting salary of $6.20 an hour and each year increases $1.00 an hour. Job B starts at $6.00 an hour and also increases $1.00 an hour per year.
   a. When does Job A pay more?
   b. Show how algebra can be used to represent this situation.

2. a. Add $-2x$ to both sides of the sentence $2x + 10 < 2x + 8$. What sentence results?
   b. What should you write to describe the solutions to this sentence?

3. a. Add $5y$ to both sides of the sentence $-5y + 9 = 3 - 5y + 6$. What sentence results?
   b. What should you write to describe the solutions to this sentence?

In 4–7, solve.

4. $2(2y - 5) \leq 4y + 6$

5. $3x + 5 = 5 + 3x$

6. $-2m = 3 - 2m$ 7. $2A - 10A > 4(1 - 2A)$

8. The population of City 1 is about 200,000 and growing at about 5000 people a year. City 2 has a population of about 200,000 and is growing at about 4000 people a year.
   a. In y years, what will be the population of City 1?
   b. In y years what will be the population of City 2?
   c. When will their populations be the same?

Applying the Mathematics

In 9–11, consider the following information. Apartment A rents for $375 per month with a $500 security deposit. Apartment B rents for $315 per month with a $400 security deposit, but the renter must pay $60 per month for utilities.

9. a. What sentence would you solve to find out when apartment A is cheaper?
   b. Solve this sentence.

10. a. What sentence would you solve to find out when apartment B is cheaper?
    b. Solve this sentence.

11. If Naomi wanted to rent an apartment for two years, which one is cheaper?

Review

12. Solve $\begin{cases} 4x - 3y = 12 \\ 8x - 6y = 24 \end{cases}$ *(Lesson 11-8)*

13. A retailer sold 220 records at $10 each. What can the sale price be of the 110 remaining records to have an overall average price of $8? *(Lesson 11-7)*

14. Solve the system by graphing. $\begin{cases} y = -x \\ y = x + 3 \end{cases}$ *(Lesson 11-1)*

15. Solve: $-(a - 4) = 2a + 3$. *(Lessons 6-5, 6-6)*

16. Anna spends $1.30 on breakfast, then d dollars on lunch, then $2 on flowers. *(Lessons 2-1, 3-3)*
   a. How much does she spend altogether?
   b. She had $10 to start with, and now has $1.25 left. If she spent no other money that day, how much did she spend on lunch?

17. A book has an average of 25 lines per page, w words per line, and 21 pages per chapter. Estimate the number of words per chapter in this book. *(Lesson 4-2)*

18. Simplify in one step. *(Lesson 9-8)*
   a. $\left(\dfrac{2}{3}\right)^2$ **b.** $(5d^2g)^2$ **c.** $\dfrac{1}{4}\left(\dfrac{c}{a}\right)^2$

19. A person wants 10,000 square meters of floor space for a store. Walking by a set of vacant stores he wonders if the 4 stores together will meet his requirements. It is possible to measure their widths, and they are given in the floor plan below. How deep must these stores be to give the required area? *(Lessons 4-1, 4-4)*

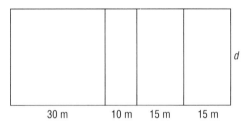

30 m   10 m   15 m   15 m

20. *Skill sequence*   Solve. *(Lesson 5-7)*
   a. $\dfrac{x}{2} = \dfrac{7}{8}$
   b. $\dfrac{3.5}{y} = \dfrac{2}{y + 2}$
   c. $\dfrac{z + 2}{2} = 4$

Exploration

21. When solving the equation $ax + b = cx + d$ for x, there may be no solution, exactly one solution, or infinitely many solutions. What must be true about a, b, c, and d to guarantee that there is exactly one solution?

11-10

Systems of Inequalities

In Lesson 7-2, you graphed linear inequalities like $x > -5$ and $y \leq 6$ on a plane. These sentences describe half-planes with boundary lines that are horizontal or vertical. In Lesson 8-9, you went on to graph half-planes with boundaries that are oblique lines. Now you will graph regions described by two or more inequalities. Solving a **system of inequalities** involves finding the common solutions of two or more linear inequalities.

Recall that the graph of $Ax + By < C$ is a half-plane. It lies on one side of the boundary line $Ax + By = C$. When systems involve linear inequalities, their solutions are intersections of half-planes.

Example 1 Graph all solutions to the system

$$\begin{cases} y \geq -3x + 2 & (1) \\ \quad y < x - 2 & (2) \end{cases}$$

Solution First graph the boundary line $y = -3x + 2$ for the first inequality. This line contains $(0, 2)$ and $(1, -1)$. The graph of $y \geq -3x + 2$ consists of points on or above the line with equation $y = -3x + 2$.

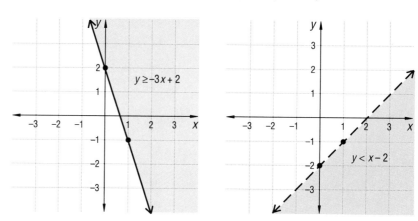

Next graph the boundary line $y = x - 2$ for the second inequality. Two parts on this line are $(0, 2)$ and $(1, -1)$. The graph of $y < x - 2$ consists of points below the line with equation $y = x - 2$. Points on the line are excluded.

We have drawn the half-planes on different axes only to make it easier to see them. You should draw them on the same axes. The part of the plane marked with both types of shading is the solution set for the system. Geometrically, it can be described as the interior of an angle with vertex (1, -1).

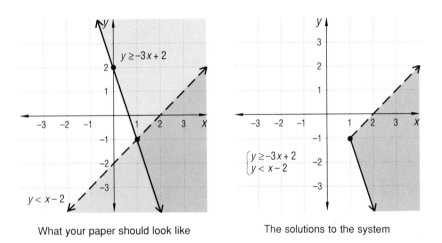

What your paper should look like   The solutions to the system

Example 2 involves a system of inequalities quite often seen in applications.

Example 2   Graph all solutions to the system

$$\begin{cases} x > 0 \\ y > 0 \end{cases}$$

Solution   The graph of the solution to the system is the intersection of the left and center sets graphed below. It is the part of the right graph that has both colors.

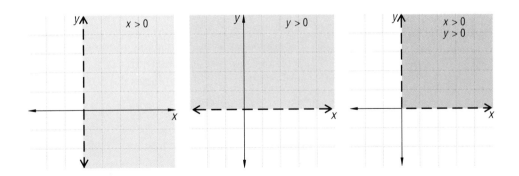

Example 3   Suppose the sum of two positive numbers x and y is less than 50 and greater than 25. The desired numbers are the solution to this system of inequalities.

$$\begin{cases} x > 0 \\ y > 0 \\ x + y < 50 \\ x + y > 25 \end{cases}$$

Solution   The graph of the solution to the system is the intersection of the four sets graphed below.

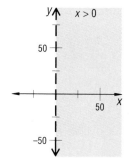 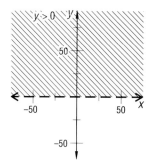 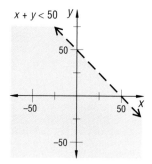 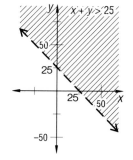

The result below shows only the intersection of the four graphs drawn on the same set of axes.

Graph of all solutions to the system.

$$\begin{cases} x > 0 \\ y > 0 \\ x + y < 50 \\ x + y > 25 \end{cases}$$

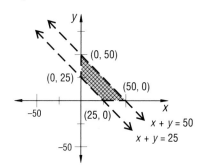

Notice that there are infinitely many solutions to the system, so they cannot be listed. But the graph is easy to describe. It is the interior of the quadrilateral with vertices (0, 50), (50, 0), (25, 0), and (0, 25).

Systems of inequalities arise from many different kinds of situations. Here is one.

Example 4 In basketball a player scores 2 points for a basket and 1 point for a free throw. Suppose a player has scored no more than 20 points in a game. How many baskets *b* and free throws *f* could the player have made?

Solution You could answer this question by trial and error, but there are a lot of possibilities. It is much easier to show the answers on a graph. The numbers *b* and *f* must be positive integers or zero.

$$b \geq 0 \text{ and } f \geq 0$$

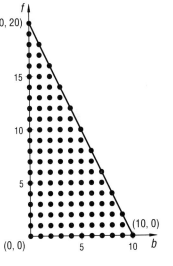

Since the total number of points is less than or equal to 20,

$$2b + f \leq 20.$$

Since this is a discrete situation, only the dotted points on or above the *b*-axis, on or to the right of the *f*-axis, and on or below $2b + f = 20$ are possible solutions.

For example, the point (5, 3) represents the possibility of the player making 5 baskets and 3 free throws. Altogether there are 121 possibilities. The above graph is the solution to this system of inequalities.

$$\begin{cases} b \geq 0 \\ f \geq 0 \\ 2b + f \leq 20 \end{cases}$$

If the situation allowed fractions, the entire triangle above would be shaded in.

In 1–4, refer to Example 1.

1. The graph of all solutions to the system
$$\begin{cases} y \geq -3x + 2 \\ y < x - 2 \end{cases}$$
is the __?__ of the graphs of $y \geq -3x + 2$ and $y < x - 2$.

2. The graph of $y < x - 2$ is a __?__.

3. Why does the graph of $y \geq -3x + 2$ include its boundary line?

4. Is (2, 1) a solution to this system?

5. The graph of $\begin{cases} x > 0 \\ y > 0 \end{cases}$ consists of all points in which quadrant?

6. Graph the system $\begin{cases} x < 0 \\ y > 0 \end{cases}$.

In 7 and 8 consider the system of Example 3.
$$\begin{cases} x > 0 \\ y > 0 \\ x + y < 50 \\ x + y > 25 \end{cases}$$

7. The graph of all solutions to this system is the interior of a quadrilateral with what vertices?

8. Is (25, 25) a solution to the system?

In 9–12, refer to Example 4.

9. Give at least three possible combinations of baskets and free throws that would total exactly 20 points.

10. For which part of the plane is $b \geq 0$?

11. For which part of the plane is $f \geq 0$?

12. Suppose the player made at least five baskets. How many possibilities are there for $2b + f \leq 20$?

In 13 and 14, describe the system graphed.

13.

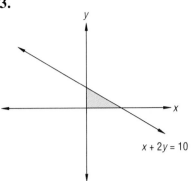

$x + 2y = 10$

14.

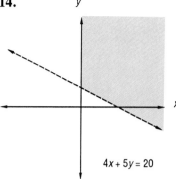

$4x + 5y = 20$

In 15 and 16, graph the solution set.

15. $\begin{cases} x > 0 \\ y > 0 \\ x + y < 6 \end{cases}$

16. $\begin{cases} x \geq -1 \\ y \leq 2 \\ y \geq x - 1 \end{cases}$

In 17 and 18, accurately graph the set of points that satisfies each situation.

17. It takes a good typist about 10 minutes to type a letter of moderate length and about 8 minutes to type a normal double-spaced page. About how many letters L and how many pages P will keep a typist busy for 1 hour or less?

18. A hockey team estimates that it needs at least 16 points to make the playoffs. A win is worth 2 points and a tie is worth 1 point. How many wins w and ties t will get the team into the playoffs?

Review

19. Solve. *(Lesson 11-9)*
    a. $6(3 - 2x) = -12(x - 3)$
    b. $6(3 - 2y) = -12(y - \frac{3}{2})$

20. Consider the equation $\dfrac{6(z - 3)}{2(z - 3)} = 3$.
    a. What value can z not have?
    b. Solve for z. *(Lesson 11-9)*

In 21 and 22, determine whether the lines are parallel, coincident, or intersecting. *(Lesson 11-8)*

21. $\begin{cases} 4x - y = 8 \\ 8x - y = 8 \end{cases}$
        22. $\begin{cases} y = 3x + 9 \\ 6x - 2y = -27 \end{cases}$

23. Solve the system using any method you wish. $\begin{cases} 3x + 5y = 4 \\ 2x - 3y = 9 \end{cases}$
    (Lessons 1-3, 11-1, 11-5)

24. Given the system $\begin{cases} w = -9z & (1) \\ z - 2w = 323 & (2) \end{cases}$, a student substituted $-9z$ for w in equation (2). The student wrote $z - 18z = 323$.
    a. Is this result correct?
    b. If it is correct, finish solving the system. If not, describe what is wrong with it. *(Lessons 11-2, 11-3)*

25. For what value of c will $x^2 + 6x + c$ be a perfect square? *(Lesson 10-7)*

26. If $x^2 = 121$, then $x = \underline{\ ?\ }$ or $\underline{\ ?\ }$. *(Lesson 7-4)*

27. Suppose a bank offers an 8.5% annual yield. What would be the amount in an account after 5 years if $1000 is invested? *(Lesson 9-1)*

28. Maureen has an ID number that consists of two letters followed by four single-digit numbers. How many different possible ID numbers can be made this way? *(Lesson 4-7)*

Exploration

29. The graph of the solution to the system of inequalities on page 576 is a special type of quadrilateral called an *isosceles trapezoid*. Look in a dictionary or geometry book for a definition of isosceles trapezoid.

Summary

A system is a set of sentences in which you want to find all the solutions common to each sentence in the set. Thus the solution set for a system is the intersection of the solutions for the individual sentences.

One way to solve a system is by graphing. By looking for intersection points on a graph you can quickly tell if there are any solutions to the system. There are as many solutions as intersection points. Graphing is also a way to describe solutions to systems that have infinitely many solutions. For instance, graphs show the solutions to systems of linear inequalities.

However, graphing does not always yield exact solutions. In this chapter you studied three strategies for finding exact solutions to systems of

linear equations. They are substitution, addition, and multiplication. Substitution is a good method to use if an equation is given in $y = mx + b$ form. Addition works if opposite terms are in two equations. Multiplication is a good method when both equations are in $Ax + By = C$ form.

Any kind of situation that leads to a linear equation can lead to a linear system. All that is needed is more than one condition to be satisfied. Problems involving mixtures were solved using weighted averages.

The graph of all solutions to a system of inequalities can be found by graphing each inequality on the same axes. The solution is the intersection of all the individual graphs.

Vocabulary

Below are the new terms and phrases for this chapter. You should be able to give a general description and a specific example for each.

Lesson 11-1
system
solution set for a system
linear systems

Lesson 11-2
substitution method for solving a system

Lesson 11-4
Generalized Addition Property of Equality
addition method for solving a system

Lesson 11-5
multiplication method for solving a system

Lesson 11-6
weighted average
weight

Lesson 11-8
coincident lines

Lesson 11-10
system of inequalities

Progress Self-Test

Take this test as you would take a test in class. Then check your work with the solution in the Selected Answers section in the back of the book.

In 1 and 2, solve.

1. $\begin{cases} a - 3b = \text{-}8 \\ \qquad b = 3a \end{cases}$

2. $\begin{cases} p = 5r + 80 \\ p = \text{-}7r - 40 \end{cases}$

3. Line 1 has equation $y = \text{-}5x - 15$. Line 2 has equation $y = x - 3$. Find the point of intersection.

In 4–6, solve.

4. $\begin{cases} m - n = \text{-}1 \\ \text{-}m + 2n = 4 \end{cases}$

5. $\begin{cases} 2x + y = \text{-}6 \\ 2x + 3y = 10 \end{cases}$

6. $\begin{cases} 7x + 3y = 1 \\ 4x - y = 6 \end{cases}$

In 7 and 8, determine whether the lines coincide, intersect, or are parallel and nonintersecting.

7. $\begin{cases} 5 = 2A + 7B \\ 10 = 4A + 14B \end{cases}$

8. $\begin{cases} \quad y = 2x + 7 \\ y - 2x = 3 \end{cases}$

In 9 and 10, solve and check.

9. $12z + 8 = 12z - 3$

10. $\text{-}19p < 22 - 19p$

11. Lisa weighs 4 times as much as her baby sister. Together they weigh 92 pounds. How much does each person weigh?

12. The Reid family goes to a restaurant and orders 3 hamburgers and 4 small salads. Without tax the bill comes to $21.30. At the same restaurant the Millers order 5 burgers and 2 small salads. Their bill without tax is $22.90. What is the cost of a small salad?

13. If the cost per unit is constant, could the following situation have happened?
   10 roses and 15 daffodils were sold for $35.
   2 roses and 3 daffodils were sold for $8.

14. South Carolina had a 1980 population of about 3,100,000 and was growing at a rate of 55,000 people each year. Louisiana had a 1980 population of about 4,200,000 and was growing at a rate of 55,000 people each year. If these rates continue, in how many years after 1980 will these two states have the same population?

15. Suppose 14 liters of a 40% salt solution are mixed with 6 liters of an 80% salt solution. What is the percent of salt in the resulting mixture?

16. A car averages 55 mph for the first 3 hours of a trip. How long can it travel at 40 mph and still maintain an overall average speed of 45 mph?

In 17 and 18, solve the system by graphing.

17. $\begin{cases} \quad y = 3x - 2 \\ x + y = 2 \end{cases}$

18. $\begin{cases} \qquad x \geq 0 \\ \qquad y \geq 0 \\ x + y \leq 20 \\ x + y \geq 10 \end{cases}$

Chapter Review

Questions on **SPUR** Objectives

SPUR stands for **S**kills, **P**roperties, **U**ses, and **R**epresentations.
The Chapter Review questions are grouped according to the
SPUR Objectives for this chapter.

SKILLS deal with the procedures used to get answers.

■ **Objective A.** *Solve systems using substitution.*
(Lessons 11-2, 11-3)

1. Determine (a, b).
$$\begin{cases} b = 3a \\ 60 = a + b \end{cases}$$

2. Determine (x, y).
$$\begin{cases} x - y = 13 \\ \quad x = 6y - 7 \end{cases}$$

3. Solve for c, p, and q.
$$\begin{cases} p = 2c \\ q = 3c \\ p + 6q = 200 \end{cases}$$

4. Solve the system.
$$\begin{cases} 10y = 20x + 20 \\ 2x + 4y = 29 \end{cases}$$

In 5–8, solve.

5. $\begin{cases} y = x + 5 \\ 300 + 10x = 35y \end{cases}$

6. $\begin{cases} q = z + 8 \\ 14z + 89 = 13q \end{cases}$

7. $\begin{cases} a = 2b + 3 \\ a = 3b + 20 \end{cases}$

8. $\begin{cases} 16 - 2x = y \\ \quad x + 4 = y \end{cases}$

In 9 and 10, two lines have the given equations.
Find the point of intersection.

9. Line 1: $y = 7x + 20$
Line 2: $y = 3x - 16$

10. Line 1: $y = \frac{2}{3}x - \frac{1}{6}$

Line 2: $y = \frac{1}{3}x + \frac{1}{3}$

■ **Objective B.** *Solve systems by addition. (Lesson 11-4)*

In 11–14, solve.

11. $\begin{cases} 3m + b = 11 \\ -4m - b = 11 \end{cases}$

12. $\begin{cases} 6a + 2c = 200 \\ 9a - 2c = 25 \end{cases}$

13. $\begin{cases} .6x - .4y = 1.1 \\ .2x - .4y = 2.3 \end{cases}$

14. $\begin{cases} \frac{1}{2}x + 3y = -6 \\ \frac{1}{2}x + y = 2 \end{cases}$

■ **Objective C.** *Solve systems by multiplying.*
(Lesson 11-5)

In 15 and 16, **a.** multiply one of the equations
by a number which makes it possible to solve
the system by adding. **b.** Solve the system.

15. $\begin{cases} 5x + y = 30 \\ 3x - 4y = 41 \end{cases}$

16. $\begin{cases} 5u + 6v = -295 \\ \quad u - 9v = 400 \end{cases}$

In 17–20, solve the system.

17. $\begin{cases} 3y - 2z = 3 \\ 2y + 5z = 21 \end{cases}$

18. $\begin{cases} 7m - 4n = 0 \\ 9m - 5n = 1 \end{cases}$

19. $\begin{cases} a + b = 3 \\ 5b - 3a = -17 \end{cases}$

20. $\begin{cases} 46 = 2t + u \\ 20 = 8t - 4u \end{cases}$

■ **Objective D.** *Recognize sentences with no solutions, one solution, or all real numbers as solutions.* *(Lesson 11-9)*

In 21–24, solve.

21. $2a + 4 < 2a + 3$

22. $12c < 6(3 + 2c)$

23. $7x - x = 12x$

24. $-10x = 15 - 10x$

■ **Objective E.** *Determine whether a system has 0, 1, or infinitely many solutions.* *(Lesson 11-8)*

In 25–28, determine whether each system describes lines that coincide, intersect, or are parallel.

25. $\begin{cases} 2x + 4y = 7 \\ 10x + 20y = 35 \end{cases}$

26. $\begin{cases} y - 2x = 5 \\ \quad\ y = 2x + 4 \end{cases}$

27. $\begin{cases} 6 = m - n \\ -6 = n - m \end{cases}$

28. $\begin{cases} a - 3b = 2 \\ a - 4b = 2 \end{cases}$

■ **Objective F.** *Use systems of equations to solve real world problems involving linear combinations.* *(Lessons 11-2, 11-3, 11-4, 11-5, 11-8, 11-9)*

29. Suppose Joe earned three times as much as Marty during the summer. Together they earned $210. How much did each person earn?

30. A punch is made by mixing 4 parts cranberry juice to 1 part club soda. If 8 gallons of punch are to be made, how many gallons of cranberry juice and how many gallons of club soda will it take?

31. In her restaurant, Charlene sells 2 eggs and a muffin for $1.80. She sells 1 egg with a muffin for $1.35. At these rates, how much is she charging for the egg and how much for the muffin?

32. A hotel offers two weekend packages. Plan A which costs $315, gives one person 3 nights lodging and 2 meals. Plan B gives 2 nights lodging and 1 meal and costs $205. At these rates, what is the charge for a room for one night?

33. If the cost per unit is constant, could the given situation have happened?

16 pencils and 5 erasers were bought for $8.00.

32 pencils and 10 erasers were bought for $16.00.

34. Renting a car from company (1) costs $39 plus $.10 a mile. Renting a car from company (2) costs $22.95 plus $.25 a mile. At what distance do the cars cost the same?

35. The starting salary on Job (1) is $7.00 an hour and every 6 months increases $0.50 an hour. For Job (2) the starting salary is $7.20 an hour and every 6 months increases $0.50 an hour. When does Job (2) pay more than Job (1)?

36. The Lorain, Ohio, area had a 1980 population of about 276,000 people and was growing at a rate of 1800 people each year. The Virginia Beach, Virginia, area had a 1980 population of about 262,000 people and was growing at a rate of 1950 people each year. If these rates continue, in how many years after 1980 will these two areas have the same population?

■ **Objective G.** *Calculate and use weighted averages.* (*Lessons 11-6, 11-7*)

37. Suppose towns A and B are 300 miles apart. If a car averages 50 mph from A to B and 40 mph on the way back, what is the average speed for the whole trip?

38. Linda had the following transcript: 4 A's, 9 B's, and 2 C's. What is her grade point average? (Assume A = 4.0, B = 3.0, C = 2.0, D = 1.0, F = 0.0)

39. Suppose 6 liters of a 25% alcohol solution are mixed with 4 liters of a 100% solution (pure alcohol). What is the percent of alcohol in the resulting mixture?

40. A retailer sold 250 T-shirts at $8 retail and has 150 left. For what sale price should the remaining T-shirts be sold to have an overall average price of $7 a T-shirt?

41. How many pounds of granola at $.99 per pound must be mixed with 5 pounds of raisins at $1.79 per pound to produce a trail mix worth $1.19 per pound?

42. A car averages 30 mph for the first 3 hours of a trip. How long must the car travel at 55 mph to have an overall average speed of 45 mph?

REPRESENTATIONS deal with pictures, graphs, or objects that illustrate concepts.

■ **Objective H.** *Find solutions to systems of equations by graphing.* (*Lessons 11-1, 11-8*)

In 43–46, solve each system by graphing.

43. $\begin{cases} y = 4x + 6 \\ y = \frac{1}{2}x - 1 \end{cases}$

44. $\begin{cases} y = x - 4 \\ y = -3x \end{cases}$

45. $\begin{cases} y = x + 3 \\ -2x + 3y = 4 \end{cases}$

46. $\begin{cases} 2y - 4x = 1 \\ y = 2x + \frac{1}{2} \end{cases}$

47. *Multiple choice*   Two straight lines *cannot* intersect in
(a) exactly one point
(b) no points
(c) exactly two points
(d) infinitely many points.

48. Parallel nonintersecting lines have the same __?__ but different *y*-intercepts.

■ **Objective I.** *Graphically represent solutions to systems of linear inequalities.* (*Lesson 11-10*)

49. Graph the solution set of this system.
$\begin{cases} y \le \frac{1}{2}x + 4 \\ y \ge -x + 1 \end{cases}$

50. Graph the solution set of this system.
$\begin{cases} x > 0 \\ y > 0 \\ x + y < 2 \end{cases}$

51. Graph all solutions to this system.
$\begin{cases} x \ge 0 \\ y \ge 0 \\ x + y \le 6 \\ x + y \ge 4 \end{cases}$

In 52 and 53, accurately graph the set of points that satisfies each situation.

52. A small elevator in a building only has a capacity of 280 kg. If a child averages 40 kg and an adult 70 kg, how many children *C* and adults *A* can you have in the elevator without its being overloaded?

53. A person wants to buy *x* pencils at 5¢ each and *y* erasers at 15¢ each and cannot spend more than 60¢.

Parabolas and Quadratic Equations

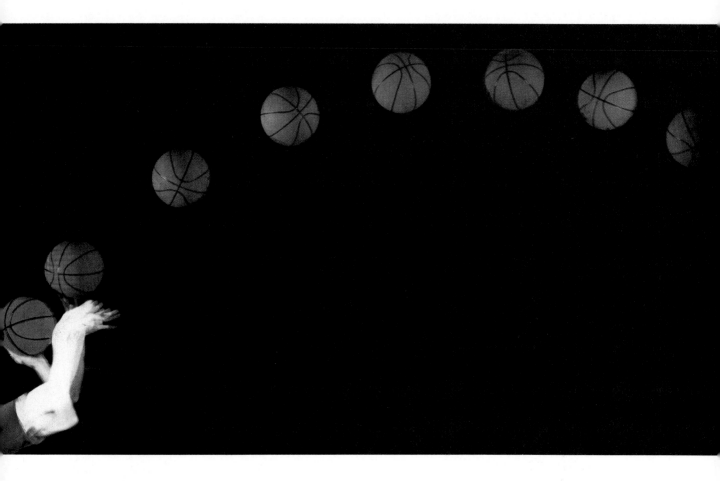

Many situations can be described by second degree polynomial equations. These are called quadratic equations.

For instance, suppose an insurance company finds that the number y of accidents per 50 million miles driven is related to a driver's age x in years for $16 \leq x \leq 74$ by the formula

$$y = .4x^2 - 36x + 1000.$$

Below is a graph of this equation.

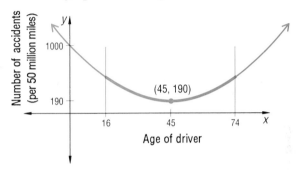

The graph shows that the insurance company predicts 45-year-olds to have the least number of accidents (190 per 50 million miles driven) while drivers both younger and older than 45 are predicted to have more. The shape of the graph is a curve called a **parabola**.

Parabolas occur often both in nature and in manufactured objects. A parabola is the shape of the path of a basketball tossed into a hoop as shown at left, the path of a football kicked for a field goal,

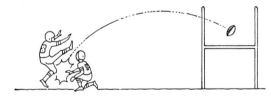

and the spray of water in a fountain.

12-1

Graphing $y = ax^2 + bx + c$

At the right is a square with side of length s. The area is described by the familiar property $A = s^2$. It is possible to graph all possibilities for s and A on the coordinate plane as shown at the left below.

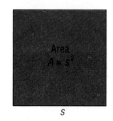

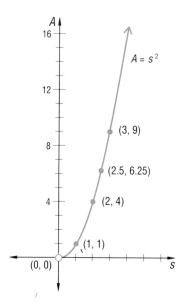

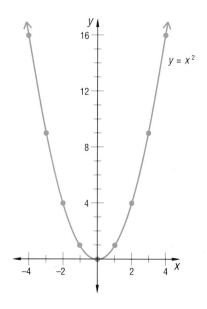

At the right above is the graph of the equation $y = x^2$. The graph has the same equation as $A = s^2$ except for the letters. In the area situation, A and s can only be positive because they are measures of area and length. In $y = x^2$, x is not limited to positive numbers. A table of values for $y = x^2$ is given below.

x	-4	-3	-2	-1	0	1	2	3	4
y	16	9	4	1	0	1	4	9	16

The graph of $y = x^2$ is a parabola that opens up. The y-axis separates the parabola into two halves. The right half is the exact same graph as the graph for the side and area of a square. The left half is a reflection or mirror image of the right half. If the parabola is folded along the y-axis, the two halves will coincide. For every point to the left of the y-axis, there is a matching point to the right. For this reason we say the parabola is **symmetric** to the y-axis. The y-axis is called the **axis of symmetry** of the parabola.

$A = s^2$ and $y = x^2$ are both equations of the form $y = ax^2$, with $a = 1$. Equations of this form all yield parabolas with the y-axis as the axis of symmetry. When a is negative, the parabola opens down.

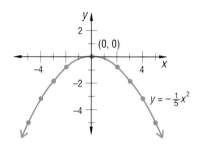

Example 1 Graph $y = -\frac{1}{5}x^2$.

Solution Make a table of values.

x	-5	-4	-3	-2	-1	0	1	2	3	4	5
y	-5	$-\frac{16}{5}$	$-\frac{9}{5}$	$-\frac{4}{5}$	$-\frac{1}{5}$	$-\frac{0}{5}$	$-\frac{1}{5}$	$-\frac{4}{5}$	$-\frac{9}{5}$	$-\frac{16}{5}$	-5

Except when $x = 0$, $-\frac{1}{5}x^2$ is negative.

So except for the origin, the parabola lies below the x-axis.

The intersection of a parabola with its axis of symmetry is called the **vertex** of the parabola. In Example 1, the vertex is also (0, 0). In summary:

The graph of $y = ax^2$ is a parabola with the following properties:
1. (0, 0) is the vertex.
2. It is symmetric to the y-axis.
3. If $a > 0$, the parabola opens up.
   If $a < 0$, the parabola opens down.

A parabola can be positioned so that the vertex is some other point in the plane. If the axis of symmetry is vertical, then it has an equation of the form $y = ax^2 + bx + c$ (with $a \neq 0$). The values of a, b, and c determine where the parabola is positioned in the plane and whether it opens up or down.

Example 2
a. Write a computer program to print a table of values for
$y = 2x^2 - 8x + 6$. Use the x-values 0, 0.5, 1, 1.5, ... , 4.
b. Draw the graph.

Solution **a.** x must take on values from 0 to 4, but must increase by 0.5 each time, instead of by 1. The command **STEP** tells the computer how much to add to x each time through the FOR/NEXT loop. The value of x increases in "steps" of 0.5.

```
10 PRINT "VALUES FOR Y = 2X^2 - 8X + 6"
20 PRINT "X", "Y"
30 FOR X = 0 TO 4 STEP .5
40    LET Y = 2 * X * X - 8 * X + 6
50    PRINT X, Y
60 NEXT X
70 END
```

b. At the left below is what the computer would print if the program is run. At the right below, the points are plotted and connected to show the parabola.

VALUES FOR Y = 2X^2 − 8X + 6

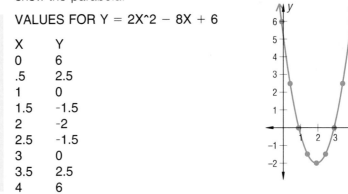

X	Y
0	6
.5	2.5
1	0
1.5	-1.5
2	-2
2.5	-1.5
3	0
3.5	2.5
4	6

The vertex of the parabola in Example 2 is at (2, -2). The graph is symmetric to the line $x = 2$, its axis of symmetry. Except for (2, -2), there are pairs of points with the same y-value. (1, 0) and (3, 0) are both on the x-axis. (1.5, 2) and (3.5, 2) are both on the horizontal line $y = 2$. There are two points where the y-coordinate is 6.

In the form $y = ax^2 + bx + c$, as with the form $y = ax^2$, if $a > 0$ the parabola opens up, and if $a < 0$ the parabola opens down.

Example 3 Graph $y = -x^2 − 12$.

Solution Form a table of values and plot.

x	-2	-1	0	1	2
y	-16	-13	-12	-13	-16

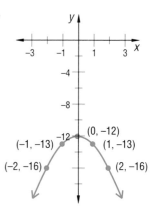

Recall that an *x-intercept* is a point where a curve crosses the *x*-axis. Look back at the parabolas in the examples of this lesson. Notice that a parabola can have two *x*-intercepts (Example 2), one *x*-intercept (Example 1) or no *x*-intercept (Example 3).

Questions

Covering the Reading

1. Name three instances in nature or in synthetic objects where parabolas occur.

2. The graph of $y = x^2$ is symmetric to the __?__.

3. What shape is the graph of $y = 47x^2$?

4. Refer to Example 2. Change lines in the computer program to print values for x and y for $y = 7x^2 - 3x + 5$, where x has values 0, .25, .50, .75, ... , 4.

In 5 and 6, consider the graph of the equation $y = ax^2$.

5. What is the vertex of this parabola?

6. **a.** If a is positive, the parabola opens __?__.
   b. If a is negative, the parabola opens __?__.

In 7–9, **a.** make a table of values for values of x between -3 and 3;
b. graph the equation.

7. $y = 3x^2$ 8. $y = x^2 - 2x - 3$ 9. $y = -3x^2 - 5$

10. Write the table of the values that this program will print.

```
10 PRINT "VALUES FOR X * X − 2X"
20 PRINT "X", "Y"
30 FOR X = -1 TO 3 STEP.5
40    LET Y = X * X − 2 * X
50    PRINT X, Y
60 NEXT X
70 END
```

Applying the Mathematics

11. **a.** In a circle of radius r, the area is described by $A = \pi r^2$. Graph all possibilities for r and A on a coordinate plane.
    b. Graph $y = \pi x^2$.

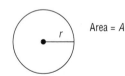

Area = A

12. On the parabola below, points A and B are symmetric to the y-axis. What are the coordinates of B? **(6, 10)**

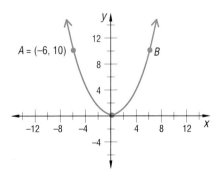

$A = (-6, 10)$

B

13. Match each graph with its equation.

(i) $y = x^2$ (ii) $y = x^2 + 1$

(iii) $y = x^2 - 1$ (iv) $y = -x^2 - 1$

a.

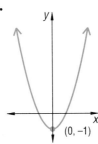

$(0, -1)$

b.

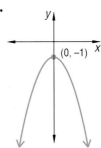

$(0, -1)$

c.

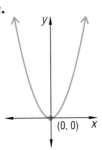

$(0, 0)$

d.

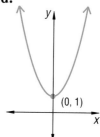

$(0, 1)$

14. Which of these could be the graph of $y = x^2 - 6x + 8$?

a.

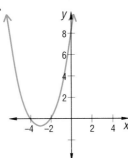

b.

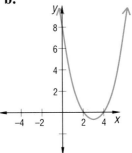

15. If (5, -11) is the vertex of a parabola with a vertical symmetry line and one of its x-intercepts is 2, find the other x-intercept.

Review

In 16 and 17, write in the form $ax^2 + bx + c = 0$. *(Lesson 10-1)*

16. $18 = 7x - 10x^2$

17. $3x + 8 = 4x^2 - 7x$

In 18–21, simplify.

18. $x^4 \cdot y \cdot x^3 \cdot y^3$ *(Lesson 9-5)*   **19.** $\dfrac{a^9}{a^6}$ *(Lesson 9-7)*

20. $\left(\dfrac{2a}{5}\right)^3$ *(Lesson 9-8)*   **21.** $\dfrac{-4 + 6x}{2}$ *(Lesson 10-4)*

22.

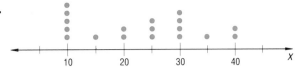

Multiple choice Which is correct for the data in the dot frequency diagram above? *(Lesson 1-2)*
(a) The mean is 4.15, median is 20, mode is 25.
(b) The mean is 24.4, median is 25, mode is 10.
(c) The mean is 25, median is 20, mode is 30.
(d) The mean is 23.05, median is 25, mode is 10.

23. *Multiple choice* The graph below is of all solutions to which inequality? *(Lesson 2-7)*

(a) $w + 6 < 5$ (b) $w - 6 < 5$
(c) $w + 5 < 6$ (d) $w - 5 < 6$

24. Calculate in your head. *(Lesson 10-9)*
 a. $(20 - 1)(20 + 1)$ **b.** $(100 - 5)(100 + 5)$
 c. $(1 - 0.1)(1 + 0.1)$

25. A house with 1800 square feet is being built on a 7500-square foot lot. The driveway will cover k square feet. Write an expression for the area left for the lawn. *(Lesson 3-2)*

26. *Skill sequence* If $a = -7$, $b = 6$, $c = 15$, find the value of each expression.
 a. $-4ac$ **b.** $b^2 - 4ac$ **c.** $\sqrt{b^2 - 4ac}$ *(Lessons 1-4, 7-4)*

Exploration

27. Draw a set of axes on graph paper. Hold a lighted flashlight at the origin so the light is centered on an axis as shown. What is the shape of the lighted area? Keep the lighted end of the flashlight in the same position and raise the other end. How does the shape change?

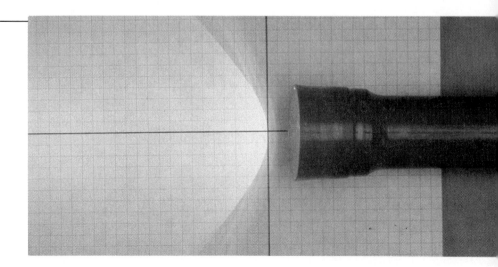

Graphs of equations are so helpful to have that there exist calculators that will automatically display part of a graph. Also, there are programs for every personal computer that will display graphs. Because computer screens are larger than calculator screens, they can more clearly show more of a graph, but graphing calculators are less expensive and more convenient.

Graphing calculators and computer graphing programs work in much the same way, and so we call them **automatic graphers** and do not distinguish them. Of course, no grapher is completely automatic. Each has particular keys to press that you must learn from a manual. Here we discuss what you need to know in order to use any of them.

The part of the coordinate grid that is shown is called a **window**. The screen at the right displays a window in which

$$-2 \le x \le 12$$
$$\text{and} \quad -3 \le y \le 7.$$

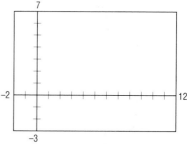

On calculators, the intervals for x and y may be left unmarked. Usually you need to pick the x-values at either end of the window. Some graphers automatically adjust and pick y-values so your graph will fit, but often you need to pick the y-values. If you don't do this, the grapher will usually pick a **default window** (perhaps $-10 \le x \le 10$, $-6 \le y \le 6$) and your graph may not appear on the screen.

On almost all graphers, the equation to be graphed must be a formula for y in terms of x.

$$y = 3x^2 + 4x - 2 \quad \text{can be handled.}$$
$$x = 4y \quad \text{cannot be handled.}$$

In order to determine a reasonable window for your graph, you may have to do some calculations, but nothing else is needed. The steps, then, are

1. Put the equation you wish to graph in y = _?_ form.
2. Calculate (perhaps by hand) the coordinates of some points you want on the graph.
3. Determine a window and key it in.
4. Give instructions to graph.

Example 1 Graph $y = 3x^2 + 4x - 2$ using an automatic grapher.

Solution 1
1. The equation is already solved for y. Key in
   y $=$ 3 $\times$ x^2 $+$ 4 $\times$ x $-$ 2 for a calculator or
   y = 3*x*x + 4*x − 2 for a computer.
2. To determine a window, you can find some points on the graph by hand. To begin it is useful to pick values of x not so close to each other.

x	y
0	-2
10	338
5	93
-5	53
1	5
-1	-3

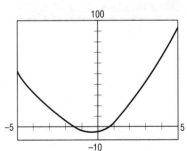

The values suggest a window from -5 to 5 for x-values and a window of -10 to 100 for the y-values. Key these values in. (Keying is different on different machines.)

3. Key in what is needed to tell the computer or calculator to graph and see a result like that shown above.

Solution 2 If you neglect step 2, the automatic grapher uses a default window (or perhaps the window of the last problem graphed). Below, the *same* equation is graphed with the default window
$$-10 \le x \le 10$$
$$-6 \le y \le 6.$$

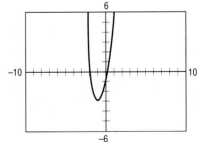

Some graphers have a **zoom** feature like those found on cameras. This feature enables you to change the window of a graph without retyping in the intervals for x and y. With such a feature, for Example 1 you could begin with Solution 2 and use it to suggest a better window for seeing the graph. In general a good window for a parabola is one in which you can estimate the coordinates of the vertex, the values of the x-intercepts (if any), and any other points you need in the problem. For these purposes, Solution 1 to Example 1 is better. The vertex seems to be at about (-.7, -3.3) and the x-intercepts are near .4 and -1.7. In the next lesson you will see why these values are important.

With most graphers more than one graph can be displayed on the same screen. This enables you to explore what happens when values in the equation are changed in some way.

Example 2   **a.** Graph $y = ax^2$ when $a = \frac{1}{2}$, 1, 2, and 3.
b. What happens to the graph as a gets larger?

Solution   **a.** The question asks to graph
$$y = \frac{1}{2}x^2$$
$$y = x^2$$
$$y = 2x^2$$
$$y = 3x^2$$

Normally, this would be quite time-consuming, but with a function grapher, it is not too difficult. We use the window $-6 \leq x \leq 6$ knowing that these parabolas are symmetric to the y-axis. The interval $-2 \leq y \leq 10$ is reasonable for comparing.

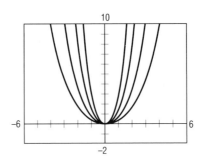

b. The parabola that looks widest is $y = \frac{1}{2}x^2$. As the value of a increases, the parabola looks thinner and thinner. The thinnest parabola is $y = 3x^2$.

Questions

Covering the Reading

1. On an automatic grapher, to what does the *window* refer?

2. Describe the window in Solution 1 to Example 1.

In 3–5, decide whether the equation is in a form in which it can be graphed with a grapher.

3. $x = 3y - 10$ **4.** $y = 1.3x^2 + 2.7x$ **5.** $y = 4 + 3x^2$

In 6–8, refer to Example 1.

6. Describe the default window.

7. Estimate the coordinates of the vertex of the parabola.

8. Estimate the larger x-intercept.

In 9–11, refer to the parabola graphed at right.

9. Describe the window.

10. Estimate the coordinates of the vertex of the parabola.

11. Estimate the x-intercepts.

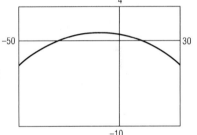

In 12–14, refer to Example 2.

12. Which value of a gives the widest parabola?

13. Which value of a gives the thinnest parabola?

14. Copy the graphs and add a sketch of the graph of $y = 4x^2$ to them.

Applying the Mathematics

15. a. Graph $y = -2x^2 + 20x - 6$ using the window $-10 \le x \le 10$ and $-100 \le y \le 100$.
  b. Graph $y = -2x^2 + 20x - 6$ using the window $-5 \le x \le 5$ and $-10 \le y \le 10$.
  c. Explain the difference between parts a and b.

16. a. Graph the equation $y = .4x^2 - 36x + 1000$ given on page 587, using the window $-10 \le x \le 90$ and $-100 \le y \le 400$.
  b. Use the graph to estimate the values of x for which $y = 400$.
  c. At about what ages are there 400 accidents per 50 million miles driven?

17. a. Graph $y = ax^2$ when $a = -\frac{1}{2}$, -1, -2, and -3.
  b. What happens to the graph as a gets smaller?

18. **a.** Graph $y = 3x + 5$ and $y = x^2$ on the same axes.
    b. Estimate the two points of intersection of these graphs. These are the solutions to the system $\begin{cases} y = x^2 \\ y = 3x + 5. \end{cases}$

19. *Skill sequence* When $a = 2$, $b = 3$, and $c = 4$, give the value of each expression. *(Lessons 1-4, 7-4)*
    a. $-4ac$ **b.** $b^2 - 4ac$ **c.** $\sqrt{b^2 - 4ac}$

20. If $\sqrt{20} = 2\sqrt{k}$, what is the value of k? *(Lesson 7-6)*

21. Calculate the distance between the origin and $(-2, -4)$. *(Lesson 7-7)*

In 22–24, simplify. *(Lesson 10-4)*

22. $\dfrac{3x + 6}{3}$ 23. $\dfrac{4 - 2\sqrt{5}}{2}$ 24. $\dfrac{3x^2 + 4x^3 - 5x^4}{x}$

25. A tortoise is walking at a rate of $3 \dfrac{\text{feet}}{\text{minute}}$. Assume this rate continues.
    a. How long will it take the tortoise to travel 20 feet?
    b. How long will it take the tortoise to travel f feet? *(Lesson 5-2)*

26. Refer to the table below. If the populations of Dallas and San Antonio continue increasing at the rate of increase from 1970 to 1984, when will they have the same population and what will that population be?
    (Lessons 6-6, 11-3)

Populations	1970	1984
Dallas	844,000	974,000
San Antonio	654,000	843,000

Dallas skyline

27. **a.** Multiply $(2x + 3)$ by $(3x + 2)$.
    b. Multiply $(6x - 1)$ by $(x - 6)$. *(Lesson 10-6)*

28. **a.** Graph $y = x^2 + bx + 2$ when $b = 0, 1, 2,$ and 3.
    b. How does the value of b affect the graph?
    c. From your answer to part b, predict where the graph of $y = x^2 + 100x + 2$ would be located.
    d. Explore the negative values of b.

Applications of Parabolas

The path of a tossed ball is a part of a parabola. Consider a quarterback who tosses a football to a receiver 40 yards downfield. If the ball is thrown and caught 6 feet above the ground, and is 16 ft above the ground at the peak of the throw (the vertex), then the path of the ball can be graphed as it is below.

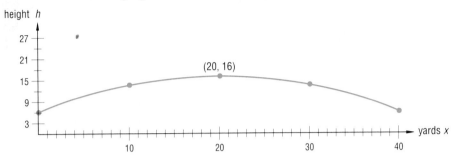

This parabola is symmetric to the vertical line $x = 20$. So the ball is at its peak 20 yards downfield. If the ball is at a height of h feet when it is x yards downfield, its path is a piece of the parabola described by the equation

$$h = -\frac{1}{40} (x - 20)^2 + 16.$$

Suppose there is a defender 3 yards in front of the receiver. This means the defender is 37 yards from the quarterback. Would he be able to deflect or catch the ball?

Example 1 Refer to the above situation. What is the height of the ball 37 yards downfield from the quarterback?

Solution Substitute 37 for x into the equation.

$$h = -\frac{1}{40} (x - 20)^2 + 16$$

$$= -\frac{1}{40} (37 - 20)^2 + 16$$

$$= -\frac{1}{40} (17)^2 + 16$$

$$= -7.225 + 16$$

$$= 8.775$$

The ball will be 8.775 feet above the ground. This is approximately 8 feet 9 inches.

To deflect or intercept the ball, the defender would have to reach a height of 8 feet 9 inches. With a tall defender and a well-timed jump, this is quite possible.

A different parabola represents the height of a throw against the time the ball has been in the air. The throw on page 599 reaches a height of h feet after t seconds, where $h = -10(t - 1)^2 + 16$. A graph of this parabola is shown below.

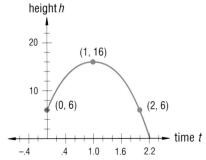

The throw on page 599

Example 2 Refer to the above situation. How high will the ball be after half a second?

Solution Substitute $t = .5$.

$$h = -10(.5 - 1)^2 + 16$$
$$= -10(-.5)^2 + 16$$
$$= -10 \cdot .25 + 16$$
$$= -2.5 + 16$$
$$= 13.5$$

In half a second, the ball will be at a height of 13.5 feet.

Example 3 Again refer to the above situation. Estimate from the graph when the ball would reach a height of 9 feet.

Solution Draw a horizontal line at $h = 9$ feet. The ball reaches this height at two points, at about 0.2 seconds and at about 1.8 seconds.

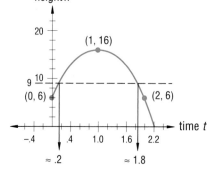

Notice that both these times are 0.8 seconds away from 1 second, where the vertex of the parabola is located.

You have seen that some parabolas describe the actual path of an object, the height at a given *distance* from the start. Other parabolas describe an object's height at a specific *time*. It is essential to read the labels of the axes to know what a graph is representing.

Questions

Covering the Reading

1. What is the shape of the path of a tossed ball?

In 2 and 3, refer to the tossed football at the beginning of the lesson.

2. If a defender is 5 yards in front of the receiver, how far is he from the quarterback?

3. What is the height of the ball 39 yards downfield from the quarterback? Write the answer: a. in feet; b. in feet and inches.

In 4 and 5, a quarterback throws the ball to a receiver 60 yards downfield. The ball is at a height of h feet, x yards downfield where
$h = -\frac{1}{45} (x - 30)^2 + 25$.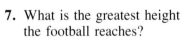

4. At what height does the receiver catch the ball?

5. A defender is 5 yards in front of the receiver. How high would he have to reach to deflect the ball?

6. When the height of a toss is plotted against time, what shape is the graph?

In 7–11, refer to the graph at right. It shows h, the height in yards of a football t seconds after it is kicked into the air.

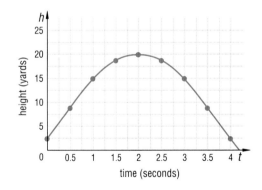

7. What is the greatest height the football reaches?

8. How high is the ball 1 second after it is kicked?

9. At what times is the height 18 yards?

10. How long is the football in the air?

11. For how many seconds is the football more than 15 yards above the ground?

In 12–15, a projectile is shot from the edge of a cliff. The projectile is y meters above the cliff after x seconds where $y = 30x - 5x^2$. This equation is graphed at right.

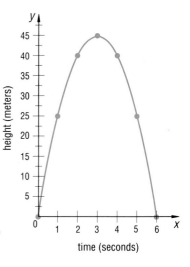

12. What is the greatest height reached?

13. How far above the cliff edge is the projectile after 5 seconds?

14. When is the projectile 40 meters above the cliff edge?

15. How far below the edge of the cliff would the projectile be after 7 seconds?

16. *True or false*   The graph of $y = 6x^2$ has a vertex at $(0, 0)$. *(Lesson 12-1)*

In 17 and 18, **a.** make a table of values; **b.** graph the equation. *(Lessons 12-1, 12-2)*

17. $y = \frac{1}{2} x^2$

18. $y = 6x - x^2$

19. Solve in your head: $2x(x - 4)(x - 3) = 0$. *(Lesson 10-10)*

20. Find the endpoints of the interval 8 ± 1.27. *(Lesson 1-3)*

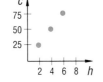

21. *Multiple choice*   The graph at the left represents: *(Lesson 2-4)*
    (a) the cost c of h pencils at 2 for 25¢;
    (b) the cost c of h pencils at 25¢ each;
    (c) the cost c of h pencils at 50¢ each.

22. *Multiple choice*   Which of the following is not equivalent to ab^2? *(Lessons 9-5, 9-8)*
    (a) $a \cdot b \cdot b$ (b) $a \cdot (b^2)$
    (c) $(a \cdot b)^2$ (d) $a \cdot (b)^2$

23. What is the image of $(-6, 0)$ after a slide 3 units to the right and 5 units down? *(Lesson 2-5)*

24. *Skill sequence*   Write as a single fraction. *(Lessons 2-2, 7-8)*
    a. $2 + \frac{7}{11}$ **b.** $2 + \frac{x}{y}$ **c.** $a + \frac{b - 1}{c + 1}$

25. A pitched ball in baseball does not always follow the path of a parabola. Why not?

26. **a.** Graph $y = 4x^2 - 20x + c$ for four different values of c of your own choosing.
    b. How does the value of c affect the graph?

12-4

The Quadratic Formula

A cliff diver in Acapulco, Mexico dives from a height of approximately 27 meters to the waters below "La Quebrada" ("the break in the rocks"). If the diver is x meters away from the cliff and y meters above the water, then under certain conditions $y = -x^2 + 2x + 27$ describes the path of the dive.

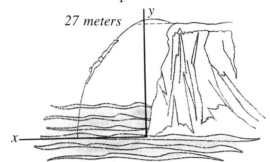

27 meters

When the diver enters the water, the height $y = 0$. So solving

$$0 = -x^2 + 2x + 27$$

or, equivalently,

$$-x^2 + 2x + 27 = 0$$

will give the number of meters from the cliff the diver enters the water.

The equation $-x^2 + 2x + 27 = 0$ is an example of a quadratic equation. A **quadratic equation** is an equation that can be simplified into the form $ax^2 + bx + c = 0$. Solutions to *any* quadratic equation can be found by using the Quadratic Formula.

Quadratic Formula:

If $ax^2 + bx + c = 0$ and $a \neq 0$, then

$$x = \frac{-b \pm \sqrt{b^2 - 4ac}}{2a}.$$

The Quadratic Formula is one of the most famous formulas in all of mathematics. *You should memorize it today.*

The Quadratic Formula uses the shorthand symbol ± means "plus or minus." That symbol signifies that there are possibly two solutions to any quadratic equation:

$$x = \frac{-b + \sqrt{b^2 - 4ac}}{2a} \text{ or } x = \frac{-b - \sqrt{b^2 - 4ac}}{2a}.$$

Caution: Many calculators have a $\boxed{+/-}$ key. That key takes the opposite of a number. It does *not* perform the two operations $+$ and $-$ required in the Quadratic Formula.

To apply the Quadratic Formula, notice that a is the coefficient of x^2, b is the coefficient of x, and c is the constant term. One side of the equation *must* be 0. Rearrange the other side in descending order of exponents. $ax^2 + bx + c = 0$ is called the **standard form of a quadratic equation**.

Example 1   Solve $3x^2 - 6x - 45 = 0$ by using the Quadratic Formula.

Solution   In general, $x = \dfrac{-b \pm \sqrt{b^2 - 4ac}}{2a}$.

Here $a = 3$, $b = -6$ and $c = -45$.

So $x = \dfrac{-(-6) \pm \sqrt{(-6)^2 - 4 \cdot 3 \cdot -45}}{2 \cdot 3}$

Use the order of operations. Do the work inside the radical sign, first powers, then multiplications then subtractions.

$$= \frac{6 \pm \sqrt{36 - (-540)}}{6}$$

$$= \frac{6 \pm \sqrt{576}}{6}$$

$$= \frac{6 \pm 24}{6}.$$

Now translate the shorthand.

$$x = \frac{6 + 24}{6} = \frac{30}{6} = 5 \quad \text{or} \quad x = \frac{6 - 24}{6} = \frac{-18}{6} = -3.$$

Check   Substitute 5 for x. $3(5)^2 - 6(5) - 45 = 75 - 30 - 45 = 0$. Substitute -3 for x. $3(-3)^2 - 6(-3) - 45 = 27 + 18 - 45 = 0$. It checks.

Working with an equation before applying the Quadratic Formula can occasionally make the calculations simpler. In Example 1, the numbers could have been made smaller by multiplying both sides of the original equation by $\frac{1}{3}$. The resulting equation, $x^2 - 2x - 15 = 0$, has the same solutions, $x = 5$ or $x = -3$.

When the term under the radical sign in the formula is not a perfect square, a calculator will give approximations.

Example 2 Solve $-x^2 + 2x + 27 = 0$ to find the distance of the diver from the cliff when he enters the water.

Solution Recall that $-x^2 = -1x^2$. So rewrite the equation as

$$-1x^2 + 2x + 27 = 0.$$

Since the left side is in descending order of exponents, apply the Quadratic Formula with $a = -1$, $b = 2$, and $c = 27$.

$$x = \frac{-2 \pm \sqrt{2^2 - 4 \cdot -1 \cdot 27}}{2 \cdot -1}$$

$$= \frac{-2 \pm \sqrt{4 - (-108)}}{2}$$

$$= \frac{-2 \pm \sqrt{112}}{-2}$$

So $x = \dfrac{-2 + \sqrt{112}}{-2}$ or $x = \dfrac{-2 - \sqrt{112}}{-2}$.

These are exact answers to the equation. Using a calculator gives approximations.

$$x \approx \frac{-2 + 10.6}{-2}$$ or $$x \approx \frac{-2 - 10.6}{-2}$$

So $x \approx -4.3$ or $x \approx 6.3$

The diver cannot land a negative number of meters from the cliff, so the solution $x \approx -4.3$ does not make sense in this situation. The diver will enter the water about 6.3 meters away from the cliff.

Check Does 6.3 work in the equation $-x^2 + 2x + 27 = 0$? Substitute 6.3 for x. $-(6.3)^2 + 2(6.3) + 27 = -39.69 + 12.6 + 27 = -.09$. This is close enough to zero, given that 6.3 is an approximation.

In Example 3, the equation has to be put into $ax^2 + bx + c = 0$ form before the formula can be applied.

■ ■ ■ ■ ■ ■ ■ ■

Example 3   Solve $m^2 - 3m = 14$. Give m to the nearest hundredth.

Solution   To apply the Quadratic Formula, one side of the equation must be 0. Add -14 to both sides.

$$m^2 - 3m - 14 = 0$$

Now the formula can be easily applied with $a = 1$, $b = -3$, $c = -14$.

$$m = \frac{-(-3) \pm \sqrt{(-3)^2 - 4 \cdot 1 \cdot -14}}{2 \cdot 1} = \frac{3 \pm \sqrt{9 - (-56)}}{2}$$

$$= \frac{3 \pm \sqrt{65}}{2}$$

Thus $m = \dfrac{3 + \sqrt{65}}{2}$ or $m = \dfrac{3 - \sqrt{65}}{2}$

$$m \approx \frac{3 + 8.06}{2} \quad \text{or} \quad m \approx \frac{3 - 8.06}{2}$$

So $m \approx 5.53$ or $m \approx -2.53$.

Check   The check is left to you as Question 7.

Questions

Covering the Reading

1. State the Quadratic Formula.

2. *True or false*   The Quadratic Formula can be used to solve any quadratic equation.

3. Refer to Example 1. Verify that $x = 5$ or $x = -3$ are solutions to $x^2 - 2x - 15 = 0$.

4. Give the two solutions to the sentence $x^2 + x - 1 = 0$, using the "$\pm$" symbol.

In 5 and 6, each sentence is equivalent to a quadratic equation of the form $ax^2 + bx + c = 0$.
  a. Give the values of a, b, and c.
  b. Give the exact solutions to the equation.
  5. $12x^2 + 7x + 1 = 0$ **6.** $3n^2 + n - 2 = 0$

7. Check both decimal answers to Example 3.

8. Suppose the cliff diver at the beginning of the lesson dove from a cliff 22 meters high. Then, under certain conditions, solving the equation $0 = -x^2 + 2x + 22$ gives the number of meters the diver is from the cliff when entering the water.
   a. Give the values of a, b, and c for use in the quadratic formula.
   b. Apply the Quadratic Formula to give the exact solutions to the equation.
   c. Approximate the solutions to the nearest tenth.
   d. How far away from the cliff does the diver enter the water?

9. a. Find a simpler equation that has the same solutions as $4m^2 - 16m - 64 = 0$.
   b. Solve the simpler equation using the Quadratic Formula.

In 10 and 11, put each equation in the form $ax^2 + bx + c = 0$. Then solve the equation using the Quadratic Formula. Round solutions to the nearest hundredth.

10. $20w^2 - 6w = 2$ **11.** $3w^2 - w = 5$

Applying the Mathematics

In 12 and 13, use this information. The solutions to $ax^2 + bx + c = 0$ are the x-intercepts of the graph of $y = ax^2 + bx + c$. Find the x-intercepts of the graph of the given equation.

12. $y = 2x^2 + 3x - 2$ **13.** $y = 12x^2 - 12x + 1$

14. A square has a side of length x. If the sum of the area and the perimeter of the square is 86.25, find x.

15. When a ball on the moon is thrown upward with an initial velocity of 6 meters per second, its approximate height y after t seconds is given by $y = -0.8t^2 + 6t$.
   a. At what *two* times will it reach a height of 10 m? Give your answers to the nearest tenth of a second.
   b. Graph $y = -0.8t^2 + 6t$. Let t take on values from 0 to 7.5.
   c. Use the graph to check your answer to part a.

16. Solve by multiplying the binomials and then using the Quadratic Formula: $(w - 11)(w + 5) = 37$.

Review

17. A rock is tossed up from a cliff with upward velocity of 10 meters per second. $h = 10t - 5t^2$ relates the height h of the rock *above the cliff* after t seconds.
   a. How high above the cliff is the rock 1 second after it is tossed?
   b. Where will the rock be after 3 seconds? *(Lesson 12-2)*

18. The x-coordinate of the vertex of a parabola is the mean of the x-intercepts of the parabola. If the x-intercepts of a parabola are 3 and -1, what is the x-coordinate of the vertex? *(Lessons 1-2, 10-1)*

19. *Multiple choice*   The graph below is the graph of
    (a) $y = 2x$ (b) $y = 2x^2$
    (c) $y = 2x^2 + 1$ (d) $y = 2x^2 - 1$. *(Lesson 12-1)*

20. On the same coordinate axes, graph $y = \frac{1}{2} x^2 + \frac{1}{2} x + 3$ and
$y = \frac{1}{2} x^2 + \frac{1}{2} x - 3$. *(Lessons 12-1, 12-2)*

21. A school begins the year with 200 reams of paper. (A ream contains 500 sheets.) The teachers are using 12 reams a week. How many reams will be left after 18 weeks? *(Lesson 6-2)*

22. Find the volume of a box that is $15a$ inches in length, $8b$ inches in width, and $6c$ inches in height. *(Lesson 4-7)*

23. Mr. Robinson is ordering a new car. He can choose from 5 models, 12 exterior colors, and 9 interior colors. How many combinations of models, interiors, and exteriors are possible? *(Lesson 4-6)*

24. Factor $x^2 - 2x + 1$. *(Lesson 10-11)*

25. *Skill sequence*   Expand. *(Lesson 10-7)*
    a. $(x + 1)^2$ **b.** $(3x - 2)^2$ **c.** $(2ax + b)^2$

Exploration

26. A team's opening batter named Nero
Squared his number of hits, the hero!
After subtracting his score,
He took off ten and two more,
And the final result was zero.
How many hits did Nero have?

12-5

Analyzing the Quadratic Formula

In the previous lesson, we introduced the Quadratic Formula without telling you why it works. There is no mystery. The formula can be found by solving the general equation

$$ax^2 + bx + c = 0$$

using only properties you already know. The idea used below was first written by the Arab mathematician al-Khowarizmi in A.D. 825 in a book entitled *Hisab al-jabr w'al muqabalah*. This book was very influential. From the second word in that title comes our modern word "algebra." From his name comes our word "algorithm." The Quadratic Formula provides an algorithm for solving any quadratic equation.

The idea is to work with the equation $ax^2 + bx + c = 0$ until the left side is a perfect square. Then the equation will have the form $t^2 = k$, which you know how to solve for t.

> If $k > 0$, then $t^2 = k$ has two real solutions, namely $t = \sqrt{k}$ or $t = -\sqrt{k}$.
> If $k = 0$, then $t^2 = 0$ and there is only one solution, $t = 0$.
> If $k < 0$, then $t^2 = k$ has no real solutions.

Examine the argument in the next paragraph closely. See how each equation follows from the preceding equation.

Multiply both sides of $ax^2 + bx + c = 0$ by $4a$. This makes the le term equal to $4a^2x^2$, the square of $2ax$.

$$4a^2x^2 + 4abx + 4ac = 0$$

When the quantity $2ax + b$ is squared, it equals $4a^2x^2 + 4abx + b^2$ The left and center terms of this trinomial match what is in the equa tion. We add b^2 to both sides of the equation to get all three terms into our equation.

$$4a^2x^2 + 4abx + b^2 + 4ac = b^2$$

The first three terms are the square of $2ax + b$.

$$(2ax + b)^2 + 4ac = b^2$$

Add $-4ac$ to both sides.

$$(2ax + b)^2 = b^2 - 4ac$$

Now the equation has the form $t^2 = k$, with $t = 2ax + b$ and $k = b^2 - 4ac$. At this point we need to ask: Is $b^2 - 4ac$ greater than, equal to, or less than 0?

If $b^2 - 4ac > 0$, then there are two real solutions. They are found by taking the square roots of both sides.

$$2ax + b = \pm \sqrt{b^2 - 4ac}$$

It is beginning to look like the formula. Now add $-b$ to each side.

$$2ax = -b \pm \sqrt{b^2 - 4ac}$$

Dividing both sides by $2a$ results in the solutions given by the Quadratic Formula.

$$x = \frac{-b \pm \sqrt{b^2 - 4ac}}{2a}$$

If $b^2 - 4ac = 0$, then you can still take the square roots, but the formula reduces to

$$x = \frac{-b \pm 0}{2a} = \frac{-b}{2a}.$$

This verifies that there is exactly one solution.

If $b^2 - 4ac < 0$, then the quadratic equation has no real number solutions. The formula still works, but you have to take square roots of negative numbers to get solutions. You will study these nonreal solutions in a later course.

Because the value of $b^2 - 4ac$ discriminates between the various possible numbers of real number solutions to a quadratic equation, it is called the **discriminant** of the equation.

Discriminant Property:

Suppose $ax^2 + bx + c = 0$ and a, b, and c are real numbers. Let $D = b^2 - 4ac$. Then

when $D > 0$, the equation has exactly two real solutions.
when $D = 0$, the equation has exactly one solution.
when $D < 0$, the equation has no real solutions.

Example 1 How many real solutions does the equation $8x^2 - 5x + 2 = 0$ have?

Solution Find the value of the discriminant $b^2 - 4ac$.
Here $a = 8$, $b = -5$, and $c = 2$.
So $b^2 - 4ac = (-5)2 - 4 \cdot 8 \cdot 2$
$$= 25 - 64$$
$$= -39.$$
Since $b^2 - 4ac$ is negative, there are no real solutions.

The discriminant can quickly give information about a situation. Recall from Lesson 12-4 that the cliff diver's height y above the water at a distance x meters from the cliff was described by the formula $y = -x^2 + 2x + 27$. Remember that the diver starts 27 meters above the water.

Example 2   Will the diver ever reach a height of 27.5 meters above the water?

Solution   Let $y = 27.5$ in the equation $y = -x^2 + 2x + 27$.

$$27.5 = -x^2 + 2x + 27$$

Add -27.5 to both sides to put the equation in standard form.

$$0 = -x^2 + 2x - 0.5$$

Now calculate the discriminant. $b^2 - 4ac = 2^2 - 4 \cdot (-1) \cdot (-0.5) = 2$. Since the discriminant is positive, there are two solutions. Use the Quadratic Formula to find them. Notice that the calculation is easy since $b^2 - 4ac$ has been found.

$$x = \frac{-2 \pm \sqrt{2}}{-2}$$

$$x \approx \frac{-2 + 1.414}{-2} \text{ or } x = \frac{-2 - 1.414}{-2}$$

$$\approx 0.293 \text{ or } x \approx 1.707$$

The diver reaches the height of 27.5 meters twice, at about 0.3 meters from the cliff (on the way up) and about 1.7 meters from the cliff (on the way down).

Example 3   Will the diver ever reach a height of 28 meters above the water?

Solution 1   Let y be 28 in the equation $y = -x^2 + 2x + 27$.

$$28 = -x^2 + 2x + 27$$

Put the equation in general form by adding -28 to both sides.

$$0 = -x^2 + 2x - 1$$

The discriminant is $2^2 - 4 \cdot (-1) \cdot (-1) = 0$. Because the discriminant is 0, there is exactly one solution. The diver will be 28 meters above the water once. This must be the vertex of the parabolic path of the dive.

Solution 2 The equation $-x^2 + 2x - 1 = 0$ from Solution 1 can be solved without using the Quadratic Formula. Multiply both sides by -1. Then the left side is a perfect square trinomial that can be factored.

$$x^2 - 2x + 1 = 0$$

Factor.

$$(x - 1)(x - 1) = 0$$

Use the Zero Product Property.

$$
\begin{array}{ccc}
x - 1 = 0 & \text{or} & x - 1 = 0 \\
x = 1 & \text{or} & x = 1
\end{array}
$$

This verifies that there is exactly one solution.

■ ■ ■ ■ ■ ■ ■

Example 4   Will the diver ever reach a height of 29 meters?

Solution   Solve $29 = -x^2 + 2x + 27$

Add -29 to both sides to put into standard form,

$$0 = -x^2 + 2x - 2.$$

Here $a = -1$, $b = 2$, and $c = -2$. $-b^2 - 4ac = 2^2 - 4 \cdot (-1) \cdot (-2) = -4$. The discriminant is negative. Since negative numbers have no square roots in the real number system, this equation has no real solution. The diver will never reach a height of 29 meters.

Examples 2, 3, and 4 can be verified by looking at the graph of $y = -x^2 + 2x + 27$.

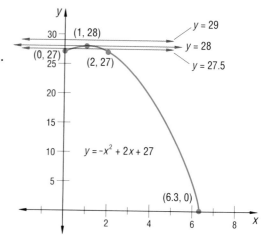

Here is a summary of the examples.

diver height	equation	discriminant	number of real solutions	graph intersects line
27.5 meters	$-x^2 + 2x + 27 = 27.5$	positive	2	$y = 27.5$ twice
28 meters	$-x^2 + 2x + 27 = 28$	zero	1	$y = 28$ once
29 meters	$-x^2 + 2x + 27 = 29$	negative	0	$y = 29$ never

Questions

Covering the Reading

1. The first person to write down an algorithm for solving quadratic equations was __?__ in the year __?__.

In 2–4, give all real solutions to the equation.

2. $x^2 = 10$ **3.** $y^2 = 0$ **4.** $z^2 = -81$

5. Here are steps in the derivation of the Quadratic Formula. Tell what was done to get each step.
$$ax^2 + bx + c = 0$$
a. $4a^2x^2 + 4abx + 4ac = 0$

b. $4a^2x^2 + 4abx + b^2 = b^2 - 4ac$

c. $(2ax + b)^2 = b^2 - 4ac$

d. $2ax + b = \pm\sqrt{b^2 - 4ac}$

e. $2ax = -b \pm \sqrt{b^2 - 4ac}$

f. $x = \dfrac{-b \pm \sqrt{b^2 - 4ac}}{2a}$

6. Define: discriminant.

7. Give the number of real solutions to a quadratic equation when its discriminant is
a. positive   **b.** negative   **c.** zero.

In 8–11, **a.** give the value of the discriminant; **b.** give the number of real solutions; **c.** find all the real solutions.

8. $w^2 - 16w + 64 = 0$ **9.** $4x^2 - 3x + 8 = 0$

10. $25y^2 = 10y - 1$ **11.** $13z = 5z^2 + 9$

12. What equation can be solved to determine how far away from the cliff the diver in this lesson will be when he is 28 meters above the water?

13. When will the diver of this lesson reach a height of 30 meters above the water?

14. When will the diver of this lesson be 10 meters above the water?

Applying the Mathematics

15. a. Solve $4x^2 + 8x = 5$ **b.** Solve $4x^2 + 8x = -1$.
c. Solve $4x^2 + 8x = -10$.

16. Suppose an equation that describes a diver's path when diving off of a platform is $d = -5t^2 + 10t + 20$, where d is the distance above the water (feet) and t is the time from the beginning of the dive (seconds).
a. How high is the diving platform? (Hint: let $t = 0$ seconds).
b. After how many seconds is the diver 25 feet above the water?
c. After how many seconds does the diver enter the water? (Round your answer to the nearest tenth.)

17. For what value of h does $x^2 + 3x + h = 0$ have exactly one solution?

Review

18. Solve $60x^2 - 190x - 110 = 0$. *(Lesson 12-4)*

19. Alan receives $50 a year on each birthday and invests it with a scale factor of z. After 3 years, he knows he has $50z^2 + 50z + 50$ dollars. He also knows he has $160.
   a. Find z. *(Lesson 12-3)*
   b. What is the interest rate on his investment? *(Lesson 9-1)*

20. Calculate $\frac{15!}{8!7!}$. *(Lesson 4-9)*

In 21 and 22, state whether the parabola opens up or down. *(Lesson 12-1)*

21. $y = -\frac{1}{5}x^2 - 3x + 2$ **22.** $x^2 + x = y$

23. Solve. *(Lesson 6-7)*
   a. $a - \frac{1}{3} < 0$ **b.** $\frac{1}{3} - b < 0$ **c.** $\frac{1}{3}c < 0$

In 24 and 25, a softball pitcher tosses a ball in practice to her catcher 50 ft away. The ball is at a height of h ft, x feet from the pitcher where

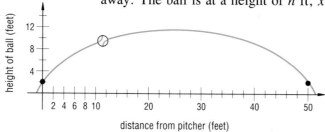

$$h = -\frac{1}{62.5}(x - 25)^2 + 12.$$

24. How high is the ball at its peak? *(Lesson 12-2)*

25. a. If the batter is 2 ft in front of the catcher, how far is she from the pitcher?
   b. How high is the ball when it reaches the batter? *(Lesson 12-2)*

26. *Skill sequence*   Add and simplify. *(Lessons 7-4, 7-6)*
   a. $\dfrac{-5 + x}{2a} + \dfrac{-5 - x}{2a}$

   b. $\dfrac{-b + y}{2a} + \dfrac{-b - y}{2a}$

   c. $\dfrac{-b + \sqrt{z}}{2a} + \dfrac{-b - \sqrt{z}}{2a}$

   d. $\dfrac{-b + \sqrt{b^2 - 4ac}}{2a} + \dfrac{-b - \sqrt{b^2 - 4ac}}{2a}$

Exploration

27. Graph the parabola $y = 4x^2 + 8x$ as accurately as you can. Explain the answers you get to Question 15 by referring to this parabola.

Rational and Irrational Numbers

A number that can be written as a ratio of two integers is called a **rational number**. Here are some examples.

$\frac{1}{2}$ is rational. It is the ratio of the integers 1 and 2.

$\frac{-6}{5}$ is rational. It is the ratio of the integers -6 and 5.

$3\frac{1}{7}$ is rational. It equals $\frac{22}{7}$, the ratio of the integers 22 and 7.

0 is rational. It is the ratio of 0 and any other integer n.

149 is rational. It equals $\frac{149}{1}$, the ratio of 149 and 1.

18.89 is rational. It equals $\frac{1889}{100}$.

$-4.\overline{2}$ is rational. It equals $-4\frac{2}{9}$, or $-\frac{38}{9}$.

In general, any integer, simple fraction, mixed number, finite decimal, or repeating decimal, whether positive or negative, represents a rational number.

Real numbers that are not rational are called **irrational**. Every infinite decimal that does not repeat represents an irrational number. Examples are

$$\pi \approx 3.141592653 \ldots$$
$$\text{and } \sqrt{2} = 1.414213562 \ldots$$

In fact, any square root of an integer that is not itself an integer is irrational. Namely, $\sqrt{3}, \sqrt{5}, \sqrt{6}, \sqrt{7}, \sqrt{8}, \sqrt{10}, \ldots$, are all irrational. (But $\sqrt{4}$ is rational, for it equals $\frac{2}{1}$.)

Thus, if a number is rational, it has a finite or repeating decimal. But if a number is irrational, you can only approximate it by a decimal.

All of this is useful in solving quadratic equations by the Quadratic Formula. Under the square root sign in the formula is the discriminant. When a, b, and c are integers, the discriminant $b^2 - 4ac$ will be an integer. If $b^2 - 4ac$ is a perfect square, then the solutions are rational. If $b^2 - 4ac$ is positive but not a perfect square, then $\sqrt{b^2 - 4ac}$ will be irrational and you can only write a decimal approximation to the roots.

If a, b, and c are integers and $b^2 - 4ac$ is a positive number but not a perfect square, then the equation $ax^2 + bx + c = 0$ has two irrational solutions.

■ ■ ■ ■ ■ ■ ■

Example 1   Determine whether the solutions to $6x^2 + 5x - 4 = 0$ are rational or irrational.

> **Solution**   Calculate the discriminant when $a = 6$, $b = 5$, and $c = -4$.
>
> $$b^2 - 4ac = 5^2 - 4 \cdot 6 \cdot -4 = 25 + 96$$
> $$= 121$$
>
> 121 is a perfect square, so the solutions are rational.
>
> **Check**   By the Quadratic Formula, $x = \dfrac{-5 \pm \sqrt{121}}{12}$
>
> $$= \dfrac{-5 \pm 11}{12}$$
>
> $$x = \dfrac{-5 + 11}{12} \text{ or } x = \dfrac{-5 - 11}{12}.$$
>
> Thus, $x = \dfrac{6}{12}$ or $x = \dfrac{-16}{12}$
>
> $$x = \dfrac{1}{2} \quad \text{ or } x = \dfrac{-4}{3}.$$
>
> These are rational numbers.

■ ■ ■ ■ ■ ■ ■

Example 2   Are the solutions to $x^2 + 8x - 5 = 0$ rational?

> **Solution**   Here $a = 1$, $b = 8$, and $c = -5$.
>
> $$b^2 - 4ac = 8^2 - 4 \cdot 1 \cdot -5 = 64 + 20$$
> $$= 84$$
>
> 84 is not a perfect square, so the solutions are irrational.
>
> **Check**   Using the Quadratic Formula, $x = \dfrac{-8 \pm \sqrt{84}}{2} \approx \dfrac{-8 \pm 9.165}{2}$.
> Thus $x \approx 0.582$ or $x \approx -8.582$. These both approximate infinite decimals.

Questions

Covering the Reading

1. Define: rational number.

In 2–9, **a.** decide whether the number is rational or irrational. **b.** If it is rational, find a ratio of integers equal to it.

2. $\sqrt{13}$ 3. $\frac{6}{8}$ 4. 1.5   5. -1.5

6. π 7. $\frac{1}{3}$ 8. 0.004   9. $\sqrt{16}$

10. If a, b, and c are integers, when does $ax^2 + bx + c = 0$ have rational solutions?

In 11–18, use the discriminant to determine the number of rational solutions the equation has. Then solve.

11. $6x^2 + 13x + 6 = 0$

12. $2x^2 - 7x - 10 = 0$

13. $y^2 - 2 = 0$

14. $5t^2 + 3t = 0$

15. $10x^2 = 3x + 1$

16. $400v^2 + 100 = 400v$

17. $t(t + 9) = 5$

18. $x^2 + \frac{1}{4}x - \frac{1}{8} = 0$

Applying the Mathematics

19. Find two different values for k that will make $3x^2 + 8x + k = 0$ have rational solutions.

20. Find a value for k so that $x^2 + kx + 5 = 0$ has rational solutions.

Review

21. Find the x-intercepts of $y = x^2 + 9x - 5$. *(Lesson 12-3)*

22. Solve $4x^2 - 12x + 9 = 0$ by factoring and using the Zero Product Property. *(Lessons 10-5, 10-10)*

In 23–25, in a vacuum chamber on Earth an object will drop d meters in approximately t seconds, where $d = 4.9t^2$. *(Lesson 12-2)*

23. How far will the object drop in 3 seconds?

24. Write an expression for how far an object would drop in $n + 1$ seconds.

25. **a.** Write an equation that could be used to find the number of seconds it takes an object to drop 10 meters.
    b. Solve the equation in part a.

26. In how many ways can you draw two prize tickets from a jar with 30 tickets? (Assume the tickets are not replaced.) *(Lesson 4-7)*

27. *Skill sequence* Simplify. All variables represent positive numbers. *(Lesson 7-6)*
    a. $\sqrt{169x^2}$
    b. $\sqrt{25y^4}$
    c. $\sqrt{169x^2 \cdot 25y^4}$

In 28–30, calculate in your head. *(Lessons 4-2, 5-4)*

28. one third of a dozen

29. $33\frac{1}{3}\%$ of 900

30. $\frac{2}{6}$ of 30

Exploration

31. **a.** Find an integer value of c so that the equation $x^2 + 6x + c = 0$ has rational solutions.
    b. How many possible values of c are there?

12-7

Factoring Quadratic Trinomials

In Chapter 10, you studied how differences of squares and perfect square trinomials can be factored into a product of two binomials. For instance,

$$x^2 - y^2 = (x + y)(x - y)$$
$$x^2 + 6x + 9 = (x + 3)(x + 3)$$
$$x^2 - 2x + 1 = (x - 1)(x - 1).$$

Other quadratic trinomials can be factored.

$$x^2 + 5x + 6 = (x + 2)(x + 3)$$
$$3m^2 + 5m - 12 = (3m - 4)(m + 3)$$
$$4y^2 - 4y - 3 = (2y - 3)(2y + 1)$$

In all the examples and questions of this lesson, we are looking for factors with integer coefficients, as in the examples above. If a, b, and c are integers, then $ax^2 + bx + c$ is factorable if and only if the discriminant $b^2 - 4ac$ is a perfect square. You will learn why this is so in the following lessons. First we discuss how to factor.

Notice the general pattern. On the left side of each equal sign is a quadratic trinomial. On the right side are two binomials. Here is the form.

$$ax^2 + bx + c = (dx + e)(fx + g)$$

The product of d and f, the first terms of the binomials, is a. The product of e and g, the last terms of the binomials, is c.

So, if you need to factor the quadratic trinomial

$$ax^2 + bx + c$$

and you know it is factorable, first write the parentheses and the variables.

$$ax^2 + bx + c = (x \quad)(x \quad).$$

The coefficients of the x's must multiply to a and the constant terms must multiply to c. The problem is to find them so that the rest of the multiplication gives b.

Example 1   Is $x^2 + 9x + 14$ factorable?

Solution   Here $a = 1$, $b = 9$, and $c = 14$. So $b^2 - 4ac = 81 - 4 \cdot 1 \cdot 14 = 25$. Since the discriminant is a perfect square, $x^2 + 9x + 14$ is factorable.

Example 2   Factor $x^2 + 9x + 14$.

Solution   The idea is to rewrite the expression as a product of two binomials.

$$x^2 + 9x + 14 = (dx + e)(fx + g).$$

Look for integers d, e, f, and g.
The coefficient of x^2 is 1, so $df = 1$. Thus $d = 1$ and $f = 1$. Now you know

$$x^2 + 9x + 14 = (x + e)(x + g).$$

The product of e and g is 14. Try each possibility.
Can $e = 1$ and $g = 14$?

$$(x + 1)(x + 14) = x^2 + 15x + 14. \quad \text{No, you want } b = 9, \text{ not } 15.$$

Can $e = 2$ and $g = 7$?
$(x + 2)(x + 7) = x^2 + 9x + 14$. This does it. The answer is $(x + 2)(x + 7)$.

Check   The problem is already checked by multiplication. Another check is to substitute a value for x, say 4. Then does

$$x^2 + 9x + 14 = (x + 2)(x + 7)?$$
$$4^2 + 9 \cdot 4 + 14 = (4 + 2)(4 + 7)?$$
$$16 + 36 + 14 = 6 \cdot 11?$$

Yes. Each side equals 66.

In Example 2, because the coefficient of x^2 is 1, and all numbers are positive, there are only a few possible factors. Example 3 has more possibilities, but the idea is still the same. Try factors until you find the correct ones.

Example 3   Factor $6y^2 - 7y - 5$. (Assume it is factorable.)

Solution   First put down the form.
$$6y^2 - 7y - 5 = (\_y + \_)(\_y + \_).$$
The coefficients of y will multiply to 6. Thus either there is $3y$ and $2y$, or y and $6y$. The constant terms will multiply to -5. So they are either 1 and -5, or -1 and 5.

Here are all the possibilities with $3y$ and $2y$.

$$(3y + 1)(2y - 5)$$
$$(3y - 1)(2y + 5)$$
$$(3y - 5)(2y + 1)$$
$$(3y + 5)(2y - 1)$$

Here are all the possibilities with y and $6y$.

$$(y + 1)(6y - 5)$$
$$(y - 1)(6y + 5)$$
$$(y - 5)(6y + 1)$$
$$(y + 5)(6y - 1)$$

At most, you need to do these eight multiplications. If one of them gives $6y^2 - 7y - 5$, then that is the correct factoring.

We show all eight multiplications. You can see that the desired one is third.

$$(3y + 1)(2y - 5) = 6y^2 - 13y - 5$$
$$(3y - 1)(2y + 5) = 6y^2 + 13y - 5$$
$$(3y - 5)(2y + 1) = 6y^2 - 7y - 5$$
$$(3y + 5)(2y - 1) = 6y^2 + 7y - 5$$
$$(y + 1)(6y - 5) = 6y^2 + y - 5$$
$$(y - 1)(6y + 5) = 6y^2 - y - 5$$
$$(y - 5)(6y + 1) = 6y^2 - 29y - 5$$
$$(y + 5)(6y - 1) = 6y^2 + 29y - 5$$

So the correct answer is $(3y - 5)(2y + 1)$.

In Example 3, notice that each choice of factors gives a product that differs only in the coefficient of y (the middle term). If the problem was to factor $6y^2 - 100y - 5$, this process would show that no factors with integer coefficients will work.

Factoring quadratic trinomials using trial and error is a skill people learn to do in their heads. But it takes practice. This skill can save you time in solving some quadratic equations, as the next lesson explains.

Questions

Covering the Reading

1. If a, b, and c are integers, when is $ax^2 + bx + c$ factorable?

2. Consider $ax^2 + bx + c = (dx + e)(fx + g)$ for all values of x.
   a. The product of d and f is ___?___.
   b. The product of ___?___ and ___?___ is c.

3. How can you check the factors of a quadratic trinomial?

In 4–11, **a.** determine whether the trinomial can be factored; **b.** if so, factor it.

4. $x^2 + 7x + 10$ **5.** $x^2 + 21x + 110$

6. $x^2 + 2x + 3$ **7.** $y^2 + 10y + 9$

8. $n^2 + 4n - 12$ **9.** $t^2 - t - 30$

10. $4x^2 - 12x - 7$ **11.** $3x^2 + 11x - 4$

Applying the Mathematics

12. a. Factor $x^3 - 5x^2 + 6x$ into the product of a monomial and a trinomial.
   b. Complete the factoring by finding factors of the trinomial in part a.

13. Find k if $8x^2 + 10x - 25 = (2x + k)(4x - k)$.

Review

In 14 and 15, determine whether the solutions to the equation are rational numbers. *(Lesson 12-7)*

14. $13x^2 - 63x - 10 = 0$ **15.** $x^2 + x = 1$

16. *Multiple choice*   Which equation has more than one real solution? *(Lesson 12-6)*
   (a) $x^2 - 22x + 121 = 0$ (b) $9p^2 + 14p + 16 = 0$
   (c) $64z^2 - 32 = 0$ (d) $w^2 + 1 = 0$

17. Solve $3x^2 - 5x + 1 = 0$ **a.** exactly; **b.** to the nearest tenth. *(Lesson 12-4)*

18. *Multiple choice*   Which equation has the graph below? *(Lesson 12-1)*
   (a) $y = \frac{1}{2}x^2$
   (b) $y = 2x^2$
   (c) $y = -2x^2$
   (d) $y = -\frac{1}{2}x^2$

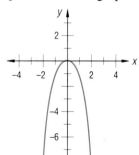

19. Here are 3 instances of a pattern. Describe the pattern using variables. *(Lesson 10-7)*

$$(72 + 100)(36 - 50) = 2(36^2 - 50^2)$$
$$(30 + 20)(15 - 10) = 2(15^2 - 10^2)$$
$$(2 + 1)(1 - 0.5) = 2(1^2 - 0.5^2)$$

In 20 and 21, give the slope and y-intercept for each line. *(Lesson 8-4)*

20. $y = \frac{1}{2}x$

21. $8x - 5y = 1$

22. State the Zero Product Property. *(Lesson 10-10)*

In 23–26, calculate the area of the shaded region. *(Lessons 4-1, 7-4)*

23.

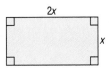

24.

25.

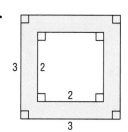

26.

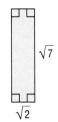

Exploration

27. The polynomial $x^3 + 6x^2 + 11x + 6$ can be factored into the form

$$(x + a)(x + b)(x + c).$$

Find a, b, and c. Hint: What is abc?

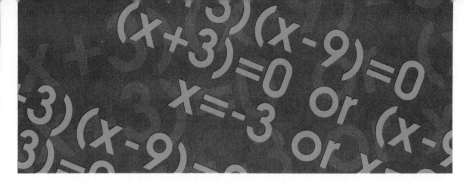

LESSON 12-8

Solving Some Quadratic Equations by Factoring

You already know that the Quadratic Formula will give you solutions to any quadratic equation you encounter — or show that the equation has no real solutions. There are lots of calculations involved in using the formula, however, so if you can factor a quadratic by sight and use the Zero Product Property to find solutions, you can save time and effort.

Example 1 Solve $x^2 - 6x - 27 = 0$.

Solution Since the coefficient of x^2 is 1, it is worth trying factoring. $1 \cdot 1 = 1$, so $x^2 - 6x - 27 = (x + \quad)(x + \quad)$. It is easy to see that 3 and -9 in the blanks yield the $-6x$ and -27 needed. Thus,

$$(x + 3)(x - 9) = 0$$

Use the Zero Product Property and solve.

$$(x + 3) = 0 \quad \text{or} \quad (x - 9) = 0.$$
$$x = \text{-3} \quad \text{or} \quad x = 9.$$

The solutions are -3 or 9.

Check
Substitute -3. Does $(-3)^2 - 6(-3) - 27 = 0$? $9 + 18 - 27 = 0.$
Substitute 9. Does $(9)^2 - 6(9) - 27 = 0$? $81 - 54 - 27 = 0.$
It checks.

Keep in mind that some equations have common monomial factors. By first factoring out the monomial, you can further factor the remaining expression.

Example 2   Solve $4x^3 - 18x^2 + 8x = 0$.

Solution   This trinomial is not a quadratic, so in this form even the Quadratic Formula is of no use. Inspection shows, however, that $2x$ is a common monomial factor.

$$4x^3 - 18x^2 + 8x = 0$$
$$2x(2x^2 - 9x + 4) = 0$$

Now factor the quadratic expression. $(x - 4)$ and $(2x - 1)$ work out, so the equation can be rewritten as

$$2x(x - 4)(2x - 1) = 0.$$

Apply the Zero Product Property. The factor 2 cannot be 0, but any of the other three could be. So,

$$x = 0 \text{ or } (x - 4) = 0 \text{ or } (2x - 1) = 0.$$
$$x = 0 \text{ or } \qquad x = 4 \text{ or } \qquad 2x = 1$$
$$x = \tfrac{1}{2}.$$

The equation has three solutions: 0, 4, or $\tfrac{1}{2}$.

Check   The check is left to you as Question 6.

The shortcut of solving equations by factoring may even help you answer some application questions more quickly.

Example 3   In a round robin chess tournament with n players, $\dfrac{n^2 - n}{2}$ names are needed. If there is time for 55 games, how many players can be entered in the tournament?

World Champion chess player Gary Kasparov

Solution   You must solve $\dfrac{n^2 - n}{2} = 55$.

This is a quadratic equation, but it is not in standard form.
Multiply both sides by 2 and then subtract 110 from both sides.

$$n^2 - n = 110$$
$$n^2 - n - 110 = 0$$

The quadratic expression can be factored.

$$(n - 11)(n + 10) = 0$$

Use the Zero Product Property and solve.

$$n - 11 = 0 \quad \text{or} \quad n + 10 = 0$$
$$n = 11 \quad \text{or} \qquad n = \text{-}10$$

A negative number of players does not make sense in this situation.
So 11 players can be entered.

Check   Substitute. Does $\dfrac{11^2 - 11}{2} = 55$? Yes, it does.

Questions

Covering the Reading

1. To solve quadratic equations by factoring, first make sure the equation is in standard form. Then __?__ the quadratic and apply the __?__ Property to solve.

In 2 and 3, solve by factoring.

2. $x^2 + 2x - 3 = 0$ 3. $y^2 - 4y - 5 = 0$

In 4 and 5, **a.** factor out the common monomial factor; **b.** solve the equation.

4. $7x^2 + 7x - 84 = 0$ 5. $9x^3 + 12x^2 - 12x = 0$;

6. Check each solution to Example 2.

7. In Example 3, if there is time for 66 games, how many players can be entered?

Applying the Mathematics

In 8–11, solve by factoring.

8. $a^2 - 5a = 50$ 9. $26 + b^2 = \text{-}15b$

10. $x^2 + 620x = 0$ 11. $20y^3 - 95y^2 - 25y = 0$

In 12 and 13, solve by any method.

12. $4t^2 + 4t + 1 = 0$ **13.** $9v = 2v^2 - 35$

14. The height h in feet of a projectile after t seconds is given by $h = -16t^2 + 128t$.
   a. When is the projectile 192 feet above the ground?
   b. When is the projectile on the ground? (Think: what is the value of h at ground level?)

Review

15. Multiply: $20(x - \frac{1}{5})(x + \frac{1}{4})$. *(Lesson 10-4)*

16. Match each graph with its equation. *(Lessons 12-1, 7-1)*
   (i) $y = (x - 1)^2$ (ii) $y - x = 1$ (iii) $-1 + x^2 = y$

a.

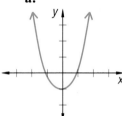

b.

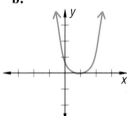

c.

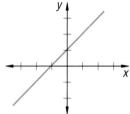

17. *True or false* $x^2 + x + 100 = 0$ has no real solutions. *(Lesson 12-4)*

In 18–23, tell whether the number is rational or irrational. *(Lesson 12-6)*

18. $\sqrt{81}$ **19.** $3 + \sqrt{2}$ **20.** 7.984

21. $\sqrt{80}$ **22.** $\sqrt{\frac{4}{9}}$ **23.** 0

In 24–26, write the reciprocal. *(Lesson 4-3)*

24. $\frac{2x}{3y}$ **25.** $\frac{a + b}{5}$ **26.** $2.8q^2$

27. Factor $\frac{1}{4}y^2 - 9$. *(Lesson 10-9)*

Exploration

28. Some trinomials that are not quadratics can be rewritten as quadratics and solved by the method of this lesson. For example, $x^4 - 3x^2 - 4 = (x^2)^2 - 3(x^2) - 4 = (x^2 - 4)(x^2 + 1)$. Use this idea to solve the following equations.
   a. $x^4 - 10x^2 + 9 = 0$ **b.** $m^4 - 13m^2 + 36 = 0$

29. Make up some examples of quadratic equations of the form $ax^2 + bx + c = 0$ that can be solved by factoring and some that cannot.

12-9

The Factor Theorem

In the last two lessons you have been factoring $ax^2 + bx + c$ by trial and error. There is another way, using the Quadratic Formula. The idea is to compare the solutions to $ax^2 + bx + c = 0$ with the factors of $ax^2 + bx + c$.

The equation
$x^2 - 13x + 40 = 0$
has solutions 5 and 8.

The expression
$x^2 - 13x + 40$
has factors $x - 5$ and $x - 8$.

Factor Theorem for Quadratic Expressions:

If r is a solution of $ax^2 + bx + c = 0$, then $(x - r)$ is a factor of $ax^2 + bx + c$.

The Factor Theorem is true for any quadratic expression.

Example 1 6 and -19 are solutions to $x^2 + 13x - 114 = 0$. Write the factors of $x^2 + 13x - 114$.

Solution Since 6 and -19 are solutions, $(x - 6)$ and $(x - -19)$ are factors of $x^2 + 13x - 114$. These simplify to $(x - 6)$ and $(x + 19)$. That is, $x^2 + 13x - 114 = (x - 6)(x + 19)$.

Check $(x - 6)(x + 19) = x^2 + 19x - 6x - 114 = x^2 + 13x - 114$

Example 2 Factor $x^2 + 2x - 15$.

Solution Solve $x^2 + 2x - 15 = 0$ using the Quadratic Formula. $a = 1, b = 2, c = -15$.

$$x = \frac{-2 \pm \sqrt{(2)^2 - 4 \cdot 1 \cdot -15}}{2 \cdot 1}$$

$$= \frac{-2 \pm \sqrt{64}}{2}$$

$$= \frac{-2 \pm 8}{2}$$

$$x = \frac{-2 + 8}{2} = \frac{6}{2} = 3 \text{ or } x = \frac{-2 - 8}{2} = \frac{-10}{2} = -5.$$

Because the solutions are 3 and -5, the Factor Theorem gives $(x - 3)$ and $(x - -5)$ as factors. So $x^2 + 2x - 15 = (x - 3)(x + 5)$.

Check $(x - 3)(x + 5) = x^2 - 3x + 5x - 15 = x^2 + 2x - 15.$

Of course you could have factored the quadratic in Example 2 in your head using the method of Lesson 12-7. For some quadratics, however, there are many possible factors and the Factor Theorem may save time.

When the coefficient of x^2 is not 1, the process requires a step to multiply by this coefficient. This step is seen in Example 3.

Example 3   Factor $12x^2 - x - 6$.

Solution   Both 12 and 6 have many pairs of factors. So solve $12x^2 - x - 6 = 0$ and use the Factor Theorem.

$$x = \frac{-(-1) \pm \sqrt{(-1)^2 - 4 \cdot 12 \cdot -6}}{2 \cdot 12}$$

$$= \frac{1 \pm \sqrt{1 + 288}}{24}$$

$$= \frac{1 \pm \sqrt{289}}{24}$$

$$= \frac{1 \pm 17}{24}$$

$$x = \frac{1 + 17}{24} = \frac{18}{24} = \frac{3}{4} \text{ or } x = \frac{1 - 17}{24} = \frac{-16}{24} = \frac{-2}{3}$$

Notice that $x - \frac{3}{4}$ and $x + \frac{2}{3}$ are factors but there is a constant factor also.

$$12x^2 - x - 6 = 12\left(x - \tfrac{3}{4}\right)\left(x + \tfrac{2}{3}\right)$$

Factor the 12 into $4 \cdot 3$ to match the denominators. (This is always possible.)

$$= 4\left(x - \tfrac{3}{4}\right) \cdot 3\left(x + \tfrac{2}{3}\right)$$
$$= (4x - 3)(3x + 2)$$

Questions

1. State the Factor Theorem for quadratic expressions.

2. -8 and 13 are solutions to $x^2 - 5x - 104 = 0$.
   a. Write the factors of $x^2 - 5x - 104$.
   b. Check your work using the FOIL algorithm.

In 3–6, factor by first solving the corresponding quadratic equation.

3. $x^2 - 7x + 10$ **4.** $y^2 - 11y - 60$

5. $5k^2 - 8k - 4$ **6.** $24t^2 + 59t + 36$

In 7 and 8, use this information. Whenever the coefficients a, b, and c of the quadratic equation $ax^2 + bx + c = 0$ sum to zero, then 1 is a solution to the equation.

7. If 1 is a solution to $ax^2 + bx + c = 0$, then what is a factor?

8. Factor $98x^2 - 99x + 1$.

In 9–12, factor using any method.

9. $x^2 + 16x + 64$

10. $y^2 + 3y - 700$

11. $16m^3 - 44m^2 + 120m$

12. $12 + 25a + 12a^2$

In 13–15, a quadratic expression is given with one of its factors. Find the other factor.

13. $x^2 + 7x + 6$; $(x + 1)$

14. $2x^2 - 17x + 8$; $(2x - 1)$

15. $3x^2 - 7x + 4$; $(x - 1)$

In 16 and 17, solve. *(Lesson 12-7)*

16. $x^2 - 15x + 14 = 0$ **17.** $5x^3 + 15x^2 = 50x$

18. How many real solutions does $x^2 - 2x + 2 = 0$ have? *(Lesson 12-4)*

In 19 and 20, a cannon is set so that the path of its cannonball is described by the equation $y = -136x^2 + 54x + 5$ where y is the cannonball's distance above the ground (in meters) and x is the corresponding distance along the ground (in kilometers).

19. How high is the cannonball when it is 300 m along the ground from the point of firing? *(Lesson 12-5)*

20. How far will the cannonball travel before it hits the ground?

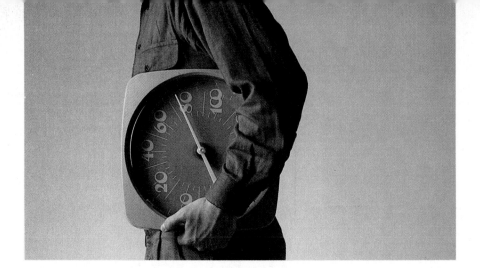

In 21 and 22, a thermometer is taken from a room temperature of 70°F to a point outside where it is 45°F. After t minutes, the thermometer reading D is approximated by $D = 24(1 - .04t)^2 + 45$.

21. What temperature will the thermometer read after 10 minutes?
(Lesson 12-2)

22. How long will it take the thermometer to get down to 45°F?
(Lessons 12-2, 7-6)

23. The total cost of a car is $3869 including 6% sales tax. What is the cost of the car before tax? *(Lesson 6-3)*

24. Simplify $(7b + 8) - (2 - 3b)$. *(Lesson 6-5)*

25. Donna has $75 and is saving $5 per week. Her brother has $48 and is saving $8 per week. In how many weeks will her brother have saved more money? *(Lesson 6-7)*

26. *Multiple choice*   Choose the equation of the line graphed below.
(Lesson 7-2)
(a) $y = x - 2$
(b) $y = x + 2$
(c) $y = 2 - x$
(d) $y = 2x$

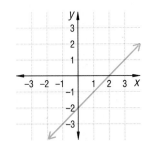

27. Graph $y = a(x - 2)(x - 8)$ for $a = 1, 2, 3,$ and 4.
   a. What do these parabolas have in common?
   b. Explore whether the same property holds when a is negative.

Summary

This chapter discusses graphs and equations involving quadratic expressions. A quadratic expression is an expression that can be put in the form $ax^2 + bx + c$. These expressions describe the shapes of trajectories and are found in area and other formulas.

A quadratic equation is an equation equivalent to $ax^2 + bx + c = 0$. The solutions to that equation are found by the Quadratic Formula.

$$x = \frac{-b \pm \sqrt{b^2 - 4ac}}{2a}$$

The discriminant of a quadratic equation $ax^2 + bx + c = 0$ is $b^2 - 4ac$. If the discriminant is positive, there are two real solutions; if zero, there is one solution; if negative, there are no real solutions. If a, b, and c are integers and $b^2 - 4ac$ is a perfect square, then the solutions are rational and the expression $ax^2 + bx + c$ can be factored.

If r is a solution of $ax^2 + bx + c = 0$, then $(x - r)$ is a factor. Thus knowing the roots, you can find the factors of a quadratic expression.

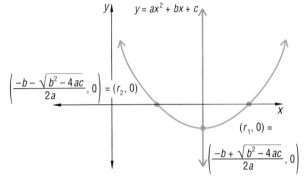

The graph of $y = ax^2 + bx + c$ is a parabola. This parabola is symmetric through the vertical line through the vertex. If $a > 0$, the parabola opens up. If $a < 0$, the parabola opens down. The Quadratic Formula gives the x-intercepts of the parabola if there are any.

Vocabulary

Below are the most important terms and phrases for this chapter. You should be able to give a general description and a specific example of each.

Lesson 12-1
parabola
symmetric
axis of symmetry
vertex
STEP command

Lesson 12-2
automatic grapher
window, default window
zoom feature

Lesson 12-4
quadratic equation
Quadratic Formula
standard form of a quadratic equation

Lesson 12-5
discriminant
Discriminant Property

Lesson 12-6
rational number, irrational number

Lesson 12-9
Factor Theorem for Quadratic Expressions

Progress Self-Test

Take this test as you would take a test in class. Then check your work with the solution in the Selected Answers section in the back of the book.

In 1 and 2, give the exact solutions to the equation.

1. $x^2 - 9x + 20 = 0$

2. $5y^2 - 3y = 11$

3. Find a simpler equation that has the same solutions as $18z^2 - 3z + 24 = 0$.

In 4–7, find all real solutions. Approximate answers to the nearest hundredth.

4. $k^2 - 6k - 3 = 0$

5. $3r^2 + 15r + 8 = 0$

6. $8x^2 - 7x = -11$

7. $z^2 = 16z - 64$

8. *True or false*   The solutions to $8x^2 - 18x - 5 = 0$ are rational.

9. *True or false*   The parabola $y = 3x^2 - 7x - 35$ opens down.

10. If -2 and 18 are solutions to $x^2 - 16x - 36 = 0$, what are the factors of $x^2 - 16x - 36$?

In 11 and 12, factor.

11. $x^2 + 8x + 19$

12. $x^2 - 12x + 35$

13. Solve $x^2 - x - 42 = 0$ by factoring.

In 14 and 15, Harry tosses a ball to Ferdinand who is 20 yards away. At x yards away from Harry, the ball is at height h feet, where
$$h = -.07(x - 10)^2 + 12.$$

14. If Melody is 2 yards in front of Ferdinand, how high is the ball when it passes by her?

15. Ferdinand misses the ball and it falls to the ground. To the nearest yard, how far did it go?

In 16 and 17, **a.** make a table of values; **b.** graph the equation.

16. $y = -2x^2$

17. $y = x^2 - 4x + 3$

18. *Multiple choice*   Which of these is the graph of $y = 2.5x^2$?

(a)
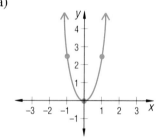

(b)
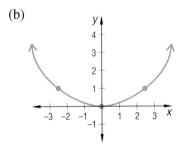

19. Define: rational number.

20. *Multiple choice* __ Which number is rational?
(a) $\sqrt{23}$ (b) $\sqrt{24}$ (c) $\sqrt{25}$ (d) $\sqrt{26}$

Chapter Review

Questions on **SPUR** Objectives

SPUR stands for **S**kills, **P**roperties, **U**ses, and **R**epresentations.
The Chapter Review questions are grouped according to the
SPUR Objectives for this chapter.

SKILLS deals with the procedures used to get answers.

■ **Objective A.** *Solve quadratic equations using the Quadratic Formula. (Lessons 12-4, 12-5)*

In 1–4, give the exact solutions to the equation.

1. $6y^2 + 7y - 20 = 0$

2. $x^2 + 7x + 12 = 0$

3. $4a^2 - 13a = 12$

4. $-q^2 - 6q + 12 = -4$

In 5 and 6, find a simpler equation that has the same solutions as the given equation. Then solve.

5. $10m^2 - 50m + 30 = 0$

6. $20y^2 + 14y - 24 = 0$

In 7–14, solve when possible. Approximate irrational answers to the nearest hundredth.

7. $k^2 - 7k - 2 = 0$

8. $2m^2 + m - 3 = 0$

9. $22a^2 + 2a + 3 = 0$

10. $0 = x^2 + 10x + 25$

11. $13k^2 + k = 2$

12. $16x^2 + 8x = -5$

13. $z^2 = 9z - 14$

14. $14y - 3 = 2y^2$

■ **Objective B.** *Factor quadratic trinomials. (Lessons 12-7, 12-9)*

In 15–20, factor.

15. $x^2 + 5x - 6$

16. $3y^2 + 2y - 8$

17. $10a^2 - 19a + 7$

18. $12m^3 + 117m^2 + 81m$

19. $x + 5x^2 - 6$

20. $-3 - 2k + 8k^2$

21. Given that -3 and 17 are solutions to $x^2 - 14x - 51 = 0$, write the factors of $x^2 - 14x - 51$.

22. Factor $x^2 + 5x + 4$ by first solving $x^2 + 5x + 4 = 0$.

■ **Objective C.** *Solve quadratic equations by factoring. (Lesson 12-8)*

In 23–26, solve by factoring.

23. $6y^2 + y - 2 = 0$

24. $z^2 + 7z = -12$

25. $x^2 - 2x = 0$

26. $0 = 16m^2 - 8m + 1$

PROPERTIES deal with the principles behind the mathematics.

■ **Objective D.** *Recognize properties of the parabola. (Lessons 12-1, 12-5)*

27. What shape is the graph of $y = -6x^2$?

28. *True or False* The parabola $y = -2x^2 + 3x + 1$ opens down.

29. What equation must you solve to find the x-intercepts of the parabola $y = ax^2 + bx + c$?

30. *True or false* The graph of $y = x^2 + 2x + 2$ intersects the x-axis twice.

Objective E. *Recognize properties of quadratic equations.* *(Lessons 12-4, 12-5)*

31. Give two solutions to the sentence $ax^2 + bx + c = 0$.

32. *True or false*   Some quadratic equations have no real solutions.

33. *True or false*   Any quadratic equation can be solved using the Quadratic Formula.

34. If a quadratic equation has solutions which are rational, the discriminant is a __?__.

In 35–38, use the discriminant to determine the number of real solutions to the equation.

35. $2x^2 - 3x + 4 = 0$

36. $a^2 = 3a + 8$

37. $9d = 40 + 8d^2$

38. $n(n + 1) = -5$

Objective F. *Determine whether a number is rational or irrational.* *(Lesson 12-6)*

In 39–44, tell whether the number is rational.

39. $\sqrt{2}$

40. 7.0707

41. $\frac{1}{3}$

42. -86

43. $\pi + 1$

44. $\sqrt{9}$

USES deal with applications of mathematics in real situations.

Objective G. *Use the parabola and quadratic equations to solve real world problems.* *(Lesson 12-2, 12-3, 12-8)*

45. Consider again the quarterback in Lesson 12-2 who tosses a football to a receiver 40 yards downfield. The ball is at height h feet, x yards downfield where $h = -\frac{1}{40}(x - 20)^2 + 16$. If a defender is 6 yards in front of the receiver,
   a. how far is he from the quarterback?
   b. Would the defender have a chance to deflect the ball?

46. When a ball is thrown into the air with initial upward velocity of 20 meters per second, its approximate height y above the ground after x seconds is given by $y = 20x - 5x^2$. When will the ball hit the ground?

47. A golf ball was hit on the moon. Suppose that the height h meters of the ball t seconds after it was hit is described by $h = -.8t^2 + 10t$. At what times was the ball at a height of 20 meters?

48. Refer to the graph below. It shows the height h in feet of a ball, t seconds after it is thrown from ground level at a speed of 64 feet per second.
   a. Estimate from the graph at what height the ball will be after 1 second.
   b. Estimate from the graph when the ball will reach a height of 35 feet.

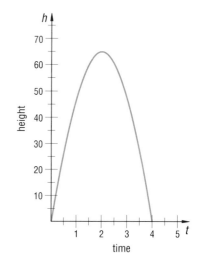

■ **Objective H.** *Graph parabolas.* *(Lessons 12-1, 12-2)*

In 49–52, **a.** make a table of values; **b.** graph the equation.

49. $y = 3x^2$

50. $y = -\frac{1}{2}x^2$

51. $y = x^2 + 5x + 6$

52. $y = x^2 - 4$

53. *Multiple choice* Which equation has the given graph?
(a) $y = -4x^2$
(b) $y = \frac{-1}{4}x^2$
(c) $y = 4x^2$
(d) $y = \frac{1}{4}x^2$

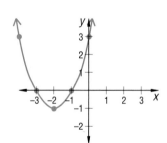

54. *Multiple choice* Which of these is the graph of $y = x^2 + 4x + 3$?

(a)

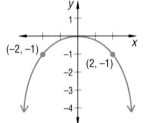

(b)

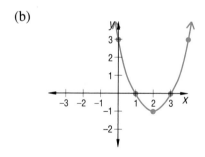

55. Describe the graphing window shown below.

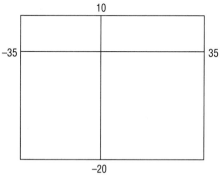

56. Consider $y = 2x^2 - 11.8x + 17$.
a. Graph this parabola using an automatic grapher.
b. Estimate the coordinates of its vertex.
c. Estimate the values of its x-intercepts.

Functions

Fahrenheit-Celsius temperatures

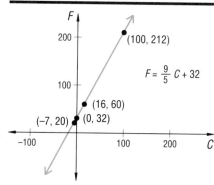

$$F = \frac{9}{5} C + 32$$

In earlier chapters, you have seen many types of graphs. Below are four of them.

These graphs picture situations that have much in common.
(1) There are two variables. (The first variable is indicated horizontally; the second vertically.)
(2) Every value of the first variable determines exactly one value of the second variable.

For instance, in the La Quebrada graph, the first variable stands for the distance from the cliff. The second variable stands for the height above the water. The distance from the cliff determines the height above the water. (But the height above the water does not always determine exactly the diver's distance from the cliff.)

Look at the graph of temperatures. If you were at the base of Mount Rainier, near Seattle, (shown on page 636), the temperature could be 60°F, or about 16°C. If you were close to the summit, the temperature could be 20°F, or about -7°C. The Fahrenheit-Celsius relationship is a linear one.

When a situation has two variables, it is natural to think of describing the situation using ordered pairs. The general name given to sets of ordered pairs like those above is *function*. You can think of a function as a particular kind of relationship between variables. The analysis of functions is extremely important in mathematics. Entire courses are often devoted to this. In this chapter, you will review many of the ideas of the earlier chapters using the language of functions, and you will encounter some functions you have not seen before.

Women's Olympic 400-meter freestyle

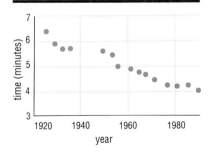

Compound interest

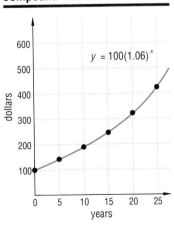

La Quebrada cliff diver

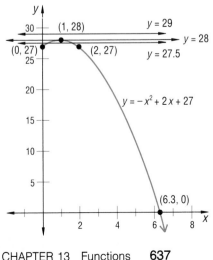

What Are Functions?

On pages 636 and 637, four functions are graphed. Below, a set of ordered pairs is graphed that is *not* a function.

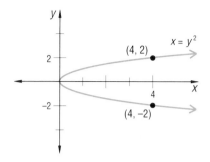

The distinction between this graph and the four on the previous pages has to do with the number of *y*-values determined by each *x*-value. The four previous graphs have the characteristic that a value of *x* gives you exactly one value of *y*. For example, there is only one Women's Olympic winning freestyle swimming time in 1972, 4.32 minutes. On the graph above, however, the value $x = 4$ corresponds to two possible values for *y*; $y = 2$ or -2.

Since the points (4, 2) and (4, -2) are on the graph, the set of ordered pairs satisfying $x = y^2$ is not a function.

Definition:

> A function is a set of ordered pairs in which each first coordinate appears with exactly one second coordinate.

*Canal in
Copenhagen, Denmark*

Red Square, Moscow, U.S.S.R.

Example 1   On page 399, latitudes and April mean high temperatures (°F) for 10 selected cities are graphed. Here that graph is repeated. Does the graph describe a function?

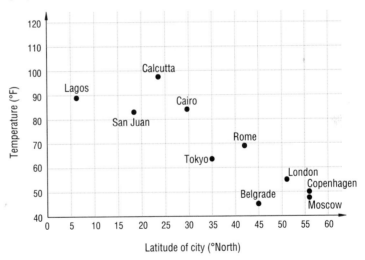

Solution   No. The latitudes for Copenhagen and Moscow are the same, but the temperatures are different. These ordered pairs have the same first coordinate but different second coordinates. The graph pictures a relation, but not a function.

In a function, the first variable determines the second. If you know the value of the first variable (often called x), then there is only one value for the second (often called y). A value of the second variable is called a **value of the function.** Many equations you have studied describe functions.

To determine whether an equation describes a function, solve it for the second variable. For instance, $y^2 = x$ becomes $y = \pm\sqrt{x}$. It is clear in $y = \pm\sqrt{x}$ that every positive value of x corresponds to two values of y. So $y^2 = x$ does *not* describe a function. It describes a relation. A **relation** is any set of ordered pairs. So a function is always a relation, but not all relations are functions.

Example 2 Does the equation $3x + 4y = 12$ describe a function?

Solution 1 Solve the equation for y.

$$4y = -3x + 12$$

$$y = \tfrac{-3}{4}x + 3$$

Since each value of x corresponds to one value of y, the equation describes a function.

Solution 2 Graph $3x + 4y = 12$. Since the graph does not contain two points with the same first coordinate, the graph describes a function.

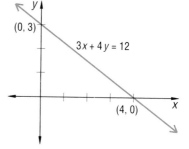

The function of Example 2 is a **linear function.** Linear functions have equations of the form $y = mx + b$. Another example is the Fahrenheit-Celsius relationship graphed on page 636. A **quadratic function** has an equation of the form $y = ax^2 + bx + c$ and its graph is a parabola. The La Quebrada graph on page 637 is a graph of a quadratic function. You studied exponential functions with the form $y = ab^x$ in Chapter 9. The compound interest graph on page 637 is a graph of an exponential function.

Functions are found in every branch of mathematics and are extremely important in applications. As a result, there are many ways to describe them:
(1) by a graph
(2) by an equation
(3) by listing all the ordered pairs (only possible if there are finitely many pairs)
(4) by a written rule.
For instance, a written rule for the Fahrenheit-Celsius function on page 637 is: To find the Celsius temperature, subtract 32 from the Fahrenheit temperature and then multiply the difference by $\tfrac{5}{9}$.

Questions

Covering the Reading

1. Define: function.

2. Define: relation.

In 3–6, two first coordinates of the relation are given. **a.** Find all corresponding second coordinates. **b.** Is the relation also a function?

3. Fahrenheit-Celsius relation on page 636; $F = 32°$, $F = 212°$.

4. Women's Olympic 400-meter freestyle swimming champion times on page 637; 1984, 1988.

5. La Quebrada cliff diver on page 637; $x = 27$, $x = 28$.

6. Latitude-high temperature relation on page 639; latitude $= 6$, latitude $= 56$.

7. Why is $x = y^2$ not an equation for a function?

8. Name four ways in which a function can be described.

9. Explain why the graph of $x - 3y = 6$ describes a function.

Applying the Mathematics

10. *Multiple choice* Which set of ordered pairs is *not* a function?
(a) $\{(0, 0), (1, 1), (2, 2)\}$ (b) $\{(3, 5), (5, 3), (4, 4)\}$
(c) $\{(0, 0), (1, 0), (0, 1)\}$ (d) $\{(\frac{1}{2}, 1), (\sqrt{7}, \sqrt{8}), (6, \frac{-9}{23})\}$

11. Explain why a vertical line cannot be the graph of a function.

In 12–14, can the graph be a graph of a function?

12. 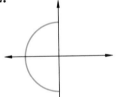 **13.** 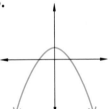 **14.**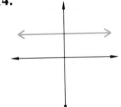

In 15 and 16, name three points of the function with the given equation. Then graph the function.

15. $y = \frac{1}{2}x^2 + 3$ **16.** $x = y + 400$

In 17 and 18, a written rule for a function is given. Name three ordered pairs of the function.

17. The cost of wrapping and mailing a particular package is three dollars plus fifty cents a pound.

18. To find the volume of a sphere, multiply the cube of the radius by $\frac{4}{3}\pi$.

Review

19. *Skill sequence*   Write as a single fraction. *(Lessons 2-3, 7-8)*

a. $\frac{1}{2} + \frac{1}{3}$ **b.** $\frac{1}{x} + \frac{1}{y}$ **c.** $\frac{1}{a + 1} + \frac{1}{b - 2}$

20. A class of 24 students contains 3% of all the students in the school. How many students are in the school? *(Lesson 5-4)*

21. When $m > n > 0$, which is larger, $\frac{1}{m}$ or $\frac{1}{n}$? *(Lesson 4-3)*

22. *True or false*   The slope of the line through (x_1, y_1) and (x_2, y_2) is the opposite of the slope of the line through (x_2, y_2) and (x_1, y_1). *(Lesson 8-2)*

23. Write $2^{-3} + 4^{-3}$ as a simple fraction. *(Lesson 9-6)*

Exploration

24. A function contains the ordered pairs $(1, 1)$ and $(2, 4)$.
   a. Find a possible equation describing this function.
   b. Find a second possible equation describing this function.
   c. Find a third possible equation describing this function. (The equations should not be equivalent.)

Function Notation

Ordered pairs in functions need not be numbers. You are familiar with the abbreviation P(E), read "the probability of E," or even shorter, "P of E." In P(E), the letter E names an event and P(E) names the probability of that event. An event can have only one probability, so any set of events and their probabilities is a function. For instance, in the tossing of a fair coin, P(heads) $= \frac{1}{2}$ and P(tails) $= \frac{1}{2}$. The function P is {(heads, $\frac{1}{2}$), (tails, $\frac{1}{2}$)}.

This kind of abbreviation is used for all functions. For instance, we could use the shorthand s(x), read "s of x," to stand for the *square of x*. Then for each number x, there is a value of the function s(x), its square. The function is named s and called the *squaring function*. In the abbreviation s(x), as in P(E), the parentheses do *not* mean multiplication. s(x) stands for the square of x. When $x = 3$, s(x) $= 3^2 = 9$. Here are some pairs of values.

x	$s(x)$
3	9
-4	16
-13	169
$\frac{1}{5}$	$\frac{1}{25}$
1.7	2.89

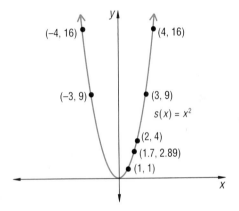

An equation or formula for this function is $s(x) = x^2$. You are familiar with this function. In previous lessons it was named $y = x^2$. Its graph is a parabola.

Example 1 Suppose A(n) is the average number of accidents per 50 million miles driven by a driver who is n years old. Using function notation, the formula on page 587 becomes

$$A(n) = 0.4n^2 - 36n + 1000.$$

Find A(16) and A(40).

Solution To find A(16), substitute 16 for n in the formula.
$A(16) = 0.4 \cdot 16^2 - 36 \cdot 16 + 1000$
$= 102.4 - 576 + 1000$
$= 526.4$
A(16) is shorthand for the average number of accidents per 50 million miles driven by 16-year-olds. A(16) is about 500.

To find $A(40)$, substitute 40 for n in the formula.
$$A(40) = 0.4 \cdot 40^2 - 36 \cdot 40 + 1000$$
$$= 640 - 1440 + 1000$$
$$= 200$$
Thus the average number of accidents per 50 million miles driven by 40-year-olds is about 200. That is quite a bit lower than the number for 16-year-olds.

In Chapter 12, we described the relationship of Example 2 with the equation $y = 0.4x^2 - 36x + 1000$. So why use function notation? One advantage of function notation is that it is shorter to write.

$A(16) > A(40)$ means   the number of accidents per 50 million miles for 16-year-olds is greater than the number for 40-year olds.

Another advantage is that the letters x and y do not always convey any meaning. But the letter A, which stands for the function, is the first letter of "accident" and could be easier to remember.

Computer programs take advantage of function notation. Remember from Chapter 7 that in BASIC,

SQR(X)	means	the square root of x.
ABS(X)	means	the absolute value of x.

In this way, BASIC uses function notation. The names of the functions are SQR and ABS.

When there is no application, the most common letter used to name a function is f.

∎ ∎ ∎ ∎ ∎ ∎ ∎∎

Example 2   What kind of function is described with the equation $f(x) = -5x + 40$?

Solution 1   Since $f(x)$ stands for the second coordinates of the points, write y in place of $f(x)$. This gives $y = -5x + 40$, which we know to be a linear function with slope -5 and y-intercept 40.

Solution 2   Use an automatic grapher (which is also called a function grapher or function plotter). Enter $y = -5x + 40$ or $f(x) = -5x + 40$. (Some graphers use function notation.) A line should appear, like that shown at the right.

(0, 40)

$f(x) = -5x + 40$

(8, 0)

Questions

Covering the Reading

In 1–3, write out how each symbol is read.

1. P(E)   **2.** s(x)   **3.** f(x) = x²

In 4–6, let s(x) = x². Give the value of:

4. s(3)   **5.** s(-8)   **6.** s($\frac{2}{5}$)

In 7–10, refer to Example 1.

7. *Multiple choice*   The function of this example is named
   (a) A   (b) n   (c) A(n)   (d) A(40)

8. The average number of accidents per 50 million miles driven by a 25-year-old is __?__.

9. A(60) stands for __?__.

10. Which is larger, A(16) or A(20), and what does the answer mean?

In 11–13, consider the function f of Example 2.

11. What kind of a function is f?

12. Calculate f(3) and f(5).

13. What is the slope of f?

14. Evaluate SQR(40).

15. Evaluate ABS(-2.5).

Applying the Mathematics

16. a. According to the function in this lesson, who has more accidents on average, a 40-year-old or a 50-year old?

   b. Verify your answer by graphing the function.

17. At what age is there an estimated 1000 accidents per 50 million miles?

18. Suppose L(x) = 17x + 10.
   a. Calculate L(5).
   b. Calculate L(2).
   c. Calculate $\frac{L(5) - L(2)}{5 - 2}$.
   d. What have you calculated in part c?

19. Let c(n) = n³.   **a.** Calculate c(1), c(2), c(3), c(4), and c(5).   **b.** What might be an appropriate name for c?

20. Let $s(p)$ = the number of sisters of a person p. Let $b(p)$ = the number of brothers of a person p.
   a. If you are the person p, give the values of $s(p)$ and $b(p)$.
   b. What does $s(p) + b(p) + 1$ stand for?

21. If $SS(n)$ = the sum of the squares of the integers from 1 to n, then
$$SS(n) = \frac{n(n + 1)(2n + 1)}{6}.$$
   a. Calculate $SS(11)$.
   b. Evaluate $1^2 + 2^2 + 3^2 + 4^2 + 5^2 + 6^2 + 7^2 + 8^2 + 9^2 + 10^2 + 11^2 + 12^2$.

Review

22. *Skill sequence* Write as a single fraction. *(Lesson 2-3)*
   a. $3 + \frac{2}{5}$ **b.** $3 + \frac{7}{5}$ **c.** $3 + \frac{k}{5}$

23. Simplify $6 \cdot 3^{-2}$. *(Lesson 9-6)*

24. *Multiple choice* From a 16″ by 16″ sheet of wrapping paper, rectangles of width x'' are cut off two adjacent sides. What is the area (in square inches) of the colored region that remains? *(Lesson 10-6)*

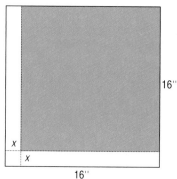

   (a) $16 - x$ (b) $256 + x^2$
   (c) $256 - x^2$ (d) $256 - 32x + x^2$

25. Give the value of $|x| + |-x|$ when $x = -2$. *(Lesson 7-3)*

26. Solve: $(t + 5)^2 = 3$. *(Lesson 7-8)*

27. A case contains c cartons. Each carton contains b boxes. Each box contains 100 paper clips. How many paper clips are in the case? *(Lesson 4-2)*

Exploration

28. Let $f(x) = \dfrac{12}{x - a}$. Using an automatic grapher, graph the function f from $x = -5$ to $x = 5$ when $a = 0$, $a = 1$, $a = 2$, and $a = 3$. What do the graphs have in common? How are they different?

13-3

Absolute Value Functions

The function with equation

$$f(x) = |x|$$

is the simplest example of an **absolute value function.** In BASIC and other computer languages, $|x|$ is written as ABS(X). By substitution, ordered pairs for this function can be found.

x	$f(x) = |x|$	Pair
2	$f(2) = |2| = 2$	(2, 2)
-8	$f(-8) = |-8| = 8$	(-8, 8)
0	$f(0) = |0| = 0$	(0, 0)

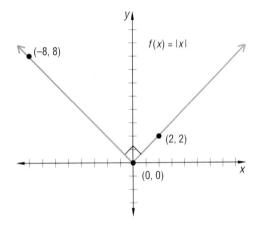

In general, when x is positive, $f(x) = x$, so the graph is part of the line $y = x$. When x is negative, then $f(x) = -x$, and the graph is part of the line $y = -x$.

The result is that the graph of the function is an angle. The angle has vertex at the origin (0, 0) and has measure 90°. It is a right angle.

You may wonder why absolute value functions are needed. (They must be needed, or computer languages would not have a special name for them and we would not have the special symbol | |.) One reason is that they occur in the study of error.

Example 1   A psychologist asked some teenagers to estimate the length of a minute. The psychologist rang a bell on a desk in front of each teenager. The teenager was to wait until he or she thought a minute was up, and then ring the bell again. The estimate x (in seconds) determines the error $|60 - x|$. Graph the function with equation

$$f(x) = |60 - x|.$$

Solution   Draw axes with units of 10, so 60, clearly a key number, can fit. Make a table.

x	error: $f(x) = |60 - x|$
60	0
50	10
40	20
70	10
0	60
100	40

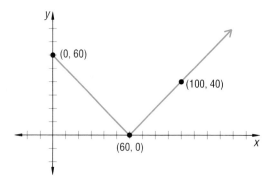

These and other points are plotted above. You can see that the graph again is a right angle, but its vertex is at (60, 0).

Another reason for absolute value functions is that they help to explain some complicated situations.

Example 2   A plane crosses a check point going due east. It then flies at 600 $\frac{km}{hr}$ for 2 hours, then returns flying due west. Is there any formula describing its distance from the check point?

Solution   Let $d(t)$ = the plane's distance from the check point t hours after crossing it. The given information tells you some points on this function.

When $t = 0$, the plane is at the check point.   So $d(0) = 0$.
When $t = 1$, the plane is 600 km east.   So $d(1) = 600$.
When $t = 2$, the plane is 1200 km east.   So $d(2) = 1200$.

All this time the plane has been going at a constant rate. Now it turns back.

When $t = 3$, the plane is again 600 km east. So $d(3) = 600$.
When $t = 4$, the plane crosses the check point again. So $d(4) = 0$.

The graph below results from graphing these and other ordered pairs. Notice how the graph describes the situation.

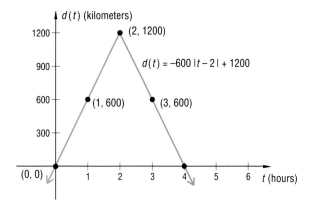

Because the graph is an angle, you should expect a formula for $d(t)$ to involve the absolute value function. And it does.

$$d(t) = -600 \, |t - 2| + 1200.$$

You should check that the formula does work for the values we found and for other values.

Without knowing the graph of the absolute value function $f(x) = |x|$, you probably would never think the situation in Example 2 could involve absolute value. The formula

$$d(t) = -600 \, |t - 2| + 1200$$

is a special case of the general absolute value function

$$f(x) = a \, |x - h| + k,$$

where $a = -600$, $h = 2$, and $k = 1200$. You are asked to explore what the graph of this function looks like for various values of a, h, and k. In your later study of mathematics, you will learn how to determine such formulas.

Earlier in this book, you would have seen the formula $d = -600|t - 2| + 1200$ without function notation. Using the $d(t)$ function notation makes it clear that the value of d depends on t. There may be other quantities that depend on the time. We could write

$a(t)$ = altitude of the plane t hours after crossing
$f(t)$ = amount of fuel used t hours after crossing
and so on.

By using $a(t)$ and $f(t)$, it is clear that the altitude and fuel used by the plane depend on time.

Questions

Covering the Reading

In 1–4, let $f(x) = |x|$. Calculate.
1. $f(-3)$ 2. $f(2)$ 3. $f(-\frac{3}{4})$ 4. $f(0)$

5. The shape of the graph of an absolute value function is __?__.

6. Name two reasons for studying absolute value functions.

In 7–9, let f be the function of Example 1.
7. If $x = 90$, $f(x) =$ __?__.

8. The graph of f has vertex __?__.

9. $f(x)$ stands for the absolute difference between the actual and estimated values of __?__.

In 10–13, let $d(t) = -600|t - 2| + 1200$.
10. Calculate $d(0)$, $d(1)$, $d(2)$, and $d(3)$.

11. Describe a situation that can lead to the function d.

12. The function d contains $(1.5, 900)$. What does this point represent?

13. The function d contains $(5, -600)$. What could this point represent?

Applying the Mathematics

In 14 and 15, graph the function with the given equation.
14. $f(x) = |3x|$ 15. $y = |x - 10| + 7$

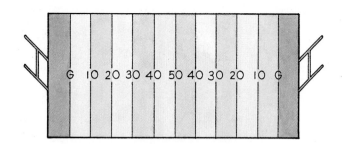

In 16 and 17, suppose you start at the goal line of a football field and walk to the other goal line. After you have walked w yards, you will be on the y yard line.

16. *Multiple choice* Which equation relates w and y?
(a) $y = w$
(b) $y = |w|$
(c) $y = |50 - w| + w$
(d) $y = -|50 - w| + 50$

17. Let f be the function relating w and y. Graph f.

Review

18. *Skill sequence* Simplify. *(Lesson 2-3)*

a. $x + \frac{x}{2}$ **b.** $\frac{x}{3} + \frac{x}{2}$ **c.** $\frac{x}{3} + \frac{y}{2}$

19. Let $A(x) =$ the April mean high temperature for city x, as shown in Lesson 13-1. What is $A(\text{Moscow})$? *(Lesson 13-1)*

20. Let $f(x) = \frac{2}{3}x + 5$.
a. Calculate $f(120)$.
b. Calculate $f(-120)$.
c. Describe the graph of f. *(Lesson 13-1)*

21. If 10 pencils and 7 erasers cost $4.23 and 3 pencils and 1 eraser cost $0.95, what is the cost of two erasers? *(Lesson 11-5)*

Exploration

In 22 and 23, consider absolute value functions of the form $f(x) = a|x - h|$.

22. Fix $a = 1$. Then vary the value of h, choosing any numbers you wish. For instance, if you let $h = 3$, then $f(x) = |x - 3|$.
a. Graph the function f for four different values of h.
b. What effect does h have on the graphs?

23. Fix $h = 2$. Now vary the value of a, choosing any numbers you wish. For instance, if you let $a = 4$, then $f(x) = 4|x - 2|$.
a. Graph the function f for four different values of a.
b. What effect does a have on the graphs?

13-4

Function Language

Every function can be thought of as a set of ordered pairs. The set of first coordinates of these pairs is called the **domain of the function.** Here are two functions with the same formula but the domains are different.

Let $f(x)$ be the price of two basketballs if one costs x dollars. Then $f(x) = 2x$.

Let $g(x)$ be the number of children in x set of twins. Then $g(x) = 2x$.

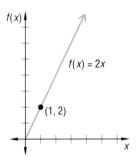

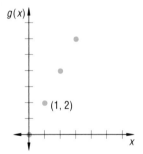

If you were to describe the above graphs by $y = 2x$, you would not be able to distinguish the functions. In this way, function language is sometimes clearer.

The domain of a function is the replacement set for the first variable. When no domain is given, the domain is assumed to be the largest set possible. For instance, if $A(t)$ stands for the area of a triangle t, then the domain is the set of all triangles. If a function is defined by the equation $f(x) = 4x + 3$ and no other information is given, we assume the domain is the set of real numbers. However, the domain of a function sometimes cannot contain particular values.

Example 1 What is the domain of the function f with rule $f(x) = \dfrac{x + 1}{x - 2}$?

Solution The domain is the set of allowable values for x. The numerator can be any number, so it can be ignored. Since the denominator cannot be 0, the domain is the set of all real numbers but 2.

Whereas the *domain* of a function f is the set of possible replacements for x, the **range of a function** is the set of possible values of $f(x)$. For instance, in the absolute value function with equation $f(x) = |x|$, $f(x)$ can be any positive number or zero. Thus the range of this function is the set of nonnegative real numbers.

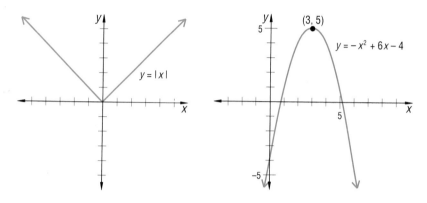

range = set of nonnegative numbers   range = set of numbers less than or equal to 5

You can often determine the range of a function by examining its graph. In the graph of the quadratic function above at right, y could be any number less than or equal to 5. So the range of that function is the set of real numbers less than or equal to 5.

If you do not have a graph of a function, then calculating the range can be more difficult.

Example 2 Determine the range of the function f with equation $f(x) = \sqrt{x} + 4$.

Solution The number $\sqrt{x}$ can be any positive number or zero. Thus $\sqrt{x} + 4$ can be any number greater than or equal to 4. So the range is the set of numbers greater than or equal to 4.

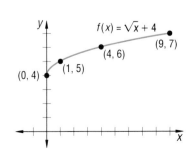

Check A graph of $f(x) = \sqrt{x} + 4$ with an automatic grapher is shown at left. The graph verifies the solution.

If the function is a finite set of ordered pairs, then you can often list the elements of the domain and range.

Example 3 Give the domain and range of the function {(1945, 51), (1965, 117), (1985, 159)} that associates a year with the number of members of the United Nations in that year. (The United Nations was formed in 1945.)

Solution Only three points are given for this function.
The domain is the set of first coordinates {1945, 1965, 1985}.
The range is the set of second coordinates {51, 117, 159}.

Questions

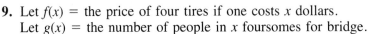

Covering the Reading

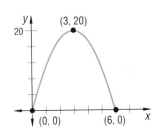

1. Define: domain of a function.

2. Define: range of a function.

In 3–5, give the domain and range of the function.

3. {(1, 2), (3, 4), (5, 7)}

4. The function graphed at left

5. The function with equation $f(x) = \sqrt{x} + 4$

6. When no domain is given for a function, what can you assume about the domain?

7. Give the domain and range of the function with equation $y = |x|$.

8. What number is not in the domain of the function g with rule
$g(x) = \dfrac{x - 2}{x - 3}$?

9. Let $f(x)$ = the price of four tires if one costs x dollars.
Let $g(x)$ = the number of people in x foursomes for bridge.
a. Give a formula for f.
b. Give a formula for g.
c. How do f and g differ?

10. How many more members did the United Nations have in 1985 than when it was founded?

11. a. Graph a function whose domain is {-1, 2, 3} and whose range is {5, 8, 0}.
  b. How many possible functions are there?

In 12–16, give the domain and range of the function.

12. $y = 3x + 1$

13. $y = 2x^2 - 3$

14. The exponential function with equation $y = 2^x$ graphed below at the left

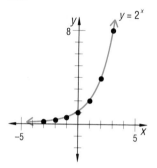
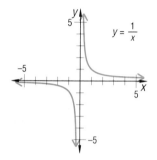

15. The function with equation $y = \dfrac{1}{x}$ graphed above at the right

16. The football field walking function of Questions 16 and 17 of Lesson 13-3

17. The graph of $y = |x| + 3$ is __?__ with vertex __?__ . *(Lesson 13-3)*

18. A poll in an election for senator predicts that candidate A will get 46% of the vote. If instead A gets x% of the vote, by how far off was the poll? *(Lesson 13-3)*

19. If $f(x) = (x - 1)(x + 2)$, find $f(1)$, $f(2)$, and $f(\frac{7}{3})$. *(Lesson 13-2)*

20. In 1986, Cuba had a population of about 10.2 million living on about 44,218 square miles of land. Mexico had a population of 81.7 million living on 761,604 square miles. Which country was more densely populated? *(Lesson 5-2)*

21. Solve for x: $y = 2x + 6$. *(Lesson 6-1)*

22. The price of an item was increased by $\frac{1}{3}$ its former price. The new price is $10.00.
  a. What equation can be solved to find the former price?
  b. What was that price? *(Lesson 5-6)*

23. Consider the absolute value functions of the form $f(x) = |x| + k$.
  a. What is the effect of k on the graph of f? (You may need to graph f for different values of k.)
  b. What is the range of f in terms of k?

Probability Functions

In Monopoly™ and many other board games, two dice are thrown and the sum of the numbers that appear is used to make a move. Since the outcome of the game depends on landing or not landing on particular spaces, it is helpful to know the probability of obtaining each sum. The following picture, which first appeared in Lesson 1-7, is helpful. It shows the 36 possibilities for two fair dice. (A fair die is one on which each side is equally likely to appear up.)

If the dice are fair, then each of the 36 outcomes has a probability of $\frac{1}{36}$. Let $P(n)$ = the probability of getting a sum of n. The domain of P is the set of possible values for n, namely $\{2, 3, 4, 5, 6, 7, 8, 9, 10, 11, 12\}$. By counting, $P(n)$ has the values given on page 657. The range of the function P is thus $\{\frac{1}{36}, \frac{2}{36}, \frac{3}{36}, \frac{4}{36}, \frac{5}{36}, \frac{6}{36}\}$. This *probability function* is graphed on page 657.

n	2	3	4	5	6	7	8	9	10	11	12
$P(n)$	$\frac{1}{36}$	$\frac{2}{36}$	$\frac{3}{36}$	$\frac{4}{36}$	$\frac{5}{36}$	$\frac{6}{36}$	$\frac{5}{36}$	$\frac{4}{36}$	$\frac{3}{36}$	$\frac{2}{36}$	$\frac{1}{36}$

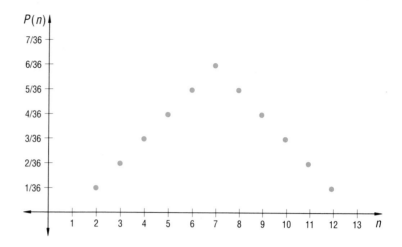

The graph is part of an angle. This suggests that there is a formula for $P(n)$ involving absolute value. In fact, there is.

$$P(n) = \tfrac{-1}{36}|n - 7| + \tfrac{1}{6}$$

For instance, using the formula, the probability of a sum of 3, $P(3) = \tfrac{-1}{36}|3 - 7| + \tfrac{1}{6} = \tfrac{-1}{36}|\text{-}4| + \tfrac{1}{6} = \tfrac{-4}{36} + \tfrac{1}{6} = \tfrac{2}{36}$. Although you could substitute any real number for n, in this situation the formula has meaning only when n is in the domain of the function P.

In general, a **probability function** is a function whose domain is a set of outcomes in a situation and in which each ordered pair contains an outcome and its probability. Many probability functions have simpler formulas than that of the dice example.

Example 1 Consider the spinner at left. Assume all regions have the same probability of the spinner landing in them. Let P(n) = the probability of landing in region n. Graph the function P.

Solution Since there are five regions, each with the same probability, each has probability $\frac{1}{5}$. So $P(1) = \frac{1}{5}$, $P(2) = \frac{1}{5}$, and so on. The graph is shown at right. An equation for the function is $P(n) = \frac{1}{5}$.

Events do not have to be numerical. Below are graphs for two functions related to the birth of boys and girls. There were about 1,928,000 boys and 1,833,000 girls born in the U.S. in 1985. The function at left assumes the two are equally likely. The function at right gives probabilities calculated from the relative frequencies of boy and girl births in 1985. A basic question for statisticians is "Could you expect relative frequencies like those at right if the probabilities at left are true?"

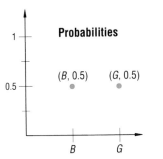

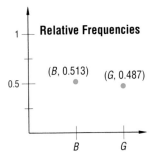

The answer in this case is "No." It is so unlikely to have such relative frequencies (with so many births) that it is possible to conclude that a baby is more likely to be a boy than a girl. In fact, in the U.S. about 105 boys are born for every 100 girls born. But women live longer, so the population contains more women than men.

Questions

Covering the Reading

1. Define: probability function.

In 2–5, let P(n) = the probability of getting a sum of n when two fair dice are thrown.

2. Give an algebraic formula for P(n).

3. Verify that P(12) = $\frac{1}{36}$ using the algebraic formula.

4. The graph of P is part of what geometric figure?

5. If P(n) = $\frac{5}{36}$, then n = __?__ or __?__.

6. Why can't 2 be in the range of a probability function?

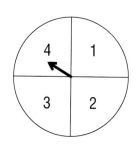

7. Consider the spinner at left. Assume the spinner can land in each region with the same probability. Let P(n) = the probability of landing in region n. Graph the function P.

In 8–10, *true or false*.

8. In 1985, more boys than girls were born in the U.S.

9. In 1985, more men than women were living in the U.S.

10. In a single birth, the probability that a baby is a boy is $\frac{1}{2}$.

Applying the Mathematics

11. At the right is a graph of a probability function for a weighted (unfair) 6-sided die.

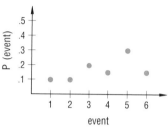

a. $P(3) = $ __?__
b. P(a number > 3 appears) $= $ __?__
c. What would the graph look like if the die were fair?

12. Graph the probability function suggested by this situation. If two brown-eyed people marry, then P(a particular child will be brown-eyed) $= \frac{3}{4}$.

13. A letter is mailed Saturday. P(it arrives Monday) $= \frac{1}{2}$. P(it arrives Tuesday) $= (\frac{1}{2})^2$. P(it arrives Wednesday) $= (\frac{1}{2})^3$. P(it arrives Thursday) $= (\frac{1}{2})^4$. P(it arrives Friday) $= (\frac{1}{2})^5$. Calculate P(it does not arrive by Friday) and graph the appropriate probability function.

Review

14. If the probability that a boy is born is about 0.513, what is the probability that a family with two children has two boys? *(Lesson 4-8)*

15. Give the domain and range of the function of Example 1 in this lesson. *(Lesson 13-4)*

16. a. Graph $y = |x + 2| - 3$.
b. Graph $y = (x + 2)^2 - 3$.
c. Compare and contrast these graphs. *(Lessons 12-2, 13-3)*

17. Find a value of x for which $2x^2 + 3x - 20$ is a prime number. (Hint: Factor the trinomial.) *(Lesson 12-8)*

18. Do this problem in your head. Since one thousand times one thousand equals one million, $1005 \cdot 995 = $ __?__. *(Lesson 10-9)*

19. Simplify $\sqrt{12} + \sqrt{3}$. *(Lesson 7-6)*

Exploration

20. a. Toss two dice a large number of times (at least 60).
b. Do you think your dice are fair?

21. Imagine tossing *three* fair dice.
a. What are the possible outcomes?
b. What is the probability of each possible sum?
c. Is the graph part of an absolute value function?

LESSON

13-6

The Tangent of an Angle

Consider right triangle *ABC* below. Its *legs* (the sides forming the right angle) have lengths 3 and 4. ∠*C* is a right angle and is thus a 90° angle. Since the measures of the three angles of a triangle add up to 180°, m∠*A* + m∠*B* = 90.

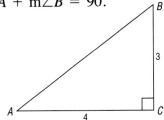

It is possible to determine the measures of the angles of this triangle using a function called the **tangent function**. The **tangent of angle** *A* in a right triangle, abbreviated **tan A,** is defined as a ratio of legs in the triangle.

Definition:

In a right triangle with acute angle *A*,
$$\tan A = \frac{\text{length of the leg opposite angle } A}{\text{length of the leg adjacent to angle } A}.$$

Above, *BC* is the leg opposite ∠*A*, and *AC* is the leg adjacent to ∠*A*. So $\tan A = \frac{BC}{AC}$. In BASIC, TAN(*A*) stands for tan *A*. Most scientific calculators have a ⟨tan⟩ key. In this lesson, the domain of the tangent function is the set of acute angles in right triangles. In later courses you will study this function with a larger domain.

■ ■ ■ ■ ■ ■ ■ ■

Example 1 In △*ABC* above, (a) find tan *A*; (b) find m∠*A*.

Solution
(a) Substitute the lengths for △*ABC* in the definition.

$$\tan A = \frac{\text{length of the leg opposite angle } A}{\text{length of the leg adjacent to angle } A} = \frac{BC}{AC} = \frac{3}{4}$$

(b) Use your calculator. Be sure it is set to degrees. When $\tan A = \frac{3}{4} = .75$, the key sequence

$$0.75 \;\boxed{\text{INV}}\; \boxed{\text{tan}}$$

undoes the tangent and gives the result. m∠*A* ≈ 37°.

One of the wondrous sights in Seattle is the peak of Mt. Rainier.

Example 2

From Seattle, which is at sea level, it is 58 miles to Mount Rainier. Mount Rainier towers 14,410 ft above sea level. When viewed from Seattle, how far is the peak of Mount Rainier above the horizon?

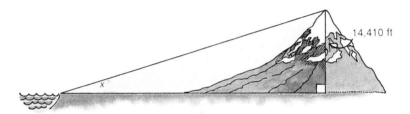

14,410 ft

Solution The picture above, though not drawn to scale, represents the situation. Since the tangent is a ratio, the sides must be measured in the same units. Change miles to feet.

$$58 \text{ miles} \cdot \frac{5280 \text{ ft}}{1 \text{ mile}} = 306,240 \text{ feet}$$

Thus,

$$\tan x = \frac{14,410 \text{ ft}}{306,240 \text{ ft}}$$

$$\approx 0.047$$

The key sequence .047 [INV] [tan] reveals that $x \approx 2.7°$.

The solution in Example 2 ignores the curvature of the earth, which lowers the perceived height by about 2240 feet. Even with this correction, the peak of Mount Rainier is 2.3° above the horizon. The diameter of the moon is about 0.5°. Since $2.3° > 4 \cdot 0.5°$, Mount Rainier is over 4 moon diameters above the horizon.

In Examples 1 and 2, the calculator sequence [INV] [tan] gives you an angle when given the tangent. When you know an angle with measure x, the sequence X [tan] gives you the tangent. This sequence is used in Example 3.

Example 3 Nancy had to look up 50° to see the top of a tree 5 meters away. If her eyes are 1.5 meters above the ground, how tall is the tree?

Solution If h is the height of the tree *above eye level,* then the height of the tree is $h + 1.5$. First use the triangle to find h, then add 1.5. To find h, note that

$$\tan 50° = \frac{h}{5}.$$

To compute tan 50° use the key sequence 50 [tan].

$$\tan 50° \approx 1.19.$$

Substituting,

$$\frac{h}{5} \approx 1.19.$$
$$h \approx 5 \cdot 1.19$$
$$h \approx 6.0.$$

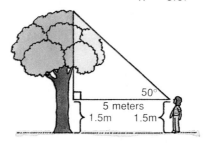

So the full height of the tree is about $h + 1.5 \approx 6.0 + 1.5 = 7.5$ m.

You may not realize it, but you have already calculated tangents on the coordinate plane. Consider the line $y = \frac{2}{3}x - 4$. This is a line with a slope of $\frac{2}{3}$. It crosses the x-axis at $(6, 0)$ and also goes through $(7, \frac{2}{3})$.

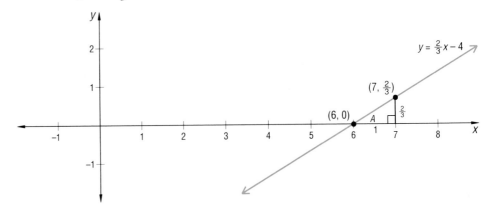

Let A be the angle of the upper half of the line with the positive ray of the x-axis. From the graph on page 662 you can see that

$$\tan A = \frac{\frac{2}{3}}{1} = \frac{2}{3}.$$

This is the slope of the line!

If A is the angle formed by the upper half of the oblique line $y = mx + b$ and the positive ray of the x-axis, then

$$\tan A = m.$$

The tangent function is quite a function; it combines the concepts of slope, graphing, angles, ratios, and triangles.

Questions

Covering the Reading

In 1 and 2, use right triangle *DEF* below.

1. a. $m\angle D + m\angle E = \underline{\ ?\ }$.
   b. The relationship in part a means that $\angle D$ and $\angle E$ are $\underline{\ ?\ }$ angles.

```
D
|\
| \
|  \
|   \
F|___\E
```

2. a. Name the side opposite $\angle E$.
   b. Name the side adjacent to $\angle E$.
   c. What is tan E?

3. What is the domain of the tangent function in this lesson?

In 4 and 5, estimate $m\angle A$ to the nearest degree for the given value of tan A.

4. tan $A = 1.963$ **5.** tan $A = 0.5$

6. Refer to $\triangle GHI$ below.
   a. Find tan I.   **b.** Find $m\angle I$.

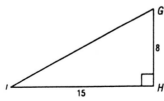

In 7 and 8, use your calculator. Give a three-place decimal approximation.

7. tan 57°;   **8.** tan 3°.

9. Lester had to look up 65° to see the top of a tree 6 meters away. If his eyes are 1.7 meters above the ground, how tall is the tree?

10. From sea level in Tacoma, Washington it is 42 miles to Mount Rainier. Find the measure of the angle needed to see the peak of Mount Rainier from Tacoma.

11. What is the relationship between the tangent function and the slope of a line $y = mx + b$?

12. Refer to $\triangle ABC$ at right.
   a. Find AC.
   b. Find m$\angle A$.
   c. Find m$\angle B$.

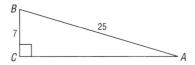

13. To the nearest degree, find the measure of the angle formed by the upper half of the line $y = 4x - 3$ and the positive ray of the x-axis.

14. Line P goes through the origin and the upper half makes an angle of 140° with the positive ray of the x-axis.
   a. Find the slope of line P.
   b. Find an equation for line P.

15. A meter stick casts a shadow 0.6 meters long. Find the measure of the angle needed to see the sun. (This is called the *angle of elevation* of the sun.)

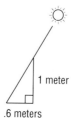

16. Joan had to look up 40° to see the top of a flagpole 20 feet away. The situation is pictured below. If her eyes are 5 feet above the ground, how high is the flagpole?

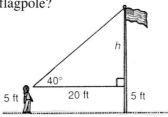

17. The Sears Tower in Chicago is about 443 meters tall. If a person two meters tall is 100 meters from the front door, about what is the measure of the angle needed to see the top?

18. In right triangle *DEF*, ∠*D* is a right angle. If m∠*E* is four times m∠*F*, find m∠*E*. *(Lesson 3-8)*

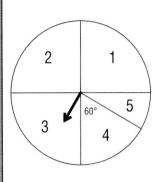

19. In the spinner at left, you receive the number of points indicated when the spinner lands in the region. The two diameters are perpendicular and the angle of region 4 has a measure of 60°.
   a. What is the measure of the angle in region 5?
   b. Give P(landing in region *n*) for *n* = 1, 2, 3, 4, and 5.
   c. Graph the probability function *P*. *(Lesson 13-5)*

20. Graph $y = -|x|$. *(Lesson 13-3)*

21. What number is not in the domain of the function *f* if $f(x) = \dfrac{x - 1}{2x + 4}$? *(Lesson 13-4)*

22. **a.** If $y = 3(x + 1)^2 - 4$, what is the smallest possible value of *y*?
   b. If $y = 3|x + 1| - 4$, what is the smallest possible value of *y*?
   c. If $y = 3\sqrt{x + 1} - 4$, what is the smallest possible value of *y*?
   (Lessons 12-2, 13-2, 13-4)

23. Solve for *x*: $ax^2 + bx + c = 0$. *(Lesson 12-4)*

24. If $\dfrac{(a^{11})^{12} \cdot a^{13}}{a^{14}} = a^t$, what is the value of *t*? *(Lessons 9-5, 9-7)*

25. When $x = \dfrac{\pi A}{180°}$, $\tan A \approx \dfrac{2x^5 + 5x^3 + 15x}{15}$.
   a. Let *A* = 10°. How close is the polynomial approximation to the calculator value of tan *A*?
   b. Repeat part a when *A* = 70°.

26. What became of the man who sat on a beach along the Gulf of Mexico?

13-7

Functions on Calculators

Calculators use the idea of functions. Rules programmed into calculators are designed to give you single answers when you enter a value of the domain. A calculator will indicate an error message when you enter a value not in the domain. Here are some familiar keys and their domains.

Key	Function	Domain	Errors
the square root key $\boxed{\sqrt{}}$	$SQR(x) = \sqrt{x}$	set of nonnegative reals	$x < 0$
the factorial function key $\boxed{!}$	$FACT(n) = n!$	set of nonnegative integers	nonintegers
reciprocal function key $\boxed{\frac{1}{x}}$	$f(x) = \dfrac{1}{x}$	set of nonzero reals	$x = 0$
the squaring key $\boxed{x^2}$	$s(x) = x^2$	set of all reals	none

In Lesson 13-6, you used the $\boxed{\text{tan}}$ key. This key defines the tangent function. Its equation is $y = \tan x$. Below is a graph of this function for values of x between $0°$ and $90°$.

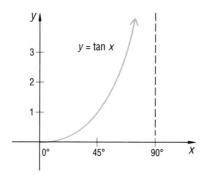

You have learned some applications for the tangent function. In this lesson, we introduce you to some of the other keys on your scientific calculator.

Two keys related to the tangent key are the sine key $\boxed{\text{sin}}$ and the cosine key $\boxed{\text{cos}}$. These keys also give values of ratios of sides in right triangles. Again consider a right triangle ABC.

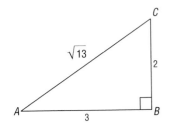

$$\mathbf{sin}\ A = \frac{\text{length of the leg opposite angle } A}{\text{hypotenuse}}$$

$$\mathbf{cos}\ A = \frac{\text{length of leg adjacent to angle } A}{\text{hypotenuse}}$$

So in this case, $\sin A = \dfrac{2}{\sqrt{13}}$ and $\cos A = \dfrac{3}{\sqrt{13}}$.

It is possible to define these functions and the tangent function for any degrees. A surprise comes when the sine or cosine function is graphed. Below is the graph of $y = \sin x$.

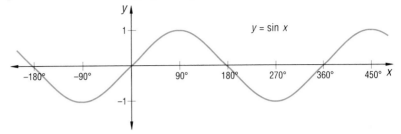

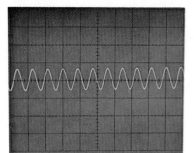

A sine function shown on an oscilloscope

The curve is called a *sinusoidal curve*, and it has the same shape as sound waves and radio waves. The sine, cosine, and tangent functions are part of the area of mathematics called trigonometry. These functions are so important that most high schools offer a full course in trigonometry devoted to studying them and their applications.

Another function key on almost all scientific calculators is the **common logarithm key** [log]. This key defines a function $y = \log x$, read "y equals the common logarithm of x." The common logarithm of a number is the power to which 10 must be raised to equal that number. So, since 1 million $= 10^6$, log (1 million) $= 6$. Logarithms provide a way to deal easily with very large or very small numbers. A graph of the common logarithm function is given below.

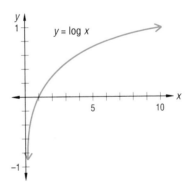

This graph pictures the kind of growth often found in learning. At first, one learns an idea quickly, so the curve increases quickly. But after a while it is more difficult to improve, so the curve increases more slowly.

Most scientific calculators have other function keys. These keys would not be there unless many people needed to get values of that function. Calculators have made it possible for people to obtain values of these functions more easily than most people ever imagined. The algebra that you have studied this year gives you the background to understand these functions and to deal with them.

Questions

Covering the Reading

In 1–6, approximate each value to the nearest thousandth using a calculator.

1. tan 11° **2.** sin 45° **3.** cos 47°

4. log (10^6) **5.** $(-3.489)^2$ **6.** $\sqrt{0.5}$

In 7–9, consider the $\boxed{\sqrt{}}$, $\boxed{!}$, $\boxed{\frac{1}{x}}$, and $\boxed{x^2}$ function keys on your calculator. Which produce error messages when the given is entered?

7. 3.5 **8.** -4 **9.** 0

10. Which function has a graph that is the shape of a sound wave?

11. Which function has a graph that is sometimes used to model learning?

12. Use △*ABC* at right. Give the value of:
   a. *AC*
   b. sin *A*
   c. cos *A*

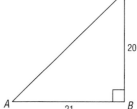

In 13–15, refer to the graph of *y* = sin *x* in this lesson.
a. Estimate the value from the graph.
b. Use a calculator to check your estimate.

13. sin 90° **14.** sin 360° **15.** sin(-70°)

Applying the Mathematics

16. a. Make a table of values for *y* = cos *x* for values of *x* from -90° to 360° in increments of 10°.
   b. Carefully graph this function.
   c. What graph in this lesson does the graph of *y* = cos *x* most resemble?

17. Many computers use the name LOG to refer to a different logarithm function than the one in the lesson.
  a. Run this program or use your calculator to determine what is printed.

```
10 PRINT "X", "LOG X"
20 FOR X = 1 TO 10
30    Y = LOG(X)
40    PRINT X, Y
50 NEXT X
60 END
```

  b. Graph the ordered pairs that are printed.
  c. Is your graph like that in the lesson, or is it different? If it is different, how does it differ?

18. a. Graph $y = \tan x$ on an automatic plotter with x set for degrees. Use the graph to estimate $\tan (-10°)$.
  b. Verify the value of $\tan (-10°)$ on your calculator.

19. Graph the reciprocal function $f(x) = \dfrac{1}{x}$.

20. The graph of the squaring function $s(x) = x^2$ is what curve?

Review

21. a. In tossing two fair dice, what number is the most likely sum to appear?
  b. What is its probability of occurring? *(Lesson 13-5)*

22. In $\triangle ABC$ at the left, find the measure of $\angle A$ to the nearest degree. *(Lesson 13-6)*

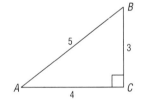

23. Graph $f(x) = 2|x - 3| + 4$. *(Lesson 13-3)*

24. The explorers were 13 km from home base at 2 P.M. and 10 km from home base at 3:30 P.M. At this rate, when will they reach home? *(Lesson 8-6)*

In 25 and 26, $S(x) = x + \frac{x^2}{20}$ gives the number of feet a car traveling at x miles per hour will take to stop. $B(x) = \frac{x^2}{20}$ gives the number of feet after brakes are applied.

25. About how many feet does it take to stop at 40 mph? *(Lesson 13-2)*

26. If skid marks in an accident are 100 feet long, at least how fast was the car going? *(Lesson 13-2)*

27. If you read 17 pages of a 300-page novel in 45 minutes, about how long will it take you to read the entire novel? *(Lesson 5-7)*

In 28–30, solve.

28. $\sqrt{v - 6} = 4$ *(Lesson 7-4)*

29. $\frac{m}{2} = \frac{m + 36}{11}$ *(Lesson 5-7)*

30. $3x + 9 > x$ *(Lesson 6-7)*

31. Here are the total numbers of votes (to the nearest million) cast for major candidates in the presidential elections since 1940.

Year	Number of Votes (millions)	Winner
1940	50	Franklin D. Roosevelt
1944	48	Franklin D. Roosevelt
1948	48	Harry S Truman
1952	61	Dwight D. Eisenhower
1956	62	Dwight D. Eisenhower
1960	68	John F. Kennedy
1964	70	Lyndon B. Johnson
1968	73	Richard M. Nixon
1972	76	Richard M. Nixon
1976	83	Jimmy Carter
1980	85	Ronald Reagan
1984	92	Ronald Reagan
1988	88	George Bush

a. Graph the ordered pairs.
b. Use the graph to predict how many votes will be cast for major candidates in the presidential election of 2000. *(Lesson 8-7)*

Exploration

32. List all the function keys of a scientific calculator to which you have access. Separate those you have studied from those you have not. Identify at least one situation in which each function you have studied might be used.

Summary

A function is a set of ordered pairs in which each first coordinate appears with exactly one second coordinate. A function may be described by a graph, by a written rule, by listing the pairs, or by an equation. The key idea in functions is that knowing the first coordinate of a pair is enough to determine the second coordinate of the pair. For this reason, functions exist whenever one variable determines another.

If a function f contains the ordered pair (a, b), then we write $f(a) = b$. We say that b is the value of the function at a. The set of possible values of a is the domain of the function. The set of possible values of b is the range of the function. If a and b are numbers, then the function can be graphed on the coordinate plane and values of the function can be found or approximated by reading the graph. The purpose of automatic graphers is to graph functions. Though convenient, it is not necessary to use $f(x)$ notation for functions; y is often used to stand for the second coordinate. Many of the graphs you studied in earlier chapters describe functions.

Equation	Graph	Type of function
$f(x) = mx + b$	line	linear
$f(x) = ax^2 + bx + c$	parabola	quadratic
$f(x) = b^x$	exponential curve	exponential
$f(x) = a\lvert x - h \rvert + k$	angle	absolute value

When the domain of a function is a set of outcomes in a situation and the range of the function the probabilities of these outcomes, then the function is a probability function. Graphs of probability functions take on various shapes.

Calculator keys determine values of functions. The ⟨tan⟩ key evaluates the tangent function, which can be used to determine lengths of sides and measures of angles in right triangles. You will encounter this function and many others in your future work in mathematics.

Vocabulary

Below are the most important terms and phrases for this chapter. You should be able to give a general description and a specific example of each.

Lesson 13-1
function
relation
value of a function
linear function
quadratic function

Lesson 13-2
$f(x)$ notation

Lesson 13-3
absolute value function

Lesson 13-4
domain of a function
range of a function

Lesson 13-5
probability function

Lesson 13-6
tangent function
tangent of an angle in a right triangle, tan A
⟨tan⟩, ⟨INV⟩ ⟨tan⟩

Lesson 13-7
function key on a calculator
sine of an angle in a right triangle, sin A
cosine of an angle in a right triangle, cos A
⟨sin⟩, ⟨cos⟩
logarithm key ⟨log⟩

Progress Self-Test

Take this test as you would take a test in class. You will need graph paper and a calculator. Then check your work with the solutions in the Selected Answers section in the back of the book.

1. If $f(x) = 3x + 5$, then $f(2) = $? .

2. If $g(t) = t^2 + 4t$, solve for t: $g(t) = 5$.

3. Estimate tan 82° to the nearest thousandth.

4. Give the value of sin 30°.

5. Define: function.

6. a. Give an example of a quadratic function.
   b. The graph of the quadratic function in part (a) is a ? .

7. Explain why the equation $x = |y|$ does not describe a function.

8. If the set $\{(10, 4), (x, 5), (30, 6)\}$ is a function, what values can x not have?

In 9 and 10, tell whether the graph represents a function. If so, give its domain and range. If not, tell why not.

9.

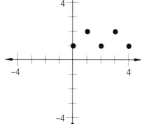

10.

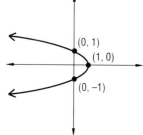

11. In the tossing of two fair dice, what is the probability of obtaining a sum of 10?

12. In the plane crossing a check point example of Lesson 13-3, $d(t) = -600|t - 2| + 1200$. Calculate $d(2.5)$ and tell what that means.

In 13 and 14, assume each of ten regions in the circle below has the same probability that the spinner will land on it. Let $P(n) = $ the probability of landing in region n.

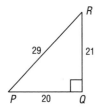

13. Calculate $P(2)$.

14. Graph the function P.

15. In $\triangle PQR$ below, determine the measure of $\angle P$ to the nearest degree.

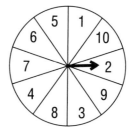

In 16 and 17, graph the function for values of x between -4 and 6.

16. $f(x) = 2|x - 3|$ **17.** $g(x) = 10 - x^2$

Chapter Review

Questions on **SPUR** Objectives

SPUR stands for **S**kills, **P**roperties, **U**ses, and **R**epresentations.
The Chapter Review questions are grouped according to the
SPUR Objectives for this chapter.

SKILLS deal with the procedures used to get answers.

▧ **Objective A.** *Find values of functions from their formulas. (Lessons 13-2, 13-3)*

In 1–4, $f(x) = x^2 - 3x + 8$. Calculate:

1. $f(2)$ **2.** $f(3)$

3. $f(-7)$ **4.** $f(0)$

5. If $A(t) = 2|t - 5|$, calculate $A(1)$.

6. If $g(n) = 2^n$, calculate $g(3) + g(4)$.

7. If $f(x) = -x$, what is $f(-1.5)$?

8. If $h(t) = 64t - 16t^2$, find $h(4)$.

▧ **Objective B.** *Find and analyze values of functions. (Lessons 13-6, 13-7)*

9. If $f(x) = |x + 3|$, solve $f(x) = 5$.

10. Let $N(t)$ = the number of chirps of a cricket in a minute at a temperature $t°$ Fahrenheit. If $N(t) = \frac{1}{4}t + 37$, for what value of t is $N(t) = 60$?

11. What is the largest possible value of the function f, where $f(x) = -x^2 + 10$?

12. What is the smallest possible value of the function A with equation $A(n) = 5|n - 3| - 9$?

In 13–20, approximate to the nearest thousandth.

13. $\frac{1}{17}$ **14.** $10!$

15. $\sqrt{11469}$ **16.** 0.8^{-3}

17. $\tan 30°$ **18.** $SQR(6.5)$

19. $\sin 82.4°$ **20.** $\log 5$

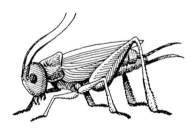

PROPERTIES deal with the principles behind the mathematics.

▧ **Objective C.** *Determine whether a set of ordered pairs is a function. (Lesson 13-1)*

In 21–24, tell whether or not the equation determines a function.

21. $x = |y + 1|$

22. $x^2 = \sqrt{y}$

23. $3x - 5y = 7$

24. $y = \tan x$

In 25–27, tell whether or not the set of ordered pairs is a function.

25. $\{(0, 1), (1, 2), (2, 3), (3, 4)\}$

26. $\{(1, 8), (1, 9), (1, 10), (1, 11)\}$

27. the set of pairs (students, age) for students in your class.

Objective D. *Classify functions.* *(Lessons 13-1, 13-3)*

In 28–33, classify the function as linear, quadratic, exponential, absolute value, or other.

28. $f(x) = 4^x$

29. $g(x) = \dfrac{x^2}{5} + \dfrac{x}{3}$

30. $x + y = 1$

31. $y = |3x + 4| - 2$

32. $y = mx + b$

33. $f(t) = t^{10} - t^9 + t^8 - t^7$

Objective E. *Find the domain and range of a function from its formula, graph, or rule.* *(Lesson 13-4)*

34. If the domain of a function is not given, what should you assume?

35. Give the domain and range of this population function for Los Angeles, California. $\{(1850, 1610), (1900, 102479), (1950, 1970358), (1960, 2479015), (1970, 2811801), (1980, 2966850)\}$.

36. *Multiple choice* The domain of a function is $\{1, 2, 3\}$. The range is $\{4, 5, 6\}$. Which of these could not be a rule for the function?
(a) $y = x + 3$ (b) $y = 7 - x$
(c) $y = x - 3$ (d) $y = |x| + 3$

37. What is the range of the function $A(x) = |x - 2|$?

In 38 and 39, determine the domain and range of the function from its graph.

38.

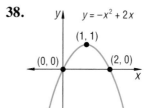

39.

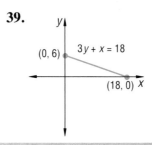

USES deal with applications of mathematics in real situations.

Objective F. *Determine values of probability functions.* *(Lesson 13-5)*

40. If the spinner below has the same probability of landing in any direction, find each of the following:
a. $P(1)$
b. $P(2)$
c. $P(3)$

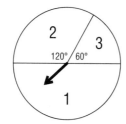

41. What is the probability of tossing a sum of 12 with two fair dice?

42. If you guess on three multiple-choice questions with four choices each, the probability you will get exactly n correct is $\dfrac{3!}{n!(3 - n)!} \cdot \left(\frac{1}{4}\right)^n \cdot \left(\frac{3}{4}\right)^{3-n}$. Calculate the probability you will get exactly 2 correct.

43. A letter is mailed. Suppose P (the letter arrives the next day) $= 0.75$. What is P(the letter does not arrive the next day)?

Objective G. *Find lengths and angle measures in triangles using the tangent function.* *(Lesson 13-6)*

44. Find m∠A to the nearest degree in the triangle below.

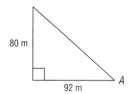

80 m
92 m
A

45. Two friends are 200 feet apart. At the same time one of the friends looks straight up at a kite, the other has to look up 20°. How high is the kite?

46. On a field trip, a girl whose eyes are 150 cm above the ground sights a nest high on a pole 25 meters away. If she has to raise her eyes 50° to see the nest, how high is the nest?

47. What is the slope of line ℓ graphed below?

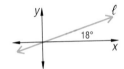

y
ℓ
18°
x

REPRESENTATIONS deal with pictures, graphs, or objects that illustrate concepts.

Objective H. *Determine whether or not a graph represents a function.* *(Lesson 13-1)*

In 48–51, tell whether or not the set of ordered pairs graphed is a function.

48.

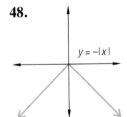

$y = -|x|$

49.

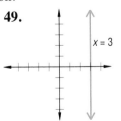

$x = 3$

50.

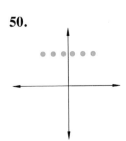

51.

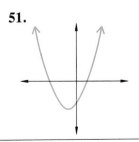

Objective I. *Graph functions.* *(Lessons 13-2, 13-3, 13-5)*

In 52–55, graph each function over the domain $\{x: -5 \leq x \leq 5\}$.

52. $f(x) = 3|2x + 1|$

53. $g(t) = t^2 - 10$

54. $y = \frac{1}{5}x$

55. $f(n) = 2^n$, n an integer

56. A weighted die has the following possibilities of landing on its sides. P(1) = 0.12; P(2) = 0.19; P(3) = 0.09; P(4) = 0.21; P(5) = 0.15. Find P(6) and graph the probability function.

57. Graph the probability function for a fair die.

Scientific Calculators

You will be using a scientific calculator for many lessons in this book so it is important for you to know how your calculator works. Use your calculator to do all the calculations described in this appendix. Some of the problems are very easy. They were selected so that you can check whether your calculator does the computations in the proper order.

Suppose you want to use your calculator to find 3 + 4. Here is how to do it:

	Display shows
Press 3	3
Now press $+$	3
Now press 4	4
Now press $=$	7

Pressing calculator keys is called **entering** or **keying in.** The set of instructions in the left column is called the **key sequence** for this problem. We write the key sequence for this problem using boxes for everything pressed except the numbers.

$$3 \quad \boxed{+} \quad 4 \quad \boxed{=}$$

Sometimes we put what you would see in the calculator display underneath the key presses.

Key sequence	3	$\boxed{+}$	4	$\boxed{=}$
Display	3	3	4	7

Next consider 12 + 3 · 5. In Lesson 1-4, you learned that multiplication should be performed before addition. Perform the key sequence below on your calculator. See what your calculator does.

Key sequence	12	$\boxed{+}$	3	$\boxed{\times}$	5	$\boxed{=}$
Display	12	12	3	3	5	27

Different calculators may give different answers even when the same buttons are pushed. If you have a calculator appropriate for algebra, the calculator displays 27. **Scientific calculators** which follow the order of operations used in algebra should be used with this book. If your calculator gave you the answer 75, then it has done the addition first and does not follow the algebraic order of operations. Using such a calculator with this book may be confusing.

Example 1 Evaluate $ay + bz$ when $a = 0.05$, $y = 2000$, $b = 0.06$ and $z = 9000$. (This is the total interest in a year if $2000 is earning 5% and $9000 is earning 6%.)

Solution
Key sequence: $a \boxed{\times} y \boxed{+} b \boxed{\times} z \boxed{=}$

Substitute in the key sequence:

0.05 $\boxed{\times}$ 2000 $\boxed{+}$ 0.06 $\boxed{\times}$ 9000 $\boxed{=}$
Display:

$\boxed{0.05}$ $\boxed{0.05}$ $\boxed{2000}$ $\boxed{100}$ $\boxed{0.06}$ $\boxed{0.06}$ $\boxed{9000}$ $\boxed{640}$

(The total interest is $640.)

Most scientific calculators have parentheses keys, $\boxed{(}$ and $\boxed{)}$. To use them just enter the parentheses when they appear in the problem. Remember to use the $\boxed{\times}$ key every time you do a multiplication, even if $\times$ is not in the expression.

Example 2 Use the formula $A = 0.5h(b_1 + b_2)$ to calculate the area of the trapezoid at the right.

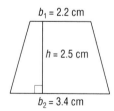

$b_1 = 2.2$ cm

$h = 2.5$ cm

$b_2 = 3.4$ cm

Solution
Remember that $0.5h(b_1 + b_2)$ means $0.5 \cdot h \cdot (b_1 + b_2)$.

Key sequence: $0.5 \boxed{\times} h \boxed{\times} \boxed{(} b_1 \boxed{+} b_2 \boxed{)} \boxed{=}$

Substitute: $0.5 \boxed{\times} 2.5 \boxed{\times} \boxed{(} 2.2 \boxed{+} 3.4 \boxed{)} \boxed{=}$
Display:

$\boxed{0.5}$ $\boxed{0.5}$ $\boxed{2.5}$ $\boxed{1.25}$ $\boxed{1.25}$ $\boxed{2.2}$ $\boxed{2.2}$ $\boxed{3.4}$ $\boxed{5.6}$ $\boxed{7.}$

The area of the trapezoid is 7 square centimeters.

Some numbers are used so frequently that they have special keys on the calculator.

Example 3   Find the circumference of the circle at the right.

Solution
The circumference is the distance around the circle, and is calculated using the formula $C = 2\pi r$, where $C =$ circumference, and $r =$ radius. Use the π key.

4.6 miles

Key sequence: $2 \boxed{\times} \boxed{\pi} \boxed{\times}$ r $\boxed{=}$

Substitute: $2 \boxed{\times} \boxed{\pi} \boxed{\times} 4.6 \boxed{=}$

$\boxed{2} \quad \boxed{2} \quad \boxed{3.1415927} \quad \boxed{6.2831853} \quad \boxed{4.6} \quad \boxed{28.902652}$

Rounding to the nearest tenth, the circumference is 28.9 miles.

As a decimal, $\pi = 3.141592653\ldots$ and the decimal is unending. Since it is impossible to list all the digits, the calculator rounds the decimal. Some calculators, like the one in Example 3, round to the nearest value that can be displayed. Some calculators truncate or round down. If the calculator in the example had truncated, it would have displayed 3.1415926 instead of 3.1415927 for π.

On some calculators you must press two keys to display π. If there is a small π written next to a key, two keys are probably needed. Then you should press $\boxed{\text{INV}}$, $\boxed{\text{2nd}}$, or $\boxed{F}$ before pressing the key next to π.

Negative numbers can be entered in your calculator using the plus-minus key $\boxed{\pm}$ or $\boxed{+/-}$. For example, to enter -19, use the following key sequence:

Key sequence: 19 $\boxed{\pm}$

Display: $\boxed{19} \quad \boxed{-19}$

You will use powers of numbers throughout this book. The scientific calculator has a key $\boxed{y^x}$ (or $\boxed{x^y}$) which can be used to raise numbers to powers. The key sequence for

3^4 is   3 $\boxed{y^x}$ 4 $\boxed{=}$

You should see displayed $\boxed{3}$ $\boxed{3}$ $\boxed{4}$ $\boxed{81}$.

$3^4 = 81$.

■ ■ ■ ■ ■ ■
Example 4   A formula for the volume of a sphere is $V = \dfrac{4\pi r^3}{3}$, where r is the radius. The radius of the moon is about 1080 miles. Find its volume.

Solution
Key sequence: $4 \boxed{\times} \boxed{\pi} \boxed{\times}$ r $\boxed{y^x} 3 \boxed{\div} 3 \boxed{=}$

Substitute: $4 \boxed{\times} \boxed{\pi} \boxed{\times} 1080 \boxed{y^x} 3 \boxed{\div} 3 \boxed{=}$

You see
$\boxed{4}$ $\boxed{4}$ $\boxed{3.1415927}$ $\boxed{12.566371}$ $\boxed{1080}$ $\boxed{1080}$ ...

... $\boxed{3}$ $\boxed{1.583 \quad 10}$ $\boxed{3}$ $\boxed{5.2767 \quad 09}$

The display shows the answer in scientific notation. If you do not understand scientific notation, read Appendix B.

The volume of the moon is about $5.28 \cdot 10^9$ cubic miles.

Note: You may be unable to use a negative number as a base on your calculator. Try the key sequence $2 \boxed{\pm} \boxed{y^x} 5$ to evaluate $(-2)^5$. The answer should be -32. However, some calculators will give you an error message. You can, however, use negative *exponents* on scientific calculators.

Questions

Covering the Reading

Covering the Reading

1. What is meant by the phrase "keying in"?

2. To calculate $28.5 \cdot 32.7 + 14.8$, what key sequence should you use?

3. Consider the key sequence $13.4 \boxed{-} 15 \boxed{\div} 3 \boxed{=}$. What arithmetic problem does this represent?

4. a. To evaluate $ab - c$ on a calculator, what key sequence should you use?
   b. Evaluate $297 \cdot 493 - 74{,}212$.

5. Estimate 26π to the nearest thousandth.

6. What number does the key sequence $104 \boxed{+/-}$ yield?

7. a. Write a key sequence for -104 divided by 8.
   b. Calculate -104 divided by 8 on your calculator.

8. Calculate the area of the trapezoid below.

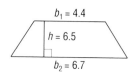

9. Find the circumference of a circle with radius 6.7 inches.

10. Which is greater, $\pi \cdot \pi$ or 10?

11. What expression is evaluated by $5 \boxed{y^x} 2 \boxed{=}$?

12. A softball has a radius of about 1.92 in. What is its volume?

13. What kinds of numbers may not be allowed as bases when you use the $\boxed{y^x}$ key on some calculators?

Applying the Mathematics

14. Use your calculator to help find the surface area $2LH + 2HW + 2LW$ of the box at right.

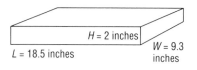

15. Remember that $\frac{2}{3} = 2 \div 3$.
   a. What decimal for $\frac{2}{3}$ is given by your calculator?
   b. Does your calculator *truncate* or *round to the nearest*?

16. Order $\frac{3}{5}$, $\frac{4}{7}$, and $\frac{5}{9}$ from smallest to largest.

17. Use the clues to find the mystery number y.
   Clue 1: y will be on the display if you alternately press 2 and $\boxed{\times}$ again and again …
   Clue 2: $y > 20$.
   Clue 3: $y < 40$.

18. $A = \pi r^2$ is a formula for the area A of a circle with radius r. Find the area of the circle in Example 3.

19. What is the total interest in a year if $350 is earning 5% and $2000 is earning 8%? (Hint: Use Example 1.)

20. To multiply the sum of 2.08 and 5.76 by 2.24, what key sequence can you use?

Scientific Notation

The first three columns in the chart on page 683 show three ways to represent integer powers of ten: in exponential notation, with word names, and as decimals. The fourth column describes a distance or length in meters. For example, the top row tells that Mercury is about ten billion meters from the sun.

You probably know the quick way to multiply by 10, 100, 1000, and so on. Just move the decimal point as many places to the right as there are zeros.

$$84.3 \cdot 100 = 8430 \qquad 84.3 \cdot 10,000 = 843,000$$

It is just as quick to multiply by these numbers when they are written as powers.

$$489.76 \cdot 10^2 = 48,976 \qquad 489.76 \cdot 10^4 = 4,897,600$$

The general pattern is as follows.

To multiply by a positive power of 10, move the decimal point to the *right* as many places as indicated by the exponent.

The patterns in the chart on the following page help to explain powers of 10 where the exponent is negative. Each row describes a number that is $\frac{1}{10}$ of the number in the row above it. So 10^0 is $\frac{1}{10}$ of 10^1.

$$10^0 = \frac{1}{10} \cdot 10 = 1$$

To see the meaning of 10^{-1}, think: 10^{-1} is $\frac{1}{10}$ of 10^0 (which equals 1).

$$10^{-1} = \frac{1}{10} \cdot 1 = \frac{1}{10} = 0.1$$

Remember that to multiply a decimal by 0.1, just move the decimal point one unit to the left. Since $10^{-1} = 0.1$, to multiply by 10^{-1}, just move the decimal point one unit to the left.

$$435.86 \cdot 10^{-1} = 43.586$$

To multiply a decimal by 0.01, or $\frac{1}{100}$, you move the decimal point two units to the left. since $10^{-2} = 0.01$, the same is true for multiplying by 10^{-2}.

$$435.86 \cdot 10^{-2} = 4.3586$$

The following pattern emerges.

> To multiply by a negative power of 10, move the decimal point to the *left* as many places as indicated by the exponent.

Integer Powers of Ten

Exponential Notation	Word Name	Decimal	Something about this length in meters
10^{10}	ten billion	10,000,000,000	distance of Mercury from Sun
10^{9}	billion	1,000,000,000	radius of Sun
10^{8}	hundred million	100,000,000	diameter of Jupiter
10^{7}	ten million	10,000,000	radius of Earth
10^{6}	million	1,000,000	radius of Moon
10^{5}	hundred thousand	100,000	length of Lake Erie
10^{4}	ten thousand	10,000	average width of Grand Canyon
10^{3}	thousand	1,000	5 long city blocks
10^{2}	hundred	100	length of football field
10^{1}	ten	10	height of shade tree
10^{0}	one	1	height of waist
10^{-1}	tenth	0.1	width of hand
10^{-2}	hundredth	0.01	diameter of pencil
10^{-3}	thousandth	0.001	thickness of window pane
10^{-4}	ten thousandth	0.000 1	thickness of paper
10^{-5}	hundred thousandth	0.000 01	diameter of red blood corpuscle
10^{-6}	millionth	0.000 001	mean distance between successive collisions of molecules in air
10^{-7}	ten millionth	0.000 000 1	thickness of thinnest soap bubble with colors
10^{-8}	hundred millionth	0.000 000 01	mean distance between molecules
10^{-9}	billionth	0.000 000 001	size of air molecule
10^{-10}	ten billionth	0.000 000 000 1	mean distance between molecules in a crystal

Example 1   Write $68.5 \cdot 10^{-6}$ as a decimal.

Solution
To multiply by 10^{-6}, move the decimal point six places to the left. So $68.5 \cdot 10^{-6} = .0000685$.

The names of the negative powers are very similar to those for the positive powers. For instance, 1 billion $= 10^9$ and 1 billionth $= 10^{-9}$.

Example 2   Write 8 billionths as a decimal.

Solution
8 billionths $= 8 \cdot 10^{-9} = .000000008$.

Most calculators can display only the first 8, 9, or 10 digits of a number. This is a problem if you need to key in a large number like 455,000,000,000 or a small number like .00000000271. However, powers of 10 can be used to rewrite these number in **scientific notation.**

$$455,000,000,000 = 4.55 \cdot 10^{11}$$
$$.00000000271 = 2.71 \cdot 10^{-9}$$

Definition:

In scientific notation, a number is represented as $x \cdot 10^n$, where $1 \leq x \leq 10$ and n is an integer.

Scientific calculators can display numbers in scientific notation. The display for $4.55 \cdot 10^{11}$ will usually look like one of these shown here.

$$\boxed{4.55 \quad E \quad 11} \qquad \boxed{4.55 \qquad 11} \qquad \boxed{4.55 \times 10 \quad 11}$$

The display for $2.71 \cdot 10^{-9}$ is usually one of these

$$\boxed{2.71 \quad E \ -0.9} \qquad \boxed{2.71 \qquad -09} \qquad \boxed{2.71 \times 10 \ -09}$$

Numbers written in scientific notation are entered into a calculator using the (EXP) or (EE) key. For instance, to enter $6.0247 \cdot 10^{23}$ (known as Avogadro's number), key in

$$6.0247 \ \boxed{EE} \ 23.$$

You should see this display.

$$\boxed{6.0247 \qquad 23}$$

In general, to enter $x \cdot 10^n$, key in x $\boxed{\text{EE}}$ n.

■ ■ ■ ■ ■ ■ ■
Example 3   The total number of hands possible in the card game bridge is about 635,000,000,000. Write this number in scientific notation.

Solution
First, move the decimal point to get a number between 1 and 10. In this case the number is 6.35. This tells you the answer will be:

$$6.35 \cdot 10^{\text{exponent}}$$

The exponent of 10 is the number of places you must move the decimal point in 6.35 to get 635,000,000,000. You must move it 11 places to the right, so the answer is $6.35 \cdot 10^{11}$

■ ■ ■ ■ ■ ■ ■
Example 4   The charge of an electron is .00000000048 electrostatic units. Put this number in scientific notation.

Solution
First move the decimal point to get a number between 1 and 10. The result is 4.8. To find the power of 10, count the number of places you must move the decimal to change 4.8 to .00000000048. The move is 10 places to the left, so the charge of the electron is $4.8 \cdot 10^{-10}$ electrostatic units.

■ ■ ■ ■ ■ ■ ■
Example 5   Enter 0.00000000123 into a calculator.

Solution
Rewrite the number in scientific notation.
$0.00000000123 = 1.23 \cdot 10^{-9}$.

Key in 1.23 $\boxed{\text{EE}}$ 9 $\boxed{+/-}$.

Questions

Covering the Reading

1. Write one million as a power of ten.

2. Write 1 billionth as a power of 10.

In 3–5, write as a decimal.

3. 10^{-4} **4.** $28.5 \cdot 10^7$ **5.** 10^0

6. To multiply by a negative power of 10, move the decimal point to the __?__ as many places as indicated by the __?__.

7. Write $2.46 \cdot 10^{-8}$ as a decimal.

8. Why is $38.25 \cdot 10^{-2}$ not in scientific notation?

9. Suppose $x \cdot 10^y$ is in scientific notation.
   a. What is the domain of x?
   b. What is the domain of y?

In 10–14, rewrite the number in scientific notation.

10. 5,020,000,000,000,000,000,000,000,000 tons, the mass of Sirius, the brightest star

11. 0.0009 meters, the approximate width of a human hair

12. 763,000   **13.** 0.00000328   **14.** 754.9876

15. One computer can do an arithmetic problem in 2.4×10^{-9} seconds. What key sequence can you use to display this number on your calculator.

Applying the Mathematics

In 16 and 17, write in scientific notation.

16. 645 billion   **17.** 27.2 million

In 18–22, use the graph below.

Write the world population in the given year: **a.** as a decimal; **b.** in scientific notation.

18. 10,000 B.C.   **19.** A.D. 1

20. 1700   **21.** 1970

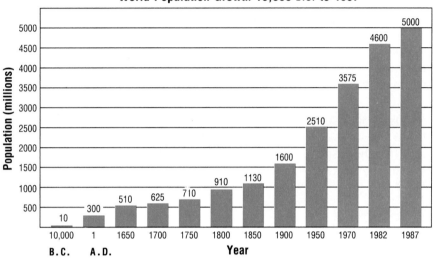

World Population Growth 10,000 B.C. to 1987

22. How can you enter the world population in 1987 into your calculator?

23. How many digits are in 1.7×10^{100}?

In 24–26, write the number in scientific notation.

24. 0.00002   **25.** 0.0000000569

26. 400.007

In 28–30, write as a decimal.

28. $3.921 \cdot 10^5$ **29.** $3.921 \cdot 10^{-5}$

30. $8.6 \cdot 10^{-2}$

Exploration

31. a. What is the largest number you can display on your calculator?
   b. What is the smallest number you can display? (Use scientific notation and consider negative numbers.)
   c. Find out what key sequence you could use to enter -5×10^{-7} in your calculator.

BASIC

In BASIC (Beginner's All-Purpose Symbolic Instruction Code), the arithmetic symbols are: + (for addition), − (for subtraction), ∗ (for multiplication), / (for division), and ^ (for powering). In some versions of BASIC, ↑ is used for powering. The computer evaluates expressions according to the usual order of operations. Parentheses ( ) may be used. The comparison symbols =, >, < are also used in the standard way, but BASIC uses <= instead of ≤, >= instead of ≥, and <> instead of ≠.

Variables are represented by letters or letters in combination with digits. Consult the manual for your version of BASIC for restrictions on the length or other aspects of variable names. Examples of variable names allowed in most versions are N, X1, and AREA.

Commands

The BASIC commands used in this course and examples of their uses are given below.

LET ...   A value is assigned to a given variable. Some versions of BASIC allow you to omit the word LET in the assignment statement.

LET X = 5	The number 5 is stored in a memory location called X.
LET N = N + 2	The value in the memory location called N is increased by 2 and then restored in the location called N.

PRINT ...   The computer prints on the screen what follows the PRINT command. If what follows is a constant or variable, the computer prints the value of that constant or variable. If what follows is in quotes, the computer prints exactly what is in quotes.

PRINT X	The computer prints the number stored in memory location X.
PRINT "X-VALUES"	The computer prints the phrase X-VALUES.

INPUT ...   The computer asks the user to give a value to the variable named, and stores that value.

INPUT X	When the program is run, the computer will prompt you to give it a value by printing a question mark, and then store the value you type in memory location X.
INPUT "HOW OLD?"; AGE	The computer prints HOW OLD? and stores your response in memory location AGE.

REM ...   This command allows remarks to be inserted in a program. These may describe what the variables represent, what the program does or how it works. REM statements are often used in long complex programs or programs others will use.

REM PYTHAGOREAN THEOREM   The statement appears when the LIST command is given, but it has no effect when the program is run.

FOR ...
NEXT ...
STEP ...

The FOR command assigns a beginning and ending value to a variable. The first time through the loop, the variable has the beginning value in the FOR command. When the computer hits the line reading NEXT, the value of the variable is increased by the amount indicated by STEP. The commands between FOR and NEXT are then repeated. When the incremented value of the variable is larger than the ending value in the FOR command, the computer leaves the loop and executes the rest of the program. If STEP is not written the computer increases the variable by 1 each time through the loop.

```
10 FOR N = 3 TO 6 STEP 2
20    PRINT N
30 NEXT N
40 END
```

The computer assigns 3 to N and then prints the value of N. On reaching NEXT, the computer increases N by 2 (the STEP amount), and prints 5. The next N would be 7 which is too large. The computer executes the command after NEXT, ending the program.

IF ... THEN ...   The computer performs the consequent (the THEN part) only if the antecedent (the IF part) is true. When the antecedent is false, the computer *ignores* the consequent and goes directly to the next line of the program.

```
IF X > 100 THEN END
PRINT X
```

If the X-value is less than or equal to 100, the computer ignores "END," goes to the next line, and prints the value stored in X. If the X-value is greater than 100, the computer stops and the value stored in X is not printed.

GO TO ...   The computer goes to whatever line of the program is indicated. GOTO statements are generally avoided because they interrupt program flow and make programs hard to interpret.

GOTO 70   The computer goes to line 70 and executes that command.

END ...   The computer stops running the program. No program should have more than one end statement.

The following built-in functions are available in most versions of BASIC. Each function name must be followed by a variable or constant enclosed in parentheses.

ABS   The absolute value of the number that follows is calculated.

> LET X = ABS (-10)   The computer calculates $|-10| = 10$ and assigns the value 10 to memory location X.

SQR   The square root of the number that follows is calculated.

> C = SQR (A * A + B * B)   The computer calculates $\sqrt{A^2 + B^2}$ using the values stored in A and B and stores the result in C.

A program is a set of instructions to the computer. In most versions of BASIC every step in the program must begin with a line number. We usually start numbering at 10 and count by tens, so intermediate steps can be added later. The computer reads and executes a BASIC program in order of the line numbers. It will not go back to a previous line unless told to do so.

To enter a new program type NEW, and then the lines of the program. At the end of each line press the key named RETURN or ENTER. You may enter the lines in any order. The computer will keep track of them in numerical order. If you type LIST the program currently in the computer's memory will be printed on the screen. To change a line re-type the line number and the complete line as you now want it.

To run a new program after it has been entered, type RUN, and press the RETURN or ENTER key.

Programs can be saved on disk. Consult your manual on how to do this for your version of BASIC. To run a program already saved on disk you must know the exact name of the program including any spaces or punctuation. To run a program called TABLE SOLVE, type RUN TABLE SOLVE, and press the RETURN or ENTER key.

The following program illustrates many of the commands used in this course.

10 PRINT "A DIVIDING SEQUENCE"	The computer prints A DIVIDING SEQUENCE.
20 INPUT "NUMBER PLEASE" X	The computer prints NUMBER PLEASE? and waits for you to enter a number. You must give a value to store in the location X. Suppose you use 20. X now contains 20.
30 LET Y = 2	2 is stored in location Y.
40 FOR Z = -5 TO 4	Z is given the value -5. Each time through the loop, the value of Z will be increased by 1.

50	IF Z = 0 THEN GOTO 70	When Z = 0 the computer goes directly to line 70. When Z ≠ 0 the computer executes line 60.
60	PRINT (X * Y) / Z	On the first pass through the loop, the computer prints -8 because $(20 \cdot 2)/(-5) = -8$.
70	NEXT Z	The value in Z is increased by 1 to -4 and the computer goes back to line 50.
80	END	After going through the FOR … NEXT … loop with Z = 4, the computer stops.

The output of this program is

```
A DIVIDING SEQUENCE
NUMBER PLEASE? 20
-  8
-10
-13.3333
-20
-40
 40
 20
 13.3333
 10
```

LESSON 1-1 (pp. 4–8)
1. 1.6 miles **3.** Yes **5.** Yes **7.** 5 and 9 **9. a.** $m > 3$ **b. See below. 11.** b **13.** $\frac{5}{6} < \frac{17}{20}$ or $\frac{17}{20} > \frac{5}{6}$ **15.** $q < 15$ **17.** c
19. 3.375 **21.** 112.9$\overline{3}$ **23. a.** sample: 4 (any number greater than 3) **b.** sample: 1 (any number less than 3) **c.** 3
25. -3, -2 and 7 **27.** $\frac{2}{3}, \frac{7}{10}, \frac{3}{4}$

9. b.

LESSON 1-2 (pp. 9–13)
1. a. 1 student **b.** 12 **c.** 79.875, or about 80 **3.** median
5. a. 66.25 **b.** 66.5 **c.** 80 **d.** mode **7. a.** 8 days **b.** 24 days
c. 9° and 13° **d.** 6° **11. a.** $C > 100$ **b. See below. 13.** -15, -10, -4, -2, 0, 5, 6.8 **15.** -4 **17.** $350 **19.** 40 **21.** $\frac{6}{35}$ **23.** 4

11. b.
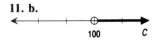

LESSON 1-3 (pp. 14–19)
3. a. 47% ± 5% **b.** 42% and 52% **5. a.** $5 \le a \le 12$
b. closed **c.** 7 **7. a.** $5 \le n \le 35$ **b. See below. c.** from 5 to 35 hours **9. a.** $20 \le g \le 20.5$ **b.** 20.25 **11. See below.**
13. $\frac{3}{8} \le x \le \frac{5}{8}$ **15. a.** $d > 1800$ **b. See below. 17. a. See below. b.** 0 **c.** 1 **19. a.** 3.6 **b.** 25.2 **c.** 54.75 or, in leap years, 54.9 **21. a.** $\frac{13}{5}$ **b.** $\frac{17}{10}$ **c.** $\frac{89}{55}$ **23. a.** -56 **b.** -56 **c.** 56

7. b.
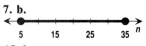
11.

15. b.

17. a.
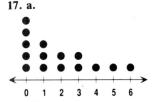
Number of swimming trips

LESSON 1-4 (pp. 20–24)
1. numerical **3.** algebraic **5.** 4 **7.** 6 **9.** 8
11. 10 PRINT ''ANSWER TO QUESTION 11''
20 LET M = 2.4
30 LET N = .2
40 PRINT ((M + 12.5)/(N + 3))^10
50 END

13. 5 * X + 3 * Y ^10 **15.** 110 mm **17.** $(1.4x - 2.3y)$,
$4xy$ **19. See below. 21. a.** $0 < P < 7.98$ **b. See below.**
23. a. See below. b. 80 **c.** 80 **d.** 72 **e.** 70 **25. a.** -85.7
b. -85.7 **c.** 85.7 **27.** a and c **29.** $\frac{11}{16}$

19.
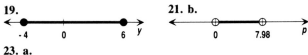
21. b.

23. a.

30 40 50 60 70 80 90 100
Scores

LESSON 1-5 (pp. 25–29)
1. L, d and s **3.** 634 **5.** $H = .8(200 - A)$ **7.** d **9.** pause and wait for a value to be typed in **11.** ≈ 22.9 in. **13. a.** 18.75 lb
b. It increases **c.** a heavy person **15. a.** 0 lb **b.** 40 in. is outside the domain of most adult heights, and the answer has no meaning. (No one has an ideal weight of 0.) **17.** 82°F
19. 3125 **21.** ANSWER TO QUESTION 21
15.24
23. 10 PRINT ''ANSWER TO QUESTION 23''
20 LET X = 5.7
30 LET Y = 2.006
40 LET Z = 51.46
50 PRINT X + Y + 3 * Z
60 END

25. a. 1 **b.** $\frac{5}{8}$ **c.** $\frac{11}{20}$ **27.** 29,360,000

LESSON 1-6 (pp. 30–35)
3. a. discrete **b.** sample: sister **5. a.** continuous **b.** sample: any real number except 2 **7.** I or R **9.** R **11. a.** {0, 1, 2}
b. 3 **15.** sample: $\frac{a + b}{2}$ **17. a.** the set of real numbers
b. $0 \le E \le 6194$ **c.** the set of real numbers between 0 and 6194 (closed interval) **c. See below. 21.** discrete **23.** {2, 4}
25. See below. 27. See below. 29. {5, 7, 9} **31.** $7\frac{1}{2}$
33. $35 - (20 - 7) = 22$ **35. a.** -6 **b.** 9 **c.** 8 **d.** -9

17. c.
25.
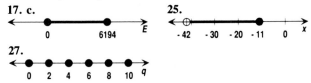
27.

LESSON 1-7 (pp. 36–41)
1. $\frac{1}{400}$ **5.** $\frac{2}{6} = \frac{1}{3}$ **7.** $\frac{1}{52}$ **9. a.** $\frac{2}{5}$ **b.** $\frac{3}{5}$ **c.** 1 **11. a.** (4, 6), (6, 4), (5, 5), (5, 6), (6, 5), (6, 6) **b.** $\frac{6}{36} = \frac{1}{6}$ **13.** 8 prizes

15. a. $\frac{1}{4}$ **b.** $\frac{2}{4} = \frac{1}{2}$ **17. a.** You cannot have more successes than possible outcomes. **b.** You cannot have a negative number of outcomes. **c.** The largest probability possible is 1, when all possible outcomes are successes. **19.** $\frac{11}{26}$

21. $10 \cdot (-3)^2 = 90$; $10 \cdot 0^2 = 0$; $10 \cdot 5^2 = 250$

23. sample: $\frac{x + y}{2}$ **25.** b **27. a.** 221.34 **b.** 22.134

c. 2.2134 **29.** 15%

LESSON 1-8 (pp. 42–47)
1. a. about .505 **b.** 50.5% **c.** It's smaller **d.** 49.5%

3. $\approx .052$ or 5.2% **7. a.** $\frac{5}{17}$ **b.** $\frac{1}{10}$ **c.** $\frac{4}{y}$ **9. a.** 60 people

b. 300 people **c.** 3 people **11.** No, the event did not occur.
13. c **15. a.** X **b.** No, but it probably would. **17.** 1 **19.** a

21. 3 **23.** 40% **25. a.** $\frac{1}{52}$ **b.** $\frac{13}{52} = \frac{1}{4}$ **c.** $\frac{4}{52} = \frac{1}{13}$

27. a. 8.85 **b.** .885 **c.** $38.43 **29.** 1 **31.** $\frac{7}{2}$

CHAPTER 1 PROGRESS SELF-TEST (p. 49)

1. $2(a + 3b)$ when $a = 3$ and $b = 5$; $2(3 + 3 \cdot 5) = 2(18) = 36$ **2.** $5 \cdot 6^n = 5 \cdot 6^4 = 5 \cdot 1296 = 6480$

3. $\frac{p + t^2}{p + t}$ when $p = 5$, $t = 2$ $\frac{5 + 2^2}{5 + 2} = \frac{9}{7}$ **4.** $\frac{3}{5}$

5. $\frac{7 \cdot 9abc}{7 \cdot 4b} = \frac{9ac}{4}$ **6.** $\frac{50.7}{4} \approx 12.7$ **7.** $10, $6, $6, $5, $5,

$4.50; $\frac{\$6 + \$5}{2} = \$5.50$

8. ANSWER TO QUESTION 8
8.82

9. See below.
10. $S > 25$ **11.** positive real numbers **12.** integers, real numbers **13.** 67% + 3% = 70%; 67% − 3% = 64% **14.** 8
15. sample: 0, 2, 4 **16.** $c = 20(4 - 1) + 25$; $c = 60 + 25$; $c = 85¢$ **17.** $A = \pi r^2 = 3.14159(3)^2 \approx 28$ m²
18. a. 3 bars **b.** 8 bars **19.** $8 \leq c \leq 31$ **20.** $\frac{2284}{100,000}$ or $\approx .02$ **21.** $\frac{8}{300}$ or $\frac{2}{75}$ **22.** $\frac{4}{26}$ or $\frac{2}{13}$ **23.** Donna **24.** See below. **25.** 11

9.

240   336

24.
0 1 2 3 4 5 6 7 8 x

The chart below keys the **Progress Self-Test** questions to the objectives in the **Chapter Review** on pages 50–53 or to the **Vocabulary** (Voc.) on page 48. This will enable you to locate those **Chapter Review** questions that correspond to questions you missed on the **Progress Self-Test**. The lesson where the material is covered is also indicated in the chart.

Question	1–3	4	5	6–7	8	9	10	11	12
Objective	C	A	A	H	D	K	K	G	Voc.
Lesson	1-4	1-1	1-8	1-2	1-4	1-3	1-1	1-6	1-6

Question	13	14–15	16–17	18–19	20	21–22	24	25
Objective	K	B	I	L	J	F	K	E
Lesson	1-3	1-1	1-5	1-2	1-8	1-7	1-6	1-6

CHAPTER 1 REVIEW (pp. 50–53)

1. $\frac{1}{6} > \frac{3}{20}$ or $\frac{3}{20} < \frac{1}{6}$ **3.** $\frac{3}{4}$ **5.** $\frac{11q}{12m}$ **7.** $\frac{7}{60}$ **9.** 4 **11.** sample:
75, 10^2, 387.2 **13.** -7 **15.** 45 **17.** 33.6 **19.** 250 **21.** 10
23. ANSWER TO QUESTION 23
ENTER A NUMBER
?2
6
25. a. -1, 0, 1, 2, 3 **b.** 5 **27.** True **29.** 50 **31.** 5 **33.** $\frac{4}{52} =$

$\frac{1}{13}$ **35. a.** B **b.** A **37.** whole numbers, discrete **39.** positive real numbers, continuous **41.** True **43.** True **45.** ≈ 1357 cm³
47. $159.73 **49.** 240 women **51.** $\approx 82\%$ **53.** a
55. a. $45 \leq s \leq 65$ **55. b.** See below. **57. a.** See below.
57. b. See below. **59.** b **61. a.** 3 states **b.** 16 states **c.** 118°
55. b.
45   65 s
57. a.
19 y
57. b.

19 20 21 22 y

CHAPTER 2 REFRESHER (p. 53)

1. 7.8 **2.** 133.56 **3.** 11.0239 **4.** $3\frac{3}{4}$ **5.** 9 **6.** $\frac{209}{210}$ **7.** 18%

8. 31.2% **9.** 11.03 cm **10.** 2.3 km **11.** 3′ **12.** 7′3″
13. 9 lb 14 oz **14.** 7 lb 9 oz **15.** 24 **16.** -15 **17.** 1 **18.** 13
19. -4 **20.** -16 **21.–29.** See below. **30.** $x = 8$ **31.** $z = 31$
32. $w = 407$ **33.** $m = 5$ **34.** $n = 539$ **35.** $s = 299$

21.–29.
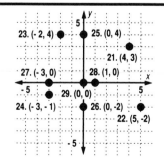
23. (-2, 4) 25. (0, 4) 21. (4, 3) 27. (-3, 0) 28. (1, 0) 29. (0, 0) 24. (-3, -1) 26. (0, -2) 22. (5, -2)

LESSON 2-1 (pp. 56–61)

3. 1988 **5.** Let E be the amount available for other expenses. Then $52.50 + 23.75 + 20 + E = 115$. **7. a.** 17 cm **b.** 15 cm **11. a 13. b 15.** $50 + x = 54$ **17.** There is overlap; zero is in both sets. **19.** $p + q + 26$ **21.** $49.95 + .05 + 59.28 + .72 = (49.95 + .05) + (59.28 + .72) = 50 + 60 = 110$ **23. a.** No **b.** Yes **c.** No **d.** Sample: brushing your teeth followed by combing your hair; yes **25.** $\frac{35}{68}$ or $\approx .515$ **27. a.** $\frac{2}{3}$ **b.** $\frac{2y}{3z}$ **c.** $\frac{2a}{3n}$ **29. a.** $\frac{26}{3}$ **b.** $-\frac{26}{3}$ **c.** $\frac{53}{10}$

LESSON 2-2 (pp. 62–67)

1. a. See below. b. gain of 4 yards **3. a.** -23 **b.** -2x **c.** -143 **5.** $y + -11$ **9.** any negative value, such as -6 **11.** Property of Opposites **13.** Adding Like Terms Property **15.** $3 + 8 \cdot 4 \neq 44$ **17.** $x + -3 + 5$ **19.** $40 - n + 2n$ or $40 + n$ **21.** 18 **23.** $-c$ **25.** 0 **27.** $14a - 4$ **29.** $(87 + k)$ sq cm **31. a.** $F = W * A/G$ **b.** 12.6 pounds **33.** 9

1. a.

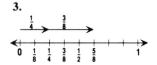

LESSON 2-3 (pp. 68–73)

3. See below. 5. -.625, -.625 **7.** $4\frac{1}{2}$ **9.** 3 **11. a.** $\frac{8x}{15}$ **b.** $\frac{8}{x}$ **c.** $\frac{5f + 3g}{5g}$ **d.** $\frac{bx + 3c}{cx}$ **13.** $\frac{17k}{12}$ **15.** $f + \frac{1}{3}$ or $\frac{3f + 1}{3}$ **17.** c **19.** $60\% = .60 = \frac{3}{5}$; $5\% = .05 = \frac{1}{20}$; $10\% = .10 = \frac{1}{10}$ **21.** sample: The temperature ranged from -3 to 4 degrees. **23.** 11 **25.** About 11.1 **27.** perpendicular lines

3.

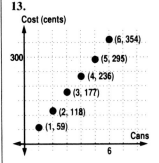

LESSON 2-4 (pp. 74–79)

3. a. 2 **b.** Beth, 60 minutes; Gary, 90 minutes **5.** to make the graph easier to use and to read. **7.** $126 billion less **9.** 30 mph **13.** 31 minutes **15.** c **17. a.** 4 **b.** 4m **19.** $\frac{1}{8}$ **21.** $\frac{17}{5a}$ **23. a.** -17.3 **b.** Property of Opposites **25.** $x + 7 + y = 140$ **27.** 1460 **29.** 8.232

LESSON 2-5 (pp. 80–84)

3. (-3.5, 5) **5. See below. 7. a.** $(x + 3, y + -7)$ **b.** sample: $x = 2, y = 0$; $(2 + 3, 0 + -7) = (5, -7)$ **See below. 9.** (4, -10) **11.** samples: 2N, 4E, 2N; 4E, 4N; 4N, 4E **13. See below. 15.** $\frac{6 + 7x}{21}$ **17.** -$1.8 million **19.** $1\frac{1}{3}$ **21.** 40 **23. a.** Commutative Property of Addition **b.** $x + y$ **25. See below.**

5.

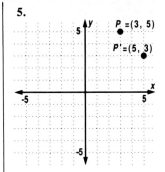

13.

Cost (cents)

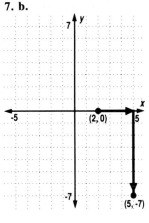

Cans

7. b.

25.

LESSON 2-6 (pp. 85–90)

1. My friend and I are the same age. **3.** Six years ago we were also the same age. **7.** Additive Identity Property **9. a.** 12 **b.** $y = -229$ **c.** Does $-12 + -229 = -241$? Yes. **11. a.** $a + b + c = p$ **b.** $b = p + -a + -c$ **c.** $b = 33$ **13.** b **15. a.** $C + 43 = 120$ **b.** $C = 77$. Does $77 + 43 = 120$? Yes. **17. a.** $7\frac{1}{4}$ **b.** Does $3\frac{1}{4} + 7\frac{1}{4} = 10\frac{1}{2}$? $3 + 7 + \frac{1}{4} + \frac{1}{4} = 10\frac{1}{2}$? Yes. **19.** $d = c + -a$ **21.** $25 + c = -12$, $c = -37$, fallen 37°, $25 + -37 = -12$ **23. a.** 2 right, 4 down; **b.** $D' = (2, -4)$; $E' = (3, 0)$ **25.** B **27.** Merry Berry **29.** 18.5¢

LESSON 2-7 (pp. 91–95)

1. $x < y$ **3.** $x + -6 < y + -6$ **5.** $y \geq 0.19$ **7.** d **9. a.** $z + 2 \geq 18$ **b.** $z \geq 16$ **11.** True **13. a.** $129 + p \leq 162$; $p \leq 33$ **b. See below. 15.** $x \geq 1\frac{3}{4}$. Step 1: Does $\frac{1}{2} + 1\frac{3}{4} + \frac{3}{4} = 3$? Yes. Step 2: Try 3. Is $\frac{1}{2} + 3 + \frac{3}{4} \geq 3$? Yes, $4\frac{1}{4} \geq 3$. **17.** 0 **19.** $-5\frac{1}{2}$ **21.** $B = -20.6$ **23.** 60° **25.** (7, 12) **27. a.** $(0, -2), (1, -1), (2, 0), (3, 1), (4, 2), (5, 3)$ **b. See below. 29.** $\frac{3}{16}$

13. b.

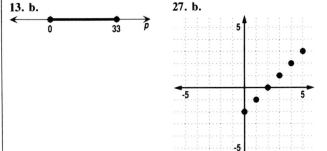

27. b.

1. $10 + d$ **5. a.** $4 + 5 > x, x + 4 > 5, x + 5 > 4$
b. $x < 9, x > 1, x > -1$ **c.** 9 **d.** 1 **e.** 1, 9 **7.** 0.5 km,
2.1 km **9.** because $1 + 2 < 4$ **11.** $2.6 < m < 20$ light-years

13. $\frac{7}{6}$ or $1\frac{1}{6}$ **15.** 8 **17.** 0 **19. a.** 15 **b.** 9 **c.** 15 **21.** They are
equal. **23.** $-50 + c + -20 = 210; c = 280$

CHAPTER 2 PROGRESS SELF-TEST (p. 101)

1. Associative Property of Addition **2.** $x + y + 35 = 6 \cdot 12$
3. $-10 + d = t$ **4.** 27 **5.** $x + -3$ **6.** $\frac{2 + m}{n}$ **7.** $\frac{3}{4} + \frac{3}{5} +$
$\left(\frac{-3}{10}\right) = \frac{3}{4} + \frac{3}{10} = \frac{15 + 6}{20} = \frac{21}{20}$ or $1\frac{1}{20}$ **8.** $-11\frac{5}{6}$ **9.** $-p$
10. $-2x + 4y$ **11.** d **12. a.** $-4 + 1\frac{1}{2} + -3 + L = -10$
b. $-5\frac{1}{2} + L = -10; L = -4\frac{1}{2}$; she must lose $4\frac{1}{2}$ lb.
13. Addition Property of Equality **14.** 16 **15.** $a > 13.7$ **16.** $\frac{7}{4}$
17. $1 > z + -10, 11 > z, z < 11$ **18.** $b = 100 + -a$ **19.** Is
$15 \leq -100 + 87$? Is $15 \leq -13$? No **20.** $-13 \geq x$; See below.
21. $m + 47.50 \geq 150$ **22.** $9 + b + 21 = 43; b + 30 =$
$43; b = 13$ **23.** $C + -7$ **24.** $(5 + -4, -2 + 5) = P' =$
$(1, 3)$ **25.** $-4 + a = 6, a = 10; 7 + b = 5, b = -2;$
$(-5 + 10, -2 + -2) = B' = (5, -4)$ **26.** $x \leq 30 + 12,$

$x + 12 \geq 30$, and $x + 30 \geq 12$ gives $x \leq 42$ and $x \geq 18$
$(18 \leq x \leq 42)$ **27.** 18 km **28.** $b + g = 45$ **29.** (10, 35),
(20, 25), (23, 22), (44, 1) **30.** See below.

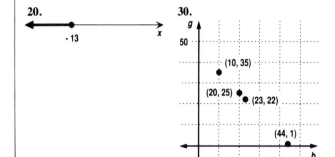

20.
30.

The chart below keys the **Progress Self-Test** questions to the objectives in the **Chapter Review** on pages 102–105 or to the **Vocabulary** (Voc.) on page 100. This will enable you to locate those **Chapter Review** questions that correspond to questions you missed on the **Progress Self-Test.** The lesson where the material is covered is also indicated in the chart.

Question	1	2	3	4, 5	6–8	9	10	11	12	13	14	15
Objective	E	H	H	B	A	E	B	A	H	F	C	D
Lesson	2-1	2-1	2-2	2-2	2-3	2-2	2-2	2-2	2-6	2-6	2-6	2-7

Question	16	17	18	19	20	21	22, 23	24, 25	26, 27	28	29, 30
Objective	C	D	C	D	J	H	H	L	G, I	H	K
Lesson	2-6	2-7	2-6	2-7	2-7	2-7	2-6	2-5	2-8	2-6	2-4

CHAPTER 2 REVIEW (pp. 102–105)

1. $\frac{13}{12}$ **3.** $-\frac{21}{20}$ **5.** $\frac{x + y}{3}$ **7.** $\frac{mq + np}{nq}$ **9.** 5 **11.** $12a +$
$2b + -5$ **13.** $3t + 3$ **15.** $6b$ **17.** $m = 10$; does $2 + 10 =$
12? Yes **19.** $t = -0.6$; does $2.5 = -0.6 + 3.1$? Yes
21. $m = 7487$; does $21,625 + 7487 = 29,112$? Yes
23. $y = -1$; does $-2 + -1 = -3$? Yes **25.** $r = 5p$ **27.** $x > 5$
29. $x > 1$ **31.** $15 + 35 = 50$; not correct **33.** Additive
Identity Property **35.** Op-op Property **37.** Adding Like Terms
Property **39.** Addition Property of Equality **41.** c
43. $a + b > c, b + c > a, c + a > b$ **45.** $0.2 < y < 4.6$
47. 24° **49.** $5.4 + d + 7.50 < 26$ **51.** $T_1 + C > T_2$
53. $+\frac{5}{8}$ **55.** It would take more than 10 minutes and less than

50 minutes. **57.** See below. **59.** See below. **61.** halfway up
63. Ohio **65.** 1950 to 1960 **67.** $(x + 4, y + -10)$ **69.** See
below.

57.
59.

69.

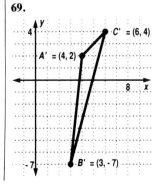

CHAPTER 3 REFRESHER (p. 105)
1. -160 **2.** -3 **3.** -8 **4.** -1 **5.** 199 **6.** 1 **7.** b **8.** c **9.** 75°
10. 120° **11.** See below. **12.** $x = 51$ **13.** $909 = y$
14. $w = 113$ **15.** $502 = z$

11.

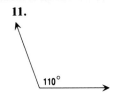

110°

CHAPTER 3 OPENER (p. 107)
1.–6. See below.

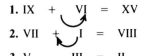

1. IX + VI = XV

2. VII + I = VIII

3. V – III = II

4. VII – IV = III

5. XI – V = VI

6. VII – VI = I

LESSON 3-1 (pp. 108–112)
1. a. $12 + -15$ **b.** $12 - 15$ **3.** $-2 + -7$ **5.** $x + d$
7. a. $\frac{3}{5} + \frac{7}{10}$ **b.** $\frac{13}{10}$ **9. a.** False **b.** Associative Property of
Subtraction **11.** -57 **13.** $10p + 6q + 4$ **15.** -22 **17.** 0
19. a. $-4 + -3 + -3 + 5$ **b.** $-4 - 3 - 3 + 5$ **c.** -5 lb
21. -1, 1 **23. a.** $-15x$ **b.** $-14y$ **c.** $2x + 32$ **d.** $4x$
25. a. $x = 4$ **b.** $y = 0.7$ **c.** $z = \frac{1}{2}$ **27. a.** 17 **b.** 10 **c.** 800
29. $y = 21x + 3$

LESSON 3-2 (pp. 113–117)
5. $18 - A$ **7.** $1600 - b^2$ **9. a.** Bernie **b.** 7 years
11. a. 0.5° **b.** -1.5° **13.** 47° **15.** $(x - 3)$ feet **17.** $4x$
19. -24.73 **21.** $-13ab - 2a - 4b$ **23. a.** -10 **b.** $-6 + n$
25. 37 **27.** $x = 3, y = 1$ **29.** $\frac{E}{G}$ **31. a.** $\frac{1}{2}b^2$ **b.** $6h^2$ **c.** $\frac{1}{16}$

LESSON 3-3 (pp. 118–122)
3. a. $s + -1240 = 20,300$ **b.** $s = 21,540$ **5. a.** $x + 60 <$
140 **b.** $x < 80$ **7. a.** Definition of subtraction **b.** Addition
Property of Inequality **c.** Property of Opposites **d.** Additive
Identity Property **9.** $A - 2768 \geq 1000$ **11.** $x \leq y + 35$
13. a. Definition of subtraction **b.** Addition Property of
Equality **c.** Associative Property of Addition **d.** Property of
Opposites **e.** Additive Identity Property **15.** 3 **17.** $100\frac{1}{2}$
19. $q > 31$ See below. **21.** $2a + a = 3a$ **23.** $2000 - n$
25. $\$5500 - 4500 = \$1000$ **27.** -14 **29. a.** 200,000 **b.** 10
c. 5

19.

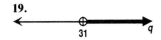

31   q

LESSON 3-4 (pp. 123–127)
3. {6, 12} **5.** See below. **7.** $\frac{1}{3}$ **11. a.** See below. **b.** See
below. **13.** $\frac{4}{26}$ **15.** $\frac{15}{30}$ or $\frac{1}{2}$ **17. a.** {1, 7} **b.** {1}
19. a. $T + S - B$ **21.** $t = 141$ **23.** $b = 4$ **25.** -4 **27. a.** -3
b. $-\frac{3}{5}$ **c.** $\frac{-3}{x}$

5.

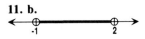

-196   °C

11. a.

-1   2

11. b.

-1   2

LESSON 3-5 (pp. 128–132)
1. The Greens beat the Reds and the Blues beat the Purples.
3. {2, 3, 4, 6, 8, 9, 10, 12, 15} **5. a.** I **b.** III **c.** II
7. a. $\approx 6\%$ **b.** $\approx 56\%$ **9. a.** {3} **b.** {1, 2, 3, 5, 6, 7, 9}
11. See below. **13.** See below. **15.** See below. **17. a.** $x =$
7.4 **b.** $y = 2\frac{1}{2}$ **c.** $z = -1.5$ **19.** $\frac{7}{5}x = -5$ **21. a.** $\frac{1}{2} - \frac{2}{3} = \frac{-1}{6}$
b. $\frac{x}{2} + \frac{-2x}{3} = \frac{-x}{6}$ **c.** $\frac{x}{2a} + \frac{2x}{3a} = \frac{7x}{6a}$ **23.** $0 < s < 24$ where
s is length of third side **25. a** **27. a**

11.

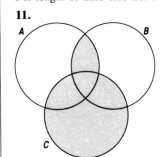

A   B

C

13.

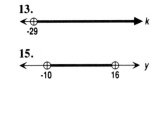

-29   k

15.

-10   16   y

LESSON 3-6 (pp. 133–137)
5. 25 **9.** .87 **11.** 6 **13. a.** 6 **b.** See below. **15.** $\frac{7}{12}$
17. a. $x = 9$ **b.** $y = 6$ **c.** $z = 2$ **19. a.** $y - 12$ **b.** $y + 38$
21. $\approx .5$ **23.** $\frac{4}{9}$; slightly above par

13.

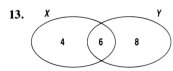

X   Y
4   6   8

LESSON 3-7 (pp. 138–142)
5. a. See below. **5. b.** See below. **7. a.** (1, 7), (2, 6), (3, 5),
(4, 4), (5, 3), (6, 2), (7, 1) **b.** 3 **9.** (0, 3), (1, 2), (2, 1),
(3, 0); c **11.** $y = x + 5$ **13.** 22.6 **15.** 2326 **17. a.** 95%
b. 430 **19.** samples: violin, piano, kettle drum **21. a**
23. $9.50 **25. a.** 1000 **b.** 2000 **c.** 125

5. a.

x Xandra	y Yvonne	(x, y) Ordered Pairs
1	6	(1, 6)
2	7	(2, 7)
3	8	(3, 8)
4	9	(4, 9)
5	10	(5, 10)
6	11	(6, 11)

5. b.

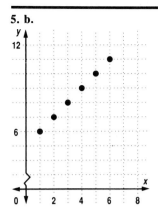

(10, 80); (30, 60); (45, 45); (70, 20); (85, 5) **b. See below.**
c. See below. 13. m∠C = 25°, m∠T = 65° **15.** 50°
17. m∠C = 180 − (m∠A + m∠B) **19.** z < 108
21. x = 10 **23.** 3279 **25.** 961 **27.** EVALUATE A * B

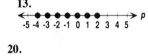

29. b **31.** $-\frac{7}{12}x$

11. b. 11. c.

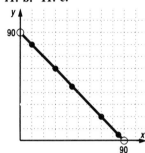

LESSON 3-8 (pp. 143–147)
1. 138° **5.** 65° **7.** m = 48 **9.** True **11. a.** sample: 10° and 80°; 30° and 60°; 45° and 45°, 70° and 20°; 85° and 5°

CHAPTER 3 PROGRESS SELF-TEST (p. 149)

1. subtracting 7 **2.** $1n + -16 + -2n + 12 = -1n + -4 = -n - 4$ **3.** $-8x + -2x + x = -10x + 1x = -9x$ **4.** $\frac{3}{4} + -\frac{7}{8} = \frac{6}{8} + -\frac{7}{8} = -\frac{1}{8}$ **5.** $\frac{2m}{4a} + -\frac{3m}{4a} = \frac{-1m}{4a} = -\frac{m}{4a}$ **6.** $y + -13 = -7; y + -13 + 13 = -7 + 13; y + 0 = 6; y = 6$
7. $m + -2 < 6; m + -2 + 2 < 6 + 2; m + 0 < 8; m < 8$
8. $b + -a = 100; b + -a + a = 100 + a; b + 0 = 100 + a; b = 100 + a$ **9.** $m + -7n = -22n; m + -7n + 7n = -22n + 7n; m + 0 = -15n; m = -15n$ **10.** {-3, 3}
11. { -6, -3, -1, 0, 1, 3, 5, 6, 7, 9} **12. See below. 13. See below. 14.** $(321 + 215) - 480 = 56$ **15.** $V - H$
16. $B - S - 3 + N$ **17.** $15 + 11 - 20 = 6, \frac{6}{30} = \frac{1}{5}$ **18.** c
19. See below. 20. See below. 21. c

12.

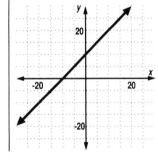

13.

19.

20.

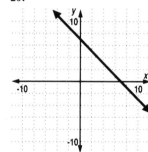

The chart below keys the **Progress Self-Test** questions to the objectives in the **Chapter Review** on pages 150–152 or to the **Vocabulary** (Voc.) on page 148. This will enable you to locate those **Chapter Review** questions that correspond to questions you missed on the **Progress Self-Test.** The lesson where the material is covered is also indicated in the chart.

Question	1	2–5	6, 7	8, 9	10	11	12	13	14	15, 16	17
Objective	E	D	B	B	A	A	H	H	G	F	G
Lesson	3-1	3-1	3-3	3-3	3-4	3-5	3-5	3-4	3-6	3-2	3-6

Question	18	19, 20	21
Objective	C	J	I
Lesson	3-8	3-7	3-5

CHAPTER 3 REVIEW (pp. 150–152)

1. {15, 25} **3.** {8, 9} **5.** x = 45; Does 45 − 47 = -2? Yes
7. $y = -\frac{1}{2}$ Does $\frac{3}{2} + -\frac{1}{2} - \frac{1}{4} = \frac{3}{4}$? Yes **9.** z < 23 Does 23 − 12 = 11? Yes; try 3: Is 3 − 12 < 11? Yes **11.** Does 10 − 30 = 40? No. He is wrong. **13.** -z = y **15.** 27.5° **17.** x + z = 90 **19.** $-1\frac{7}{15}$ **21.** $-25\frac{1}{3}$ **23.** 0 **25.** $1 + -3z^3$
27. -8 + -v = 42 **29.** d **31.** p + 80 > L where

p = weight of other passengers; $p > L - 80$ **33.** $S - 40 <$ 3; $S < 43$ **35.** 31,500 feet **37.** $5000 - H$ **39.** $\frac{22}{25}$ **41.** 91 **43.** See below. **45.** See below. **47.** {-11, -1, 2} **49.** See below. **51.** See below. **53.** See below.

43.

-2   0   →y

45.

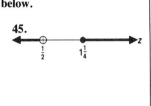
$\frac{1}{2}$ $1\frac{1}{4}$ →z

49.

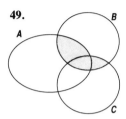

51.

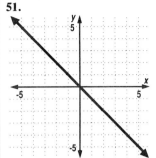

53.

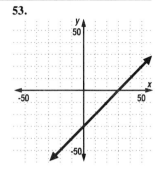

CHAPTER 4 REFRESHER (p. 153)

1. 15.087 **2.** 12.48 **3.** 0.00666 **4.** 0.0034 **5.** 20 **6.** $27\frac{1}{2}$ **7.** $\frac{1}{6}$
8. $\frac{1}{24}$ **9.** $2\frac{21}{32}$ **10.** $33\frac{1}{3}$ **11.** $39.27 **12.** $93.75 **13.** $5\frac{1}{4}$ inches
14. 36 **15.** $180 **16.** 78.75 **17.** -6 **18.** -121 **19.** -24 **20.** 25
21. 0 **22.** -1820 **23.** 3540 **24.** 6 **25.** $x = 4$ **26.** $y = 22$

27. $z = \frac{1}{2}$ **28.** $\frac{1}{9} = w$ **29.** $\frac{3}{25} = a$ **30.** $b = \frac{2}{3}$ **31.** $c = \frac{1}{2}$
32. $3\frac{1}{2} = d$ **33.** 180 in.² **34.** 14.4 cm² **35.** 8 m² **36.** 8200 ft²
37. 96 cm³ **38.** 162 in.³ **39.** $\frac{1}{8}$ ft³ **40.** 30,000 m³

LESSON 4-1 (pp. 156–161)

3. square inches **5.** $8y$ in.² **7.** 3984 **9.** $21xy$ cubic units
11. $45ab$ **13. a.** kn **b.** $2k + 2n$ **15.** $12x^3$ cm³ **17. a.** $\frac{1}{2}$ in. •
$\frac{3}{4}$ in. **b.** $\frac{3}{8}$ **c.** $\frac{3}{8}$ **19.** 576 m² **21.** samples: 3 by 4 by 7, 2 by 6
by 7, 14 by 2 by 3, 28 by 1 by 3, $\frac{1}{2}$ by $\frac{1}{2}$ by 336 **23.** $240x^2$
25. 17 **27.** 175 **29. a.** If p pounds of meat are eaten, then
$17 \le p \le 23$ **29. b. See below.**

29. b.

17   23 p

LESSON 4-2 (pp. 162–167)

1. $\frac{10}{33}$ **3.** $\frac{7m^2}{3n}$ **5.** $6k$ **7. a.** A typist types 70 words per minute **b.** 70 words/minute **c.** $70\frac{\text{words}}{\text{minute}}$ **9.** 17.55 miles
11. a. -1.5 inches **b.** -1.5 **13.** (b) **15. a.** 30 **b.** A printer is
printing 15-page documents at the rate of $\frac{1}{2}$ page per minute.
How many minutes will it take to print each document?
17. $\frac{60ab}{x}$ **19. a.** 1536 oz **b.** $128c$ oz **21.** 3,503,000 babies
23. Ds dollars **25.** $\approx$ 11.4 sec **27.** $\frac{1}{2}$ in.²
29. 5   12   534.072 **31. a.** True **b.** True **c.** False

LESSON 4-3 (pp. 168–172)

5. a. opposites or additive inverses, zero **b.** reciprocals, one
7. $\frac{1}{10}$; $10 \cdot \frac{1}{10} = 1$ **9.** $\frac{5}{23}$; $\frac{23}{5} \cdot \frac{5}{23} = 1$ **11.** $-\frac{7}{6}$; $-\frac{6}{7} \cdot -\frac{7}{6} = 1$
13. a. 10 in. • $2.54\frac{\text{cm}}{\text{in.}} = 25.4$ cm **b.** x in. • $2.54\frac{\text{cm}}{\text{in.}} =$
$2.54x$ cm **15.** $abcde$ **17.** a and d **19.** Yes **21.** $a = 1$,
$b = 0$ **23.** $10x$ **25. a.** $32p$ **b.** -125 **c.** x^2 **27.** 33 math
problems/night **29. a.** 3080 ft² **b.** 513 people **31.** $T - \frac{3}{2}$ hours

LESSON 4-4 (pp. 173–177)

3. a. -32 **b.** $-\frac{1}{32}$ **c.** -13 **5. a.** $\frac{3}{32}$ **b.** $\frac{32}{3}$ **c.** 8 **7.** -.05 **11.** No
13. 200 cm from the turning point **15.** 12.41 cm
17. a. $DV = M$ **b.** 648.96 pounds **19.** (a) **21.** 16.2 minutes
23. $\approx$ 171 cm² **25. a.** $a = -7$ **b.** $a \le 48$ **c.** $a > 166 - c$
27. a. 24% **b.** 30% **c.** 140%

LESSON 4-5 (pp. 178–181)

1. 0 has no reciprocal. **3. a.** There is no solution. **b.** { } or
$\varnothing$ **7.** $x = -40$ **9.** $z = 0$ **11.** $v = 3.74$ **13.** $x = 89$ seconds **15. a.** $y = m - 25$ **b.** $2 = 27 - 25$ **17.** $N = 0$
19. $p = -3$ **21.** $r = 30$ **23.** -238 **25. a.** 1 **b.** -1 **c.** When
the number of factors is even, the value is 1. When the
number of factors is odd, the value is -1. **27.** $4 + -3$
29. $-432 \cdot -175$

LESSON 4-6 (pp. 182–185)

1. $120 < 180$ **3.** $-80 > -120$ **7. a.** No **b.** No **c.** Yes **d.** No
e. No **f.** No **9.** $y > -100$ **11.** $\frac{13}{2} > z$ **13.** $-0.01 < c$
15. $m > -8$ **17.** $t > 0$ **19.** 30 rows **21.** $n < 272$ **23.** $x = 4$
25. $b = 10$ **27.** $d = 7$ **29.** $30 = d$ **31.** 57 cubic units **33. a.** $\frac{1}{2}$
b. $\frac{1}{2}$ **c.** $\frac{1}{3}$ **d.** $\frac{2}{3}$ **35.** $A = \frac{1}{2}s^2$ **37. a.** $1\frac{1}{5}$ **b.** 6 **c.** 7.5 **d.** $\frac{1}{6}$

LESSON 4-7 (pp. 186–190)

5. See below. **7.** $5m^2n^4$ **9. a.** $2^{20} = 1,048,576$
b. $\frac{1}{1,048,576}$ **11.** Takeshi and Reiko, Takeshi and Akiko,
Takeshi and Kimiko, Satoshi and Reiko, Satoshi and Akiko,
Satoshi and Kimiko, Izumi and Reiko, Izumi and Akiko,
Izumi and Kimiko, Mitsuo and Reiko, Mitsuo and Akiko,
Mitsuo and Kimiko **13.** 144 **15.** 72 **17.** 2º **19.** $k = -\frac{1}{50}$
21. $m < \frac{1}{40}$ **23.** $d < 13$ **25.** 5.9 **27.** $(2w + 5)w$
29. Commutative Property of Multiplication **31.** Property of
Opposites **33.** $\approx$ 38.8 **35. a.** $\frac{1}{4}$ **b.** $\frac{11}{2}$ **c.** -1 **37.** 70%

5.

		South Gate		
		I	F	E
	A	(A, I)	(A, F)	(A, E)
North Gate	B	(B, I)	(B, F)	(B, E)
	C	(C, I)	(C, F)	(C, E)

LESSON 4-8 (pp. 191–196)

1. b. $\frac{2}{30}$ or $\frac{1}{15}$ **5.** $\frac{3}{70} \approx .043$ or 4.3% **7.** $\frac{19}{34}$ **9.** $\approx 18\%$
11. (b) **13.** 30% **15. a.** $\frac{1}{2}$ **b.** $\frac{7}{19}$ **c.** $\frac{7}{38}$ **17.** $\frac{5m^2}{9n^2}$ **19. a.** $\frac{1}{4}$
b. $\frac{1}{8}$ **21.** 17,576,000 **23.** $cr = d$ **25.** $\frac{6}{5}$ **27.** $\frac{1}{a}$ **29.** $x = -\frac{12}{m}$
31. a. (1, -3) **b.** (-7, $r + 2$) **c.** (-7 + m, 2 − n)

LESSON 4-9 (pp. 197–201)

1. Bird, Jordan, Johnson; Bird, Johnson, Jordan; Jordan, Johnson, Bird; Jordan, Bird, Johnson; Johnson, Bird, Jordan; Johnson, Jordan, Bird **5. a.** 1 **b.** 2 **c.** 6 **d.** 24 **e.** 120 **f.** 720 **g.** 5040 **h.** 40,320 **7.** 2.6525×10^{32} **9.** 100 **13.** 720 **15. a.** $8 \cdot 7 = 56$ **b.** $8! = 40,320$ **17. a.** 10
b. $\frac{n!}{(n-1)!} = n$ **19. a.** $\frac{12}{25}$ **b.** $\frac{2}{6}$ or $\frac{1}{3}$ **c.** $\frac{6}{25} \cdot \frac{2}{6} = \frac{2}{25}$ **21.** 72.3
23. $y < \frac{10}{3}$ **25.** $t \geq$ -8 **27.** 0 **29.** $\approx$ 691,200,000 grains
31. $6.6a$ **33.** $64c^2$

CHAPTER 4 PROGRESS SELF-TEST (p. 203)

1. $\frac{5 \cdot 4 \cdot 3 \cdot 2 \cdot 1}{3 \cdot 2 \cdot 1 \cdot 2 \cdot 1} = \frac{5 \cdot 4}{2} = \frac{20}{2} = 10$
2. $7(2.4 + 2.9)(2 \cdot 2.4 + 3.1)(0) = 0$ **3.** $\frac{5}{3} \cdot \frac{x}{5} = \frac{x}{3}$
4. a. sample: $(2 \cdot 5) \cdot 8 = 2 \cdot (5 \cdot 8)$ **b.** sample: $6 \cdot \frac{1}{6} = \frac{1}{6} \cdot 6 = 1$ **5. See below. 6.** $\frac{1}{30} \cdot 30x = \frac{1}{30} \cdot 10$; $x = \frac{10}{30}$; $x = \frac{1}{3}$
7. $4 \cdot \frac{1}{4}k = 4 \cdot$ -24; $k =$ -96 **8.** $\frac{1}{3} \cdot 15 \leq \frac{1}{3} \cdot 3m$; $\frac{15}{3} \leq m$; 5 ≤
m **9.** $-y \leq$ -2; $y \geq 2$ **10.** $1.46 = 2.7 + $-$t$; $2.7 + 1.46 =$
$-2.7 + 2.7 + $-$t$; $-1.24 = 0 + $-$t$; $1.24 = t$ **11.** $\frac{1}{3m}$ **12.** $n \cdot$
$1 = n$ **13.** $11 \cdot (14 + 6 + 4) - (5 \cdot 6) = (11 \cdot 24) -$
$30 = 264 - 30 = 234$ **14.** $4 \cdot 5 = 20$ **15.** $5! = 120$

16. $\$15.50s$ **17.** 300 mi. $\cdot \frac{1}{55}\frac{\text{hour}}{\text{mile}} = 5.\overline{45}$ hours, or about
$5\frac{1}{2}$ hours **18.** 8 inches $\cdot 2.54\frac{\text{cm}}{\text{in.}} = 20.32$ cm **19.** $\frac{3}{7} \cdot \frac{2}{6} = \frac{6}{42} =$
$\frac{1}{7}$ **20. a.** 16 cm $\cdot w \cdot 5$ cm = 1000 cm³ **b.** $80w = 1000$ cm;
$\frac{1}{80} \cdot 80w = \frac{1}{80} \cdot (1000 \text{ cm})$; $w = 12.5$ cm **21.** $\frac{14}{40} \cdot \frac{1}{2} = \frac{7}{40}$
5.

	$\frac{1}{5}$	$\frac{1}{5}$	$\frac{1}{5}$	$\frac{1}{5}$	$\frac{1}{5}$
$\frac{1}{2}$					
$\frac{1}{2}$					

The chart below keys the **Progress Self-Test** questions to the objectives in the **Chapter Review** on pages 204–207 or to the **Vocabulary** (Voc.) on page 202. This will enable you to locate those **Chapter Review** questions that correspond to questions you missed on the **Progress Self-Test.** The lesson where the material is covered is also indicated in the chart.

Question	1	2	3	4	5	6	7	8, 9	10	11, 12
Objective	E	F	A	F	L	B	B	D	C	F
Lesson	4-9	4-3	4-2	4-3	4-1	4-4	4-5	4-6	4-5	4-3

Question	13	14	15	16, 17, 18	19	20	21
Objective	G	H	J	G	I	K	I
Lesson	4-2	4-9	4-2	4-2	4-8	4-4	4-8

CHAPTER 4 REVIEW (pp. 204–207)

1. $\frac{27}{40}$ **3.** $\frac{c}{d}$ **5.** $m = 150$; Does $2.4m = 2.4 \cdot 150 = 360$? Yes
7. $f =$ -2.3; Does $-10f = -10 \cdot$ -2.3 $= 23$? Yes **9.** $A = \frac{15}{2}$;
Does $\frac{4}{25}A = \frac{4}{25} \cdot \frac{15}{2} = \frac{30}{25} = \frac{6}{5}$? Yes **11.** $x =$ -12; Does
$31 - $-$12 = 31 + 12 = 43$? Yes **13.** $z = $-$\frac{1}{5}$; Does
$\frac{1}{5} - $-$\frac{1}{5} = \frac{1}{5} + \frac{1}{5} = \frac{2}{5}$? Yes **15.** $m \leq 2$; Step 1: Does $8 \cdot 2 =$
16? Yes; Step 2: Try 0. Is $8 \cdot 0 \leq 16$? Yes **17.** $u >$ -2;
Step 1: Does $6 \cdot$ -2 = -12? Yes; Step 2: Try 0. Is
$6 \cdot 0 >$ -12? Yes **19.** $g \leq 10$; Step 1: Does $\frac{1}{2} \cdot 10 = 5$?
Yes; Step 2: Try 4. Is $\frac{1}{2} \cdot 4 \leq 5$? Yes **21.** 30 **23.** 240

25. 20 $\boxed{x!}$ $\boxed{\div}$ 15 $\boxed{x!}$ $\boxed{=}$ $\boxed{\div}$ 5 $\boxed{x!}$ $\boxed{=}$ 15,504 **27. a.** $2200x$
b. Associative and Commutative Properties of Multiplication
29. $n \cdot \frac{1}{n} = 1$ **31.** $\frac{1}{0.6}$ or $\frac{5}{3}$ **33. a.** $(-6.2 + 3.8)(4.3 - 6.2)$
$(0) = 0$ **b.** Multiplication Property of Zero **35.** opposite
37. sample: $0x = 5$ **39.** 48 ft² **41.** $\$350k$ **43.** 1,045,440 sq ft
45. C bookcases $\cdot 24\frac{\text{ft}}{\text{bookcase}} \cdot B\frac{\text{books}}{\text{ft}} = 24 \cdot B \cdot C$ books
47. 9,765,625 **49.** $\frac{1}{3} \cdot \frac{2}{8} = \frac{1}{12}$ **51. a.** $\frac{2}{10}$ or $\frac{1}{5}$ **b.** $\frac{10}{30} \cdot \frac{2}{10} =$
$\frac{2}{30}$ or $\frac{1}{15}$ **53. a.** $\frac{13}{52} \cdot \frac{12}{51} = \frac{3}{51}$ **b.** $\frac{4}{52} \cdot \frac{3}{51} = \frac{1}{221}$ **55. a.** $9! =$
362,880 **b.** 3.6288×10^5 **57.** 8.22 hr or $\approx$ 8 hr 15 min

59. $d \leq 12.2$; at most 12 days **61.** Commutative Property of Multiplication **63. See below. 65.** $\frac{3}{4} \cdot \frac{2}{5} = \frac{6}{20}$ or $\frac{3}{10}$
67. $20q$ tiles **69.** $24k^3$

63.

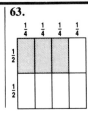

CHAPTER 5 REFRESHER (p. 207)
1. 0.40 **2.** 2.5 **3.** 2.4 **4.** 1,600 **5.** .2 **6.** .2 **7.** .0016
8. ≈ 4.8 **9.** 6 **10.** $\frac{4}{5}$ **11.** $\frac{2}{3}$ **12.** $1\frac{7}{8}$ **13.** 3 ft **14.** $2.5m$
15. $14\frac{2}{7}$ kg **16.** .24 lb **17.** .5; 50% **18.** .75; 75% **19.** .025;
2.5% **20.** .73; 73% **21.** .14 **22.** 6.67 **23.** 6.47 **24.** 3.14
25. 220% **26.** 27% **27.** 89% **28.** 18% **29.** .3; $\frac{3}{10}$

30. .01; $\frac{1}{100}$ **31.** 3.0; $\frac{3}{1}$ **32.** .0246; $\frac{123}{5000}$ **33.** .0003; $\frac{3}{10,000}$
34. .0025; $\frac{1}{400}$ **35.** \$240 **36.** 68 questions **37.** 2942 voters
38. 12,000 square miles **39.** 0 **40.** 2 **41.** -8 **42.** -.5
43. -.025 **44.** $\frac{1}{2}$ **45.** -100 **46.** divisor 3; dividend 21; quotient
7 **47.** divisor 100; dividend 20; quotient 0.2 **48.** divisor 7;
dividend 56; quotient 8

LESSON 5-1 (pp. 210–214)
1. $32 \div 6 = 5.\overline{33}$; $32 \cdot \frac{1}{6} = 5.\overline{33}$ **3. a.** n **b.** $\frac{1}{n}$ **5. a.** 2
b. 2 **c.** $\frac{1}{2}$ **d.** 100 **7.** $\frac{25}{2n}$ **9. a.** $\frac{6}{25} \div \frac{10}{7}$ **b.** $\frac{6}{25} \cdot \frac{7}{10}$ **c.** $\frac{42}{250} =$
$\frac{21}{125}$ **11. a.** $-\frac{1}{3}$ **b.** -3 **13.** 50 **15.** $\frac{1}{2}$ **17.** $\frac{1}{6}$ **19. a.** -3.5, -3.5, and
-3.5 **b.** $-.\overline{27}$, $-.\overline{27}$, and $-.\overline{27}$ **c.** $\frac{-x}{y} = \frac{x}{-y} = -\frac{x}{y}$ **21.** A
positive number divided by a positive number is positive; a
negative number divided by a negative number is positive; a
negative number divided by a positive number is negative; a
positive number divided a negative number is negative.
23. $t + b = 5$; $t \geq 0$; $b \geq 0$; **See below. 25.** 120
27. a. 3 **b.** 10 **29.** 4.5 m²

23.

x (turkey)	y (beef)
0	5
1	4
2	3
3	2
4	1
5	0

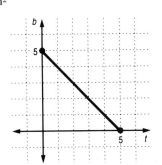

LESSON 5-2 (pp. 215–220)
1. a. 37.5 mph **b.** $\frac{2}{75}$ or $\approx .027$ hour **3. a.** degrees, hours
b. -2.2 degrees per hour **5.** \$55 per day **7.** ≈ 34.6 miles per
gallon **9.** -1 **11.** so that the program will divide by the result
of $K - 5$ **13. a.** 1544.1 people per sq mi **b.** .06 person per
sq mi **15.** (b) **17.** An error message (E) appears. **19.** $\frac{x}{3} \cdot \frac{1}{5}$
21. $\frac{1}{2}$ **23.** 12 **25.** $x = -300$ **27.** $-\frac{10}{t}$ **29.** ≈ 738 francs
31. 13; Does $8 = 13 - 5$? Yes **33.** $\frac{15}{4}$; Does $13 - 15 + \frac{15}{4} =$
$2\frac{1}{2} + \frac{-3}{4}$? Yes

LESSON 5-3 (pp. 221–225)
1. Yes **5.** $\approx 7\%$ **9.** b **11.** $\frac{w}{h} > 1$ **13.** ≈ 148 **15. a.** $\frac{.64}{16} = \frac{1}{25}$

b. 4% **17.** $\frac{6}{4} = \frac{3}{2}$ **19.** $\frac{y}{x}$ **21.** $-\frac{16}{21}$ **23.** .12 **25. a.** 13.5 ft²
b. $\frac{1}{3}$ m² **c.** $6x^2$ **27. a.** \$65,000 **b.** \$67,500 **29.** 11

LESSON 5-4 (pp. 226–230)
5. 1100 **7.** $\approx 17\%$ **9. a.** $.124x = 70,000$ **b.** $x = 564,516$
11. 33% **13.** 33% **15.** 25% **17.** \$88,000 **19.** $\approx 166\%$
21. $66\frac{2}{3}\%$ **23.** \$8.50 **25.** -6 **27.** 130.5 **29.** $y = \frac{14x}{3}$
31. $c > \frac{2}{5}$

LESSON 5-5 (pp. 231–235)
1. a. $\approx .08$ **b.** $\approx 8\%$ **3.** .5 **5.** $\frac{13}{36}$ **7. a.** 100 **b.** 100 −
$25\pi \approx 21.5$ **c.** $\frac{\pi}{4} \approx 78.5\%$ **d.** 100%; $p + q$ is the area of
the whole region **9. a.** $\approx .28$ **b.** $\approx .21$ **11.** $\frac{ab}{pq}$
13. $\approx 26.3\%$ **15.** $\frac{98}{135}$ **17.** ad **19.** more than $\frac{85}{6}$ (≈ 14.2) cm
21. a. 3,628,800 **b.** 132 **c.** 56 **23. See below.**
23.

LESSON 5-6 (pp. 236–241)
5. 6.4 in. by 9.6 in. **9.** $(8x, 8y)$ **11. a.** $J' = (-3, 0)$; $K' =$
$(-3, 6)$, $L' = (0, 10.5)$; $M' = (9, 6)$, $N' = (9, 0)$ **b. See
below. 13.** \$7.80/hr **15.** 150 **17. a.** The image is the same
as the original figure. **b.** Multiplicative Identity Property of 1
19. $\approx 14\%$ **21. a.** 70% **b.** $c = .7(23.95)$ **c.** \$16.77
23. $18 \leq x$ **25.** $y = \frac{53}{x}$ **27. a.** $\frac{5}{8} + x = 80$ **b.** $79\frac{3}{8}$ **29.** ≈ 100
31. ≈ 8 **33. a.** positive **b.** negative **35.** 144

11. b.

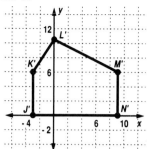

LESSON 5-7 (pp. 242–246)

1. True **3.** Yes **5.** No **7.** ≈328.7 miles **9. a.** $21 = 5n$

b. $15 = 7n$ **c.** $21 = 5n$ **d.** b **11. a.** $11x = 21$ **b.** $x = \frac{21}{11}$

13. $\frac{13}{40}$ or .325 of a picture **15.** 2.25 inches **17.** $b = \frac{2}{7}$

19. equal **21. a.** (5, -10) **b.** expansion **23.** $\frac{4x - 3y}{4x}$ or $1 - \frac{3y}{4x}$

25. a. $\frac{3,120,348}{5,232,113}$ **b.** ≈59% **27.** -3

LESSON 5-8 (pp. 247–252)

3. a. $\overline{BG}$ **b.** $\overline{IG}$ **5. a.** $\frac{2}{0.8} = \frac{3}{x}$ or (equivalent) **b.** $x = 1.2$

c. .4 or 2.5 **7.** sample: $\frac{p}{t} = \frac{a}{s} = \frac{r}{e} = \frac{k}{w}$ **9. a. See below.**

b. $\frac{t}{6} = \frac{25}{10}$ **c.** 15 ft **11. a.** $\frac{1.5 \text{ in.}}{30 \text{ ft.}}$ **b.** $\frac{1.5}{30} = \frac{1}{x}$ **c.** 20 ft

13. $A = \frac{1}{4}$ **15.** $C = \frac{15}{2}$ **17.** 90 students per year **19.** $\frac{1}{38}$

21. ≈ 19% **23.** $\frac{15}{y^2}$ **25.** 42 **27.** $a \le 15$ **29.** $b \le 75$

9. a.

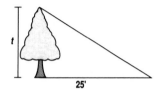

CHAPTER 5 PROGRESS SELF-TEST (pp. 254–255)

1. $15 \cdot \frac{-2}{3} = -10$ **2.** $\frac{x}{9} \cdot \frac{3}{2} = \frac{x}{6}$ **3.** $\frac{2b}{3} \cdot \frac{3}{b} = 2$ **4.** $\frac{4}{7} \cdot \frac{m}{3} = \frac{4m}{21}$

5. $23y = 22;\ y = \frac{22}{23}$ **6.** $60 = 15y,\ y = 4$ **7.** $.14 \cdot b = 60$,

$b \approx 428.6$ **8.** $.25n = 4.5,\ n = 18$ **9.** $\frac{1}{2} = x \cdot \frac{4}{5};\ x = 62.5\%$

10. $\frac{1}{8}$ **11.** 3 and x **12.** $v + 1 = 0;\ v = -1;\ v$ cannot be -1.

13. False **14. a.** $\frac{36}{30} = 1.2 = 120\%$ **b.** 20% **15.** $n \cdot c = d$,

$n = \frac{d}{c}$ **16.** p pages in $7y$ minutes **17.** $40p = 16,\ p = .4 =$

40% **18.** $\frac{1}{12}$ **19.** $\frac{35 \text{ days}}{3 \text{ chapters}} \cdot 13$ chapters ≈ 152 days

20. $\frac{250 \text{ miles}}{12 \text{ gallons}} \cdot 14$ gallons ≈ 292 miles **21.** $B = 5$

22. a. See below. b. See below. 23. $\frac{9}{12} = \frac{4}{z},\ 9z = 48$,

$z = \frac{48}{9} = 5\frac{1}{3}$ **24.** $\frac{2x}{7} = \frac{5}{10},\ 20x = 35,\ x = \frac{35}{20} = \frac{7}{4} = 1\frac{3}{4}$

25. $25 \cdot \frac{20}{27} \cdot 39 = \frac{500}{1053} \approx .47$

22. a.

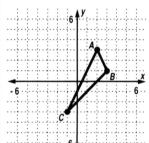

22. b.

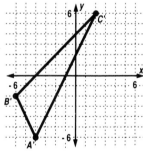

The chart below keys the **Progress Self-Test** questions to the objectives in the **Chapter Review** on pages 256–259 or to the **Vocabulary** (Voc.) on page 253. This will enable you to locate those **Chapter Review** questions that correspond to questions you missed on the **Progress Self-Test.** The lesson where the material is covered is also indicated in the chart.

Question	1–4	5, 6	7–9	10	11	12	13	14	15
Objective	A	C	B	A	E	D	E	G	F
Lesson	5-1	5-7	5-4	5-1	5-7	5-2	5-7	5-3	5-2

Question	16	17	18	19, 20	21	22	23, 24	25
Objective	H	I	J	F	L	M	K	J
Lesson	5-4	5-7	5-5	5-2	5-2	5-6	5-8	5-5

CHAPTER 5 REVIEW (pp. 256–259)

1. 125 **3.** $\frac{6}{7}$ **5.** $\frac{4}{3}$ **7.** $\frac{ad}{cb}$ **9.** 2 **11.** 24 **13.** 3.6 **15.** 33.33%

17. $\frac{5}{2}$ **19.** 156 **21.** $-\frac{15}{2}$ **23.** $-\frac{0.44}{2.3} \approx -0.19$ **25.** 0 **27.** .2

29. a. 8 and 15 **b.** 5 and 24 **31.** $\frac{3}{2}$ **33.** Sample: What was the Johnson's average speed? 40 miles per hour **35.** Sample:

What was the puppy's average weight loss? $-\frac{3}{4}$ kg per week **37.** $\frac{1}{11}$ hours per mile **39.** $\frac{1}{4}$ hr (15 min) **41.** $6w$ words in $2m$ min **43.** $\frac{17}{10}$ **45.** 40% **47. a.** $133\frac{1}{3}$% **b.** $33\frac{1}{3}$% **c.** 25% **49.** 44,590 **51.** 25% **53.** 2610 cm **55.** 48 **57.** $\approx$ \$59.11 **59.** $\frac{150}{360} = \frac{5}{12}$ **61.** $\frac{1}{2}$ **63.** 12.5 **65. a.** 5.5 **b.** 30 **67.** $\frac{27}{n}$ **69.** See below. **71.** (2, 3) **73.** See below.

69. 10 PRINT "EVALUATE (K + 1)/(K − 2)"
20 PRINT "ENTER VALUE OF K"
30 INPUT K
40 IF K = 2 THEN PRINT "IMPOSSIBLE"
50 IF K<>2 THEN PRINT (K + 1)/(K − 2)
60 END

73.

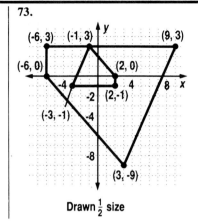

Drawn $\frac{1}{2}$ size

23. $z < 77$ **25. a.** January, February, March, April, May, June, October, November **b.** July and August **c.** zero inches **d.** June and July **27. a.** 3 **b.** 9 **c.** 21 **d.** 16

LESSON 6-1 (pp. 262–266)
1. a. If B equals the weight of each box, then $2B + 4 = 12$. **b.** One: remove 4 kg from each side of the balance. Two: remove half the contents of each side of the balance. **3. a.** Add 61 to each side. **b.** Divide each side by 55. **5.** $y = 9$; Does $7 \cdot 9 − 11 = 52$? Does $63 − 11 = 52$? Yes **7.** $z = 128$; Does $2 \cdot 128 + 32 = 288$? Does $256 + 32 = 288$? Yes **9.** $w = 7$; Does $312 = 36 \cdot 7 + 60$? Does $312 = 252 + 60$? Yes **11.** $t = 25$; Does $7 + 8 \cdot 25 = 207$? Does $7 + 200 = 207$? Yes **13.** b **15.** $x = \frac{-15}{7}$; Does $7 \cdot \frac{-15}{7} + 11 = -4$? Does $-15 + 11 = -4$? Yes **17.** $266\frac{2}{3}$ days **19.** $11\frac{2}{3}$ in. **21. a.** $3x + 16$ **b.** $x = 10$ **23.** -312 **25.** -85 **27. a.** $x = -15$ **b.** $y = 15$ **c.** $z = 45$ **29.** 11.25 feet

LESSON 6-2 (pp. 267–270)
3. $x = 18$; Does $\frac{2}{3} \cdot 18 + 15 = 27$? Does $12 + 15 = 27$? Yes **5.** $C = 20°$ **7. a.** Add 3.5 and 5.6. **b.** $x = .45$; Does $3.5 + 2 \cdot .45 + 5.6 = 10$? Does $3.5 + .9 + 5.6 = 10$? Yes **9.** $n = -2.7$; Does $6 − 30 \cdot -2.7 − 18 = 69$? Does $6 + 81 − 18 = 69$? Yes **11.** $x = 28$; Does $-4 \cdot 28 + 12 = -100$? Does $-112 + 12 = -100$? Yes **13.** $f \approx 42.5$ cm **15.** $x = \frac{c − b}{a}$ **17.** $x = 0$ **19.** $y = -37$ **21. a.** \$35,000 **b.** $x \approx 51.6$ or 52 concerts **23.** 2.0625 in.²

LESSON 6-3 (pp. 271–275)
1. $h = 20.5$ **3.** $g = 8400$ **5.** $h + h + \frac{1}{4}h = 90,000$ Two heirs receive \$40,000 each; third receives \$10,000. **7.** \$411.76 **9.** \$11.00 **11.** .7F **13. a.** $s + .06s = 2650$ **b.** \$2500 **15.** ≈ 52% **17.** $x = \frac{17}{5}$ **19.** $y = 4.5$ **21.** $c = -2.81$ **23.** $C = 0$ **25.** 4.98×10^{-23} **27.** $(x + 2, y − 4)$

LESSON 6-4 (pp. 276–281)
1. a. t is the number of hours the car is parked. **b.** The \$0.80 is the charge for each additional hour beyond the first. **c.** \$40.70 **3.** 43 hr **5.** 82nd **7.** 9th **9.** 517 **11. a.** $1000 − 3(n − 1)$ **b.** 835 **c.** 201st **13. a.** i **b.** ii **c.** iii **d.** They are discrete points which lie on a line. **15.** $\frac{1}{5} + \frac{1}{6}p$ **17.** $x = 2$ **19.** $m\angle A = 30$, $m\angle B = 60$, $m\angle C = 90$ **21.** $x = \frac{3}{5}$

LESSON 6-5 (pp. 282–283)
1. $a = -1$ **3.** $c = \frac{2}{3}$ **5.** $e = -\frac{3}{7}$ **7.** $g = 3$ **9.** $i = 11$ **11.** $k = -2$ **13.** $m = \frac{1}{36}$ **15.** $p = 100$ **17.** $r = 1200$ **19.** \$680.00 **21.** 56 persons **23.** \$1.75 **25.** 28 cards **27.** $3 + 8(n − 1) = 331$; 42nd term **29.** $1000 − 6n = 100$; 150 days

LESSON 6-6 (pp. 284–289)
1. a. Let B = weight of 1 box, then $5B + 6 = 3B + 10$. **b.** Remove 3 boxes from both sides, remove 6 kg from both sides, and divide the contents of both sides in half. **3.** $y = 10$ **5.** $w = 2$ **7.** $z = \frac{1}{3}$ **9.** $a = 6$ **11.** -70 **13. a.** 13 **b.** 91 **15. a.** $54.73 − .33x$ **b.** $49.36 − .18x$ **c.** ≈ 36 yr **17. a.** 16 **b.** 20 **c.** -160 **19.** samples: yxz, xzy, zyx, zxy, yzx **21.** $1.05c = 10$ **23. a.** $\frac{2}{3}A = 9 \cdot 12$ **b.** $A = 162$ square feet **25.** $z = \frac{6}{7}$

LESSON 6-7 (pp. 290–295)
1. $ax + b = cx + d$ **3.** $x < 5$; See below. **5.** $-1 \le b$; See below. **7.** 59 **9.** 8 **11.** packages weighing more than 31.25 ounces **13.** $x = -3$ **15.** $m = 6$ **17.** $-(a − b) = b − a$ **19.** 7800 **21.** 9 **23.** 25% **25.** 9 **27.** $\frac{1}{3}$ **29.** 7.25×10^{11} **31.** $\frac{mn}{48}$

3.   **5.**

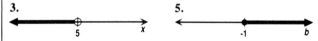

LESSON 6-8 (pp. 296–300)
3. $12k + 60$ **5.** $21a − 35ab$ **7.** $2xy + 3x$ **9.** $5(7.96) = 5(8.00) − 5(.04) = 40.00 − .20 = \$39.80$ **11.** \$21.00 **13.** $m = 73$; Does $4(73 + 7) = 320$? Does $4(80) = 320$? Yes **15.** $x = 1.8$; Does $2(1.8 + 3.1) = 9.8$? Does $2 \cdot 4.9 = 9.8$? Yes **17.** 5,999,994 **19. a.** $\frac{1}{2}x = \frac{11}{15} + -\frac{2}{3}$; $\frac{1}{2}x = \frac{1}{15}$; $x = \frac{2}{15}$ **b.** $15x + 20 = 22$; $x = \frac{2}{15}$ **21.** $b_2 = 15$ cm **23.** $n \le -\frac{1}{8}$ **25.** $\frac{9}{4}$ **27.** $x < -\frac{1}{2}$ **29.** $4b$ **31.** $\frac{1}{3}$ lb **33.** 2,657,638

LESSON 6-9 (pp. 301–305)

3. a 5. $-4n + 3m$ **7.** $-2y - 5$ **9.** -23 **11.** $D - (L + 10)$, $D - L - 10$ **13. a.** $16 - 2y$ **b.** $y = -1$ **15.** $x = -\frac{13}{3}$ **17.** $x =$ $\frac{15}{2}$ **19.** $96 - (R + S)$ or $96 - R - S$ **21.** No **23.** c **25.** $x = -1$ **27.** $z = 10.4$ **29.** \$100 **31. a.** 720 **b.** 120 **c.** 857,280 **33. a.** \$80 **b.** \$6

PROGRESS SELF-TEST (p. 307)

1. $.77m$ **2.** $4k + 9$ **3.** $10v + 250 - \frac{5}{2}w$ **4.** $14t - 18$ **5.** $r =$ 7.5 **6.** $q = -3$ **7.** $w = -36$ **8.** $x = 2$ **9.** $m = 2$ **10.** $x \le -6$ **11.** $v < -1$ **12.** $h = 0$ **13.** $5x + 3 = 12$ **14.** $6(3 - .01) =$ $18 - .06 = 17.94$ **15.** \$550 **16.** \$18.90 **17.** term $= -10 + 12n$; $470 = -10 + 12n$; $n = 40$; 40th term; **See below.** **18.** term $= 1100 - 6n$; $350 = 1100 - 6n$; $n = 125$ min; **See below.** **19.** $C = 10°$ **20.** $137.25 + 2.50w$ **21.** Let $x =$ amount Jill will receive; $x + 2x = 19.50$; $x = \$6.50$ **22.** Let $n =$ number of years; $25,000 + 1200n = 37,000 - 200n$; $n = 12$ years **23.** The picture shows that the area can be calculated in two ways: $c(a + b)$ or $ca + cb$ **24.** 62.6 years **25.** $y - (c + 15)$; $y - c - 15$

17.
term no.	term
0	-10
1	$2 = -10 + 12 \cdot 1$
2	$14 = -10 + 12 \cdot 2$
3	$26 = -10 + 12 \cdot 3$
4	$38 = -10 + 12 \cdot 4$

18.
term no.	term
0	1100
1	$1100 - 6 \cdot 1$
2	$1100 - 6 \cdot 2$

The chart below keys the **Progress Self-Test** questions to the objectives in the **Chapter Review** on pages 308–311 or to the **Vocabulary** (Voc.) on page 306. This will enable you to locate those **Chapter Review** questions that correspond to questions you missed on the **Progress Self-Test.** The lesson where the material is covered is also indicated in the chart.

Question	1	2, 3	4	5	6	7	8	9	10, 11
Objective	A	A	A	B	B	B	B	B	C
Lesson	6-3	6-8	6-9	6-5	6-1	6-6	6-8	6-9	6-7

Question	12	13	14	15, 16	17	18	19	20	21, 22
Objective	B	D	E	F	G	G	H	I	I
Lesson	6-2	6-1	6-3	6-3	6-4	6-6	6-2	6-3	6-6

Question	23	24	25
Objective	J	H	J
Lesson	6-8	6-2	6-8

CHAPTER 6 REVIEW (pp. 308–311)

1. $\frac{3}{2}c$ **3.** $.92m$ **5.** $3a + 6b$ **7.** $17x + 10$ **9.** $2 - z$ **11.** $n = 3$ **13.** $x = 68$ **15.** $m = -5$ **17.** $A = 2$ **19.** $w = 142$ **21.** $a = \frac{3}{2}$ **23.** $g = 4$ **25.** $x < 95$ **27.** $-\frac{1}{9} < g$ **29.** $y > -13$ **31.** $y \ge -1$ **33.** Distributive Property **35.** Addition Property of Inequality **37.** $2 < a + 4$ **39.** $B - 6B + 4B + 8 \ge B - 5$ **41.** $7(3 + 0.4) = \$21.28$ **43.** $9(20.00 - 0.01) = \$179.91$ **45.** \$55.50 **47.** \$9.00 **49.** 50 weeks **51.** 20 months **53. a.** 5507 **b.** 35th term **55.** $12\frac{1}{3}$ in. **57.** ≈ 38.2 cm **59.** up to 5 ounces **61.** more than 41.3 hr **63. a.** $3B + 2 = 8$ **b.** 2 kg **65.** $1.5(.5 + 2.5)$; $1.5(.5) + 1.5(2.5)$

LESSON 7-1 (pp. 314–318)

5. 2 in. **7. a.** 0 **b.** Yes **9.** 40 LET Y $= 8 * X - 3$ **11.** $\frac{2}{3}$ **13.** b **15. a.–b.** See below. **c.** (0, 0) **17.** $3.35h - .14y$ **19.** $x = 62$ **21.** $-\frac{9}{2}$ **23.** 60 **25.** $3.326 \cdot 10^6$ **27. a.** .25 **b.** 2.5 **c.** 25 **d.** 250,000

15. a.–b.

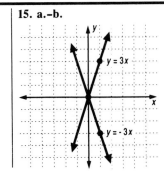

LESSON 7-2 (pp. 319–323)

3. True **5. a.–b.** See below. **7. a.–b.** See below.
9. $x = 1$ **11.** $x \geq$ -3 **13. a.** See below. **b.** See below.
15. $x = $ -6 **17.** See below. **19. a.** $y = 0$ **b.** $x = 0$
21. See below. **23.** $t = 25 + 2w$ **25.** $16x^2 + 45x - 33$
27. $25 \cdot 9.8 + 25 \cdot 14.2$; $25(9.8 + 14.2)$ **29.** $w \approx 13.48$
31. $x = .75$

5. a.

7. a.

7. b.

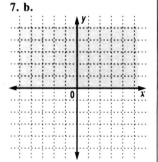

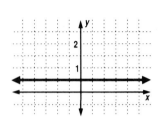

13. a.

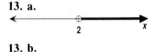

13. b.

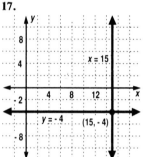

17.

21. 10 PRINT "TABLE OF (X, Y) VALUES"
20 PRINT "X VALUES", "Y VALUES"
30 FOR X = 0 TO 5
40 LET Y = -0.63 + 19 * X
50 PRINT X, Y
60 NEXT X
70 END

LESSON 7-3 (pp. 324–330)

1. a. 49 **b.** 25 **c.** $|487 - G|$ **3.** 72 **5.** 0 **7.** 1 **9.** 39
11. $|17.5 - x|$ **13.** THE DISTANCE IS 38 **15.** 8 **17.** $x =$
290 or $x = $ 310; See below. **19. a.** 4 **b.** -6 **21. a.** a
23. 36 **25. a.** $m + 4$ **b.** $m - 4$ **c.** 4 **d.** 8 **27.** See below.
29. See below. **31.** $108x^2$ sq cm

17.

27.

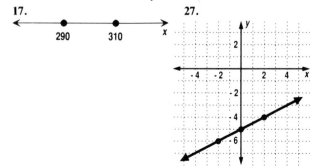

29.

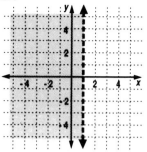

LESSON 7-4 (pp. 331–336)

1. square root **3.** square **5. a.** 10 **b.** -9 **c.** 144 **7.** 3 and 4
11. $\sqrt{301}$ or $- \sqrt{301}$, $\approx$ 17.35, or $\approx$ -17.35 **13.** 400
15. no real solutions
17. TYPE IN A NUMBER
CALCULATE SQUARE ROOTS
NO REAL SQUARE ROOTS
19. 1.732 | 3.464
2 | 3.606
2.236 | 3.742
2.449 | 3.873
2.646 | 4
2.828 | 4.123
3 | 4.243
3.162 | 4.359
3.317 | 4.472
21. 0.6561 **23.** 90 **25.** $x = 7$ or $x = $ -1 **27. a.** $x = 0$
b. y-axis **29.** -7 **31.** -16 **33.** $\approx$ .215

LESSON 7-5 (pp. 337–341)

5. a. 29 **b.** 41 **7.** 25 **9.** $\sqrt{119}$ ft $\approx$ 10.9 ft **11.** 7
13. a. 9 cm **b.** $\sqrt{162} \approx$ 12.73 cm **15.** $2\frac{1}{2}$ seconds **17.** >
19. 177 **21. a.** 11 or -11 **b.** 14,641 **c.** 121 or -121
23. a. $y = -\frac{7}{2}x + 7$ **b.** See below. **25.** w words in
m minutes **27.** b **29.** 300

23. b.

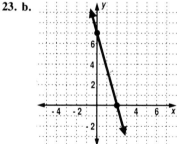

LESSON 7-6 (pp. 342–346)

3. a. 2.65 **b.** 2.24 **c.** 5.92 **d.** 5.92 **5. a.** $\sqrt{25} \cdot \sqrt{2}$
b. $5\sqrt{2}$ **7.** $\sqrt{128} \approx$ 11.314, $8\sqrt{2} = $ 11.314 **9.** $10\sqrt{2}$
11. $3\sqrt{3}$ **13.** $a = \sqrt{6}$ or $-\sqrt{6}$ Does $(2 \cdot \pm\sqrt{6})^2 = $ 24?
Does $4 \cdot 6 = $ 24? Yes **15. a.** $5\sqrt{3}$ **b.** $2\sqrt{3}$

c. $7\sqrt{3}$ **17.** $2w\sqrt{5}$ **19.** $\frac{10\sqrt{3}}{5} = 2\sqrt{3}$ **21.** 540 **23. a. See below. b. See below. 25. a.** {1, 2, 3, 4, 5, 6, 7, 8, 10, 12} **b.** $\varnothing$ **c.** {1, 2, 3, 4, 5, 6, 7, 8} **27. a.** 12.25 **b.** 12.25 **c.** 12.25 **29.** $.25q + .10d \geq 5.20$ **31.** 3

23. a.

23. b.

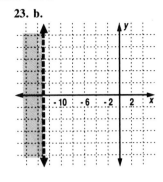

LESSON 7-7 (pp. 347–352)
1. a. 11 **b.** $\sqrt{73} \approx 8.5$ **3. a.** 8 **5.** 5 **c.** ≈ 9.4 **5.** $\sqrt{173} \approx$ 13.15 **7.** 7 miles **9.** $\sqrt{5} \approx 2.2$ miles **11.** 1000 m **13.** 29.732137 **15. a.** 20 **b.** 100 **c.** 195 **d.** 10 **17.** $\sqrt{180}$ or $6\sqrt{5}$ **19.** $2(2x + 1 - x) = (4x + 6) - 20; 8$ **21.** $\frac{38}{3}$ or

$12\frac{2}{3}$ **23.** $a = 3$ **25.** $m = 0$ **27. a.** $m = wa$ **b.** $t = \frac{s}{z}$ **c.** $k = b(x + 1)$

LESSON 7-8 (pp. 353–357)
3. 46 **5.** 12 or -18 **7.** 7, -8 **9.** $\frac{x - 8}{x + 8}$ **11.** $12\frac{1}{2}$ **13.** 2, -2 **15.** -3.45, -18.55 **17.** $7\sqrt{5}$ **19. a.** 36 **b.** 18 **c.** 90 **21.** $\frac{12}{5}$ **23. a.** 10 **b.** 15 **25.** 3.8 **27. See below. 29.** $T = sd$ **31. a. See below. b.** preimage: $4x + 7y + 10$, image: $32x + 56y + 80$

27.

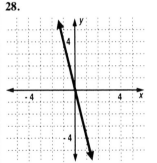

31. a.

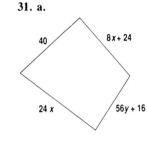

CHAPTER 7 PROGRESS SELF-TEST (p. 359)

1. 30 **2.** 72 **3.** 27 **4.** $|13x|$ or $13|x|$ **5.** 14.49 **6.** $\sqrt{29}$ cm **7.** 9 **8.** $8ab$ **9.** $3\sqrt{5}$ **10.** $2\sqrt{11}$ **11.** 14.5 **12.** $n = 24$ or -24 **13.** no solutions **14.** $y = 16$ **15.** $y = 4$ or -4 **16.** $x = 7\sqrt{2}$ or $-7\sqrt{2}$ **17.** 10 or -1 **18.** $x < -4$ or $x > 10$ **19.** $-2 \leq y \leq 6$ **20.** 17 ft **21.** $\sqrt{609} \approx 24.7$ **22.** $y = 6$ **23.** 800 m **24.** $200\sqrt{10}$ m ≈ 632 m **25.** $AB = \sqrt{145}$ **26.** (-2, 8), (-1, 5), (0, 2), (1, -1), (2, -4) **27. See below. 28. See below. 29. See below.**

27.

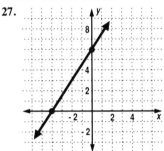

28.

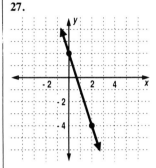

29.

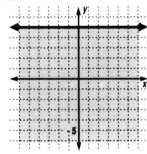

The chart below keys the **Progress Self-Test** questions to the objectives in the **Chapter Review** on pages 360–363 or to the **Vocabulary** (Voc.) on page 358. This will enable you to locate those **Chapter Review** questions that correspond to questions you missed on the **Progress Self-Test.** The lesson where the material is covered is also indicated in the chart.

Question	1, 2	3, 4	5	6	7	8–10	11	12–16	17–19
Objective	A	G	B	F	B	E	D	C	D
Lesson	7-3	7-3	7-4	7-4	7-4	7-6	7-8	7-4	7-8

Question	20	21	22	23–25	26–28	29
Objective	F	J	I	K	H	L
Lesson	7-6	7-5	7-2	7-7	7-1	7-3

1. 43 **3.** 6 **5.** -5 **7.** 10 **9.** d = 16 or -16 **11.** 277 or 323
13. 4 and 5 **15.** -10 **17.** 33 **19.** $x = \pm 2\sqrt{10}$ **21.** $q =$
$\pm 4\sqrt{5}$ **23.** no solution **25.** 25 **27.** 67.6 **29.** 16 or -8
31. $39\frac{2}{3}$ **33.** no solution **35.** $-1 < x < 7$ **37.** $10\sqrt{5}$

39. $\frac{5\sqrt{6}}{5} = \sqrt{6}$ **41.** -6 **43.** $x\sqrt{5}$ **45.** $7y\sqrt{3}$ if $y > 0$

47. $\approx$ 316 ft **49.** $\approx$ 16.6 m + Megan's height **51.** 24
53. $|j - d|$ or $|d - j|$ **55. a. See below. b.** -9 **57. a.** $t = 15 +$
$2w$ **b. See below. 59.** True **61. See below. 63.** $y \le -2$ **65.** 37
67. $2\sqrt{21} \approx 9.2$ **69. a.** 8 **b.** 11 **c.** $\sqrt{185} \approx 13.6$ **71.** 6
73. $\sqrt{241}$ **75. a.** (-2, -7); (8, 13) **b.** $10\sqrt{5}$ **77.** 14 miles
79. See below. 81. See below 83. See below.

55. a.

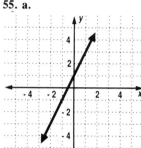

57. b.

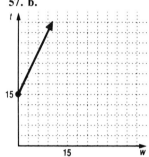

61.

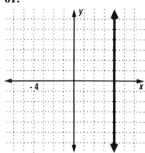

79.

81.

83.

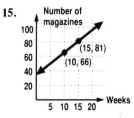

---

3. a. .75 inches per year **b.** from age 9 to 11 **5.** increase
7. -11,094.8 people per year **11. a. See below. b.** from age
10 to 12 **c.** 4 inches per year **13.** 0 meters per second

15. positive **17.** $\frac{4'3'' - h}{2 \text{ yrs}}$ **19. a.–b. See below.**

c. (0, 2) **21.** $-\frac{1}{2}$ **23. a.** 5.25 **b.** $\frac{3}{4}x - 3$ **c.** $\frac{a}{4}x - 3$ **25. a.** $\frac{1}{a}$

b. $\frac{1}{3 + 4x}$ **c.** $x = -\frac{4}{5}$

11. a.

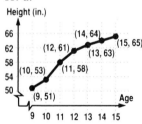

19. a.–b.

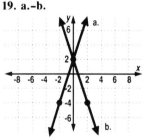

3. a. See below. b. $\frac{1}{3}$ ft per minute **5.** sample: using (1, 2)
and (-1, 0), $\frac{0 - 2}{-1 - 1} = 1$; using (0, 1) and (-1, 0), $\frac{0 - 1}{-1 - 0} =$
1 **7. a.** sample (9, 6) and (6, 8) **b.** $-\frac{2}{3}$ **9.** 8 **11.** $5\frac{1}{5}$ or 5.2
13. C; D **15. a.** 0 **b.** 0 **c.** The slope is 0. **17. a.** r **b.** q
c. p **19. a.** 1980–85 **b.** 1960–65 and 1965–70 **c.** .5 cent per
year **d.** The cost of stamps does not gradually rise over each
5-year period. For instance, the cost was never 4.3 cents.

21. See below. 23. $100 - 4x$ **25. a.** $\frac{5}{3}$ **b.** $-\frac{1}{3}$ **c.** $\sqrt{15}$,
$-\sqrt{15}$ **27. a.** 60 **b.** $5b$, if $b \ge 0$ **c.** $3\sqrt{2}$

3. a.

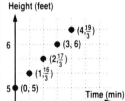

21.

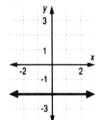

1. -0.5 **5. See below. 7.** $c = .25 + .20(9 - 1)$; $c = 1.85$
9. 1 **11.** sample: (-4, 9) **13.** sample: $\left(1, \frac{5}{4}\right)$ or (4, 5)
15. See below. 17. $-\frac{5}{3}$ **19.** 3 **21. a.** 1982–1983
b. 1983–1984 **23.** -10 **25.** 7 or -9 **27. a.** $2T - 8$
b. $\frac{2T - 8}{T}$ **c.** 12

5.

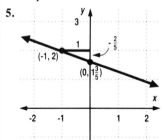

15.

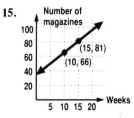

LESSON 8-4 (pp. 381–386)

3. a. 4 **b.** 2 **5. a.** ii **b.** 8 **c.** 100 **7. a.** iii **b.** 4 **c.** -350
11. $y = \frac{2}{3}x + -1$ or $y = \frac{2}{3}x - 1$ **13. a.** $y = -\frac{1}{6}x + \frac{7}{6}$ **b.** $-\frac{1}{6}$
15. a. 1 **b.** 0 **17. See below. 19. a.** u **b.** v **c.** t **d.** s
21. a. $y = 9 - 6x$ **b. See below. 23.** $y = 3x - 7$ or $y = 3x + -7$ **25.** (b) **27.** $\sqrt{176} \approx 13.3$ feet **29.** $-\frac{11}{17}$ **31.** 2

17.
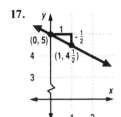

21. b.
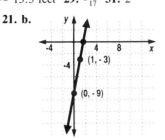

LESSON 8-5 (pp. 387–390)

3. $y = -2x - 17$ **5.** $y = -\frac{1}{2}$ **7.** $y = -3x + 110$
9. 10,143,00 **11. a.** $\frac{1}{2}$ **b.** $y = \frac{1}{2}x + 2$ **13. a.** q **b.** n **c.** p
d. m **15.** No **17. a.** A **b.** C **c.** All sections show some change. **19. See below. 21. a.** $2\sqrt{6}$ or $-2\sqrt{6}$ **b.** 576
c. $n = 4$ **23. a.** $\frac{3}{2}$ **b.** $\frac{3}{2}$ **c.** $\frac{c}{b}$

19.
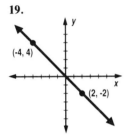

LESSON 8-6 (pp. 391–397)

1. $y = -x + 10$ **3.** $y = \frac{-11}{13}x + 11$ **5.** \$8.88 **9.** about 132
11. $S = 180n - 360$ **13. a.** $C = \frac{5}{9}F - \frac{160}{9}$ **b.** $\approx 65.6°C$
c. 302°F **15. See below. 17. a.** $\frac{15}{2}$ **b.** $\frac{2}{5}$ or $\frac{5}{2}$ **19. a.** 93
b. any score greater than or equal to 88 **c.** 88 **21. a.** $\frac{1}{12}$ **b.** $\frac{1}{3}$
c. $\frac{1}{d}$

15.
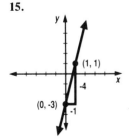

LESSON 8-7 (pp. 398–403)

9. b, c **13.** $y = \frac{3}{5}x + \frac{1}{5}$ **15. a. See below. b. See below.**
17. about 91 **19.** ≈ 74 miles **21. a.** 6 **b.** 6 **c.** -6

15. a.

15. b.
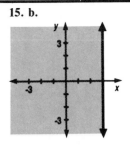

LESSON 8-8 (pp. 404–409)

7. $A = 1$, $B = -8$, $C = 2$ **9.** $-4x + y = 0$; $A = -4$, $B = 1$, $C = 0$ or $4x - y = 0$; $A = 4$, $B = -1$, $C = 0$ **11.** (0, 6) and (10, 0); **See below. 13.** sample: (4, 6), (6, 5), (8, 4)
15. $2x + 4y = 100$ **17.** $y + 2x = 4$ **19. a. See below.**
b. samples: Using (1900, 35) and (1960, 60), we get $y = .42x - 763$. **c.** ≈ 72 meters **d.** sample: The line is an estimate. **21. See below. 23.** Yes **25.** $\frac{1}{12}$ **27. a.** 2 **b.** 10
c. See below.

11.
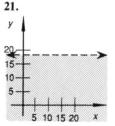

19. a.
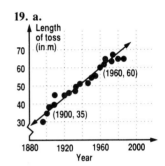

21.
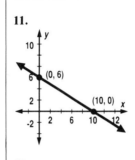

27. c.
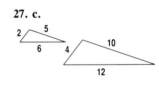

LESSON 8-9 (pp. 410–415)

7. See below. 9. See below. 11. a. $2W + T \geq 20$ **b. See below. 13. a. See below. b.** The graph is the strip of infinite length that lies between the two boundary lines.
15. $25c + 5f = 55,000$ **17.** ≈ 83.8 yd **19. a.** 0.1 **b.** .03
c. 0.00001 **21.** 3,603,266

7.
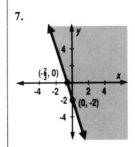

9.
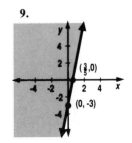

11. b.

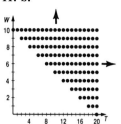

13. a.

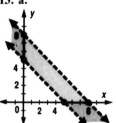

CHAPTER 8 PROGRESS SELF-TEST (p. 417)

1. $\frac{5 - 0}{0 - 2} = -\frac{5}{2}$ **2. a.** 5 **b.** 2 **3.** $\frac{1 - -5}{-2 - 4} = \frac{6}{-6} = -1$;

$\frac{-20 - 1}{20 - -2} = \frac{-21}{22}$ Since $-1 \neq \frac{-21}{22}$, the points do not lie on the same line. **4.** Rewrite as $y = -4x + 8$, so the slope is -4 and the y-intercept is 8. **5.** Rewrite as $y = \frac{-5}{2}x + \frac{1}{2}$, so the slope is $\frac{-5}{2}$ and the y-intercept $\frac{1}{2}$. **6.** Substitute $m = -2$ and $(x, y) = (-5, 6)$ into $y = mx + b$ and get $y = \frac{3}{4}x + 13$. **7.** Substitute $m = -2$ and $(x, y) = (5, 6)$ into $y = mx + b$ to get $y = -2x - 4$. **8.** Add $-5x$ to both sides; $-5x + y = -2$ with $A = -5, B = 1$, and $C = -2$. (You could also write $5x - y = 2$ with $A = 5, B = -1$, and $C = 2$.) **9.** Slope is 0. **10.** up $\frac{3}{5}$ unit

11. Do a quick estimate. Between 1942 and 1944 the increase was from about 3 million to about 8 million, or 5 million personnel. From 1943 to 1945 the increase was a little over 1 million. From 1944 to 1946 was a decrease. So the greatest 2-year increase was from 1942 to 1944.

12. $\frac{1,889,690 - 3,074,184}{1946 - 1942} \approx 296,123$ personnel per year

13. $2x + y = 67$ **14.** slope = 5, y-intercept = 100

15. First find the rate of increase of weight; $\frac{50 - 43}{14 - 12} = \frac{7}{2}$ kg per year. Next substitute $m = \frac{7}{2}$ and $(x, y) = (14, 50)$ into $y = mx + b$ and solve for b: $50 = \frac{7}{2} \cdot 14 + 6$, $50 = 49 + b$, $b = 1$. Substitute $m = \frac{7}{2}$ and $b = 1$ into $y = mx + b$ to get $y = \frac{7}{2}x + 1$. **16.** See below. **17.** First graph $y = x + 1$. It has y-intercept 1 and slope 1. Then test $(0, 0)$ in $y < x + 1$; $0 < 0 + 1$, $0 < 1$. So $(0, 0)$ is a solution and the re-

gion below the line is shaded. **See below.** **18.** about 115 feet
19. a. See below. b. Using (60, 70) and (120, 140) from the graph, $m \approx \frac{140 - 70}{120 - 60} = \frac{70}{60} = \frac{7}{6}$ **20.** $y = \frac{7}{6}x$

21. About 117 ft

16.

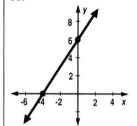

17.

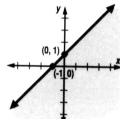

19. a.

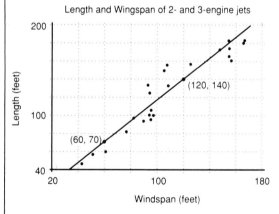

The chart below keys the **Progress Self-Test** questions to the objectives in the **Chapter Review** on pages 418–421 or to the **Vocabulary** (Voc.) on page 416. This will enable you to locate those **Chapter Review** questions that correspond to questions you missed on the **Progress Self-Test.** The lesson where the material is covered is also indicated in the chart.

Question	1–3	4, 5	6, 7	8	9	10	11, 12	13–15
Objective	A	F	B	E	D	D	G	H
Lesson	8-2	8-4	8-5	8-8	8-2	8-3	8-1	8-8

Question	16	17	18, 19	20	21
Objective	I	K	J	C	H
Lesson	8-4	8-9	8-7	8-6	8-8

1. $-\frac{1}{2}$ **3.** $\frac{3}{5.5} = .\overline{54}$ **5.** $y = 4x + 3$ **7.** $y = -2x - 7$ **9.** $y = 30x - \frac{359}{4}$ **11.** $y = \frac{1}{2}x - \frac{9}{2}$ **13.** height; right **15. a. See below.**
b. 0.46 **17.** $x - 5y = 22$; $A = 1$, $B = -5$, $C = 22$ or $-x + 5y = -22$; $A = -1$, $B = 5$, $C = -22$ **19.** $y = -2x + 4$
21. $m = 7$, $b = -3$ **23.** $m = -1$, $b = 0$ **25.** 4.3 kg per yr
27. -0.5° per hour **29.** $m = 0.25$, $b = 15$ **31.** $y = 3x + 50$ **33.** $w = .2d + 37.2$ **35.** Using (21, 1976), (20, 1972), $y = 4n + 1892$ **37. See below. 39. See below. 41. See below. 43. See below. 45.** Answers will vary. **a. See below. b.** Using (1960, 4.84) and (1980, 4.15), $m \approx -.035$. **c.** $y = -.035x + 73.45$ **d.** 3.73 minutes **47.** $.05x + .25y < 2.00$; **See below. 49. See below. 51. See below.**

15. a.

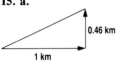

0.46 km
1 km

37.

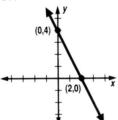

(0,4)
(2,0)

39.

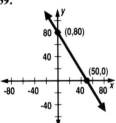

80 (0,80)
40
(50,0)
-80 -40 40 80
-40

41.

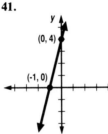

(0, 4)
(-1, 0)

43.

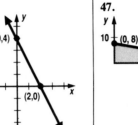

(0, 2.3)

45. a.

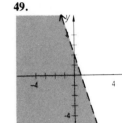

Women's 400-Meter Freestyle Olympic Winners

Time (minutes)
7
6
5
4
3
1920 1940 1960 1980
Year

47.

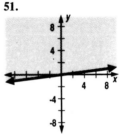

10 (0, 8)
(40, 0)
40

49.

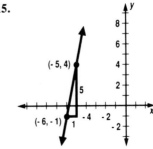

-4
4 X
-4

51.

8
4
4 8
-4
-8

LESSON 9-1 (pp. 424–428)
3. B **5.** $P(1.06)$ **7.** 573 $\boxed{\times}$ 1.063 $\boxed{y^x}$ 24 $\boxed{=}$ **9. a.** $123.88
b. $23.88 **11.** $232.18 **13. a.** Susan, $10.25; Jake, $21.00
b. No **15.** in about 9 years **17.** 79 **19. a.** $T = 7 + 2W$
b. See below. 21. $\approx .29$, ≈ -2.29

19. b.

T
10
(1, 9)
(0, 7)
5
5 W

LESSON 9-2 (pp. 429–433)
1. a. 14 **b.** 409,600 **3.** 1,476,225 **5.** (1) $1510
(2) $1,917,510.60 **7. a.** $2000 **b.** There is no time for the money to earn interest. **9.** 638 **11. a.** 20 minutes **b.** 6
c. 1,458,000 **d.** 1,062,882,000 **13.** 1605 **15.** $\approx$ $2.92

trillion **17.** 80,000 **19.** 343 **21.** $1\frac{4}{9}$ **23. a.** $x \cdot 1.07^2 = 915.92$ **b.** $800 **25.** $\frac{3}{26}$ **27.** $14n + 44$ **29.** $\frac{1}{128}$ **31. a.** $\frac{1}{2}$
b. $\frac{1}{3}$ **c.** $\frac{a}{c}$

LESSON 9-3 (pp. 434–437)
1. ≈ 48 **3.** 63,650 **5.** .937 **7.** $\approx$ 1.4 in. by 1.8 in.
9. a. 5832 **b.** $\approx .01$ **c.** 100,000, 1.02, and 10 **11. a.** 96
b. 24,576 **13.** n^x **15. See below. 17.** $k(36 - k)$
19. a. $5.88 **b.** $60.20 **c.** $6.00 **21.** 256 **23. a.** $\frac{7}{9}$ **b.** $\frac{11}{9}$
c. 27

15.

y
8
6
(- 5, 4)
4
2
5
-4 -2
(- 6, -1)
1
-2

LESSON 9-4 (pp. 438–443)

3. The graph of linear decrease is a line. The graph of exponential decay is a curve. **7.** ≈ 1300 **9. a.** 27 **b.** 20 **11.** a, c, d **13.** sample: **See below.** **15.** ≈ 2156 **17.** $k = 30(1.05)^n$ **19.** -13,824 **21. a.** 100 * 1.06 ^ YEAR
b. 20   320.71355 **c.** 30 FOR YEAR = 1 TO 100
d. 100   33930.208 **23.** $6a^3 + 12a^2 - 2a$ **25. a.** $180
b. $160 **c.** $134

13.

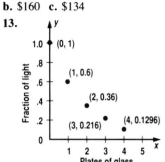

LESSON 9-5 (pp. 444–447)

1. a. $x^6 = x \cdot x \cdot x \cdot x \cdot x \cdot x$ **b.** x^6 is the growth factor after 6 years if there is growth by a factor x in each one-year interval. **3.** The expression has two different bases. **5. a.** $2000 \cdot 2^{11}$ **b.** $(2000 \cdot 2^{11}) \cdot 2^6$ or $2000 \cdot 2^{17}$ **9.** n^{12} **11.** x^{55} **13.** $(1.5)^{10}$ **15.** $2x^7$ **17.** $15m^6$ **19.** 12^{101} **21.** $a^5 + 4a^6$ **23. a.** -1; 1; -1; 1; -1; 1; -1; 1 **b.** 1 **25. See below.**
27. a. $7x + 7y = 26$ or $-7x - 7y = -26$ **b.** $y = -x + \frac{26}{7}$
29. 2.439×10^6 **31.** 8.97×10^{-1}

25.

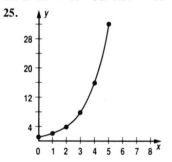

LESSON 9-6 (pp. 448–452)

1. $4^{-1} = \frac{1}{4}$, $4^{-2} = \frac{1}{16}$, $4^{-3} = \frac{1}{64}$ **3.** $\frac{1}{25}$ **5.** $\frac{1}{7}$ **7.** $\frac{x^5}{y^2}$ **9.** $\frac{1}{2^n}$
11. a. 0.000000001 **b.** $\frac{1}{1,000,000,000}$ **13. a.** $3434.58
b. $3209.89 **15.** -11 **17.** t^{-6} **19. a. See below. b. See below.** **21.** ≈ 4.60 **23.** a^7 **25.** $12c^7$ **27.** c **29.** $y = -5x + 9$
31. $\frac{3}{4}$ **33.** $\frac{4}{21}$

19. a.

x	y
2	9
1	3
0	1
-1	$\frac{1}{3}$
-2	$\frac{1}{9}$
-3	$\frac{1}{27}$

19. b.

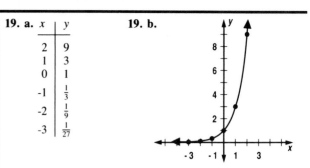

LESSON 9-7 (pp. 453–457)

1. $2^6 \cdot 2^8 = 2^{14}$ **3.** $\frac{9.29 \cdot 10^8}{2.27 \cdot 10^8} = \frac{9.29}{2.27} \approx 4 \frac{\text{five-dollar bills}}{\text{person}}$
5. $\frac{1}{36} = \frac{1}{729}$ **7.** $\frac{z^3}{42}$ **9.** bases are different **11.** $\approx .018 \frac{\text{pound}}{\text{person}}$
13. x^{9n} **15.** 1 **17.** $5p$ **19.** $\approx .32 \frac{\text{people}}{\text{sq km}}$ **21.** .015625
23. y^{12} **25.** x^{15}; try $x = 3$; $3^5 \cdot 3^5 \cdot 3^5 = 243 \cdot 243 \cdot 243 =$ 14,348,907; $3^{15} = 14,348,907$ **27.** $\frac{4}{9}$ **29. a.** $4x + 32$
b. $24x + 48$ **c.** 11 **31.** $x \geq -\frac{25}{4}$ **33.** $x \leq -6$

LESSON 9-8 (pp. 458–462)

1. a. $125x^3$ **b.** $(5 \cdot 2)^3 = 10^3 = 1000$; $125(2)^3 = 125 \cdot 8 = 1000$ **3. a.** 1953.125 cu ft **b.** 15,625 cu ft
5. a. $\approx 23,350,000,000$ km^3 **b.** $\approx 2.335 \cdot 10^{10}$ km^3
7. $\frac{343}{1000}$ **9.** $512y^3$ **11.** $\frac{k^2}{L}$ **13.** True **15.** a; x^{13} **17.** b; k^3
19. c; v^{-6} **21.** $\frac{a^{-3}}{b^{15}}$ or $\frac{1}{a^3 b^{15}}$ **23.** 1 **25.** $18k^7 q^{13}$
27. a. k^3 in.^{3} **b.** $125k^3$ in.^{3} **29.** $\frac{1}{5}$ or .2 **31. a.** $2x^5$ **b.** x^9
c. x^{10} **33.** $29.66 **35.** 414 sq units

LESSON 9-9 (pp. 463–468)

1. a. $\frac{8}{125}$ **b.** .064 **3.** d; sample: $\left(\frac{3}{4}\right)^2 = \frac{9}{16}$; $\frac{3^2}{4^2} = \frac{9}{16}$
5. a. Yes **b.** Yes **c.** Yes **d.** No **9.** No, plan (2) yields more. **11.** The pattern is false. Sample: let $x = 0$: $(2 \cdot 0)^3 = 0^3 = 0$, $2 \cdot 0^3 = 2 \cdot 0 = 0$, and the pattern holds. Let $x = 1$: $(2 \cdot 1)^3 = 2^3 = 8$, $2 \cdot 1^3 = 2 \cdot 1 = 2$, and the pattern does not hold. **13.** b; sample: $a = 25$;
$\frac{25}{b} = b$; $b^2 = 25$; $b = 5$; Does $\sqrt{25} = 5$? Yes **15.** $\frac{36}{625x^4}$
17. 9 **19.** $2^{((14)^{92})}$ **21.** The longest segment connects opposite corners of the card. If its length is d, then $d^2 = 3^2 + 5^2$, $d^2 = 9 + 25$, $d^2 = 34$, $d = \sqrt{34} \approx 5.8$ in. **23. a.-d.:** $y = -\frac{2}{3}x + \frac{4}{3}$ **25.** 30% **27.** $\frac{193}{225} \approx .86$

CHAPTER 9 PROGRESS SELF-TEST (p. 470)

1. $b^7 \cdot b^{11} = b^{7+11} = b^{18}$ **2.** $\frac{5x^2 y}{15x^{10}} = \frac{5y}{5 \cdot 3x^{10-2}} = \frac{y}{3x^8}$
3. $70 = 3 \cdot 1 + 6 \cdot 0.1 = 70 + 3 + 0.6 = 73.6$ **4.** $-\frac{1}{32.768}$
5. $\frac{3z^{6-4}}{3 \cdot 4} = \frac{z^2}{4} = 4^{-1}z^2 = .25z^2$ or $\frac{z^2}{4}$ **6.** $-5^4 = -5 \cdot 5 \cdot 5 \cdot 5 =$ -625 **7.** $\frac{3^2}{x^2} \cdot \frac{x^4}{3^4} = \frac{x^{4-2}}{3^{4-2}} = \frac{x^2}{3^2} = \frac{x^2}{9}$ **8.** $3^4 (y^2)^4 =$

$81y^{2 \cdot 4} = 81y^8$ **9.** $6(11)^0 = 6 \cdot 1 = 6$ **10.** $3^{-1} = \frac{1}{3} \neq -3$
11. a. Sample: Does $3^2 - 2^2 = (3 - 2)(3 + 2)$? $9 - 4 = 1 \cdot 5$? Yes; Does $(-4)^2 - (3)^2 = (-4 - 3)(-4 + 3)$? $16 - 9 = (-7)(-1)$? Yes; Does $(-10)^2 - (-9)^2 = (-10 + 9)$ $(-10 - 9)$? $100 - 81 = (-1)(-19)$? Yes **b.** True **12.** Product Property of a Power, of $b^j \cdot b^k = b^{j+k}$ **13.** $6500 \cdot

$(1 + .0571)^5 = \$6500(1.0571)^5 \approx \$8580.13$
14. $\$1900(1 + .058)^3 = \$1900(1.058)^3 \approx \$2250.15$,
$2250.15 - 1900 \approx \$350.15$ **15.** $P \cdot (1 + .03)^{15} =$
$P(1.03)^{15} \approx 1.56P$ **16.** $T - 100,000(1.03) \approx 94,260$ people
17. a and d **18. See below. 19.** $(1.30)^3 = 2.197$ times as
large **20.** $v = \frac{4}{3}(\pi)(433.10^5)^3 = \frac{4}{3}\pi(4.33)^3 \cdot (10^5)^3 =$
$\frac{4}{3}\pi(81.183) \cdot 10^{15} \approx 340.058 \times 10^{15} \approx 3.40058 \times$
10^{17} miles^3

18.

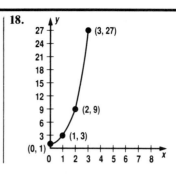

The chart below keys the **Progress Self-Test** questions to the objectives in the **Chapter Review** on pages 471–473 or to the **Vocabulary** (Voc.) on page 469. This will enable you to locate those **Chapter Review** questions that correspond to questions you missed on the **Progress Self-Test**. The lesson where the material is covered is also indicated in the chart.

Question	1	2	3	4, 5	6	7, 8	9	10, 11
Objective	B	B	A	B	A	C	E	D
Lesson	9-5	9-7	9-2	9-7	9-2	9-8	9-2	9-9

Question	12	13, 14	15	16	17	18	19	20
Objective	E	F	G	G	Voc.	I	H	H
Lesson	9-5	9-1	9-2	9-3	9-2	9-4	9-6	9-8

CHAPTER 9 REVIEW (pp. 471–473)

1. a. 81 **b.** -81 **c.** 81 **3.** 4 **5.** 703.04 **7.** $-\frac{1}{125}$ **9.** $\frac{8}{343}$
11. 81 **13.** x^{11} **15.** x^3y^{12} **17.** 1 **19.** $5 \cdot 1m^4 = .2m^4$ **21.** $\frac{x}{y^2}$
23. $\frac{y^3}{x^3}$ **25.** $\frac{a^5}{b^5}$ **27.** $\frac{4k^3}{27}$ **29.** a, b **31.** Does $(3 + 4)^3 =$
$3^3 + 4^3$? Does $7^3 = 27 + 64$? No **33.** Power of a Product
Property **35.** Zero Exponent Property **37.** Power of a Quotient Property **39.** Negative Exponent Property **41.** $3123.11
43. $1348.32 **45.** $11.41 **47. a.** 128,000 **b.** After 4 hours
there will be 128,000 bacteria. **49.** 1,106,136 **51. a.** $\frac{9}{4}$ **b.** $\frac{243}{32}$
53. ≈ 34 **55. See below. 57.** $y = 56 \cdot (1.04)^x$ **59. See
below. b. See below. c. See below.**

55.

x	y
-4	$\frac{1}{16}$
-3	$\frac{1}{8}$
-2	$\frac{1}{4}$
-1	$\frac{1}{2}$
0	1
1	2
2	4
3	8
4	16

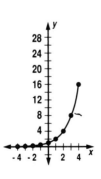

59. a.

year	0	5	10	15
amount	200	200	200	200

59. b.

year	0	5	10	15
amount	200	255.26	325.78	415.79

59. c.

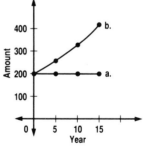

LESSON 10-1 (pp. 476–480)
1. $1150.15 **3.** $47.75 **5. a.** 10 dollars **b.** $10x + 20$ dollars
c. $10x^2 + 20x + 15$ dollars **d.** $10x^3 + 20x^2 + 15x +$
30 dollars **7. a.** $x^5 - 90x^3 + 4x^2 + 3$ **b.** 5' **9.** False
11. $160y^3 + 220y^2 + 180y + 250$ **13. a. See below.**
b. Change line 10 to
PRINT "VALUES of POLYNOMIAL X ^ 2 − 8X + 16''
Change line 30 to
30 FOR X = -6 TO 6
and line 40 to
40 LET V = X ∗ X − 8 ∗ X + 16
15. a. $110 **b.** $99 **17. a.** $\approx .51$ **b.** $\approx .49$ **19.** $6x + 13$
21. $\frac{1}{2}$ **23.** 3 **25.** $\frac{1}{5}$

13. a.

X	VALUE
-3	0
-2	-3
-1	-4
0	-3
1	0
2	5
3	12

LESSON 10-2 (pp. 481–485)
3. $16y^2 + y - 17$ **5.** $7k^2 - 5k - 7$ **7. a.** $8p(4p + 2) =$
$32p^2 + 16p$ **b.** $3p(p + 1) = 3p^2 + 3p$ **c.** $29p^2 + 13p$
9. $-2x^2 - 16x - 7$ **11.** $120y^4 + 50y^3 + 160y^2 +$
$150y + 24$ **13. a.** $2n^2 + 20n$ **b.** $18n^2 + 120n$
c. $16n^2 + 100n$ **15. a.** simplified **b.** x^{m+n} **c.** x^{m-n}
17. 98 **19.** $\frac{5}{4}$ **21.** 38, 22, or 18 **23.** 20 **25.** ab^2c

LESSON 10-3 (pp. 486–489)
1. $2h(L_1 + L_2 + L_3 + L_4), 2hL_1 + 2hL_2 + 2hL_3 + 2hL_4$
3. $zq + zr - 13z$ **5.** $-12wy^2 + 6w^2y + 3wy$ **7.** $50x^8 +$
$70x^7 + 60x^6$ dollars **9.** $-x^2 + 6x$ **11.** $2m^4 + 2m^3 + 6m^2$
13. a. $9r^2$ **b.** πr^2 **c.** $9r^2 - \pi r^2$ **d.** $5(9r^2 - \pi r^2)$ or
$45r^2 - 5\pi r^2$ **15.** $4n^4 + 15n^3 - 2n^2 - 9n + 25$ **17.** $10x$
19. a. $25x^2 + 25x$ **b.** $17x^2 + 24x + 31$ dollars
c. $42x^2 + 49x + 56$ dollars **21.** $y = -2x + 5$
23. $y = 2x + 18$ **25. a.** $\frac{28c}{39}$ **b.** $\frac{c + 5cq}{13}$ **c.** $\frac{12c + 65cq}{156q}$

LESSON 10-4 (pp. 490–494)
3. x^2; 5 **5.** $7xy$; 3; x^2 **7. a.** a^2 **b.** $a^2(23a + 1)$ **9. a.** $6y^3$
b. $6y^3(5xy - 2 + 3x^3y^2)$ **11.** $\frac{3n(4n + 5)}{3n} = 4n + 5$
13. $20y^2 - 11y + 6$ **15.** True **17.** $-x + 2 - 3y$
19. $550.19 **21.** $5x^2 + x + 9$ **23.** $4v^3 + 8v^2 + 4v$ **25.** 9
27. $\approx -.31$ kg per day **29.** $b = \frac{d - c}{x}$

LESSON 10-5 (pp. 495–498)
1. a. $(w^2 + 5w + 4)(w + 6)$ **b.** $w^3 + 11w^2 + 34w + 24$
3. $y^3 + 13y^2 + 44y + 12$ **5.** $3m^4 + 26m^3 - 33m^2 -$
$32m - 6$ **7.** $(5 \cdot 2^2 + 4 \cdot 2 + 3)(2 + 7) = 31 \cdot 9 = 279;$
$5 \cdot 2^3 + 39 \cdot 2^2 + 31 \cdot 2 + 21 = 40 + 156 + 62 +$
$21 = 279$ **9.** $x^4 + 64$ **11.** $49 - 9x^2$ **13.** $x^2 + 32$ **15. a.** Let
$x = 3$; Does $(5 \cdot 3 + 3)(5 \cdot 3 - 2) = 25 \cdot 3^2 = 6$? Does
$18 \cdot 13 = 225 - 6$? No **b.** 0 **17.** $2v(2v - 1)$ **19.** $3xy^3(x + 1)$
21. $-x^2 - 3x + 25$ **23.** $\sqrt{128}$ or $8\sqrt{2}$ km **25.** $\frac{7a^2}{3b}$

LESSON 10-6 (pp. 499–503)
1. a. $(2w + 6)(w + 7)$ **b.** $2w^2 + 20w + 42$ **3.** $ac + ad +$
$bc + bd$ **5.** $y^2 + 11y + 30$ **7.** $9k^2 - 19k - 24$ **9.** $a^2 -$
b^2 **11. a.** 892,500 **b.** $(f + p)(j + s)$ or $fj + fs + pj + ps$
13. a. $x^5 + 2x^3 - x^2 - 2$ **b.** $(4^2 + 2)(4^3 - 1) =$
$18 \cdot 63 = 1134$; $4^5 + 2 \cdot 4^3 - 4^2 - 2 = 1024 + 128 -$
$16 - 2 = 1134$ **15.** $n^3 + 3n^2 + 2n$ **17.** $300q - 225r$
19. $4w^3z^4 - 3w^2z^5 + 2wz^3$ **21.** $14x^2 + 10x - 60$
23. 150 feet **25.** $xy(3x^2 + 2y^2)$ **27.** $y = \frac{1}{2}x + \frac{17}{2}$

LESSON 10-7 (pp. 504–507)
3. a. $(3x + 7)^2$ **b.** $9x^2 + 42x + 49$ **5.** $x^2 - 2xy + y^2$
7. $-24p$ **9.** $(1 + .02)^2 = 1 + .04 + .0004 = 1.0404$
11. $(40 - 3)^2 = 1600 - 240 + 9 = 1369$ **13.** $324 +$
$72y + 4y^2$ **15. a.** $(2 + 7)^2 = 81, 2^2 + 7^2 = 53$ **b.** 28
c. 28; It's the same. **17.** $(32t + 20)(t + 3)$ or $32t^2 +$
$116t + 60$ feet **19.** $z^2 - 3z - 88$ **21. a.** $52x^4 - 8x^3$
b. $52x^4 - 8x^3 + 78x^2 + x - 2$ **c.** $52x^3 - 12x^2 + 7x - 1$
23. $60xy(x + y + 1)$ **25.** $-\frac{1}{2}$

LESSON 10-8 (pp. 508–512)
3. No **5.** No **7. a.** $(x + 5)$ **b.** -2 or 8 **9. a.** i or iii **b.** No,
$576 = 24^2$, but 776 is not a perfect square. **11.** Yes
13. a. $4x - 2y$ **b.** THEY ARE NOT EQUAL.
15. $\frac{1}{2}m^2 + 3mn + \frac{9}{2}n^2$ **17.** $a^2 - b^2$ **19.** $6x^2 + 8x$
21. $\frac{1}{11,250,000} \approx 8.9 \cdot 10^{-8}$ **23. a.** II **b.** I **c.** III

LESSON 10-9 (pp. 513–516)
3. $25y^2 - 49$ **5.** b **7.** $(90 - 5)(90 + 5) = 90^2 - 5^2 =$
8075 **9.** $(x + 1)(x - 1)$ **11.** $(6m - 5n)(6m + 5n)$
13. $9^2 - \left(\frac{1}{2}\right)^2 = 80\frac{3}{4}$ **15.** $\frac{1}{4} - 9m^4$ **17. a.** $(x^2 - 9)(x^2 + 9)$
b. $(x - 3)(x + 3)(x^2 + 9)$ **19.** False **21. a.** 441 **b.** 1.21
23. $3x^2(-3 + 8x + 6x^2)$ **25. a.** $-28x^2 + 6x - 8$ **b.** 2
27. See below. **29. a.** $x = 0$ **b.** $x < 0$

27.

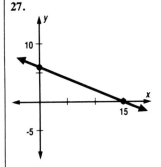

LESSON 10-10 (pp. 517–520)
3. $0, \frac{5}{3}$ **5.** $15, \frac{8}{9}$ **7.** 2.8125 **9.** $0, 11, -\frac{7}{2}$ **11.** 5 **13.** c
15. a. $t^2 - 6t = 0; t(t - 6) = 0; t = 0$ or $t = 6$ **b.** $(t - 3)^2 =$
$9; t - 3 = 3$ or $t - 3 = -3; t = 6$ or $t = 0$ **17.** -3, -2, -1, 0
19. $(7p - q)(7p + q)$ **21.** $9x^2 - 49$ **23.** $(4e + 1)^2$ or $16e^2 +$
$8e + 1$ **25. See below.** **27.** 9 **29.** 6 or -6

25.

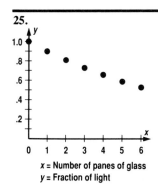

x = Number of panes of glass
y = Fraction of light

3. a. $(x + 10)(x - 1) = x^2 + 9x - 10$ **b.** Sample: let $x = 3$; $(13)(2) = 26$, and $3^2 + 9 \cdot 3 - 10 = 26$. Let $x = -4$; $(6)(-5) = -30$, and $(-4)^2 + 9(-4) - 10 = -30$. **7.** c **9.** e
11. c **13.** b **15. a.** 2 **b.** 0, 2 **c.** $\frac{-4}{3}$, 2 **d.** 5, $\frac{-4}{3}$, 2
17. $100(x - y + w - z)$

CHAPTER 10 PROGRESS SELF-TEST (p. 526)

1. a. $4x^2 + 9x^2 - 7x - 12 - 11 = 13x^2 - 7x - 23$
b. 2nd **2.** $4k^3 + 7k^3 - k^2 - 4k^2 + 6k + 8 - 2 = 11k^3 - 5k^2 + 6k + 6$ **3.** $(9v^2 + 2) - (2v^2 + 5) = 9v^2 + 2 + -2v^2 + -5 = 7v^2 - 3$ **4.** $4(3v^2 - 9 + 2v) = 12v^2 - 36 + 8v$ **5.** $18x^2 - 12x + 6 - 12x + 32 = 18x^2 - 12x - 12x + 6 + 32 = 18x^2 - 24x + 38$
6. $-5z(z^2) - 5z(-7z) - 5z(8) = -5z^3 + 35z^2 - 40z$
7. $x \cdot x + 9 \cdot x - 7 \cdot x - 7 \cdot 9 = x^2 + 2x - 63$
8. $4y \cdot 3y - 4y \cdot 16 - 2 \cdot 3y + 2 \cdot 16 = 12y^2 - 70y + 32$ **9.** $q^2 - 2 \cdot 1 \cdot 7 \cdot q + 7^2 = q^2 - 14q + 49$
10. $(3x)^2 - (8)^2 = 9x^2 - 64$ **11.** $2a \cdot a^2 + 2a \cdot 5ab + 2a \cdot 7b^2 - 3 \cdot a^2 - 3 \cdot 5ab - 3 \cdot 7b^2 = 2a^3 + 10a^2b + 14ab^2 - 3a^2 - 15ab - 21b^2$ **12.** No; $(2x - 5)^2 = 4x^2 - 20x + 25$, not $4x^2 - 10x + 25$. **13.** $6 \cdot 2m^3 - 2m^3 \cdot m^2 = 2m^3(6 - m^2)$ **14.** $50y \cdot 10x^2 + 50y \cdot 2x + 50y \cdot 1 = 50y(10x^2 + 2x + 1)$ **15.** $z^2 - 9^2 = (z + 9)(z - 9)$

16. $y^2 + 2 \cdot 6 \cdot y + 6^2 = (y + 6)(y + 6)$
17. $\frac{4c(2c + 1)}{c} = 4(2c + 1)$ **18.** $(30 - 1)(30 + 1) = 900 - 1 = 899$ **19.** a **20.** $z = 0$ or $5z - 2 = 0$, $z = 0$ or $\frac{2}{5}$ **21.** $x - 13 = 0$ or $2x + 15 = 0$, $x = 13$ or $-\frac{15}{2}$
22. $80x^2 + 60x + 90$ **23.** $8t^2 - 5t^2 = 0$ at ground level. So $t(8 - 5t) = 0$, $t = 0$ or $8 - 5t = 0$, $t = 0$ or $\frac{8}{5}$, after $\frac{8}{5}$ seconds. **24. See below. 25.** $2y(5y - 3)(y + 9) = 2y(5y^2 + 42y - 27) = 10y^3 + 84y^2 - 54y$
24.

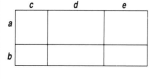

The chart below keys the **Progress Self-Test** questions to the objectives in the **Chapter Review** on pages 527–529 or to the **Vocabulary** (Voc.) on page 525. This will enable you to locate those **Chapter Review** questions that correspond to questions you missed on the **Progress Self-Test.** The lesson where the material is covered is also indicated in the chart.

Question	1, 2, 3	4	5, 6	7, 8	9	10	11	12	13, 14
Objective	A	B	B	B	B	F	B	C	D
Lesson	10-1, 10-2	10-3	10-3	10-6	10-7	10-9	10-5	10-8	10-4
Question	15	16	17	18	19	20, 21	22	23	24, 25
Objective	E	E	D	F	H	G	I	I	J
Lesson	10-9	10-8	10-4	10-7, 10-9	10-11	10-10	10-3	10-10	10-3

CHAPTER 10 REVIEW (pp. 527–529)

1. a. $7x^2 + 4x + 1$ **b.** 2nd **3.** $3.9x^2 + 1.7x + 19$
5. $-k^2 + k - 5$ **7.** $8x^2 + 10x - 36$ **9.** $4x - 4x^3$
11. $-6m^2n + 4mn^2 - 8mn$ **13.** $x^2 + 4x - 21$ **15.** $ac - ad - bc + bd$ **17.** $d^2 - 2d + 1$ **19.** $a^2 - 225$ **21.** $a^3 - 1$
23. Yes **25.** No **27.** No **29.** x^3; 7 **31.** $m^2(14m^2 + 1)$
33. $6z^2 - 1$ **35.** $(m + 8)^2$ **37.** $(a + 2)(a - 2)$
39. $(2x + 1)(2x - 1)$ **41.** 1681 **43.** 3591 **45.** For any two numbers a and b, if $ab = 0$, then $a = 0$ or $b = 0$.

47. There is no product. **49.** 3, 1 **51.** 3, $\frac{1}{2}$, $-\frac{1}{2}$
53. a. $x^2 + 5x - 6$ **b.** Sample: $3^2 + 5 \cdot 3 - 6 = 18$, $(3 + 6)(3 - 1) = 9 \cdot 2 = 18$. $4^2 + 5 \cdot 4 - 6 = 30$, $(4 + 6)(4 - 1) = 10 \cdot 3 = 30$ **55.** b **57.** $811.60
59. $25x^6 + 50x^5 + 75x^4$ **61.** $\frac{25}{8}$ **63. a.** $xy + 3y + 2x + 6$
b. $(x + 3)(y + 2)$ **c.** Yes **65.** $21x^2 - \pi x^2$ or $x^2(21 - \pi)$
67. $x^3 - x$

LESSON 11-1 (pp. 532–535)

5. a. $\begin{cases} y = 2x + 4 \\ x + 2y = 3 \end{cases}$ **b.** (-1, 2) **7. a.** $x + y = 18$, $x - y = 8$
b. See below. c. 13 and 5 **9.** Any two of: $(x, y) = (-1, -9)$;
$x = -1$ and $y = -9$; or $\{(-1, -9)\}$. **11. See below.**
13. $-0.185(2013) + 416.7 \approx 44.3$; $-0.313(2013) +$
$674.41 \approx 44.3$ **15.** $x = 2$ **17.** $12x - 3y$ **19. a.** 20 quarts

b. $4n$ quarts **21.** 10¢ per second **23. a.** sample: $\frac{325}{1} = \frac{x}{4\frac{3}{4}}$
b. ≈ 1544 **25. a.** $\frac{-20}{3} = -6\frac{2}{3}$ **b.** -10 **c.** $\frac{24}{7} = 3\frac{3}{7}$

7. b.

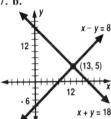

11.

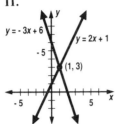

LESSON 11-2 (pp. 536–540)
1. $135 for the boys, $45 for the girls **3.** 15 quarts of orange
juice and 5 quarts of ginger ale **5.** Does $\frac{24}{11} + \frac{86}{11} = \frac{110}{11} = 10$?
Yes.; Does $\frac{32}{11} + \frac{11}{11} = \frac{43}{11}$? Yes. **7. a.** (0, -6) **b.** Does
$12(0) - 5(-6) = 30$? Yes.; Does $-6 = 2(0) - 6$? Yes.
9. a. (-12, -3) **b.** Does $2(-12) - 3(-3) = -15$? Yes.; Does
$-12 = 4(-3)$? Yes. **11. a.** $\begin{cases} J = 3M \\ J + M = 11{,}000 \end{cases}$ **b.** Mary gets
$2750 and John gets $8250. **13.** $L = $800{,}000$ and $T =$
$1,000,000 **15. a. See below. b.** There is no answer to
check. **17.** $2x^2 - 2x + 1$ **19. See below. 21.** 21

15. a.

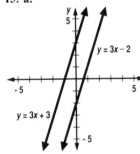

19.

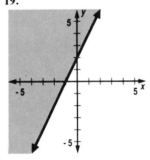

LESSON 11-3 (pp. 541–544)
1. False **3.** (-55, -610) **5.** $4.25 **7.** distances greater than
8 tenths of a mile **9. See below. 11.** (-31, -17); Does $-31 =$
$2(-17) + 3$? Yes. Does $-31 = 3(-17) + 20$? Yes. **13. a.** 30
b. $440 **15. a.** 12.2 years **b.** 546,000 **17.** $b = -1.9$;
$a = -4.6$ **19. a.** -2 **b.** 7 **c.** -7 **21. See below. 23. a.** 9
b. 3 or -3 **c.** 0 or 9

9.

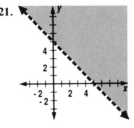

21.

LESSON 11-4 (pp. 545–550)
1. to eliminate one variable **3.** $\left(-\frac{14}{3}, 2\right)$ **5.** plane speed is
108 mph; wind speed is 12 mph **7.** $\left(2, -\frac{1}{3}\right)$ **9.** $49 per night
and $8 per meal **11.** $\left(6, -\frac{5}{3}\right)$ **13.** 45¢ **15.** 2 and -3
17. (-1, -3) **19.** sample: $P = 20$, $H = 0$; $P = 14$, $H = 5$;
$P = 8$, $H = 10$ **21.** $2b^4$ **23. See below. 25.** $8x + y = -15$
or $-8x - y = 15$ **27.** $4x - 7y = -8$ or $-4x + 7y = 8$

23.

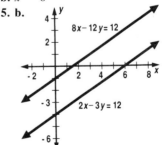

LESSON 11-5 (pp. 551–556)
1. the Multiplication Property of Equality **3. a.** -2
b. $(u, v) = \left(\frac{3}{4}, \frac{1}{2}\right)$ **5. a.** $10b = -20$ **b.** $15a = 80$
c. $\left(\frac{16}{3}, -2\right)$ **7.** $(m, n) = (57, 48)$ **9.** $(x, y) = (7, -5)$
11. Yes; make 7 pentagons and 8 rectangles. **13.** $m = 17$
and $t = 33$ **15.** $(x, y) = (1, -1)$ **17.** $(x, y) = (-30, -12)$
19. $\pm\sqrt{8} = \pm2\sqrt{2}$ **21.** ≈ 9 days **23.** $\frac{5}{36}$

LESSON 11-6 (pp. 557–561)
1. a. $\frac{6}{8}, \frac{2}{8}$ **b.** 78.25 **5.** $3.73 **7.** ≈ 34.4 mph
9. ≈ 1.7 siblings **11.** ≈ 444 mph **13. a.** ≈ 7.6 hours
b. $\left(\frac{5}{7}x + \frac{2}{7}y\right)$ hours **15.** No **17.** (4, -3)
19. a. $\begin{cases} 2n + 3m = 22 \\ n - m = 31 \end{cases}$ **b.** 23 and -8 **21.** $\pm\sqrt{32}$ or $\pm4\sqrt{2}$

LESSON 11-7 (pp. 562–565)
1. With a $9 average, more than half are $10; the balance
must be less than $9. **3.** $9.29 **5.** $(q, z) = (20, 11)$
7. 0.75 hours or 45 minutes **9.** 76 **11.** 17.5 **13.** 48%
15. No **17.** $\frac{8}{30} = \frac{4}{15}$ **19.** $0 < x < 12$

LESSON 11-8 (pp. 566–569)
1. sample: $\begin{cases} 2x + 3y = -6 \\ 4x + 6y = 24 \end{cases}$ **3.** slope **5. a.** nonintersecting
b. See below. 11. nonintersecting **13.** intersecting **15.** Yes
17. 2 hours **19.** $\left(\frac{7}{2}, -\frac{7}{3}\right)$ **21.** $(x - 7)^2$ **23. a.** $y = -12$
b. $x = 8$
5. b.

LESSON 11-9 (pp. 570–573)
1. a. always **b.** Let $y =$ number of years employed. $6.20 +$
$1.00y > 6.00 + 1.00y$; $6.20 > 6.00$ **3. a.** $9 = 3 + 6$

b. y may be any real number. **5.** x may be any real number.
7. no solution **9. a.** $375m + 500 < 315m + 400 + 60m$
b. $500 < 400$; Apartment A is never cheaper. **11.** B
13. $4.00 **15.** $\frac{1}{3}$ **17.** $525w$ **19.** ≈ 143 m

LESSON 11·10 (pp. 574–579)
1. intersection **3.** An equal sign means the boundary is included. **5.** first **7.** (0, 50), (50, 0), (25, 0) and (0, 25)
9. sample: 20 free throws and no baskets; 10 free throws and 5 baskets; 10 baskets and no free throws. **11.** on or above the b-axis **13.** $\begin{cases} x \geq 0 \\ y \geq 0 \\ x + 2y \leq 10 \end{cases}$ **15. See below. 17. See below.**

19. no solution **b.** y may be any real number.
21. intersecting **23.** (3, -1) **25.** 9 **27.** $1503.66
15. **17.**

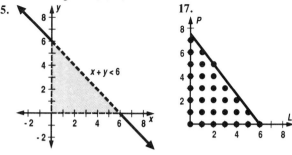

CHAPTER 11 PROGRESS SELF-TEST (p. 581)

1. $a - 3(3a) = -8$, $a - 9a = -8$, $-8a = -8$, $a = 1$;
so $b = 3(1) = 3$. The solution is $(a, b) = (1, 3)$. **2.** $5r + 80 = -7r - 40$, $12r = -120$, $r = -10$; so $p = 5(-10) + 80$, $p = -50 + 80 = 30$. The solution is $(p, r) = (30, -10)$.
3. $-5x - 15 = x - 3$, $-12 = 6x$, $x = -2$; so $y = -2 - 3 = -5$. The solution is $(x, y) = (-2, -5)$. **4.** Add the equations and get $n = 3$; so $m - 3 = -1$, $m = 2$. The solution is $(m, n) = (2, 3)$. **5.** Multiply one equation by -1 and add. We use the first equation and get $2y = 16$, $y = 8$; so $2x + 8 = -6$, $2x = -14$, $x = -7$. The solution is $(x, y) = (-7, 8)$. **6.** Multiply the second equation by 3 and add.
$19x = 19$, $x = 1$; so $4(1) - y = 6$, $4 - y = 6$, $-y = 2$, $y = -2$. The solution is $(x, y) = (1, -2)$. **7.** Multiply the first equation by -2 $\left(\text{or the second by } -\frac{1}{2}\right)$ and add. The result is $0 = 0$, so the lines are coincident. **8.** $2x + 7 - 2x = 3$, $7 = 3$ is not true, so the lines are parallel. **9.** Add $-12z$ to both sides to get $8 = -3$. This is not true, so there is no solution. **10.** Add $19p$ to both sides to get $0 < 22$. This is always true, so p may be any real number. **11.** Let $L =$ Lisa's weight and $B =$ baby's weight. Then $L = 4B$ and $L + B = 92$. So $4B + B = 92$, $5B = 92$, $B = 18.4$. Then $L = 4(18.4) = 73.6$. Lisa weighs 73.6 pounds and the baby weighs 18.4 pounds. **12.** Let $h =$ cost of a burger and $s =$ cost of a salad. Then $3h + 4s = 21.30$ and $5h + 2s =$

22.90. Multiply the second equation by -2 and add to get $-7h = -24.50$, $h = 3.50$. Then $3(3.50) + 4s = 21.30$, $4s = 10.80$, $s = 2.70$. A salad costs $2.70. **13.** Let $r =$ cost per rose and $d =$ cost per daffodil. Then $10r + 15d = 35$ and $2r + 3d = 8$. Multiply the second equation by -5 and add to get $0 = -5$. The situation could not happen.
14. Never. Since the populations are not equal in 1980 and the growth rates are the same, South Carolina population will never catch up to Louisiana population. **15.** $\frac{14}{20} \cdot 40\% + \frac{6}{20} \cdot 80\% = 28\% + 24\% = 52\%$ salt **16.** Let $x =$ time traveled at 40 mph and $y =$ total time traveled. Then $\frac{3}{y} \cdot 55 + \frac{x}{y} \cdot 40 = 45$, and $x + 3 = y$. Simplify the first equation to get $165 + 40x = 45y$, $33 + 8x = 9y$. Substitute $x + 3$ for y to get $33 + 8x = 9x + 27$, $6 = x$; it can travel 6 hours at 40 mph. **17. See below. 18. See below.**

17. **18.**

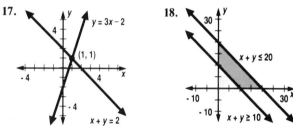

The chart below keys the **Progress Self-Test** questions to the objectives in the **Chapter Review** on pages 582–585 or to the **Vocabulary** (Voc.) on page 581. This will enable you to locate those **Chapter Review** questions that correspond to questions you missed on the **Progress Self-Test.** The lesson where the material is covered is also indicated in the chart.

Question	1, 2	3	4	5, 6	7, 8	9, 10	11	12	13, 14
Objective	A	A	B	C	E	D	F	F	D
Lesson	11-2	11-3	11-4	11-5	11-8	11-9	11-2	11-5	11-9

Question	15	16	17	18
Objective	G	G	H	I
Lesson	11-6	11-7	11-1	11-10

CHAPTER 11 REVIEW (pp. 583–585)

1. $(a, b) = (15, 45)$ **3.** $c = 10$, $p = 20$, $q = 30$
5. $(x, y) = (5, 10)$ **7.** $(a, b) = (-31, -17)$ **9.** $(x, y) =$
$(-9, -43)$ **11.** $(m, b) = (-22, 77)$ **13.** $(x, y) = (-3, -7.25)$
15. Different numbers could be used to multiply. This is one approach. **a.** Multiply $5x + y = 30$ by 4, $20x + 4y = 120$.
b. $(x, y) = (7, -5)$ **17.** $(y, z) = (3, 3)$ **19.** $(a, b) = (4, -1)$
21. no solutions **23.** 0 (one solution) **25.** coincide
27. coincide **29.** Marty earned $52.50. Joe earned $157.50.

31. $.45 for an egg and $.90 for a muffin **33.** Yes
35. always **37.** $\approx$ 44.4 mph **39.** 55% **41.** 15 pounds
43. See below. **45.** See below. **47.** c **49.** See below.
51. See below. **53.** See below.

43.

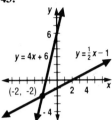

$y = 4x + 6$ $y = \frac{1}{2}x - 1$

$(-2, -2)$

45.

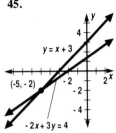

$y = x + 3$

$(-5, -2)$

$-2x + 3y = 4$

49.

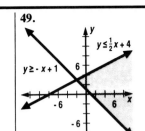

$y \le \frac{1}{2}x + 4$

$y \ge -x + 1$

51.

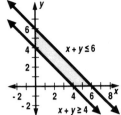

$x + y \le 6$

$x + y \ge 4$

53.

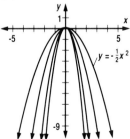

LESSON 12-1 (pp. 588–593)

3. a parabola that opens upward, vertex at (0, 0)

7. a.

x	-3	-2	-1	0	1	2	3
y	27	12	3	0	3	12	27

b. See below.

9. a.

x	-3	-2	-1	0	1	2	3
y	-32	-17	-8	-5	-8	-17	-32

b. See below. **11. a.** See below. **b.** See below. **13. a.** iii
b. iv **c.** i **d.** ii **15.** 8 **17.** $4x^2 - 10x - 8 = 0$ or
$-4x^2 + 10x + 8 = 0$ **19.** a^3 **21.** $-2 + 3x$ **23.** c
25. $7500 - 1800 - k$ or $5700 - k$

7. b.

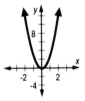

9. b.

11. a.

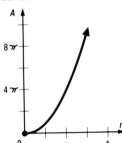

11. b.

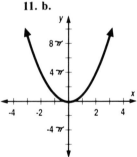

LESSON 12-2 (pp. 594–598)

3. No **5.** Yes **7.** $\approx$ (-.7, -3.3) **9.** $-50 \le x \le 30$; $-10 \le y \le$
4 **11.** $\approx$ -35 and 12 **13.** 3 **15. a.** See below. **b.** See below. **c.** The graph in part a gives you a "bigger" picture. It lets you see what is happening over more points. The graph in part b zooms into the graph in part a. You see more detail.

---

17. a. See below. **b.** As a gets smaller, the parabola appears thinner. **19. a.** $-4ac = -4(2)(4) = -32$ **b.** $b^2 - 4ac = 3^2 - 4(2)(4) = 9 - 32 = -23$ **c.** $\sqrt{b^2 - 4ac} = \sqrt{-23}$

21. $\approx$ 4.47 **23.** $2 - \sqrt{5}$ **25. a.** 20 ft $\cdot \dfrac{1 \text{ min}}{3 \text{ ft}} \approx 6.67$ min

b. f ft $\cdot \dfrac{1 \text{ min}}{3 \text{ ft}} = \dfrac{f}{3}$ min **27. a.** $(2x + 3)(3x + 2) = 6x^2 + 4x + 9x + 6 = 6x^2 + 13x + 6$

b. $(6x - 1)(x - 6) = 6x^2 - 36x - x + 6 = 6x^2 - 37x + 6$

15. a.

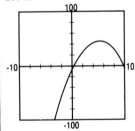

15. b.

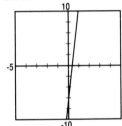

17. a. $y = -\frac{1}{2}x^2$

x	-3	-2	-1	0	1	2	3
y	-4.5	-2	$\frac{1}{2}$	0	$\frac{-1}{2}$	-2	-4.5

$y = -x^2$

x	-2	-1	0	1	2
y	-4	-1	0	1	-4

$y = -2x^2$

x	-2	-1	0	1	2
y	-8	-2	0	-2	-8

$y = -3x^2$

x	-1	0	1
y	-3	0	-3

17. a.

$y = -\frac{1}{2}x^2$

b. As a gets smaller, the parabola appears thinner

LESSON 12-3 (pp. 599–602)

3. a. 6.975 feet **b.** about 6 feet $11\frac{7}{10}$ inches **5.** ≈ 11.1 feet

7. 20 yards **9.** ≈ 1.5 and 2.5 seconds **11.** 2 seconds

13. 25 meters **15.** 35 meters

17. a. sample:

x	-2	-1	0	1	2
y	2	$\frac{1}{2}$	0	$\frac{1}{2}$	2

b. See below. 19. 0, 4, 3 **21.** a **23.** (-3, -5)

17. b.

LESSON 12-4 (pp. 603–608)

3. Substitute $x = 5$; does $5^2 - 2 \cdot 5 - 15 = 0$? Yes.
Substitute $x = -3$; does $(-3)^2 - 2 \cdot -3 - 15 = 0$? Yes.

5. a. $a = 12$, $b = 7$, $c = 1$ **b.** $-\frac{1}{4}$ or $-\frac{1}{3}$ **7.** Substitute

$m = 5.53$; $(5.53)^2 - 3(5.53) - 14 = -0.01$. Substitute
$m = -2.53$; $(-2.53)^2 - 3(-2.53) - 14 = -0.01$. Both are
close enough to zero to give the approximations. **9. a.** The
simplest is $m^2 - 4m - 16 = 0$. **b.** $2 \pm 2\sqrt{5} \approx -2.47$, 6.47

11. $a = 3$, $b = -1$, $c = -5$; 1.47 or -1.14 **13.** $\dfrac{12 \pm \sqrt{96}}{24} \approx$

.908 or .092 **15. a.** 2.5, 5 **b.–c. See below. 17. a.** 5 meters

above **b.** 15 meters below **19.** d **21.** none **23.** 540

25. a. $x^2 + 2x + 1$ **b.** $9x^2 - 12x + 4$ **c.** $4a^2x^2 + 4abx + b^2$

15. b.–c.

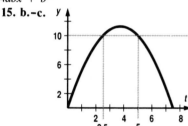

LESSON 12-5 (pp. 609–614)

3. 0 **9. a.** -119 **b.** 0 **c.** no real solution **11. a.** -11 **b.** 0

c. no real solution **13.** never **15. a.** $\frac{1}{2}$, $-\frac{5}{2}$ **b.** $\dfrac{-8 \pm \sqrt{48}}{8} \approx$

-.13 or -1.9 **c.** no real solution **17.** 2.25 **19. a.** ≈ 1.07

b. about 7% **21.** down **23. a.** $a < \frac{1}{3}$ **b.** $b > \frac{1}{3}$ **c.** $c < 0$

25. a. 48 feet **b.** ≈ 3.5 feet

LESSON 12-6 (pp. 615–617)

3. rational, sample: $\frac{3}{4}$ **5.** rational, sample: $\frac{-3}{2}$ **7.** rational,

sample: $\frac{2}{6}$ **9.** rational, sample: $\frac{4}{1}$ **11. a.** two rational **b.** $-\frac{2}{3}$ or

$-1\frac{1}{2}$ **13. a.** no rational **b.** ≈ 1.414 or ≈ -1.414 **15. a.** two

rational **b.** .5 or -.2 **17. a.** no rational **b.** $\approx .525$ or ≈ -9.525

19. samples: $3x^2 + 8x + -k = 0$ (with rational solutions);
$b^2 - 4ac =$ a perfect square; $64 - 4(3)(k) = 1$; $64 -$

$12k = 1$; $-12k = -63$; $k = \frac{63}{12} = \frac{21}{4}$, $64 - 12k = 25$;

$-12k = -39$; $k = \frac{39}{12}$ **21.** ≈ -9.525 or $\approx .525$ **23.** 44.1 ft

25. a. $d = 4.9t^2$ **b.** $1.43 \approx t$ **27. a.** $13x$ **b.** $5y^2$ **c.** $65xy^2$

29. 300

LESSON 12-7 (pp. 618–622)

3. Multiply the two binomials or substitute a number.

5. a. yes; **b.** $(x + 10)(x + 11)$ **7. a.** yes; **b.** $(y + 9)(y + 1)$

9. a. yes; **b.** $(t + 5)(t - 6)$ **11. a.** $11^2 - 4(3)(-4) = 169$;

yes; **b.** $(3x - 1)(x + 4)$ **13.** $k = 5$ **15.** No

17. a. $\dfrac{5 \pm \sqrt{13}}{6}$ **b.** 1.4 or 0.2 **19.** $(2a + 2b)(a - b) =$

$2(a^2 - b^2)$ **21.** $m = \frac{8}{5}$, $b = \frac{-1}{5}$ **23.** $2x^2$ **25.** 5

LESSON 12-8 (pp. 623–626)

1. factor, Zero Product **3.** 5, -1 **5. a.** $3x(3x^2 + 4x - 4) = 0$;

b. 0, $\frac{2}{3}$, -2 **7.** 12 **9.** -13, -2 **11.** 0, $-\frac{1}{4}$, 5 **13.** $v = -\frac{5}{2}$ or 7

15. sample: $20x^2 + x - 1$ **17.** True **19.** irrational

21. irrational **23.** rational **25.** $\dfrac{5}{a + b}$

27. $\left(\frac{1}{2}y - 3\right)\left(\frac{1}{2}y + 3\right)$

LESSON 12-9 (pp. 627–630)

3. $x = \dfrac{-(-7) \pm \sqrt{49 - 4(10)}}{2} = 5$ or 2; $(x - 2)(x - 5)$

5. $k = \dfrac{-(-8) \pm \sqrt{64 - 4(5)(-4)}}{10} = \frac{-2}{5}$ or 2; $5k = -2$ or $k = 2$;

$5k + 2 = 0$ or $k - 2 = 0$; $(5k + 2)(k - 2)$ **7.** $(x - 1)$

9. $(x + 8)^2$ **11.** $4m(4m^2 - 11m + 30)$ **13.** $x + 6$

15. $3x - 4$ **17.** 0, -5, 2 **19.** 8.96 meters high **21.** $\approx 53.6°F$

23. $3650 **25.** as soon as he passes 9 weeks

CHAPTER 12 PROGRESS SELF-TEST (p. 632)

1. $x = \dfrac{-(-9) \pm \sqrt{(-9)^2 - 4 \cdot 1 \cdot 20}}{2 \cdot 1} = \dfrac{9 \pm \sqrt{81 - 80}}{2} =$

$\dfrac{9 \pm \sqrt{1}}{2}$; so $x = 4$ or $x = 5$ (This could also be factored to

$(x - 4)(x - 5) = 0$ and solved using the Zero Product

Property.) **2.** $5y^2 - 3y - 11 = 0$;

$y = \dfrac{-(-3) \pm \sqrt{(-3)^2 - 4 \cdot 5 \cdot -11}}{2 \cdot 5} = \dfrac{3 \pm \sqrt{9 + 220}}{10} =$

$\dfrac{3 \pm \sqrt{229}}{10}$ **3.** Factor out 3; $6z^2 - z + 8 = 0$.

4. $k = \dfrac{-(-6) \pm \sqrt{(-6)^2 - 4 \cdot 1 \cdot -3}}{2 \cdot 1} = \dfrac{6 \pm \sqrt{36 + 12}}{2} =$

$\dfrac{6 \pm \sqrt{48}}{2}$; $k = 6.46$ or $k = -0.46$

5. $r = \dfrac{-15 \pm \sqrt{15^2 - 4 \cdot 3 \cdot 8}}{2 \cdot 3} = \dfrac{-15 \pm \sqrt{225 - 96}}{6} =$

$\dfrac{-15 \pm \sqrt{129}}{6}$; $r = -0.61$ or $r = -4.39$ **6.** $8x^2 - 7x + 11 =$

0; the value of the discriminant is $(-7)^2 - 4 \cdot 8 \cdot 11 = -303$,
so the equation has no real solutions. **7.** $z^2 - 16z + 64 = 0$;
factor to $(z - 8)^2 = 0$. $z = 8$ **8.** True **9.** False

10. $(x + 2)(x - 18)$ **11.** Solve $x^2 + 8x + 19 = 0$; the
discriminant is $8^2 - 4 \cdot 1 \cdot 19 = -12$, so the equation has no
solutions and the expression does not factor.

12. $(x - 7)(x - 5)$ **13.** $(x - 7)(x + 6) = 0$; $(x - 7) = 0$
or $(x + 6) = 0$; $(x - 7) = 0$ or $(x + 6) = 0$; so $x = 7$ or
$x = -6$ **14.** $h = -.07(2 - 10)^2 + 12 \approx 7.5$ ft **15.** $0 = -.07(x - 10)^2 + 12$; 23 yd

16. a. sample:

x	-2	-1	0	1	2
y	-8	-2	0	-2	-8

b. See below.

17. a. sample:

x	0	1	2	3	4
y	3	0	-1	0	3

b. See below.

18. a **19.** A number that can be written as a fraction, using whole numbers in the numerator and denominator. A number that has a repeating pattern or decimal is a rational number.
20. c

16. b.

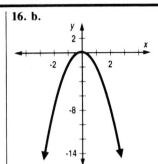

17. b.

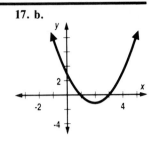

The chart below keys the **Progress Self-Test** questions to the objectives in the **Chapter Review** on pages 633–635 or to the **Vocabulary** (Voc.) on page 631. This will enable you to locate those **Chapter Review** questions that correspond to questions you missed on the **Progress Self-Test.** The lesson where the material is covered is also indicated in the chart.

Question	1	2	3	4	5	6	7	8	9	10	11
Objective	A	A	B	A	A	A	B	E	D	B	E
Lesson	12-4	12-4	12-7	12-4	12-4	12-5	12-7	12-5	12-1	12-7	12-5

Question	12	13	14	15	16	17	18	19	20
Objective	B	C	G	G	H	H	H	F	F
Lesson	12-7	12-8	12-3	12-3	12-1	12-1, 12-2	12-1	12-6	12-6

CHAPTER 12 REVIEW (pp. 633–635)

1. $-\frac{5}{2}, \frac{4}{3}$ **3.** $4, -\frac{3}{4}$ **5. a.** $m^2 - 5m + 3 = 0$ **b.** $\frac{5 \pm \sqrt{13}}{2} \approx 4.3$ or 0.7 **7.** 7.27, -0.27 **9.** no real solution **11.** 0.36, -0.43 **13.** 2, 7 **15.** $(x + 6)(x - 1)$ **17.** $(5a - 7)(2a - 1)$ **19.** $(5x + 6)(x - 1)$ **21.** $(x + 3)(x - 17)$ **23.** $\frac{1}{2}, \frac{2}{3}$ **25.** 0, 2 **27.** a parabola **29.** $ax^2 + bx + c = 0$ **31.** $\frac{-b \pm \sqrt{b^2 - 4ac}}{2a}$

33. True **35.** no real solution **37.** no real solution **39.** irrational **41.** rational **43.** irrational **45. a.** 34 yards **b.** Yes **47.** 2.5 and 10 sec **49. a.** sample:

x	-2	-1	0	1	2
y	12	3	0	3	12

b. See below. **51. a.**

x	-4	-3	-2	-1	0
y	2	0	0	2	6

b. See below. **53.** b **55.** $-35 \le x \le 35$; $-20 \le y \le 10$

49. b.

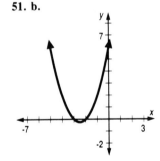

51. b.

LESSON 13-1 (pp. 638–642)

3. a. $C = 0°C, C = 100°C$ **b.** Yes **5. a.** $y = -648$; $y = -701$ **b.** Yes **9.** $y = \frac{1}{3}x - 2$; Each value of x corresponds to one value of y, so the equation describes a function. **11.** One x value is paired with infinitely many y values. **13.** Yes

15. typical points

x	-2	0	2
y	5	3	5

; **See below.**

17. $y = 3.00 + .50x$; See below. **19. a.** $\frac{5}{6}$ **b.** $\frac{x + y}{xy}$

c. $\frac{a + b - 1}{(a + 1)(b - 2)}$ **21.** $\frac{1}{n}$ **23.** $\frac{9}{64}$

15.

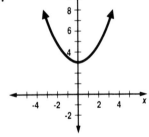

17. samples:

x	y	
0	3.00	3.00 + .50(0)
1	3.50	3.00 + .50(1)
2	4.00	3.00 + .50(2)

(0, 3.00), (1, 3.50), (2, 4.00)

LESSON 13-2 (pp. 643–646)

3. f of x equals x squared **5.** 64 **9.** the average number of accidents per 50 million miles driven by 60-year olds **13.** -5 **15.** 2.5 **17.** age 18 **19. a.** 1, 8, 27, 64, 125 **b.** cubing function **21. a.** 506 **b.** 650 **23.** $\frac{2}{3}$ **25.** 4 **27.** 100bc

LESSON 13-3 (pp. 647–651)

1. 3 **3.** $\frac{3}{4}$ **7.** 30 **9.** the length of a minute **11.** describing the distance of a plane from a check point **13.** When $t = 5$ hours, the plane is 600 kilometers west of the checkpoint.

15.

x	12	11	10	9	8	7
y	9	8	7	8	9	10

; **See below.**

17. See below. 19. 45 **21.** $0.58

15.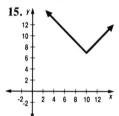

17.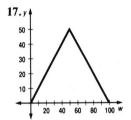

LESSON 13-4 (pp. 652–655)

1. the replacement set for the first variable **3.** {1, 3, 5}; {2, 4, 7} **5.** $x \geq 0$; $f(x) \geq 4$ **7.** all real numbers; $y \geq 0$ **9. a.** $f(x) = 4x$ **b.** $g(x) = 4x$ **c.** The domain for $f(x)$ can be any real number ≥ 0. The domain for $g(x)$ can only be whole numbers. **11. a. See below. b.** 6 **13.** all real numbers; $y \geq -3$ **15.** all nonzero real numbers; all nonzero real numbers **17.** an angle; (0, 3) **19.** 0; 4; $\frac{52}{9}$ **21.** $x = \frac{1}{2}y - 3$

11. a. sample:

b. 6

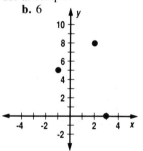

LESSON 13-5 (pp. 656–659)

3. $P(12) = -\frac{1}{36}|12 - 7| + \frac{1}{6} = -\frac{1}{36}|5| + \frac{1}{6} = -\frac{1}{36}(5) + \frac{1}{6} = -\frac{5}{36} + \frac{6}{36} = \frac{1}{36}$ **5.** 6; 8 **7. See below. 9.** False **11. a.** .2 **b.** .6 **c.** Each point would be graphed at $P = .\overline{16}$. **13.** $1 - \frac{31}{32} = \frac{1}{32}$; **See below. 15.** {1, 2, 3, 4, 5}; $\{\frac{1}{5}\}$ **17.** $x = 3$ **19.** $3\sqrt{3}$

7.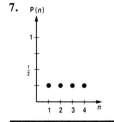

13.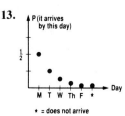

LESSON 13-6 (pp. 660–665)

1. a. 90 **b.** complementary **5.** 27° **7.** 1.540 **9.** 14.57 m **11.** tan A = m **13.** 76° **15.** ≈ 59° **17.** ≈ 77° **19. a.** 30° **b.** $\frac{1}{4}, \frac{1}{4}, \frac{1}{4}, \frac{1}{6}, \frac{1}{12}$ **c. See below. 21.** $x = -2$

23. $x = \dfrac{-b \pm \sqrt{b^2 - 4ac}}{2a}$

19. c.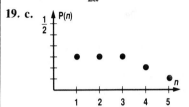

LESSON 13-7 (pp. 666–670)

1. 0.194 **3.** 0.682 **5.** 12.173 **7.** ! **9.** $\frac{1}{x}$ **11.** common logarithm **13. a.** 1 **b.** 1 **15. a.** -0.9 **b.** -0.940

17. a.

X	LOG X	or	LOG X
1	0		0
2	0.301029995		.69315
3	0.477121254		1.0986
4	0.602059991		1.3863
5	0.698970004		1.6094
6	0.77815125		1.7918
7	0.84509804		1.9459
8	0.903089987		2.0794
9	0.954242509		2.1972
10	1		2.3026

b. See below. c. almost the same, none of the graph is below the x-axis **19. See below. 21. a.** 7 **b.** $\frac{1}{6}$

23.

x	0	1	2	3	4	5	6	7
$f(x)$	10	8	6	4	6	8	10	12

See below.

25. 120 ft **27.** 794.12 min, or 13 hr 14.12 min **29.** $m = 8$ **31. a. See below. b.** Sample: 120 million

17. b.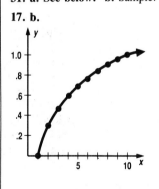

31. a.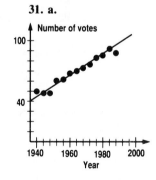

19.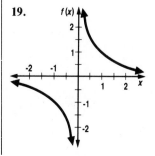

23.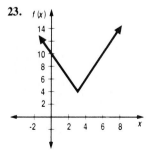

CHAPTER 13 PROGRESS SELF-TEST (p. 672)

1. $3(2) + 5 = 6 + 5 = 11$ **2.** $5 = t^2 + 4t; t^2 + 4t - 5 = 0; (t + 5)(t - 1) = 0; t + 5 = 0$ or $t - 1 = 0; t = -5$ or $t = 1$ **3.** 7.115 **4.** 0.50 **5.** A function is a set of ordered pairs in which each first coordinate appears with exactly one second coordinate. **6. a.** Answers will vary. **b.** parabola **7.** One positive value of x yields two values of y. **8.** 10 or 30 **9.** Yes; domain: {0, 1, 2, 3, 4}; range: {1, 2} **10.** No; $x = 0$ is paired with 1 and -1 for y. **11.** (6, 4), (4, 6), (5, 5); $\frac{3}{36}$ **12.** $d(2.5) = -600|2.5 - 2| + 1200 = -600|0.5| + 1200 = -600(0.5) + 1200 = -300 + 1200 = 900$; When $t = 2.5$, the plane is 900 km east. **13.** $\frac{1}{10}$ **14.** See below. **15.** $\tan P = \frac{21}{20} = 1.05$, 1.05 INV TAN $\approx 46°$

16.

x	-4	-3	-2	-1	0	1	2	3	4	5	6
$f(x)$	14	12	10	8	6	4	2	0	2	4	6

See below.

17.

x	-4	-3	-2	-1	0	1	2	3	4	5	6
$g(x)$	-6	1	6	9	10	9	6	1	-6	-15	-26

See below.

14.

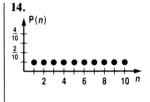

16.

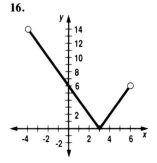

17.

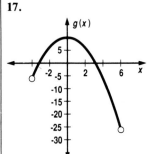

The chart below keys the **Progress Self-Test** questions to the objectives in the **Chapter Review** on pages 673–675 or to the **Vocabulary** (Voc.) on page 671. This will enable you to locate those **Chapter Review** questions that correspond to questions you missed on the **Progress Self-Test.** The lesson where the material is covered is also indicated in the chart.

Question	1–2	3	4	5	6	7	8	9–10	11
Objective	A	G	B	C	D	I	E	H	F
Lesson	13-2	13-6	13-7	13-1	13-1	13-3	13-4	13-1	13-5

Question	12	13	14	15	16	17
Objective	I	F	I	G	I	I
Lesson	13-3	13-5	13-5	13-6	13-3	13-2

CHAPTER 13 REVIEW (pp. 673–675)

1. 6 **3.** 78 **5.** 8 **7.** 1.5 **9.** 2 or -8 **11.** $f(x) = 10$ **13.** 0.059 **15.** 107.093 **17.** 0.577 **19.** 0.991 **21.** No **23.** Yes **25.** Yes **27.** Yes **29.** quadratic **31.** absolute value **33.** other **35.** domain {1850, 1900, 1950, 1960, 1970, 1980}; range {1610, 102,479, 1,970,358, 2,479,015, 2,811,801, 2,966,850} **37.** set of nonnegative real numbers **39.** domain {$0 \le x < 18$}; range {$0 \le y \le 6$} **41.** $\frac{1}{36}$ **43.** 0.25 **45.** $\approx$ 73 ft; See below. **47.** $\tan 18° \approx 0.325$ **49.** No **51.** Yes

53.

t	-5	-4	-3	-2	-1	0	1	2	3	4	5
$g(t)$	15	6	-1	-6	-9	-10	-9	-6	-1	6	15

See below.

55.

n	-5	-4	-3	-2	-1	0	1	2	3	4	5
$f(n)$	$\frac{1}{32}$	$\frac{1}{16}$	$\frac{1}{8}$	$\frac{1}{4}$	$\frac{1}{2}$	1	2	4	8	16	32

See below. **57.** See below.

53.

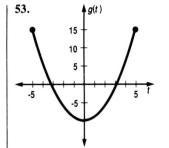

55.

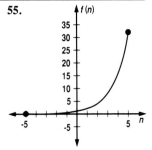

57.

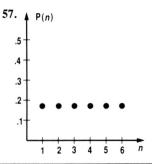

Table of Symbols

=	is equal to
≠	is not equal to
<	is less than
≤	is less than or equal to
≈	is approximately equal to
>	is greater than
≥	is greater than or equal to
±	plus or minus
{ . . . }	the symbol used for a set
Ø, { }	the empty or null set
$A \cap B$	the intersection of sets A and B
$A \cup B$	the union of sets A and B
W	the set of whole numbers
I	the set of integers
R	the set of real numbers
⌐	symbol for 90° angle
%	percent
√	square root symbol; radical sign
$\lvert x \rvert$	the absolute value of x
N(E)	the number of elements in set E
P(E)	the probability of an event E
P(A and B)	the probability that A and B occur
P(B given A)	the probability that B occurs given A
$n!$	n factorial

$f(x)$	function notation "f of x"; the second coordinates of the points of a function
tan A	tangent of $\angle A$
sin A	sine of $\angle A$
cos A	cosine of $\angle A$
%	calculator percent key
1/x	calculator reciprocal key
y^x	calculator key for powering
x^2	calculator squaring function key
√	calculator square root function key
x!	calculator factorial function key
tan	calculator tangent function key
INV tan	calculator inverse tangent keys
sin	calculator sine function key
cos	calculator cosine function key
log	calculator logarithm function key
ABS (X)	in BASIC, the absolute value of X
SQR (X)	in BASIC, the square root of X
X * X	in BASIC, X • X
X ^ Y	in BASIC, X^Y

GLOSSARY

absolute value If $n < 0$, then the absolute value of n equals $-n$; if $n \geq 0$, then the absolute value of n is n. The absolute value of a number is its distance from the origin.

absolute value function A function of the form $f(x) = a|x - b| + c$.

adding fractions property For all real numbers a, b, and c, with $c \neq 0$, $\frac{a}{c} + \frac{b}{c} = \frac{a + b}{c}$.

adding like terms property For any real numbers, a, b, and x, $ax + bx = (a + b)x$.

addition method for solving a system The method of adding the sides of two equations to yield a third equation that contains solutions to the system.

addition property of equality For all real numbers a, b, and c: if $a = b$, then $a + c = b + c$.

addition property of inequality For all real numbers a, b, and c: if $a < b$, then $a + c < b + c$.

additive identity The number 0, because if 0 is added to any number, that number keeps its "identity."

additive identity property For any real number a: $a + 0 = a$.

additive inverse The additive inverse of any real number x is $-x$. Also called *opposite*.

algebraic definition of division For any real numbers a and b, $b \neq 0$: $a \div b = a \cdot \frac{1}{b}$.

algebraic expression An expression that includes one or more variables.

and The word used to speak about the intersection of sets. If A and B are sets, $A \cap B$ is read "A and B."

annual yield The percent the money in an account earns per year.

area model (discrete version) The number of elements in a rectangular array with x rows and y columns is xy.

area model for multiplication The area of a rectangle with length ℓ and width w is ℓw.

associative property of addition For any real numbers a, b, and c: $(a + b) + c = a + (b + c)$.

associative property of multiplication For any real numbers a, b, and c: $(ab)c = a(bc)$.

"as the crow flies" The straight line distance between two points.

automatic grapher A graphing calculator or computer graphing program.

axes The perpendicular number lines in a coordinate graph.

axis of symmetry The line over which a graph could be folded to yield two coinciding halves.

base In the power x^n, x is the base.

BASIC A type of computer language standing for Beginner's All-Purpose Symbolic Instruction Code.

bar graph A way of displaying data using different sized rectangles or bars.

binomial A polynomial with two terms.

boundary line A line that separates two regions in a plane.

changing the sense (direction) of an inequality Changing from $<$ to $>$, or from $\leq$ to $\geq$, or vice-versa.

chunking The process of grouping some small bits of information into a single piece of information. In algebra, viewing an entire algebraic expression as one variable.

closed interval An interval that includes its endpoints.

coefficient In the term ax, a is the coefficient of x.

coincident lines The name given to two lines that have the same graph or coincide.

common denominator   The same denominator for two or more fractions.

common monomial factoring   Isolating the largest common factor from each individual term of a polynomial.

commutative property of addition   For any real numbers a and b: $a + b = b + a$.

commutative property of multiplication   For any real numbers a and b: $ab = ba$.

comparison model for subtraction   The quantity $x - y$ tells how much quantity x differs from the quantity y.

complementary angles   Two angles the sum of whose measures is 90°. Also called *complements*.

compound interest   A form of interest payment in which the interest is placed back into the account so that it too earns interest.

compound interest formula $T = P(1 + i)^n$, where T is the total after n years if a principal P earns an annual yield of i.

conditional probability   The probability that one event occurs given that another event occurs.

conditional probability formula $P(A$ and $B) = P(A) \cdot P(B$ given $A)$.

constant decrease   A negative rate of change that is the same between any two points of a line.

constant term   A term without a variable in a polynomial; a term in a polynomial that is a number.

continuous   A situation in which numbers between any two numbers have meaning.

contraction   A size change in which the size change factor k is between -1 and 1 ($k \neq 0$).

coordinate graph   A graph displaying points as ordered pairs of numbers.

cosine of an angle in a right triangle (cos *A*)   A ratio of sides in a right triangle given by
$$\cos A = \frac{\text{length of leg adjacent to angle } A}{\text{hypotenuse}}.$$

counterexample   A special case for which a pattern is false.

default window   The window automatically picked by an automatic grapher if a particular window is not chosen.

definition of subtraction   For all real numbers a and b: $a - b = a + \text{-}b$.

density property   Between any two real numbers are many other real numbers.

degree of the polynomial   The largest exponent in a polynomial with one variable, the largest sum of exponents in a polynomial with terms containing more than one variable.

difference of squares   An expression of the form $x^2 - y^2$.

difference of two squares pattern $(a + b)(a - b) = a^2 - b^2$.

dimensions   The number of rows and columns of an array. The lengths of sides of a polygon.

discount   The percent by which the original price of an item is lowered.

discrete   A situation in which numbers between given numbers do not have meaning.

discriminant   In the quadratic equation $ax^2 + bx + c = 0$, $b^2 - 4ac$.

discriminant property   Suppose $ax^2 + bx + c = 0$ and a, b, and c are real numbers. Let $D = b^2 - 4ac$. Then when $D > 0$, the equation has exactly two real solutions. When $D = 0$, the equation has exactly one real solution. When $D < 0$, the equation has no real solutions.

distance formula on a number line   If two points on a line have coordinates x_1 and x_2, the distance between them is $|x_1 - x_2|$.

distributive property   For any real numbers a, b, and c: $ac + bc = (a + b)c$ and $ac - bc = (a - b)c$.

domain   The values that may be meaningfully substituted for a variable.

domain of a function   The set of possible replacements for the first coordinate in a function.

dot frequency diagram   A way of displaying data using dots above a horizontal number line.

element An object that is in a set.

eliminate parentheses To use the Distributive Property to rewrite an expression without parentheses.

empty set A set that has no elements in it.

endpoints The smallest and largest numbers in an interval. The points A and B in the segment $\overline{AB}$.

equal fractions property If $k \neq 0$ and $b \neq 0$: then $\dfrac{a}{b} = \dfrac{ak}{bk}$.

equal sets Two sets that have the same elements.

equally-likely outcomes Outcomes in a situation where each outcome is assumed to occur as often as every other outcome.

equation A sentence with an equal sign.

expanding a power of a polynomial Using the Extended Distributive Property to multiply a polynomial by itself. Example: $(5a + 1)^2 = 25a^2 + 10a + 1$ in expanded form.

expansion A size change in which the factor k is greater than 1 or less than -1.

exponent In the power x^n, n is the exponent.

exponential decay A situation in which the original amount is repeatedly multiplied by a growth factor smaller than one.

exponential growth A situation in which the original amount is repeatedly multiplied by a nonzero growth factor.

extended distributive property To multiply two sums, multiply each term in the first sum by each term in the second sum.

extremes The numbers a and d in the proportion $\dfrac{a}{b} = \dfrac{c}{d}$.

fitting a line to data Finding a line that closely describes data points which themselves may not all lie on a line.

FOIL algorithm A method for multiplying two binomials; the sum of the product of the First terms, plus the product of the Outside terms, plus the product of the Inside terms, plus the product of the Last terms.
$(a + b)(c + d) = ac + ad + bc + bd$.

formula A sentence in which one variable is given in terms of other variables and numbers.

FOR/NEXT loop A BASIC command that tells the computer the number of times to execute a loop.

function A set of ordered pairs in which each first coordinate appears with exactly one second coordinate.

function key on a calculator A key that produces the value of a function when a value of the domain is entered.

fundamental principle of counting Let A and B be finite sets. Then
$N(A \cup B) = N(A) + N(B) - N(A \cap B)$.

general form of a quadratic equation A quadratic equation in which one side is 0 and the other side is arranged in descending order of exponents: $ax^2 + bx + c = 0$, where a $\neq$ 0.

general linear equation An equation of the form $ax + b = cx + d$.

generalized addition property of equality For all numbers or expressions a, b, c, and d: if $a = b$ and $c = d$, then $a + c = b + d$.

growth factor In exponential growth, the nonzero number that is repeatedly multiplied by the original amount.

growth model for powering When an amount is multiplied by g, the growth factor in each of x time periods, then after the x periods, the original amount will be multiplied by g^x.

half-plane In a plane, the region on either side of a line.

horizontal line A line with an equation of the form $y = k$, where k is a fixed real number.

hypotenuse The longest side of a right triangle.

IF . . . THEN   A BASIC command that tells the computer to perform the THEN part only if the IF part is true.

image   The final position of a figure resulting from a transformation.

inequality   A sentence with one of the following signs: "$\neq$", "$<$", "$>$", "$\leq$", "$\geq$", or "$\approx$".

INPUT   A BASIC statement that makes the computer pause and wait for you to type in a value.

integers   The whole numbers and their opposites.

interest   The money the bank pays on the principal in an account.

intersection of sets   The set of elements in both set A and set B, written $A \cap B$.

interval   The set of numbers between two numbers a and b, possibly including a and b.

irrational number   A real number that is not rational. An infinite decimal that is not repeating.

leg of a right triangle   One of the sides forming the right angle of a triangle.

LET   A BASIC command that assigns a value to a given variable.

like terms   Terms in which the variables and corresponding exponents are the same.

linear equation   An equation in which the variable or variables are all to the first power and none multiply each other.

linear function   A function that has an equation of the form $y = mx + b$.

linear inequality   A linear sentence with an inequality symbol.

linear sentence   A sentence in which the variable or variables are all to the first power and none multiply each other.

magnitude of the size change   The size change factor.

markup   A percent by which the original price of an item is raised.

mean   The average of a set of numbers.

means   The numbers b and c in the proportion $\frac{a}{b} = \frac{c}{d}$.

means-extremes property   For all real numbers a, b, c, and d (b and d nonzero): if $\frac{a}{b} = \frac{c}{d}$, then $ad = bc$.

median   The middle value of a set of numbers written in increasing order.

mode   The most frequently occurring value in an ordered set of numbers.

monomial   A polynomial with one term.

multiplication counting principle   If one choice can be made in m ways and a second choice can be made in n ways, then there are mn ways of making the choices in order.

multiplication method for solving a system   Multiplying both sides of an equation in a system by a number so that the addition method can be used on the result.

multiplication property of -1   For any real number a: $a \cdot -1 = -1 \cdot a = -a$.

multiplication property of equality   For all real numbers a, b, and c: if $a = b$, then $ca = cb$.

multiplication property of inequality   If $x < y$ and a is positive, then $ax < ay$. If $x < y$ and a is negative, then $ax > ay$.

multiplication property of zero   For any real number a: $a \cdot 0 = 0 \cdot a = 0$.

multiplicative identity   The number 1, because if any number is multiplied by 1, that number keeps its "identity."

multiplicative identity property of one   For any real number a: $a \cdot 1 = 1 \cdot a = a$.

multiplicative inverse   The multiplicative inverse of a number n is $\frac{1}{n}$, where $n \neq 0$. Also called *reciprocal*.

n factorial   The product of the integers from 1 to n. In symbols, $n!$

negative exponent property   For any nonzero b, $b^{-n} = \dfrac{1}{b^n}$, the reciprocal of b^n.

nth power   The nth power of a number x is the number x^n.

null set   A set that has no elements in it. Also called *empty set*.

numerical expression   An expression that includes only numbers.

op-op property   For any real number a: $-(-a) = a$.

open interval   An interval that does not include its endpoints.

open sentence   A sentence that contains at least one variable.

opposite   The opposite of any real number x is $-x$. Also called *additive inverse*.

or   The word used to speak about the union of sets. If A and B are sets, $A \cup B$ is read "A or B."

order of operations   The correct order of evaluating numerical expressions: first do operations within parentheses or other grouping symbols; within grouping symbols or if there are no grouping symbols—do all powers from left to right, do all multiplications and divisions from left to right, do all additions and subtractions from left to right.

origin   The point $(0, 0)$ on a coordinate plane.

outcome   A result of an experiment, such as the result of the toss of a die.

parabola   A graph whose equation is of the form $y = ax^2 + bx + c$, where $a \neq 0$.

P(E)   The probability of E, or "P of E." The letter E names an event and $P(E)$ names the probability of that event.

percent   %, times $\dfrac{1}{100}$, or "per 100".

percent of discount   The ratio of the discount to the selling price.

perfect square   A number that is the square of a whole number.

perfect square trinomial   A trinomial that is the square of a binomial.

perfect square trinomial patterns
$(a + b)^2 = a^2 + 2ab + b^2$ and
$(a - b)^2 = a^2 - 2ab + b^2$.

permutation   An arrangement of letters, names, or objects.

permutation theorem   There are $n!$ possible permutations of n different objects, when each object is used exactly once.

polynomial   An algebraic expression formed by adding, subtracting, or multiplying numbers and variables.

polynomial in the variable x   A polynomial whose only variable is x.

power of a power property   When $b \geq 0$, for all m and n: $(b^m)^n = b^{mn}$.

power of a product property   For all a and b, and any integer n: $(ab)^n = a^n \cdot b^n$.

power of a quotient property   For all a and nonzero b, and any integer n: $\left(\dfrac{a}{b}\right)^n = \dfrac{a^n}{b^n}$.

preimage   The original position of a figure before a transformation takes place.

principal   Money deposited in an account.

PRINT   A BASIC command that tells the computer to print what follows the command.

probability formula for geometric regions   If all points occur randomly in a region, then the probability P of an event is given by P = $\dfrac{\text{measure (area, length, etc.) of region for event}}{\text{measure of entire region}}$.

probability function   A function whose domain is a set of outcomes in a situation and in which each ordered pair contains an outcome and its probability.

probability of success   The ratio $\dfrac{S}{T}$, where S is the number of outcomes that are successes and T is the total number of outcomes.

probability of a union of events
$P(A \cup B) = P(A) + P(B) - P(A \cap B)$.

product of powers property When b^m and b^n are defined: $b^m \cdot b^n = b^{m+n}$.

property of opposites For any real number a: $a + -a = 0$.

property of reciprocals For any nonzero real number a: $a \cdot \dfrac{1}{a} = \dfrac{1}{a} \cdot a = 1$.

proportion A statement that two fractions are equal. Any equation of the form $\dfrac{a}{b} = \dfrac{c}{d}$.

putting-together model for addition A quantity x is put together with a quantity y with the same units. If there is no overlap, then the result is the quantity $x + y$.

Pythagorean distance formula The distance AB between points $A = (x_1, y_1)$ and $B = (x_2, y_2)$ is $AB = \sqrt{(x_2 - x_1)^2 + (y_2 - y_1)^2}$.

Pythagorean theorem In a right triangle with legs a and b and hypotenuse c, $a^2 + b^2 = c^2$.

quadrant One of four parts of the coordinate plane resulting from dividing it by the x-axis and y-axis.

quadratic equation An equation that can be simplified into the form $ax^2 + bx + c = 0$.

quadratic formula If $ax^2 + bx + c = 0$ and $a \neq 0$, then $x = \dfrac{-b \pm \sqrt{b^2 - 4ac}}{2a}$.

quadratic function A function with an equation of the form $y = ax^2 + bx + c$ or $y = \dfrac{k}{x}$.

quantity (in parentheses) An expression in parentheses.

quotient of powers property If b^m and b^n are defined and $b \neq 0$: $\dfrac{b^m}{b^n} = b^{m-n}$.

random outcomes Outcomes in a situation where each outcome is assumed to have the same probability.

range The length of an interval. Range = maximum value − minimum value.

range of a function The set of possible values of a function.

rate factor model for multiplication When a rate is multiplied by another quantity, the unit of the product is the product of units. Units are multiplied as though they were fractions. The product has meaning when the units have meaning.

rate of change The rate of change between points (x_1, y_1) and (x_2, y_2) is $\dfrac{y_2 - y_1}{x_2 - x_1}$.

rate model for division If a and b are quantities with different units, then $\dfrac{a}{b}$ is the amount of quantity a per quantity b.

ratio A quotient of quantities with the same units.

ratio model for division Let a and b be quantities with the same units. Then the ratio $\dfrac{a}{b}$ compares a to b.

ratio of similitude The ratio of corresponding sides for two similar figures.

rational number A number that can be written as the ratio of two integers.

real numbers Numbers that can be represented as finite or infinite decimals.

reciprocal The reciprocal of a number n is $\dfrac{1}{n}$, where $n \neq 0$. Also called *multiplicative inverse*.

reciprocal rates Equal rates in which the quantities are compared in reverse order.

rectangular array A two-dimensional display of numbers or symbols arranged in rows and columns.

rectangular solid A box.

relation Any set of ordered pairs.

relative frequency The ratio of the number of times an event occurred to the total number of possible occurrences.

REM A BASIC statement for a remark or explanation that will be ignored by the computer.

replacement set The values that may be meaningfully substituted in a formula.

rule for multiplication of fractions For all real numbers a, b, c, and d, with b and d not zero: $\dfrac{a}{b} \cdot \dfrac{c}{d} = \dfrac{ac}{bd}$.

ruling out possibilities   Arriving at the right answer by eliminating wrong ones.

scale factor   The amount by which interest changes in a polynomial expression.

scattergram   A two-dimensional coordinate graph that shows data.

sentence   Two algebraic expressions connected by "$=$", "$\neq$", "$<$", "$>$", "$\leq$", "$\geq$", or "$\approx$".

set   A collection of objects called elements.

similar figures   Two or more figures that have the same shape.

simplifying radicals   Rewriting a radical with a smaller integer under the radical sign.

sine of an angle in a right triangle (sin A)   In a right triangle, $\sin A = \dfrac{\text{length of the leg opposite angle } A}{\text{hypotenuse}}$.

size change factor   A number that multiplies other numbers to change their size.

size change model for multiplication   If a quantity x is multiplied by a size change factor k, $k \neq 0$, then the resulting quantity is kx.

slide model for addition   If a slide x is followed by a slide y, the result is the slide $x + y$.

slope   The name for the constant rate of change between points on a line. The amount of change in the height of the line as you go 1 unit to the right. The slope of the line through (x_1, y_1) and (x_2, y_2) is $\dfrac{y_2 - y_1}{x_2 - x_1}$.

slope-intercept form   An equation of a line in the form $y = mx + b$, where m is the slope and b is the y-intercept.

slope-intercept property   The line with equation $y = mx + b$ has slope m and y-intercept b.

solution   A replacement of the variable in a sentence that makes the statement true.

solution set   The set of numbers that are solutions to a given open sentence.

solution set to a system   The intersection of the solution sets for each of the conditions in the system.

square root   If $A = s^2$, then s is the square root of A.

square root of a product property   If $a \geq 0$ and $b \geq 0$, then $\sqrt{ab} = \sqrt{a} \cdot \sqrt{b}$.

stacked bar graph   A way of displaying data using different sized rectangles or bars stacked on top of each other.

standard form for an equation of a line   An equation of the form $Ax + By = C$.

standard form of a quadratic equation   An equation of the form $ax^2 + bx + c = 0$, where $a \neq 0$.

statistics   Numbers that represent sets of data.

substitution method for solving a system   A method in which one variable is written in terms of other variables, and then this expression is used in place of the original variable in subsequent equations.

supplementary angles   Two angles, the sum of whose measures is 180°. Also called *supplements*.

system   A set of conditions separated by the word *and*.

system of inequalities   A system in which the conditions are inequalities.

take-away model for subtraction   If a quantity y is taken away from an original quantity x, the quantity left is $x - y$.

tangent function   A function defined by $y = \tan x$.

tangent of an angle A in a right triangle (tan A)   A ratio of sides given by $\tan A = \dfrac{\text{length of leg opposite angle } A}{\text{length of leg adjacent to angle } A}$.

term   A single number or a product of numbers and variables.

testing a special case   A strategy for finding a pattern by trying out a specific instance.

tree-diagram   A tree-like way of organizing the possibilities of choices in a situation.

triangle inequality   The sum of the lengths of two sides of any triangle is greater than the length of the third side.

triangle-sum   In any triangle with angle measures a, b, and c: $a + b + c = 180$.

trinomial   A polynomial with three terms.

two-dimensional slide   A movement that can be broken into a horizontal and a vertical slide. A transformation in which the image of (x, y) is $(x + h, y + k)$.

union of sets   The set of elements in either set A or set B, written $A \cup B$.

value of function   A value of the second variable (often called y) in a function.

variable   A letter or other symbol that can be replaced by any numbers (or other objects).

Venn diagram   A diagram used to show relationships among sets.

vertex   The intersection of a parabola with its axis of symmetry.

vertical line   A line with an equation of the form $x = h$, where h is a fixed real number.

weighted average $W_1 x_1 + W_2 x_2 + \ldots + W_n x_n$, where n quantities $x_1, x_2, \ldots, x_n$ having weights $W_1, W_2, \ldots, W_n$ with $W_1 + W_2 + \ldots + W_n = 1$.

weights   The parts or percents of the total accounted for by each amount.

whole numbers   The set of numbers $\{0, 1, 2, 3, \ldots\}$.

window   The part of the coordinate grid that is shown on an automatic grapher.

x-axis   The horizontal axis in a coordinate graph.

x-coordinate   The first coordinate of a point.

x-intercept   The x-coordinate of a point where a graph crosses the x-axis.

y-axis   The vertical axis in a coordinate graph.

y-coordinate   The second coordinate of a point.

y-intercept   The y-coordinate of a point where a graph crosses the y-axis.

zero exponent property   If g is any nonzero real number, then $g^0 = 1$.

zero product property   For any real numbers a and b, if the product of a and b is 0, then $a = 0$ or $b = 0$.

zoom feature   A feature on an automatic grapher by which a window can be changed without retyping the intervals for x and y.

estimating cars on a highway, 25
force of acceleration, 28
gravity, 517
height of a rocket, 503
life expectancy, 307
normal adult weight, 28
percent discount, 51
perimeter of a rectangle, 28
perimeter of a triangle, 187
planets supporting intelligent life, 155
pricing packaged nuts, 141
probability for geometric regions, 232
probability of success, 38
profit, 121
Pythagorean distance, 349
quadratic, 603
rate of growth, 251
relating Fahrenheit and Celsius temperatures, 267
shoe size, 35, 265
sum of consecutive integers, 479
surface area of prism, 484
volume of a cone, 132
volume of a cylinder, 52
volume of a rectangular solid, 158
volume of a sphere, 459

fraction(s)
adding, 68–71
addition property for, 68
common denominator of, 68
comparing, 47
decimal form of, 6, 109, 214
dividing, 185, 210–212
equal property of, 43
multiplication rule for, 162
multiplying, 158, 162
ordering, 8
in percent form, 214
simplifying, 43–44, 455, 491
subtracting, 109

function(s)
on calculators, 666–668
cosine, 666
definition, 638
described by equation, 439–440
described by graph, 639
described by written rule, 640
domain, 652
linear, 640
notation, 643–644
probability, 656–658
quadratic, 640
range, 653
sine, 666
squaring, 643
tangent, 660
value of, 639
ways to describe, 640

fundamental counting principle, 133

general linear equation, 285
generalized addition property of equality, 545
geometry
acute angle, 72
area formulas *See* area formula(s).
betweenness on a segment, 96
circumference of a circle, 27
coincident lines, 567
complementary angles, 144
congruent figures, 214
contraction, 236
diameter, 224
expansion, 236
hypotenuse, 337
isosceles trapezoid, 579
leg of a right triangle, 337
length of segment, 96
line segment, 96
measure angles, 141
optical illusion, 313
parallel lines, 566
perimeter, 28, 87, 214
perpendicular lines, 73
point, 96
Pythagorean theorem, 337
right triangle, 144
similar figures, 238, 247–249
size change, 236
sum of angles in quadrilateral, 145
surface area, 167, 484
supplementary angles, 143
triangle inequality, 96
triangle-sum property, 144
volume *See* volume.

Gosselin, Guillaume, 214 *See* history of mathematics.

graph(s)
and absolute value, 325, 330, 354
of absolute value function, 647–649
area of a square, 588
with automatic grapher, 386, 409, 439, 489, 594–596
bar, 56
of coincident lines, 567
common logarithm function, 667
of constant sum or difference situation, 138–139
continuous, 32, 77, 138, 315
coordinate, 74
of discrete domain, 32, 314, 577
dot frequency, 9
of exponential decay, 440
of exponential growth, 438
of finite set, 138
of functions, 638–640
hyperbola, 640
of image, 80
of inequalities, 5, 92, 320

of interval, 15
of a line, 377, 383, 406
of linear decrease, 440
of linear equations, 314–316
of linear increase, 438
of linear inequalities, 410–412
on a number line, 5
of open sentences, 5
of ordered pairs, 53, 73, 74, 82, 138
parabola, 588–589, 594–596, 599–601
of preimage, 80
of probability functions, 657–658
rate of change, 366
of relations, 640
scattergram, 75–76
of set intersection, 124
of set unions, 129
of sine function, 667
of a slide, 62–65, 80–82
of solution set, 31–32, 92, 119
stacked bar, 56
symmetric, 142, 588
of systems of inequalities, 574
of systems of linear equations, 533

graph systems of linear inequalities *See* problem-solving strategies.
graphing calculator *See* automatic grapher.
graphing method, 532
grouping symbols
with order of operations, 20–21
using computer, 22
growth factor, 429
growth model for powering, 429

half-plane, 320, 411
Hamilton, Sir William Rowan, 59 *See* history of mathematics.
history of mathematics, 3, 29, 58, 59, 112, 214, 226, 235, 337, 392, 609
horizontal line, 319
equation of, 319, 407
slope of, 374
hyperbola, 640
hypotenuse, 337

identity
for addition, 63
for multiplication, 168
identity property
for addition, 63
for multiplication, 168
image, 80
after two-dimensional slide, 80
under size change, 237

model
for addition
putting-together, 56–59
slide, 63
for division
rate, 216
ratio, 221
for multiplication
area, 156
area (discrete version), 157
rate factor, 164
size change, 236
for powering
growth, 429
for subtraction
comparison, 114
take-away, 113
monomial(s), 477
dividing, 455, 491
dividing polynomial by, 490–492
multiplying, 159, 444
multiplying a polynomial by,
486–487
powers of, 458
monomial factor, 490
multiplication
of binomials, 499
of binomials containing radicals,
515
differences of squares, 513
FOIL algorithm, 500
of fractions, 158, 162
identify for, 168
model(s)
area, 156, 157
rate factor, 164
size change, 236
of monomials, 159
of a polynomial by a monomial,
447, 486
of polynomials, 495
properties
associative, 158
commutative, 156
of equality, 173
of inequality, 182, 183
of negative one, 169
of zero, 170
of radical expressions, 334
of real numbers, 8, 11, 13, 19, 24,
29, 41, 47
rules for positive and negative,
11, 164
of squares of binomials, 504
multiplication counting principle,
187
**multiplication method for solving
a system,** 551
multiplication property of equality,
173
**multiplication property of
inequality (part 1),** 182

**multiplication property of
inequality (part 2),** 183
multiplication property of –1, 169
multiplication property of zero,
170
multiplicative identity, 168
**multiplicative identity property of
1,** 168
multiplicative inverse, 169
mutually-exclusive sets, 133

negative exponent(s), 448
negative exponent property, 449
negative number, 62
absolute value of, 324
negative rate of change, 367
negative square root, 332
notation
function, 643–644
scientific, (Appendix B)
nth power, 424
null set, 33
number(s)
comparing, 6
directed, 62
graphing, 5, 32
integers, 31
irrational, 615
negative, 62
opposite, 63
positive, 62
prime, 37
rational, 615
real, 31
square, 331
whole, 30
number line
addition slides on, 62
graphing inequalities on, 5
graphing intervals on, 15
intersection of sets on, 124
union of sets on, 129
numerator, 68
numerical expression, 20

oblique line, 407
open interval, 15
open sentence, 4
operations
computer, 21
order of, 20
op-op property, 64
opposite(s)
property of, 64
of a number, 63
changing a number to, 169
of a sum, 301
optical illusion, 313
or (in sets), 128
order
of operations, 20

of numbers in scientific notation,
47
of rational numbers, 8, 13
symbols for, 4
ordered pairs
coordinates of, 73
graph of, 53, 74, 138
origin, 74
outcome
equally-likely, 36
random, 36

parabola(s), 587
applications of, 599–601
on automatic grapher, 594–596
axis of symmetry, 588
properties of, 589
vertex, 589
x-intercept, 590
parallel lines, 566
parentheses, 20–21
pattern(s), 293
difference of two squares, 513
perfect square trinomial, 505
percent, 226
decimal form of, 44, 72, 226
of discount, 51, 223
fraction form of, 44, 72, 137, 226
problems, 19, 29, 40, 41, 45, 99,
226–228
solve problems, using equations,
226–228
perfect square(s), 331, 512
recognizing, 508
trinomial, 505
perimeter formulas and problems,
23, 28, 214, 329
permutation, 198
permutation theorem, 198
perpendicular lines, 73
pi (π), 29
place value, 24, 29, 67
plus or minus
in describing interval, 14
in quadratic formula, 604
point(s), 96
distance between two, 326, 349
graphing, 53, 73, 74, 82, 138
of intersection, 532
polynomial(s), 473
addition and subtraction, 301, 481
degree of, 477
equations, 517
monomial factors, 490
multiplication, 486, 495
recognizing factors of, 521
simplifying, 481
in the variable x, 477
positive number, 62
absolute value of, 324
positive rate of change, 367
positive square root, 332

230, 234, 241, 246, 251, 265,
270, 274, 281, 288, 293, 300,
304, 318, 322, 330, 336, 340,
345, 351, 356, 370, 375, 379,
386, 389, 396, 403, 408, 413,
428, 433, 436, 442, 447, 452,
456, 462, 467, 480, 485, 488,
494, 498, 502, 507, 511, 516,
520, 523, 534, 540, 544, 549,
555, 561, 564, 569, 573, 580,
598, 602, 607, 614, 617, 621,
626, 629, 642, 646, 651, 655,
659, 665, 669 *See* chapter
review.
right triangle, 144
  Pythagorean theorem, 337
  and trigonometric ratios, 225
roman numerals, 112
root(s) *See* solution(s).
rounding numbers
  decimals, 41
  in quotients, 19
rule(s)
  for adding to both sides of
    inequality, 92
  for dividing positive and negative
    numbers, 211
  for multiplying fractions, 162
  for multiplying positive and
    negative numbers, 11, 164
  for order of operations, 20
rule for multiplication of fractions,
  162
ruling out possibilities, 521 *See*
  problem-solving strategies.

scale factor, 477
scale drawing(s), 251
scattergram, 75–76, 399
scientific notation, (Appendix B),
  67, 122, 200, 275, 295, 318,
  459, 461
segment, 96
sentence, 4
  linear, 261
  open, 4
sequence, 277–279
Servois, François, 58 *See* history of
  mathematics.
set(s), 30–33
  continuous, 32
  discrete, 32
  domain of, 26
  element of, 30
  empty, 33
  equal, 30
  fundamental counting principle of,
    133
  graphing, 32
  of integers, 31
  intersection of, 123
  mutually-exclusive, 133

null, 33
  number of elements in, 33
  of real numbers, 31
  replacement, 32
  solution, 31
  symbols of, 30, 31, 33, 123
  union of, 128
  Venn diagram of, 123, 128
  of whole numbers, 31
sides
  of an equation, 85
  of an inequality, 91
  of a triangle, 96
similar figures, 238, 247–249
similar terms *See* like terms.
simplifying
  expressions, 65, 110
  expressions involving subtraction,
    301–302
  fractions, 44, 211, 455, 491
  polynomials, 481
  radicals, 342
sine function, 666
  graph of, 667
sine ratio, 225
sinusoidal curve, 667
size change(s), 236–238
  contraction, 236
  expansion, 236
  factor, 236
  magnitude of, 236
  model for multiplication, 236
  negative factors, 237
  similar figures, 238
slide(s)
  on a number line, 62
  on a plane, 80–82
slide model for addition, 62–65
slope
  of horizontal line, 374
  of a line, 371
  of parallel lines, 566
  properties of, 376–378
  of vertical line, 376, 383
slope-intercept form of a line, 381
slope-intercept property, 381
solution(s)
  of an open sentence, 4
  of a quadratic equation, 610
  of a system, 532, 566–568
solution set, 31, 179
  to a system, 533
solve equations
  absolute value, 326, 354
  by factoring, 510, 623–625, 626
  having variables in both sides,
    284–286
  involving squares, 333, 339, 344,
    353
  with like terms, 271
  with parentheses, 296–298, 302
  plan for, 86, 119, 174, 267

polynomial, 517–519
  radical, 333
  with rational numbers, 267–269
  systems of linear, 533, 536, 541,
    545, 551, 566
  by trial and error, 5
  using chunking, 353
  using properties of equality,
    85–88, 118–119, 173–175, 179,
    212, 267–269
solving inequalities
  involving absolute value, 354
  plan for, 91, 119
  system of linear, 574–577
  using chunking, 354
  using properties of inequalities,
    91–93, 118–119, 182–184,
    290–292
  with variable on both sides,
    290–292
solving for a variable, 87
sphere, volume of, 459
spread of data *See* range.
SPUR *See* chapter review.
square(s)
  of binomials, 504
  difference of, 513
  of a number, 13, 331
  perfect, 331
square root(s), 331–334
  approximation, using calculator,
    331
square root of a product property,
  342
stacked bar graph, 56
standard form
  of a linear equation, 404
  of linear inequality, 411
  of quadratic equation, 604
statistics
  bar graph, 56
  data gathering, 79
  data, organizing, 9
  dot frequency diagram, 9
  fitting a line to data, 399
  intervals and estimates, 14–17
  mean, 10–11
  measures of central tendency, 10
  median, 10
  mode, 10
  range, 16
  scattergram, 75–76, 399
  stacked bar graph, 56
  table, 74
substitution method, 536, 541
subtraction, 110
  algebraic definition of, 110
  of fractions, 447
  of polynomials, 301, 481
  of real numbers, 108–110
sum
  of angles in a triangle, 144

of angles in quadrilateral, 145
of complementary angles, 144
of supplementary angles, 143
See addition.
supplement, 143
supplementary angles, 143
surface area
of a cylinder, 167
of a prism, 484
survey, 79
symbol(s)
absolute value, 324
braces, 30
of computer operations, 21
cursor, 27
empty or null set, 33
of equality and inequality, 4
factorial, 197
integers, set of, 30
intersection of sets, 123
number of elements in set, 33
plus or minus, 14
real numbers, set of, 30
right angle, 144
square root, 331
union of sets, 128
whole numbers, set of, 30
symmetry
axis of, 588
in a graph, 142
in a parabola, 588
system, 532
systems of inequalities, 574–577
systems of linear equations,
532–533
with infinitely many solutions, 567
with no solutions, 566
solution of, 532
solving by addition method, 545
solving by the graphing method,
533
solving by the multiplication
method, 551
solution by the substitution
method, 536, 541

take-away model for subtraction,
113
tangent, 225
of an angle, 660
function, 660
term(s), 65
coefficient of, 477
constant, 477
like, 65
of a polynomial, 477
of a sequence, 277
test a special case *See* problem-
solving strategies.
testing a special case, 464
tests *See* progress self-tests.
tolerance, 18

transformations *See* solving
equations using properties of
equality.
translating
verbal expressions to algebraic
expressions, 73, 114, 115
verbal sentences to algebraic
sentences, 5, 15, 91
translations *See* slides.
trapezoid, area of, 299
tree diagram, 187
trial and error *See* problem-solving
strategies.
triangle(s)
area of, 185
area of right, 117
inequality relationships, 96
perimeter of, 87
right, 144
sum of angles in, 144
triangle inequality, 96–97
triangle-sum property, 144
trigonometry
cosine function, 666–667
ratios, 225
sine function, 666–667
tangent function, 660–663
trinomial(s), 477
two-dimensional slides, 80–82

unbiased die, 37
union of sets, 128–130
use alternate approach *See*
problem-solving strategies.
use a counterexample *See*
problem-solving strategies.
use a diagram/picture *See*
problem-solving strategies.
use a formula *See* problem-solving
strategies.
use a graph *See* problem-solving
strategies.
use mathematical models *See*
problem-solving strategies.
use a physical model *See*
problem-solving strategies.
use a proportion *See* problem-
solving strategies.
use a ratio *See* problem-solving
strategies.
use systems of linear equations
See problem-solving strategies.
use a table *See* problem-solving
strategies.
use a theorem *See* problem-solving
strategies.

value(s)
of an expression, 20
of a function, 639
of a variable, 26

variable, 4
domain, 26
value of, 26
Venn diagram, 123
intersection of sets, 123, 124
mutually-exclusive sets, 133
union of sets, 128
verbs, mathematical, 4
vertex (vertices)
of a parabola, 589
of a triangle, 96
vertical line, 319
equation of, 320, 407
slope of, 374
vertical number line, 16
volume
of a box, 158
of a cone, 132
of a cube, 167, 458
of a cylinder, 52
of a rectangular solid, 158, 206
of a sphere, 459

weight, 557
weighted average, 557–560,
562–563
whole numbers, set of, 30
window, 594
write an inequality *See* problem-
solving strategies.
write a linear equation *See*
problem-solving strategies.
write a polynomial expression *See*
problem-solving strategies.
write a quadratic equation *See*
problem-solving strategies.

x-axis, 80
x-coordinate, 80
x-intercept
of a line, 406
of a parabola, 590

y-axis, 80
y-coordinate, 80
y-intercept, 381

zero
absolute value of, 324
division by, 217
as exponent, 428, 430
identity for addition, 62
multiplication property of, 170
rate of change, 367
reciprocal of, 178
square root of, 333
zero exponent property, 430
zero product property, 517
zoom feature, 596

ACKNOWLEDGMENTS

For permission to reproduce indicated information on the following page, acknowledgment is made to:

page 35 DENNIS THE MENACE® used by permission of Hank Ketcham and © by North American Syndicate

page 415 "We're waiting to marry" by Karren Loeb, *USA Today*, September 10, 1987. Copyright © by 1987 *USA Today*. Reprinted with permission.

Illustration Acknowledgments

Phil Renaud
54–55, 91, 166, 193, 195, 215, 236, 240, 272, 281, 295, 302, 340, 369, 371, 376, 378, 389, 391, 397, 400, 408, 420, 435, 494, 603, 608, 611, 612

Jack Wallen
160, 176, 210, 218, 234, 241, 248, 249, 250, 251, 252, 256, 262, 264, 266, 284, 287, 300, 311, 339, 345, 350, 359, 361, 356, 386, 418, 436, 486, 520, 539, 563

Jill Ruter
23, 55, 56, 74, 90, 140, 143, 147, 162, 165, 178, 186, 203, 208, 209, 221, 237, 241, 246, 258, 276, 290, 296, 335, 337, 350, 356, 364, 398, 410, 415, 417, 422, 434, 448, 472, 500, 623

Unless otherwise acknowledged, all photos are the property of Scott, Foresman and Company. Page positions are as follows: (t) top, (c) center, (b) bottom, (l) left, (r) right, (ins) inset

CHAPTER 1
2-3 Milt & Joan Mann/Cameramann International, Ltd.
2-3 (INS) Translation from *The Rhind Mathematical Papyrus*, by Arnold Buffum Chase, Courtesy The National Council of Teachers of Mathematics. 4 Stuart Westmorland/Aperture Photobank 7 David Madison/Focus On Sports 12 David Black 14 Bob Daemmrich/Click/Chicago/Tony Stone 16 Mickey Pfleger 18 Norma Morrison 23 Carl Purcell 25 Brent Jones 30 Lynn M. Stone 34 Pat and Tom Lesson/Alaska Photo 36 Space Science & Engineering Center, University of Wisconsin, Madison 41 Brent Jones 46 David R. Frazier Photolibrary

CHAPTER 2
56 © Robert Perron 61 Brent Jones 62 Focus On Sports 67 Milt & Joan Mann/Cameramann International, Ltd. 70 Milt & Joan Mann/Cameramann International, Ltd. 73 Focus On Sports 76 David R. Frazier Photolibrary 79 Milt & Joan Mann/ Cameramann International, Ltd. 94 Charles Krebs/Aperture Photobank 99 Jim Shives/Aperture Photobank 104 Brent Jones

CHAPTER 3
108 David Lissy/Focus On Sports 109 David R. Frazier Photolibrary 113 Milt & Joan Mann/Cameramann International, Ltd. 115 Carl Purcell 116 Bruce Forster/Aperture Photobank 121 Courtesy Patricia Dolan 122 Travelpix/FPG 123 Ellis Herwig/Stock Boston 126 Mickey Pfleger 129 Norma Morrison 135 David R. Frazier Photolibrary 137 David Black 140 Marty Snyderman 141 Milt & Joan Mann/Cameramann International, Ltd. 147 Focus On Sports 149 Carl Purcell 151 Virginia Historical Society

CHAPTER 4
154–155 California Institute of Technology & Carnegie Institute of Washington 163 David R. Frazier Photolibrary 167 Guy Sauvage Agency/Vandystadt/ALLSPORT USA 173 James R. Rowan 177 Focus On Sports 178 David R. Frazier Photolibrary 181 Randy Wells/Aperture Photobank 184 Norma Morrison 186 Rudi Von Briel 189 Viesti Associates 190 Roy Morsch/The Stock Market 197L Focus On Sports 197CL Focus On Sports 197CR Focus On Sports 197R Focus On Sports 198 David Black 200 David R. Frazier Photolibrary 203 COMSTOCK INC.

CHAPTER 5
219 Ahmed/David R. Frazier Photolibrary 222 ANIMALS ANIMALS/Ted Levin 223 Ann Purcell 227 Brent Jones 229 Christopher Morrow/Black Star 234 NASA 238 Ann Purcell 239 Rick McIntyre/Aperture Photobank 245 Susan Copen Oken/Dot Picture Agency 252 Carl Purcell 254 ANIMALS ANIMALS/H. Ausloos 257 Steve McCutcheon/Alaska Photo

CHAPTER 6
260–261 Frans Lanting 268 Carl Purcell 270 Deutsches Museum, Munchen 274 Thomas Valentin 275 Wolfgang Bayer Productions 276 Art Pahlke 278 Mitch Kezar 283 Brent Jones 286 Alan Magayne-Roshak/Third Coast 288 Thomas Valentin 291 Grant Heilman Photography 293 Robert Frerck/ Odyssey Productions 297 Milt & Joan Mann/Cameramann International, Ltd. 303 John McGrail 310 Focus On Sports

CHAPTER 7
312–313 J. Taposchaner/FPG 315 Mark Reinstein/Stock Imagery 317 Art Pahlke 319 Milt & Joan Mann/Cameramann International, Ltd. 323 Milt & Joan Mann/Cameramann International, Ltd. 346 Bruce McAllister/Stock Imagery 347 © 1988 Color-Art, Inc., St. Louis, Mo. 356 Francis de Richemond/The Image Works 361 Into The Wind 362 Milt & Joan Mann/Cameramann International, Ltd.

CHAPTER 8
364–365 Chuck O'Rear/West Light 368 Brent Jones 376 Milt & Joan Mann/Cameramann International, Ltd. 381 Chet Hanchett/Photographic Resources, Inc. 385 David Brownell 387 DRS Productions/The Stock Market 390 G. Newman Haynes/The Image Works 393 Jeff Foott/Bruce Coleman Inc. 395 Steinhart Aquarium/Photo Researchers 400B Robert Frerck/Odyssey Productions 401T Robert Frerck/Odyssey Productions 404 Martha Swope 406 Martha Swope 413 Focus On Sports 419 Courtesy United Airlines 421 Bob Daemmrich/Stock Boston

CHAPTER 9
423 David Black 429 Barry L. Runk/Grant Heilman Photography 432 Michael Fredericks Jr./Earth Scenes 433 Alfred Pasieka/Bruce Coleman Inc. 437 Rudi Von Briel 439 Tom Raymond/Bruce Coleman Inc. 442 Robert Frerck/Odyssey Productions 447 Frank Siteman/Stock Boston 454 Milt & Joan